Contaminants in Agriculture

M. Naeem • Abid Ali Ansari
Sarvajeet Singh Gill
Editors

Contaminants in Agriculture

Sources, Impacts and Management

Springer

Editors
M. Naeem
Department of Botany
Aligarh Muslim University
Aligarh, Uttar Pradesh, India

Abid Ali Ansari
Department of Biology
University of Tabuk
Tabuk, Saudi Arabia

Sarvajeet Singh Gill
Centre for Biotechnology
Maharshi Dayanand University
Rohtak, Haryana, India

ISBN 978-3-030-41551-8 ISBN 978-3-030-41552-5 (eBook)
https://doi.org/10.1007/978-3-030-41552-5

This Springer imprint is published by the registered company Springer Nature Switzerland AG
The registered company address is: Gewerbestrasse 11, 6330 Cham, Switzerland

Foreword

Contaminants caused by industrial activities, excessive use of pesticides, herbicides, chemical fertilizers, antibiotics, and pharmaceuticals containing heavy metals, sewage sludge, and improper disposal of waste into the soil, have become a major challenge that we need to overcome for a healthy environment. Such soils with toxic pollutants, apart from causing serious threat to all life-forms, alter plant metabolism, thus resulting in the reduction in crop quality and crop yield.

The assessment of risks of polluted soil on ecosystems and humans and the ways to reduce them have become the key subjects of concern for researchers. To address the problem, serious research into remediation processes has been carried out over the last 20 years. Various technologies have been developed for remediation of such polluted soils. The principles of phytoremediation techniques, immobilization of heavy metals, and soil washing are frequently listed among the best practices.

Remediation of heavy metal polluted soils is necessary to reduce the associated risks, make the soils available for agricultural production, enhance food security, and scale down land tenure problems arising from changes in the land-use pattern. Some plant species which grow profusely in a water-saturated soil and accumulate a spectrum of contaminants have been found to address the problem efficiently.

Accomplished scientists in the field of remediation have compiled and edited some very useful articles written by experts from different parts of the world and produced this book. It comprises 21 chapters grouped into four sections: Section A outlines the overview of contaminants in agriculture; section B illustrates sources of contaminants and their impacts on agriculture; section C discusses management strategies covering utilization, applications, and bioremediation of agricultural contaminants; and Section D focuses on approaches and challenges for crop protection and production from contaminated soils. This compendium provides extensive contributions from researchers in almost all aspects of agriculture and agronomy.

The editors have offered a good foundation of the subject for students, researchers, readers, plant growers, plant lovers, agronomists, agriculturists and all those who are concerned about soil and human health. I believe the book will prove to be a rich source of practical knowledge for all those involved and associated with the farming or processing of plants.

With my good wishes to the readers, I congratulate the editors, authors, and the publisher of this compendium.

Central University of Punjab
Bathinda (Punjab), India

R. K. Kohli

Preface

Over the past few decades, the increase in global population and advancement in modern agricultural technology has amplified the demand for agricultural and exotic crops and livestock. Agricultural, horticultural, and other industrial activities are major causes of contamination in soils, sediments, and water bodies in the adjacent environments. Agricultural contaminants include organic and inorganic fertilizers, pesticides, herbicides, and insecticides, organic matter such as animal waste and decaying plant materials, irrigation residue like salts and trace metals, and microorganisms. Heavy metals are one of the most important factors of soil and water contamination.

Innovative management strategies are essential to maximize the benefits from the agricultural inputs while minimizing their off-site migration and impact on the receiving environments. In this regard, efficient management strategies and skills for the agricultural contaminants pave the way to combat the challenges to improve the production of agricultural crops. Application of targeted, sufficient, and balanced quantities of chemicals will be necessary for high yields without environmental penalty. At the same time, every effort should be made to improve the availability and use of secondary nutrients, micronutrients, organic fertilizers, and soil conservation practices to develop overall crop production in an efficient and environmentally sustainable manner, without sacrificing soil health and productivity.

Therefore, we must address these challenging issues rising day by day in the field of agriculture. Thus, we bring forth a comprehensive volume, *Contaminants in Agriculture: Sources, Impacts and Management*, highlighting the various prospects that are involved in current scenario. The book consists of 21 chapters categorized in different sections, consisting of review articles written by global experts. We are hopeful that this volume will meet the need of all researchers who are working or have great interest in that particular field. We are thankful to the Springer International Publishing AG, Switzerland, for the compilation of this scientific work. Heartfelt thanks are expressed to the team members (Eric Stannard, Nicholas DiBenedetto, Anthony Dunlap, and Arun Siva Shanmugam) for their dedication, sincerity, and friendly cooperation in producing this volume.

With great pleasure, we extend our sincere thanks to all the contributors for their timely response, their outstanding and up-to-date research contribution, and their support and consistent patience.

Lastly, thanks are also due to well-wishers, friends, and family members for their moral support, blessings, and inspiration in the compilation of this book.

Aligarh, Uttar Pradesh, India M. Naeem
Tabuk, Saudi Arabia Abid Ali Ansari
Rohtak, Haryana, India Sarvajeet Singh Gill

Contents

About the Editors

M. Naeem is an Assistant Professor in the Department of Botany at Aligarh Muslim University, India. For more than a decade, he has devoted his research to improving the yield and quality of commercially important medicinal and aromatic plants (MAPs). His research focuses on escalating the production of MAPs and their active principles using a novel and safe technique involving depolymerized polysaccharides as well as the application of potent PGRs. His research also focuses on abiotic stress tolerance in medicinal plants. To date, he has successfully run three major research projects as the Principal Investigator sponsored by national funding agencies like the Department of Science and Technology, New Delhi, and the Council of Science and Technology UP, Lucknow. He has published more than 80 research papers in reputable national and international journals as well as 7 books. He has also participated in various national and international conferences and acquired life memberships to various scientific bodies in India and abroad. Based on his research contributions, he has been awarded a Research Associateship from the Council of Scientific and Industrial Research, New Delhi; a Yuva Vaigyanic Samman (2011) from the State Government of Uttar Pradesh; a Fast Track Young Scientist Award from the Department of Science and Technology, India; a Rashtriya Gaurav Award (2016) from the India International Friendship Society, New Delhi; a Young Scientist of the Year Award (2015), a Distinguished Young Scientist of the Year Award (2017); and a Best Scientist of the Year Award (2018) from the Scientific and Environmental Research Institute, Kolkata; and UGC-BSR Research Start-Up Grant (2018) by UGC, New Delhi.

Abid Ali Ansari is an Assistant Professor in the Department of Biology, Faculty of Science, University of Tabuk, Saudi Arabia. His research work is concerned with phytoremediation and eutrophication. He has a number of research articles of national and international repute to his credit. He has written ten edited books and a number of book chapters on varied aspects of his research. He has been awarded "Environmentalist of the Year 2011" and "Scientist of the Year 2014" by the National Environmental Science Academy, India, and "Research Excellence 2016 Award" by the University of Tabuk. He has also participated in various national and international conferences and acquired memberships to various scientific bodies (Saudi Biological Society, International Phytotechnology Society, and Society for Ecological Restoration).

Sarvajeet Singh Gill Assistant Professor, Centre for Biotechnology, Maharshi Dayanand University, Rohtak, India, has made significant contributions toward abiotic stress tolerance in crop plants. His research includes abiotic stress tolerance in crop plants, reactive oxygen species signaling and antioxidant machinery, gene expression, helicases, crop improvement, transgenics, nitrogen and sulfur metabolism, and plant fungal symbiotic interactions. Together with Dr. Narendra Tuteja at the International Centre for Genetic Engineering and Biotechnology (ICGEB), New Delhi, he worked on plant helicases and discovered a novel function of plant MCM6, PDH45, and p68 in salinity stress tolerance that will help improve crop production at sub-optimal conditions. A recipient of the Junior Scientist of the Year Award 2008 from the National Environmental Science Academy, he has edited several books and has a number of research papers, review articles, and book chapters to his name.

Part I
Overview of Contaminants in Agriculture

Organic and Inorganic Fertilizer Contaminants in Agriculture: Impact on Soil and Water Resources

I. Rashmi, Trisha Roy, K. S. Kartika, Rama Pal, Vassanda Coumar, S. Kala, and K. C. Shinoji

Contents

I. Rashmi (✉)
Soil Chemistry and Fertility, IISWC, Research Centre, Kota, India

T. Roy
Soil Chemistry and Fertility, IISWC, Dehradun, India

K. S. Kartika
Soil Chemistry and Fertility, CPCRI, Research Centre, Vittal, India

R. Pal
Environmental Science, IISWC, Research Centre, Agra, India

V. Coumar
Soil Chemistry and Fertility, IISS, Bhopal, India

S. Kala
Forestry, IISWC, Research Centre, Kota, India

K. C. Shinoji
Agriculture extension, IISS, Bhopal, India

M. Naeem et al. (eds.), *Contaminants in Agriculture*,
https://doi.org/10.1007/978-3-030-41552-5_1

1 Introduction

Nutrients are essential for crop growth and production. As plant growth involves 17 essential nutrient elements along with several other beneficial elements majority of which is supplied from soil, the replenishment of the elements taken up by the plants from soil becomes necessary for sustainable agriculture management. Therefore, in order to replenish nutrients in soil, various sources of amendments like inorganic fertilizers, organic manure sources like farmyard manure, municipal solid waste, distillery effluents, and food processing industry waste can be considered as a viable option for maintaining soil fertility. Both fertilizers and organic manures are the major source of nutrients for crop production in agriculture. In inorganic fertilizers such as urea, DAP (diammonium phosphate), and MoP (muriate of potash), nutrients are in inorganic forms and are easily accessible to crops. However, in organic manure, both organic and inorganic nutrient forms are present, which are slowly available to crops over a period of time. However, if we focus on how efficiently different nutrient elements are used as fertilizer formulations, it can be noted that in majority cases the efficiency of the applied fertilizers is way below 50%. This indicates that half of the applied fertilizer in most cases and more often almost 80–90% of the applied fertilizer remain unutilized by the plants and find entry into different sources like soil, water, and environment and become a source of contaminant. In the inorganic fertilizers, nutrient content is steady and does not vary much among various companies and is, therefore, more standardized. In contrast, organic manures vary in their nutrient content and composition depending on the feedstock or raw material used for compost preparation. Application of inorganic and organic fertilizers has both pros and cons. Indiscriminate use of inorganic fertilizers often results in increased risk of soil, water, and air pollution, which directly affects living beings. The inherent capacity of soil to produce crops in sustainable way is also hampered by imbalanced use of inorganic fertilizers. Therefore, improper management of both inorganic and organic fertilizers can detrimentally affect human health and ecological systems. According to FAOSTAT (2015), more than 50% of fertilizer is utilized by the United States, China, and India, which is visible by their agricultural and financial improvements. FAO (2016) also predicted that in the coming years, fertilizer consumption of Asian and Latin American countries would rise by 89%. Estimates of FAO highlight that supply of nutrients (N, P_2O_5, K_2O content in fertilizers) was 240 million tonnes at global level; however, nutrient requirement was 284 million tonnes in 2014, resulting in wide gap of nutrient scarcity for crop production. This high usage of fertilizers has resulted in buildup of nutrient stock in soil, leading to leaching and runoff loss affecting soil and water quality.

Both manure and fertilizers are essential for crop production, but indiscriminate use, rate of application, and improper storage can upshot contamination of environment. In recent times, the agriculture sources have become the major nonpoint source of pollution across the globe (www.epa.gov/nps/nonpoint-source-agriculture). Not only unmanaged use of fertilizers but also various other farm practices make agriculture as the dominant contributor to nonpoint source pollution. Various operations which lead to the contamination of the environment are as follows:

- Absence of proper soil and water conservation measures which leads to loss of soil and sediments through runoff.
- Poorly located or managed animal feeding.
- Excessive application of manures.
- Overgrazing practices.
- Excessive tillage operations or ploughing of the field at wrong time.
- Runoff from barnyards, feedlots, and croplands.

Many studies (Hanson 2002; Almasri and Kaluarachchi 2004) have revealed that higher crop production with indiscriminate use of fertilizers has become major source of water and soil pollution putting human health and ecological balance at risk (Khan et al. 2018). Fertilizers often contain certain heavy metals (HMs) such as chromium (Cr), cadmium (Cd), lead (Pb), arsenic (As), and mercury (Hg), and therefore, excess application on long term adversely affects the physical, chemical, and biological properties of soil. Beside soil health, crop metabolism is affected by accumulation of HMs, adversly affecting biochemical process resulting in collapse of cell organelles and plant death sometimes (Nagajyoti et al. 2010; Gupta and Sandallo 2011). Contaminants from fertilizers find their entry into food chain and, thus, can affect animal and human health (De Vries et al. 2002). Even application of huge amount amount of fertilizers during lawn making and maintenance also results in exposure of humans to different contaminants (Madrid et al. 2002). Eutrophication is a major pathway of nutrient entry into water bodies risking aquatic and human life. Runoff and soil loss due to rain, irrigation, etc. often result in heavy loading of nutrients and contaminants in water bodies. Among the nutrients, N and P are the major culprits which affect adversely water bodies. Some nutrients like N, on one hand, in the form of ammonium (NH_4^+) from fertilizer or manure is used by crops, whereas, on the other hand, other forms like ammonia (NH_3) gas is lost to the atmosphere contributing to greenhouse gas emission.

Mostly, soil gets contaminated as a result of human action when the concentration of chemicals that are not originally found in nature, nutrients, or elements in the soil exceeds naturally occurring levels. Thus, the chemicals have direct impact on functioning of the associated ecosystem and environment at an unacceptable level, it is called as soil pollution. Soil pollution brings detrimental changes in various soil properties, which adversely affects crop production, soil quality, human nutrition, and surrounding environment and thereby causes huge disturbance in the ecological balance (Tao et al. 2015). Application of precise dose of fertilizers is essential to supply optimum nutrient supply for crop production. Excessive application of fertilizers often leads to accumulate contaminants, which can adversely affect natural resources like soil, air, and water. All the above-mentioned factors strongly point out the significance of understanding the contamination of soil and water by various inorganic and organic fertilizers used for agricultural production. This chapter focuseson the various contaminants present in both fertilizers and manures and their influence on soil and water quality.

1.1 Inorganic Fertilizers: Consumption and Contaminants

Over the last few years, requirement of inorganic fertilizers increased worldwide with increase in crop production. There is a shift in the consumption pattern of fertilizers in many countries over a period of time. In Asian countries like India and China, fertilizer consumption increased rapidly; however, it almost remained constant in Western Europe and North America. As per various estimates by FAO, total fertilizer requirement is predicted to grow at 1.6% per annum globally. Demand for essential nutrients like N, P_2O, and K_2O is expected to grow by 1.5%, 2.2%, and 2.4%, respectively, from 2015 to 2020. This will result in an increase in the overall production and consumption of fertilizers in the next 5 years (FAO 2017). According to IFA (2017), with adoption of best management practices by farmers, it is hopeful that more efficient use of N and P fertilizers, followed by recycling and reusing of natural or organic nutrient sources, will tend to grow in coming years and emerge as a environment friendly pollution mitigation mechanism in agriculture.

In general, three major nutrients, N, P, and K, constitute a bulk of fertilizer industry, as these nutrients are essential for crop growth and development. It is estimated that among the three nutrients, N accounts for more than 60% of total nutrient utilized by crops followed by P and K. Compared to N fertilizer which is manufactured by chemical reaction between N from atmosphere and H from natural gas, phosphate and potash fertilizers largely involve digesting and mining activities (Arovuori and Karikallio 2009) of natural resources. As cited in many literatures, fertilizers, on one hand, were crucial element for green revolution, which resulted in significant increase in fertilizer production and consumption. There is no doubt that application of fertilizer has contributed greatly in raising agricultural productivity and reducing hunger worldwide (Erisman et al. 2008). This increase is shown in a study by Lu and Tian (2017) who reported an increase of 8 and 3 times in N and P fertilizer use. They also reported that during 1961–2013, an overall increase in fertilizer consumption resulted in improved N/P ratio to the tune of 0.8 g N g^{-1} P per decade, highlighting the role of human on ecosystem services. Some of the studies indicate that indiscriminative use of fertilizers resulted in a number of environmental and ecological problems (Sutton et al. 2011; Lu and Tian 2017). Some of the common problems often encountered, such as soil acidification, salinization, ground water contamination, eutrophication, crop yield reduction, greenhouse gas emission, and air pollution, result in deterioration of natural resources, thereby hampering sustainable food production (Ju et al. 2009; Guo et al. 2010).

Imbalanced application of chemical fertilizers increases the chance of environmental contamination. Fertilizers that act as a source of macronutrients and micronutrients to crops are also rich in heavy metal, radioactive compounds, etc. and become a major source of contaminants in long run to soil and environment. For instance, inorganic fertilizer application can affect soil health by forming hard soil surface, reducing soil pH, decreasing microbial process, negatively affecting physical and chemical properties of soil, and thus indirectly influencing crop production. Similarly, with the emerging trend of organic farming in West and South East Asia,

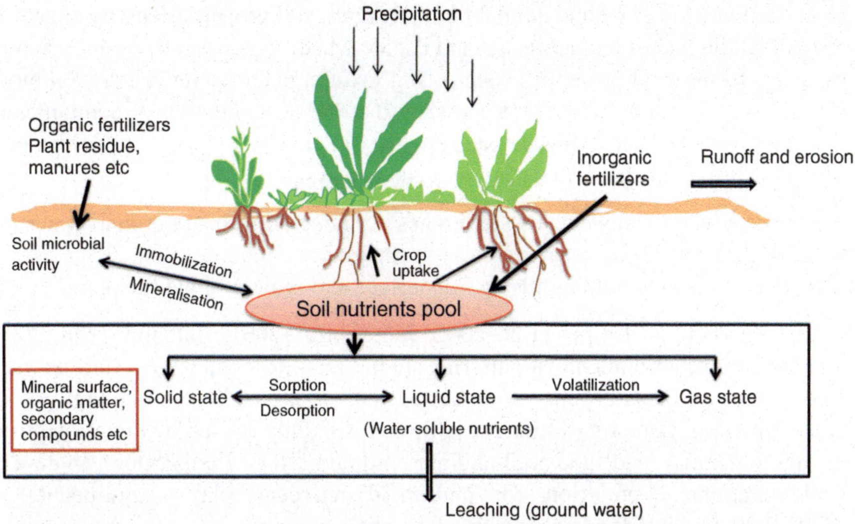

Fig. 1 Fate of organic and inorganic fertilizer behavior in soil

a lot of emphasis on the usage of organic fertilizers is promoted on larger scale across the world. But the major drawback in this system is the nonavailability of organic sources creating a wider gap between demand and supply. However, the organic fertilizers such as manures, crop residue, municipal solid waste (MSW), food processing industry waste, and others sometimes become the sink of various heavy metals, disease-causing pathogen, etc., which can have deteriorating effect on soil and water resources (Khan et al. 2018). Difference between organic and inorganic fertilizers is depicted in Fig. 1, showing the influence on soil and nutrient uptake.

Nitrogen and phosphorus are considered as backbone for any crop production system. Application of nitrogenous fertilizers in excess often results in various losses, such as leaching and volatilization, which not only reduces the nutrient use efficiency but also is an environmental threat. Nitrate is the major form of contaminants found in water bodies with excess application of N fertilizers. The second most consumed plant primary nutrient through fertilizers is phosphorus. The P is widely deficient in soil system, and 43% of world soils are scarce in available P needed for crop growth (Liu et al. 2012). The major problem with P fertilizers is the very low-use efficiency with only 10–15% of the applied fertilizer being utilized by the crop plant, while the rest remains in the soil or finds way into water bodies causing various environmental problems. In recent times, the consumption of fertilizer P increased by 3.2% from 2002 to 2010 (Lun et al. 2018). Phosphatic fertilizers manufactured from rock phosphate contain cadmium, and increased accumulation of Cd affects soil health. These contaminants might undergo some chemical changes and convert into different compound, which can be either more or less toxic to environment. In this context, sometimes HMs are easily absorbed by crops and tend to be accumulated in plant and animal body. Besides, soil properties and management also affect the fate of contaminants and decide whether they can be easily taken up by living forms. Soil properties such as soil texture, pH, organic matter, soil moisture, soil temperature, and heavy metals affect the accumulation of contaminants and their movement in soil–water system. Shayler et al. (2009) explained the mechanism of different behavior of contaminants in a system as follows:

(a) Some contaminants reach water bodies polluting surface and ground water.
(b) Some pollute air by escaping into atmosphere.
(c) Some pollutants bind tightly to soil surface and remain stable for years.

Many experts (e.g., Kolpin et al. 2002; Juhler et al. 2001; Battaglin et al. 2003) highlighted in environmental monitoring studies that EC is detected in various water body sources such as ground water, surface water, animal bodies like that of fish, and earthworm. Sometimes nutrient pollution is mainly caused by emissions from the agglomeration and industrial and agricultural sectors. Furthermore, in case of agglomerations, P emissions via household detergents play a significant role (ICPDR 2013). Nutrient discharge into water bodies can result from (i) point sources (in particular untreated/partially treated wastewaters) and/or (ii) diffuse sources (especially agriculture).

1.2 Organic Fertilizers and Contaminants

Organic fertilizers or manures are considered as biodegradable and are mostly from plant or animal origin. Most commonly used OMs include FYM, municipal solid waste, food industry waste, crop residue, different types of composts such as vermicompost, kitchen waste compost, distillery effluents, etc. According to Bruun et al. (2006) and Hargreaves et al. (2008), OM are used as fertilizers not only supply essential nutrients to plants but reduce chemical fertilizers requiement for micronutrients and eliminate the requirement of its consequent management or removal. Therefore, OMs are considered to be easiest way to recycle nutrient back to soil system. Besides acting as nutrient sources, OMs act as good soil amendments and conditioners and might reduce dependency on nonrenewable resources like fossil fuel for fertilizer production (Mondini and Sequi 2008). It has been stated that if OMs are utilized appropriately in crop production, then they are capable of supplying essential nutrients to crops. To revive barren or infertile soil, OMs are considered to be the best amendment and provide a better crop performance in agriculture (Soliva and Paulet 2001). Organic sources such as sewage sludge (SS) and animal manure are the most common organic wastes applied to soil either raw or composted. As suggested by many reports (Weber et al. 2007; Singh and Agrawal 2007), application of organic manures provides both macro- and micronutrients; improves organic matter, soil structure, bulk density, and other physical properties; and enhances microbial activities, resulting in efficient nutrient absorption by crops. Thus, proper utilization of different types of organic waste in crop production is encouraged rather than the conventional practices of applying inorganic fertilizers to carter productivity and soil health.

Generally, organic fertilizers are known for their slow release or transformation of nutrients, but the presence of contaminants such as HMs, other toxic compounds, and pathogen inoculums cannot be ignored (Petersen et al. 2003); besides, the bulky nature of OMs often results in high transportation costs. Especially, use of industrial by-products such as MSW, distillery waste, and fly ash often results in loading of HM besides PBTs (persistent, bioaccumulative, toxic chemicals) in soils and environment (cwmi.css.cornell.edu). Therefore, judicious and correct method of application of organic fertilizers should be strictly followed in order to avoid contamination of soil and water.

2 Contaminant Sources in Various Fertilizers and Manures

Contaminants from various inorganic and organic fertilizers are explained in this section. Various contaminants from fertilizer sources interfering with natural ecosystem deteriorate the soil, air, and water quality, thus directly affecting plants and animal life (Fig. 2). This section highlights the contaminants from the most commonly used N, P, and K fertilizers and organic manures and their ill effects on living forms and environment. Indiscriminate and long-term uses of inorganic and organic

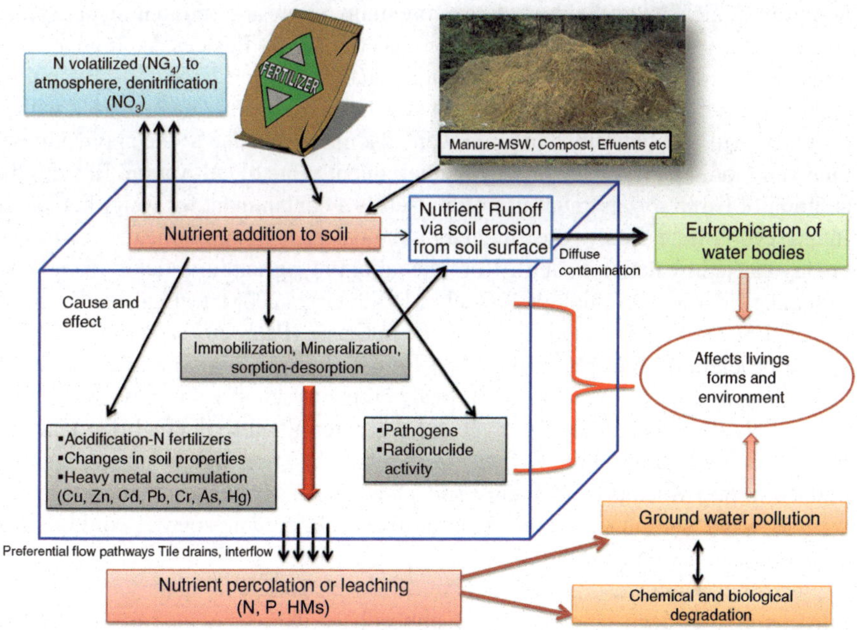

Fig. 2 Transformation of contaminants in inorganic and organic fertilizer in soil ecosystem

Table 1 Contaminants in fertilizers and manures

Inorganic fertilizers	Heavy metals	Others	References
N fertilizers	Cd	–	
P	Cd, As, Pb	Radionuclide like U, Ra, Sr	Khan et al. (2018)
K	Cl	–	
Micronutrients	As, Pb, Cd	–	MDH (2008)
Organic fertilizers			
Compost, biosolids, MSW	Cd, Zn, Ni, Pb, Hg, Cr, Cu	–	Smith (2009)

fertilizers result in accumulation of various contaminants, such as HMs and radionuclide (Table 1), thus raising environmental concerns by polluting natural resources especially soil, water, and atmosphere.

2.1 *Nitrogenous Fertilizers*

Despite the beneficial role of nitrogenous fertilizers in crop production, it is an undeniable fact that excessive N causes irreversible damage to our ecosystem. The consumption of N has increased from 12 Tg in 1960 to 113 Tg in 2010 to sustain agricultural production and feed the growing population globally (FAO 2016). Some ammonical N fertilizers such as ammonium nitrate and ammonium phosphate contain nearly 50–60 mg kg^{-1} of arsenic (As) (Li 2002). Besides As, higher accumulation of another HM like Cd was found in malt barely fertilized with ammonium nitrate especially in low N soils (Grant et al. 1996). Though barley increased with N fertilizers, Cd concentrations in grain were also increased due to the application of mono ammonium phosphate or KCl in consequent years of cultivation. Besides the addition of Cd to soil, nitrate form of N itself is a contaminant for water bodies. In India, several water bodies are affected by the nuisance of eutrophication, which is greatly caused by nutrient runoff from the adjoining agricultural fields. To name a few, Lake Udaisagar in Udaipur, Rajasthan (Vijayvirgia 2008), Dal Lake in Kashmir, Loktak Lake in Manipur, and Chilika Lake in Orissa (Patra 2012) were all affected by the problem of eutrophication. Most local water bodies adjoining agricultural fields are subjected to excess nutrient load coming as runoff from these fields. One latest study by Kritee et al. (2018) found that intermittently flooded rice farms in India emit 30–45 times more N_2O as compared to the maximum from continuously flooded farms, which predominantly emits methane. They suggest that co-management of water with inorganic nitrogen fertilizer and organic sources could reduce greenhouse emission by 90%.

Besides soil and water pollution, N fertilizers also affect air quality. The applied nitrogenous fertilizers are lost to the atmosphere in various gaseous forms like ammonia (volatilization), N_2O, and NO_x (denitrification). All these gases pollute the environment in various ways. The nitrous oxide (N_2O), in particular, is an important greenhouse gas (GHG) and has 6.2% contribution toward global warming. The persistence time of N_2O is 116 years with a global warming potential of 310 times in comparison to CO_2. Thus, with a meagre contribution of 6.2% in global warming in comparison to CO_2 (76%) and CH_4 (16%), it is one of the most potent threats to agriculture production and food security (Fagodiya et al. 2017). The nitrous oxide emission has increased steadily in the last two decades between 1990 and 2010 from 5.7 to 6.0 Tg N_2O per year with majority of the rise attributed to the intensification of agricultural practices involving increased fertilizer N application (Skiba and Rees 2014). Of the total N_2O emission, 36% is contributed by agricultural operations like fertilizer application and cultivation of farmlands (EPA 2010). The key factors, influencing nitrous oxide emission rates, are substrate availability (N), temperature (regulates the speed of the enzymatic reactions), pH, oxygen status, and also carbon for denitrification.

The National Oceanic and Atmospheric Administration (2009) has declared nitrous oxide as the second most potent ozone destroyer after chlorofluorocarbons (CFCs). In agriculture, N fertilizer is the main source of contamination, and other sources such as animal manure, sewage sludge, industrial process, and combustion reaction are major sources that release nitrous oxide to the atmosphere. Depletion of the protective ozone layer, which acts as a filter in removing the ultraviolet radiations, may cause various diseases like sunburn, cataract, genetic mutations, and cancer. However, depletion of ozone layer in the poles has resulted in more cooler springs and accentuated the contrast in temperature between poles and equator in the southern hemisphere. This contrast has caused stronger air circulation and more extreme storm events and can be considered as a direct consequence to climate change (Robinson 2014). The photochemistry of ozone was first described by Chapman (1930). Ozone is produced by the photolysis of O_2. The following reactions show the process of ozone formation.

$$O_2 + h\upsilon \rightarrow O + O \tag{1}$$

Followed by reaction of O and O_2,

$$O_2 + O + M \rightarrow O_3 + M \tag{2}$$

O and O_3 are quickly cycled between each other via

$$O_3 + h\upsilon \rightarrow O_2 + O \tag{3}$$

The ozone (O_3) again reacts with nascent oxygen and forms O_2.

$$O_3 + O \rightarrow O_2 + O_2 \tag{4}$$

Various reactive species in the atmosphere such as NO_x and ClOs react with the stratospheric ozone causing its depletion. The reactions in Eqs. 5 and 6 show how NO causes O_3 depletion.

When O_3 reacts with NO, the following reaction takes place,

$$NO + O_3 \rightarrow NO_2 + O_2 \quad (5)$$

NO_2 is again recycled back to NO, and the chain reaction leading to O_3 degeneration continues.

$$NO_2 + O \rightarrow NO + O_2 \quad (6)$$

NO is recycled back to its original form, which makes it a very potent agent for O_3 depletion. The NO can typically destroy 10^3–10^5 molecules of O_3 before it is converted to some less reactive form (Lary 1997). Comparing the ozone depletion potential (ODP) of various greenhouse gases, the ODP weighted emission due to N_2O is the highest. In the present time also, the increased concentration of N_2O routed through anthropogenic activities would be the major O_3-depleting agent compared to the halocarbons (Ravishankara et al. 2009).

2.2 *Phosphatic Fertilizers*

Compared to N fertilizers, application of phosphatic fertilizers is of significance as they accumulate not only HMs but also radioactive materials like radionuclides of ^{238}U, ^{232}Th, ^{210}Po, ^{226}Ra, and ^{40}K (FAO 2009; Sonmez and Snmez 2007; Hassan et al. 2016). Along with metal impurities, rock phosphate (RP) used in fertilizer industries contains high concentrations of radionuclides from the ^{238}U decay series, such as U, Ra, and Rn (Lopez et al. 2010). Amendments like phosphogypsum (PG) are more radioactive than P fertilizers. Potential risk by exposure to radionuclide-enriched fertilizers through food consumption in human cannot be neglected (Rehman et al. 2006; Nowak 2013). Another study by Lopez et al. (2010) widely described the radioactive impact of U-series radionuclides in phosphate rock (during the industrial process) wastes on the environment. In contrast, several studies (Saueia and Mazzilli 2006; Righi et al. 2005) suggested negligible amount of radioactive contaminants in phosphate fertilizers. Release of by-products into environment from fertilizer industries is a serious threat to aquatic life, human beings, and other life forms, creating ecological imbalance in nature. One such instance was reported by Perianez et al. (1996) where 20% of the phosphogypsum was disposed to environmental systems until 1997 and was just considered as a waste management practices by industries.

Fluoride is closely associated with RP as majority of the mineral is present in the form of fluorapatite. While manufacturing commercial fertilizers, F is released into the atmosphere, which is recycled back to the earth's surface during rainfall. Also, phosphor-gypsum, a by-product from the P fertilizer industry, also leads to F contamination (Mirlean and Roisenberg 2007). In India, Unnao district in Uttar Pradesh is fatally affected by F contamination and has seen a rise in P fertilizer consumption by 5 lakh metric tonnes in the past one decade (www.downtoearth.org.in). Lead and cadmium contamination is another problem arising from excessive use of P fertilizers. Lin (1996) detected varying amount of HMs such as Cd 9.5–96.4 mg kg^{-1}, As 19.4–273.0 mg kg^{-1}, Pb 5.6–17.2 mg kg^{-1}, and Hg 0.01–0.42 mg kg^{-1} in rock phosphate and phosphorus fertilizers. In another study, triple superphosphate fertilizer application led to increased Cd concentration compared to other fertilizers (Atafar et al. 2010). Besides the primary nutrient fertilizers, arsenic (As) concentration in soil has increased due to zinc sulfate application. Phosphate fertilizers were important carrier of heavy metals such as Zn, Cu, and Cd in agricultural soils of England and Wales (Nicholson et al. 2003). In many scenarios, long-term application of P fertilizers results in accumulation of HMs and radionuclides, which could be potential threat to environment and organisms (Huang and Jin 2008).

2.3 *Potassium Fertilizers*

Potassium fertilizer such as KCl is commonly applied by farmers, and excess application of the fertilizer often leads to accumulation of Cl^- ions in soil. Some studies have shown that chloride anions accumulate to toxic levels in legumes. These contaminants are known to enter food chain through fertilizers or other chemicals used for food production. To avoid the adverse consequences of Cl^-, K_2SO_4 would be preferred as a fertilizer source (Khan et al. 2013). In another report, Grant et al. (1996) highlighted increased concentrations of Cd in malting barley due to KCl application. Fertilizers containing high level of sodium and potassium can have negative impact on soil properties such as soil pH, microbial life, and soil physical properties like structure and bulk density, which can hamper crop production (Savci 2012).

2.4 *Organic Fertilizers*

Heavy metals are the major contaminants in organic manures. Application of organic fertilizers (i.e., compost, sludge, or manure) to fields, especially agricultural crops, provides significant input not only of nutrients (i.e., nitrogen, phosphorus, sulfur, and micronutrients) but also of some heavy metals, some of them being toxic, such as cadmium or lead (Pinamonti et al. 1997; Lipoth and Schoenau 2007). Organic manures like sewage sludge can be used in agriculture provided, HMs content should be within threshold limits for soil application. In majority of conditions, organic fertilizers are usually considered “best,” but uncontrolled use of manures may cause environmental damage due to its high content of nitrogen released into the soil. Nitrogen present in the organic fertilizers transforms rather slowly into ammonium nitrate, and thus nitrate. Similar transformation happens with inorganic fertilizers, where urea converts into nitrate a little rapidly and more rapidly with ammonium nitrate fertilizers. The chemical transformation rate depends greatly on the soil microbial activity present in soil, and environmental conditions such as warmer temperatures and humidity favor this increased rate of transformation. Under high temperature especially during spring and summer, rapid conversion occurs compared to winter and dry conditions. The formed nitrate sometimes is absorbed by crops, and the excess is lost via leaching to subsurface depth, causing pollution of ground or surface water and sometimes resulting in eutrophication of water bodies.

Contaminant concentration should be kept in mind before planning for the reuse of waste materials as organic fertilizers. It is clear that continuous and long-term application of organic fertilizers from unknown sources often favors accumulation of HMs and other contaminants in soil and water system. Interaction of HMs and other contaminants with soil components provides a direct entry into food chain, and foods directly grown on such soils adversely affect animal and human health (Khan et al. 2008; Smith et al. 2009; Zhuang et al. 2009). Other studies by Smith et al. (2009) and Lopes et al. (2011) reported that heavy metal concentration in SS is 50–90% more than compost and 20 times more than manure (especially concentration of Cd and Pb). Industrial effluents from wastewater treatment plant of sewage sludge (or compost) are major sources of HMs concentrations, and the amount of pollutants varies depending on the composition of domestic waste and country origin (Bose and Bhattacharyya 2008: Egiarte et al. 2009). Organic manures such as vegetable fruit waste from food processing industries, municipal solid waste, or sewage sludge are some of common manure sources. Another waste source is from food markets rich in nutrient levels, organic matter, and moisture content (Varma and Kalamdhad 2015). In a similar study on food market wastes in Chimborazo Region of Ecuador, Jara et al. (2015) reported mean values of OM and N–P_2O_5–K_2O as 77.3% and 2.5% ± 0.7% ± 3%, respectively. Though application of such organic waste or manures is a sustainable way to recycle nutrient and carbons into soil, precaution is needed to evaluate the possible source of contaminants present in such organic fertilizers before disposing into agriculture fields. This will not only prevent the entry of contaminants into food chain but also protect environment.

2.4.1 Heavy Metals: Major Contaminant in Organic and Inorganic Fertilizers

Soil is considered a long-term sink for toxic elements often referred to as heavy metals, such as Cu, Zn, Cd, Pb, Cr, As, and Hg. In India, heavy metal contamination in soil due to anthropogenic activity has been reported from different areas (Sachan 2007; Shanker 2005; Deka and Bhattacharyya 2009). In agriculture, soil is the major contributor of heavy metals, which includes liming materials, irrigation water, and sewage sludge as shown in Table 2. The HMs such as Cd, Cr, Zn, Pb, Cr, and As are highly contributed by inorganic fertilizers, pesticide, and organic sources in agriculture (Kelepertzis 2014; Toth et al. 2016).

Table 2 Total concentration of selected heavy metal in manures (ppm on dry weight basis)

Source	Arsenic	Cadmium	Chromium	Lead	Nickel	Copper
Cow manure	–	8	58	16	29	62
Poultry manure	0.35–110.5	–	0.6–19.6	–	–	3.5–13.5

Chhonkar (2003)

Table 3 Source of HM contaminants from inorganic and organic fertilizers in agriculture

Source	Heavy metal inputs	Contaminants	References
Inorganic fertilizers	Phosphate fertilizer	Cr, Cd, Cu, Zn, Ni, Mn, and Pb	Atafar et al. (2010), Sun et al. (2013), Toth et al. (2016), Kelepertzis (2014), etc.
	Nitrate fertilizer		
	Potash fertilizer		
	Lime		
Organic fertilizers	Animal manures	Zn, Cu, Ni, Pb, Cd, Cr, As, and Hg	Nicholson et al. (2003), Singh and Agrawal (2008, 2010), Niassy and Diarra (2012), Srivastava et al. (2015, 2016), Sharma et al. (2017), etc.
	Sewage sludge		
	Compost		
	Fly ash		

Srivastava et al. (2017)

Niassy and Diarra (2012) reported that sewage sludge, manure, and limes are major sources of cadmium enrichment. Repeated use of phosphatic fertilizers often results in deposition of HMs like Cd in soils. However, long-term application of sludge materials accumulates Cd, Cr, Ni, Pb, Cu, and Zn and builds up micronutrients like Cu, Mn, Cu, Co, and Zn (Srivastava et al. 2017). Land application of sewage sludge is one of the major contributors of heavy metal to the soil system (Srivastava et al. 2016; Sharma et al. 2017). Several studies had shown that both organic and inorganic fertilizers contribute to HM contamination as shown in Table 3. In a report by Daniel and Perinaz (2012), total urban solid waste generates nearly 68.8 million metric tonnes per year (TPY) or 1,88,500 metric tonnes per day (TPD). Out of the total, 9–10% of these wastes enter into agricultural land directly in the form of compost rich in heavy metals. This is a serious concern with the present data indicating higher accumulation of HMs by agricultural inputs directly influencing soil, water, air, and organisms. Threat or potential risks due to contamination by HMs, radionuclides, and other form of contaminants cannot be neglected in agriculture, though availability or transformation of these contaminants varies depending on soil type, input type, rate, mode of application, etc. Soil acts as a big reservoir for contaminant retention and degradation in long run protecting the environment and its ecosystem services so as to sustain several life forms. Besides negative effect on human, HMs also adversely affect soil microbial diversity, microbial-mediated process, and soil–microbe interaction (Gall et al. 2015; Rai et al. 2018). Soil faunae like invertebrates, small mammals, worms, and various agriculturally important

Table 4 Permissible HM content in soil and food material

Elements	World range of elements in nonpolluted soil (mg kg^{-1})	Maximum allowable limits of elements in fruits and vegetable (mg kg^{-1}) (*dry weight basis*)
Cd	0.07–1.1	0.2
Pb	10–70	0.3
Cu	6–60	40
Cr	5–121	2.3

Banerjee et al. (2010)

insects are affected by HM contamination (Bartrons and Penuelas 2017; Rai et al. 2018).

Soil type like those with high clay content have high buffering capacity which does not signify the effect of soil pH on bioavailability of metals and thus control metal chemistry in soils (Baldwin and Shelton 1999). The availability of metals for crop uptake from sewage sludge and other composts sometimes depends on intrinsic properties of the materials themselves. There is a need to develop more careful management scheme for experiments related for study of HM uptake by crops through compost and other sludge treatments (Smith 2009). Table 4 presents safe values for Cu, Pb, Cd, and Cr in fruits and vegetables recommended by WHO/FAO and range of heavy metals in nonpolluted soil.

Long-term application of excess organic manures with chemical fertilizers accumulates HMs like Cu, Zn, Cd, Cr, Pb, As, etc. in soils under vegetable fields of China (Huang and Jin 2008). They reported an increased accumulation of total Cu, Zn, and other heavy metals in soils with increase in vegetable production. Nicholson et al. (2003) reported the presence of high amount of HMs like Cu, Zn, Cd, other contaminants from P, and other fertilizers. In plants, Cd accumulation has a negative effect on N metabolism as it alters oxidant levels, resulting in oxidative stress with accumulation of active oxygen species (AOS), including superoxide radical (O^{2-}), hydroxyl radical (OH), and hydrogen peroxide (H_2O_2) (Gallego et al. 1996; Hassan et al. 2005). Cadmium accumulation in soil and its uptake by crops tend to induce stress in plants, thereby affecting the photosynthetic trait and its antioxidative pathway leading to growth reduction. Higher concentration of Zn usually present in sewage sludge and compost is relatively available and is easily transferred to plant tissues resulting in higher bioaccumulation (Speir et al. 2004). Bioaccumulation of HMs in plants interfere with metabolic pathways and biochemical reactions and directly affect photosynthesis, assimilation of biomolecules and elements, etc. (Kabata-Pendias and Pendias 1992), resulting in plant senescence and death.

With recent developments in agriculture, intensive cropping system is practiced by farmers, which forces excessive use of inputs such as fertilizers and pesticide in soil. These chemicals often leave residues in soil and get transported to water bodies, and thus contributes significantly to water and soil pollution (Almasri and Kaluarachchi 2004; Khan et al. 2018). Thus, pollution by contaminants has put human and animal life at risk, on the one hand, and environmental degradation, on the other hand. Fertilizers containing HMs such as Cr, Zn, Cd, Hg, and As from the raw materials contribute for higher accumulation of HMs in the soils (Huang and Jin 2008). Discriminate and blanket dose of fertilizer applied to crops results in HM buildup in soil and deteriorates soil functions, thereby adversely affecting crop growth and development. Such situations often affect both biochemical and physiological plant processes, leading to the degeneration of organelles and cells that may result in plant death (Nagajyoti et al. 2010; Gupta and Sandallo 2011). There are several studies that indicate that continuous and excessive application of inorganic and organic fertilizers will not only add nutrients to the soil but also considerable amounts of HMs in soil and plant systems. Some examples of heavy metal accumulation in soils of different experiments are shown in Table 5.

In last few decades, more emphasis has been put on the reuse of organic sources such as manure, distillery effluents, sewage/sludge water, and fly ash on agricultural field in order to reduce dependency on nonrenewable resources. Increasing population and urbanization had created more pressure for agricultural productivity with limited land use, and this has often pushed use of low-cost methods of applying such manure forms in soil, which has resulted in high buildup of HMs affecting adversely human health (Rai et al. 2019). Countries with high population like China, India, and African countries such as Zambia and Nigeria are using wastewater from sewage/sludge for irrigation without proper treatments, having direct impact on food quality and environmental issues. Long-term use of wastewater for irrigating crops in India showed accumulation of HM in plant tissues of food crops and poses health risk (Ghosh et al. 2012; Garg et al. 2014; Saha et al. 2015; Chabukdhara et al. 2016). However, in European and American countries, fertilizers, fungicides, and modern agricultural practices were responsible for HM contamination in food crops (Rai et al. 2019).

3 Fertilizers and Manures and Their Impact on Soil Health

3.1 Impact on Soil Properties

Both inorganic and organic fertilizers influence soil physical, chemical, and biological properties. The various soil properties influenced by the addition of inorganic fertilizers and organic manure are shown in Fig. 3. In this section, various soil properties influenced by inorganic and organic fertilizers application are explained.

3.1.1 Effect on Soil Physical Health

Some important physical indicators of soil are bulk density, water availability, hydraulic conductivity, compaction, pore size distribution, and soil surface cover. Soil structure is a dominant soil indicator used for crop production, which has direct influence on soil health. Application of fertilizers like $NaNO_3$, NH_4NO_3, KCl, K_2SO_4, and NH_4Cl deteriorates the structure (Savci 2012).

Table 5 Effect of various fertilizer containing HMs and their influence on soil properties

S. No.	Crop	Country	Source of fertilizer and manure, application rate	Effect of fertilizer and manure contaminants (HMs)	References
1.	Lettuce	China	Phosphate rock (PR) and triple superphosphate (TSP)	Average of 1% or less Cd was accumulated in lettuce tissue. Applications of the fertilizers at high rates could result in increased Cd accumulation in the soil over time	Huang et al. (2004)
2.	Potato–sugarbeet	Hamadan province of Iran	Pollution index was calculated for each element	Enhanced levels of As, Cr, Cu, Mn, Ni, and Pb in P-amended soils from sugar beet fields; Pb, Cr, As, and Cd for soils from potato fields; and Fe and Zn for soils from both potato and sugar beet fields	Cheraghi et al. (2012)
3.	–	Hesse, Germany	14 years of fertilizer application	Pseudo- and mobile metals (Cd, Cu, Mn, Pb, and Zn) in soils increased following 14 years of mineral fertilizer treatments (N, P, NP, and NPK). Long-term fertilizer use increased soil metal content, soil organic C, CEC, and decreased soil pH level	Czarnecki and During (2015)
4.	Land-use pattern (3): vegetable field, bare vegetable field, and grain crop field	Bejing, China	20 years of cultivation	Long-term use of excessive chemical fertilizers and organic manures in the bare vegetable field and the greenhouse vegetable field contributed to the accumulation of Cu, Zn, and other heavy metals in the soils. Cd pollution was relatively more serious in the bare vegetable field and the greenhouse vegetable field than that in the grain crop field	Huang and Jin (2008)
5	Soybean–wheat	Brazil	Phosphatic fertilizers and agricultural gypsum	Higher accumulation of uranium and Thorium radionuclides in soil where fertilizers; clayey texture retained more radionuclides than red latosol of mixed texture	Saleh et al. (2007)
6	Rice–wheat	India	Five levels of sludge, i.e., 0 (S0), 10 (S10), 20 (S20), 30 (S30), and 40 (S40) t ha^{-1}, applied to rice crop and wheat grown as residual crop with fertilizers	Improved rice and wheat yield; cd content in rice grain was above the Indian safe limit at 20 t ha^{-1} or higher levels of sludge application; significant buildup of P, S, Zn, Fe, and Mn in postharvest wheat soil at 40 t ha^{-1} sludge application	Latare et al. (2014)
7.	Animal feed and manure	England and Wales	Dairy and pig manure (85 samples)	Increment of about 5247 mg Zn, 1821 mg Cu, and 225 mg Ni per kg dry matter added to agricultural lands	Nicholson et al. (1999)

8.	Animal feed and manure	China	Pig, cattle, and chicken feed and manure (224 samples)	Long-term agricultural application of animal manure increases the potential risk of cu pollution in soil and surface water.	Zhang et al. (2011)
				Mean Cu concentrations in pig, cattle, and chicken feeds were 179.8, 16.6, and 20.8 mg kg^{-1}, respectively. Cu concentrations in manures ranged from 1.5 to 1521.2 mg kg^{-1}	
9.	Pot experiment on herbaceous plants	Shanghai, China	Sewage sludge collected from waste water treatment plant	Long-term trials of sewage sludge application show accumulation of trace metals in the soil especially Ni and Cd; sewage sludge significantly affects heavy metal uptake by herbaceous plants	Dai et al. (2006)
10.	Column experiment (Composting of MADD)	Kent, UK	Biosolids namely fresh mesophilic anaerobic digested dewatered (MADD) sludge cake used, application rate of biosolids (250 kg N ha^{-1} $year^{-1}$)	Enriched the soil with Zn: 6 mg kg^{-1}, Cu: 2 mg kg^{-1}, Pb: 5 mg kg^{-1}, and Ni: 0.2 mg kg^{-1}	Gove et al. (2001)
				The movement of metals followed order Zn = Pb > Cu > Ni; composting or drying of MADD increases contamination of ground water	
11.	Silver beet	New Zealand	Composted sewage sludge for 4 years up to a maximum rate of 200 t ha^{-1} y^{-1}	Total heavy metal concentrations in the soil were high, but basal respiration, microbial biomass C, and anaerobically mineralizable N were significantly increased in the compost-amended plots relative to the control	Speir et al. (2004)
12	Long-term study: red beet, sugar beet, or barley	Woburn, England (23-year-old experiment)	Sewage sludge or composted sludge	Zn and Cu present in compost sludge organically bound fractions in compost-amended soil; there was no evidence of an increase in bioavailability of Zn and Cd to the crop plants with time	McGrath et al. (2000)
13	Barley	Toledo, Spain (5-year study)	Compost applied at two rates of 20 t ha^{-1} and 80 t ha^{-1} with mineral fertilizers	MSW compost increased microbial biomass; activities of the intracellular enzymes increased with the rate of compost addition; urease and phosphatase activities were reduced in compost-amended soil. Heavy metal toxicity was indicated as one possible cause of the inhibition of these enzymes	Garcia-Gil et al. (2000)

(continued)

Table 5 (continued)

S. No.	Crop	Country	Source of fertilizer and manure, application rate	Effect of fertilizer and manure contaminants (HMs)	References
14	Wheat	Tunisia (Short-term study)	MSW compost applied at 40 t ha^{-1} and 80 t ha^{-1}; farmyard manure was applied at 40 t ha^{-1} with or without chemical fertilizers	At 80 t ha^{-1}, plants showed an increase of heavy metal (Cu, Cd, Zn) content in all plant parts but did not affect crop yield and growth	Lakhdar et al. (2009)
15	Maize–soybean	China	10-year study; pig manure (0, 100, 250, and 500 kg total N ha^{-1} $year^{-1}$ for 8 years and 0, 10, 25, and 50 t of pig manure fresh weight ha^{-1} $year^{-1}$ for 2 years)	Serious accumulation of Cu and Zn in soil, total Cu and Zn concentrations increased by 204% and 107% at high application rates; Cu and Zn leaching occurred in the tested soil; Cu and Zn contents in plants parts were not affected by manure	Xu et al. (2013)
16	Tomato	Tunisia	Sewage sludge was mixed with the soil at 2.5%, 5%, and 7.5% (DW) proportions	Among the three HMs (Zn, Cu, and Cr), Zn had the highest capacity for transferring from soil into plants. Low metal translocation was observed from roots to leaves. The 7.5% SS dose decreased	Elloumi et al. (2016)
				Biomass production and caused a decline in chlorophyll content and stomatal conductance	

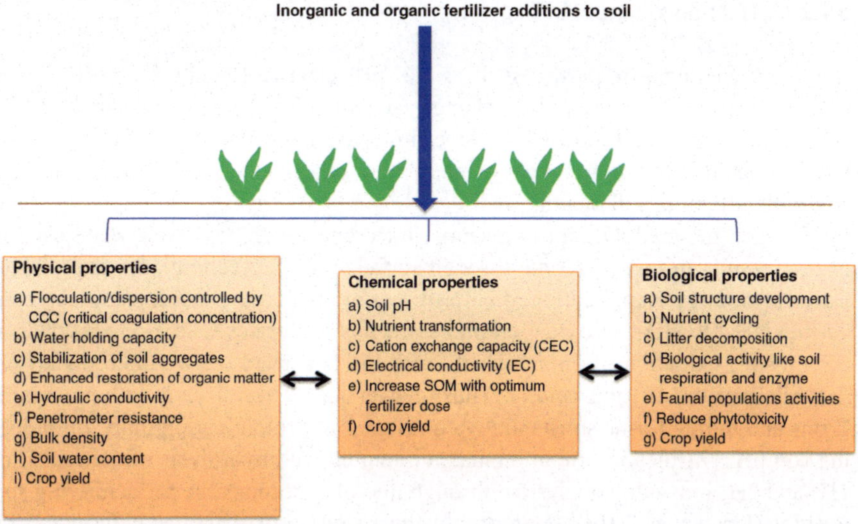

Fig. 3 Soil properties influenced by inorganic and organic fertilizers

Soil physical properties unlike chemical, are indirectly effected by fertilizers and manures. They contribute to better crop canopy, root distribution that directly effects physical properties like soil aggregation, bulk density, water movement etc. All the nutrients in the form of N, P, and K fertilizers directly play an important role in soil properties such as flocculation and dispersion depending on the composition of synthetic compounds. Both flocculation and dispersion are governed by critical coagulation concentration (CCC). As underlined by Sposito (1989), the lowest electrolyte concentration at which a soil suspension turns unstable and subjected to rapid flocculation or coagulation under a specific set of conditions is termed as CCC. Few studies indicate the benevolent impact of phosphatic fertilizers on soil physical properties (Thein 1976; Yeoh and Oades 1981a, b). Some of the properties improved by the addition of P fertilizers are lowering bulk density, enhanced water holding capacity, flocculation, etc. The addition of fertilizers is also known to improve stabilization of soil aggregates, which contributes significantly to the SOM stabilization. Similarly, application of fertilizers improved the mean weight diameter and aggregated stability of soil, which might be due to the phosphatic fertilizers binding the soil particles together (Marathe and Bharambe 2005; Selvi et al. 2005). However, Interweck et al. (1982) reported addition of ammonical fertilizers sometimes cause dispersion of soil colloids. But they did not observe any affect of ammonical fertilizer on soil physical properties like bulk density, penetration resistance, aggregate stability etc.

3.1.2 Effect on Chemical Properties

Important chemical properties such as soil pH, electrical conductivity (EC), soil nutrients interaction, and cation exchange capacity (CEC) govern soil suitability for crop production. Addition of excessive dose of chemical fertilizers to soil often leads to acidification and heavy metal and other contaminant accumulation, adversely affecting soil health.

Nitrogen in small amount is essential for crop growth; however, when used in excess, results in loss of nitrate through surface runoff, leaching (Bai et al. 2010; Lucas et al. 2011). Application of N fertilizers often increases acidification by converting NH_4^+ to NO_3^- in oxidation process, which generates H^+ and lowers the soil pH. In soil, oxidation of ammoniacal fertilizers is carried out by microorganisms that release ions causing acidity. Thus, conversion of fertilizer N to ammoniacal forms and nitrate forms substantially contributes to proton loading into soil reducing soil pH. Nitrification of ammonia, ammonium, and hydrolysis of urea releases $1H^+$ and $2H^+$ ions into soil system through microbes responsible for acidifying soil reaction (Khan et al. 2018). Interestingly, the assimilation of applied N fertilizers by crops in the form of nitrate-N and sulfate-S into organic forms, however, consumes H^+ ions, resulting in alkalinity of soil. There are number of studies illustrating the effect of N fertilizers on soil acidity (Chien et al. 2008; Jackie et al. 2011). The reduced pH increases Al solubility in soil, which becomes toxic for crop growth at pH <5.5. The study by Guo et al. (2010) in China reported that soil pH decreased significantly with application of excessive N fertilizer resulting in severe soil acidification, during 1980–2000. They also reported that application of N fertilizer released protons to the tune of 20–221 kg ha^{-1} $year^{-1}$, on the one hand, and base cation uptake contributed to additional 15–20 kg H^+ ha^{-1} $year^{-1}$ resulting in soil acidification, on the other hand. Chien et al. (2008) observed that application of various ammoniacal fertilizers changed soil pH in the following order: $(NH_4)_2SO_4 < NH_4NO_3 <$ urea<control. In contrast, study by Darusman et al. (1991) observed no significant difference in soil pH when N was applied in the form of NH_3, NH_4NO_3, urea, and urea–NH_4NO_3 at rates ranging from 0 to 224 kg ha^{-1} during 20-year period. Chen et al. (2013) also described that soil acidification induced by N fertilizers is a serious threat for organisms and various ecosystem services. Soil acidification directly influences soil pH by increasing ions of Al and Fe in soil colloids and decreasing basic cations like Ca^{2+}, Mg^{2+}, and K that are essential for crops. Acidic soil impedes P availability to crops due to higher P fixation by forming insoluble Fe and Al phosphate compounds.

Another major aspect of inorganic fertilizers addition is its influence on soil organic matter (SOM) cycling. Triberti et al. (2008) highlighted the role of inorganic and organic fertilizer addition to C sequestration in cropping system. Compared to nonfertilized plots, addition of inorganic fertilizers with manures or crop residues improved soil carbon content and sequestration efficiencies in various cropping systems (Gong et al. 2009; Huang et al. 2010). Although SOC dynamics varies with climate and soil type, optimized fertilization with inputs and proper agronomic management are key factors that improve SOC buildup in soil. In contrast, few studies by Khan et al. (2007) and Mulvaney et al. (2009) suggested that continuous application of N fertilizers could result in net loss of organic N and organic C to the tune of 92% and 74%, respectively, from temperate and tropical regions. Balanced use of chemical fertilizers help in SOM build up compared to those without fertilizers applied during cropping system (Geiseller and Scoe 2014; Korschens et al. 2013)

3.1.3 Effect on Microbial Life

Soil biota highly influences soil health and is considered a prominent indicator for improving soil quality. The soil biological indicators govern various mechanisms, such as nutrient cycling, immobilization–mineralization, residue decomposition, soil respiration, and biomass addition.

Among microbial fauna, both bacteria and fungi play important role in nutrient transformation and govern the availability to plant roots. Immobilization and mineralization of nutrients from organic matter are highly influenced by soil microbial diversity that signifies the bioavailability of N, P, S, and few micronutrients essential for plants. Allison and Martiny (2008) in a review highlighted that 84% of 38 experiments indicated that microbial biodiversity is highly sensitive to N, P, and K fertilizers. Earlier studies underlined reduction in microbial population with the application of N fertilizers due to reduction in soil pH (Bittman et al. 2005).

Even symbiotic relationship of microorganisms with plants roots is also influenced by the availability of nutrients, for example, suppress *Rhizobium* activity in legumes with excess N fertilizers (Savci 2012). Reduction in microbial biomass P with 200 mg N kg^{-1} (ammonium sulfate) was reported by Saggar et al. (2000) during 168 days of incubation. Root colonization by arbuscular mycorrhizal fungi (AMF) was reduced with P and N application in long-term experimental studies (Ryan et al. 2000). However, inorganic fertilizers do not always suppress microbial community; instead, they indirectly help build up SOM. This improves microbial population and contributes to nutrient transformation and other benefits in soil system. Geiseller and Scow (2014) in their review paper reported 15.1% increase in soil microbial biomass carbon compared to control plots under long-term fertilizer use. This is quite clear that fertilizer application improves crop or plant biomass which adds back root exudates, plant parts back to soil and enhances microbial activity in a managed ecosystem.

3.2 Water Quality Issues Associated with Organic and Inorganic Fertilizers

Almost all the countries rely heavily on ground water for drinking water supply, livestock water, irrigating crops, industrial uses, and other uses. Scope of ground water contamination in many countries is very critical. Impact of fertilizer contaminants on water quality is a serious issue, which caters immediate attention worldwide. According to Organisation for Economic Co-operation and Development (OECD) countries, agricultural water quality has been identified as a major environmental issue and as a topic for policy analysis and is an issue of relevance across all OECD countries. The primary agricultural sector is mainly responsible for nitrate, phosphorus, pesticide, soil sediment, salt, and pathogen pollution of water from crop and livestock activities, but it can also play a role under certain farm practices in terms of improving water quality through a water purification function. Surface water is primarily affected through soluble contaminants via surface runoff or insoluble contaminants carried on soil particles during erosion events. However, ground water can be contaminated with pollutants through percolation, seepage, and infiltration rate. Water pollution from agriculture has associated costs in terms of removing pollutants from drinking water supplies, as well as damage to ecosystems and commercial fishing, recreational, and cultural values associated with rivers, lakes, ground water, and marine waters (Parris 2011). This nutrient enrichment of aquatic bodies is known as eutrophication, which results in an algal bloom by undesirable plants covering the water surface and decreases the biological oxygen demand that affects living things. The decomposition of organic matter from undesirable aquatic plants causes the water body to have depleted oxygen levels and restricts water use for fisheries, recreation, industry, and drinking. In most of the developing countries, lack of consistent and comprehensive database of contaminants from agrichemicals has made difficult to link nonpoint source of pollution and human activities. In many cases, extent of agricultural ground water pollution is generally less well documented than that of surface water, chiefly due to the costs involved in sampling ground water and because most pollutants take a longer time to leach through soils into aquifers. Among the various macronutrients used as fertilizers in agriculture, in particular nitrogen (N) and phosphorus (P) can cause eutrophication of surface waters. Further, their emission and discharge into coastal areas and the marine environment can significantly impact upon the status of those ecosystems. According to ICPDR (2013) during the period of 1988–2005, an average of about 35,000 and 400,000 tonnes of inorganic P and N, respectively, is transmitted into the Black Sea through Danube river each year. Some of the major contaminants and their impact on water quality are discussed below.

3.2.1 Nitrogen Contamination in Water Bodies

Nitrogen is the most important nutrient required by crops and highly vulnerable to losses causing contamination of surface and ground water resources. Nitrogen is converted into various forms of nitrate (NO_3^-), nitrite (NO_2^-), and ammoniacal (NH_4^+) forms before assimilation by crops and organisms. Nitrate is the most common form of N available in soil and loss from soil system to nearby water bodies. However, potential of nitrate in polluting the soil depends on soil type, N application rate, rainfall, irrigation, etc. Leaching is identified as a common mechanism through which N is lost from root zone and contaminating water reservoirs on large scale. Nitrate form of N is negatively charged and, therefore, does not bind to soil surface, and thus easily leaches through soil profile and finds entry into ground water or subjected to surface runoff polluting the water reservoirs. Nitrate accumulation at the surface and bottom soil layer is reported by Wang et al. (2015) who studied leaching and accumulation of NO_3^- in a simulated rainfall experiment. However, NO_3^- concentrations initially increased but later decreased sharply and stabilized with fertilization levels. They reported that soil was able to retain 50.53% of total nitrate applied during experiment. This indicated that precipitation or irrigation in such soils would increase threat for nitrate contamination through surface and subsurface flow into water bodies. As per US environmental protection agency and health organizations, the acceptable level of NO_3^-–N levels in drinking water should be less than 10 mg L^{-1} (EPAR 2001). However, high NO_3^-–N concentration in ground water of Japan was noticed due to application of N fertilizers (Kumazawa 2002). The NO_3^-–N concentration reached 100 mg L^{-1} in some wells posing high risk to human health. There are studies which illustrate spatial and temporal variation in N loss from fertilizers. In a study by Chen et al. (2016), significant linear correlation between total nitrogen (TN) and NO_3^-–N was observed in surface runoff.

Globally, nonpoint source of N pollution has gained attention due to its serious threat to aquatic and human life. It is estimated that agriculture itself contributes to nearly 75% of nonpoint source of pollution in the United States (Line et al. 1998). In Denmark, nearly 94% of nitrogen buildup in 270 water bodies was reported due to nonpoint source of pollution (Kronvang et al. 1996). In the Netherlands, agriculture contributed to nearly 60% of N buildup in water bodies through nonpoint sources (Boers 1996). In a recent study in China, Jiao et al. (2015) reported high accumulation of total phosphorus (TP) in areas with intensive agriculture practices in catchment areas of Miyun reservoir. They reported highest TN during winter season, but high concentrations of NH_4^+–N were observed during summer, and this clearly indicates the direct effect of season variations in N forms.

In India, several water bodies are affected by the nuisance of eutrophication, which is greatly caused by nutrient runoff from the adjoining agricultural fields. To name a few, Lake Udaisagar in Udaipur, Rajasthan (Vijayvirgia 2008), Dal Lake in Kashmir, Loktak Lake in Manipur, and Chilika Lake in Orissa (Patra 2012) are all affected by eutrophication. Both USEPA and WHO (US Environmental Protection Agency (EPA) 1977; WHO 1958, 1985) have set threshold limit of NO_3–N as 10 and 50 mg L^{-1} nitrate for drinking water so as to prevent methemoglobinemia in infants. Under this condition, the NO_3^- in water enters the infant's body and is reduced to nitrite (NO_2^-), which in turn oxidizes the Fe^{2+} of hemoglobin to Fe^{3+}. The oxidized hemoglobin is incapable of binding with oxygen and thus causes anoxia and death of infants. The symptom is also referred to as "blue baby syndrome" as the blood color turns blue due to lack of oxygen. Nitrate is also known to be a carcinogen, which affects humans greatly. According to BIS (2012), the threshold limit of nitrate in drinking water is 45 mg L^{-1} in Indian situation. Almost 108.2 million people in India are exposed to more than 45 mg L^{-1} nitrate (Rai 2003). Thus, risk of nitrate contamination and diseases like cancer have become more common in India.

3.2.2 Nitrate Management in Ground Water

Several techniques of nitrate removal from ground water sources are now available with many industries and scientific organizations. The contaminated ground water is subjected to various processes to remove nitrate as explained by many authors (Khani and Mirzaei 2008; Bhatnagar and Sillanpaa 2011). Techniques such as (i) ion exchange process to remove nitrate ion from water, (ii) biological denitrification process, (iii) reverse osmosis, (iv) electrodialysis, (v) activated carbon and carbon nanotubes, (vi) montmorillonite and bentonite clays, and (vii) agricultural waste material (rice husk, wheat straw, sugarcane bagasse, etc.) are commonly utilized for nitrate removal. Other technique is management of soil organic matter to avoid nitrate leaching from fertilizers applied to soil. Thirty-year long-term experiment showed that 60–65% of applied N was assimilated by plant, 12–15% remained in soil organic pool, and 8–12% was lost to the ground water after 28 years of experiment. This shows the importance of organic matter management in soil to improve the efficiency of fertilizer N and prevent its leaching to the hydrosphere (Sebilo et al. 2013).

3.2.3 Phosphorus Contamination in Water Bodies

Unlike nitrogen, phosphorus is highly immobile in soil. Low availability of phosphatic fertilizers due to fixation to soil particles results in less leaching. Out of the total applied P, plants take up only 10–40% (Garg and Aulakh 2010). The remaining P accumulates in the soil over time and remains unused. According to Sharpley et al. (1994), soils with excessive use of animal manures, fertilizers, etc. containing more than 20 mg kg^{-1} soil solution P are subjected to surface runoff and leaching of P to water bodies. Tirado and Allsopp (2012) in a green peace report highlighted the alarming rates of P loss from field into water bodies. Due to inefficient management practices, nearly 33% of P is lost from soil by wind and water erosion. Poor P use efficiency of 15–30% of P fertilizer results in accumulation of excessive P in soil. Similarly, high rate of application and improper management of animal manure is a major pathway of P entering in water bodies. It is estimated that approximately half of manures applied to crop fields is lost to environment due to mismanagement of resources. Many studies including those of Zhang et al. (2004) and Wang et al. (2012) confirm the potential contamination of P as it is the limiting nutrient in fresh waters. Fig. 4 shows the P transformation pathways in soil water system. Surface runoff during soil erosion increases P losses from soil system to streams, rivers, lakes, and coastal regions (Eghball and Gilley 2001) and causes eutrophication from high bioavailability of soluble P in water bodies.

Water bodies enriched with P shows high algal blooms, reducing dissolved oxgyen, poor aquatic life, phenomena known as eutrophication. It is commonly observed in many developed and few developing countries with excessive application of manure and fertilizers use in agriculture. Phosphorus is considered as limiting nutrient in water bodies Correll (1998) because P concentrations below 10 and

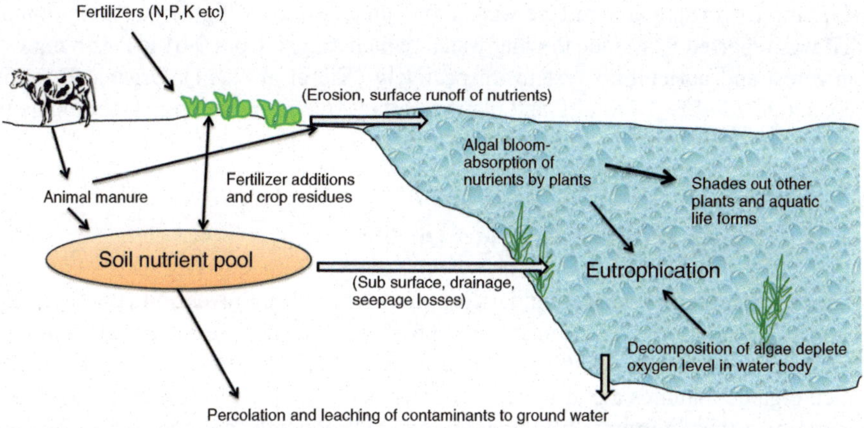

Fig. 4 Inorganic and organic fertilizers as source of contamination in water bodies

20 mg P L^{-1} supports fast growth of aquatic plants, phytoplanktons, etc. (Powlson 1998). Phosphorus had detrimental effect on drinking water quality. In a long-term fertilizer experiment of 16 years, continuous application P fertilizers led to leaching of 7–15% of applied P to ground water. Growth of microbes is a major problem in drinking water. The presence of low P levels like 1 ppm is known to increase microbial population. Miettinen et al. (1997) reported continuous growth of microbes up to 10 ppm P in water; however, the presence of other nutrients of HMs did not influence microbial growth. Higher P accumulation and downward P movement in long-term fertilizer experiments had been reported by Garg and Aulakh (2010) in 22 years of rice–wheat and 34 years of maize–wheat cropping systems in Punjab, India. They reported movement of 6–29% residual P below 30-cm soil depth and higher accumulation of P in subsurface soil layers. Another experiment on a peanut–sunflower system for 4 years showed 41% of P movement below 30-cm soil depth. Buildup and accumulation of P are significantly influenced by P management factors, such as application rate, mode of application besides soil type, depth, irrigation, and organic matter content, which govern P movement in soils.

Eutrophication conditions in water bodies develop with uncontrolled algal multiplication, and major contributor is nonpoint instead of point sources (Carpenter et al. 1998). The main reason for nonpoint sources is intensive fertilization of agricultural soils with P animal manure and fertilizers. Animal manure applied on N basis often results in high P buildup in many Western countries and is considered to be the major culprit for P loss via surface runoff (Eghball and Gilley 2001). Though P is highly susceptible to fixation by soil constituents, but many studies highlight that higher P application with animal manure or fertilizer could increase P level above threshold limit in water bodies. Such high P level causes eutrophication that cannot be reverted back within 1000 years to come (Bennett et al. 2001; Carpenter 2005). In agriculture and forest system, TP concentration varied from 0.01 to

2.47 mg L^{-1} in stagnant surface waters in China. However, higher concentration of TP was reported in surface flowing waters which ranged from 0.01 to 41.66 mg L^{-1} in forest and agriculture system, respectively (Xie et al. 2014). Organic P forms constitute 22–46% of total P and play important role in increasing algal blooms in rivers and other water bodies (Darch et al. 2014).

3.2.4 Organic Fertilizer and Water Quality

Organic manure provides essential crop nutrients and improves soil conditions by modifying aeration, water content, microbial biodiversity, organic carbon, nutrient cycling, etc., thereby significantly improving soil health. However, as a matter of fact, organic manures are also active sources that contribute to HM buildup, pathogenic microbes, nutrient accumulation, and subsequently polluting water bodies, which are serious concerns for human and environment health (Thurston-Enriquez et al. 2005; Jenkins et al. 2008). Improper and excessive application of manures often leads to high buildup of N and P that easily move out of root zone and penetrate to subsurface reaching ground water and polluting it. Surface runoff in such soils will carry N, P along with water flow during erosion causing eutrophication of water bodies nearby agriculture lands or livestock system. In agricultural fields, application of organic manures on regular basis (mostly N based) often acts as nonpoint source of water pollution. In many countries, organic fertilizers like animal manures are applied to the fields. Wastes generated from livestock, poultry, piggery, etc. are usually spread to the land leading to waste assimilation capacity and creating potential sites for ground water contamination, even when commercial fertilizers are also applied. N content in organic manure is either underestimated or ignored and is very critical for enhancing nitrate contamination in water bodies.

Municipal solid waste (MSW) and sewage sludge are other organic manures used in agriculture as soil ameliorant that improve physiochemical properties of soil and improve crop yield (Antonkiewicz and Pelka 2014). They also supply micronutrients and improve soil fertility. Heavy metals such as Cu, Cr, Zn, and Se accumulated above critical limit often interfere with living organisms at cellular levels (Zhao et al. 2012). In agricultural soils, MSW, organic manure, or industrial by-products such as distillery effluents and fly ash often accumulates HMs in large quantities (Srivastava et al. 2017; Sharma et al. 2017). These organic chemicals and HMs enter into food chain and affect biochemical pathways, resulting in reduced cell growth and death of cells subsequently risking animal life (Khaliq et al. 2017). Poultry manures (PM) are recommended for crop production due to its high N content. However, Hill et al. (2005) suggested that higher dose of poultry manure increase bacterial count and nutrient content, which is susceptible to loss of nutrients (N) under rainfall or irrigation leading to eutrophication and becoming a serious source of contaminant. Poultry manure might be good option for improving soil health, but also it poses potential risk to surface water quality. Application time of manures also poses important risk with its transport via soil erosion and runoff along with soil type, topography, and climatic conditions. In hilly terrain (10% slope), application of PM leads to highest mean annual flow weighted runoff nutrient concentrations and loadings from potato fields. In addition to high nutrient loss from field, runoff samples recorded nearly 20–230% increase in *E. coli* population in water (Rees et al. 2011). Manure application to potato crop during fall time increases nutrient load in runoff samples of water.

In most of the agriculture fields, cattle, pig, and poultry manures are common organic fertilizers used for crop application in many countries. Nutrient concentrations of N, P, etc. vary largely depending on animal type, feed, and other management practices. Pagliari and Laboski (2012) reported that cows usually managed for meat and milk production contain 4.5–14.2 g kg^{-1} P and 2.8–15.0 g kg^{-1} P, respectively, which suggests that frequent monitoring is necessary for better manure management practices. Poultry manure is rich in N content, which may give rise to nitrate leaching. Cattle, poultry, and pig manures recorded Zn (180, 400, and 500 mg kg^{-1} dm, respectively) and Cu (50, 80, and 360 mg kg^{-1} dm, respectively) contents (Nicholson et al. 1999). According to Xiong et al. (2010), Cu is the widely used animal feed additive in China and results in soil Cu pollution. Huge increase in concentration of HMs from dairy, pig, and poultry manures over a period of 18 years was reported by Wang et al. (2013). They reported that HMs such as Cu, Zn, As, Cr, and Cd increased by 212, 95, 200, 791, and 63 in dairy manure, 771, 410, 420, 220, and 63 in pig manure, and 181, 197, 1500, 261, and 196 in poultry manure. Besides buildup of HMs, animal manures sometimes transmit antimicrobial-resistant bacteria raising the issue of environmental contamination as reported by Venglovsky (2009). Therefore, judicious application of such organic and animal manures is the only way to prevent contamination of soil and water resources and will enable to dispose the waste safely.

Municipal solid waste, sewage sludge, or biosolids are another form of organic manures that are easily available source of nutrients; however, they flush high dose of HMs and other contaminants to soil with land application. Biosolids are known to contain array of HMs like Cd, Cr, Cu, Ni, Pb, and Zn as contaminants (Lavado et al. 2005). These HMs are toxic to various life forms and create a number of environmental issues that interfere with the valuable provisional ecosystem services. Besides loss of nutrients, HMs are also lost from soils due to their high buildup and subsequent dilution from soil solution system instead of binding with organic matter and clay particles (Luczkiewicz 2006). Some studies highlight the improvement in soil fertility with the application of sludges; although the presence of HMs such as Cu, Zn, and Ni is essential, high concentrations of the same become toxic for microorganism and plants. Reduction in enzymatic activity and microbial biomass due to HM accumulation has been reported by Singh et al. (2012). However, other studies have reported improvement in soil properties with the application of MSW on field crops. Improvement in soil microbial activity, aggregate associated carbon, and reduction in bulk density with short-term application of MSW were reported by Mondal et al. (2015) in cowpea–wheat cropping system. Generally, MSWs are enriched with organic matter which acts as food and energy sources for heterotrophic microorganisms accelerating biomass C in soil. Sludges from paper, oil, and sugar industries recorded 50, 70, and 312 mg l^{-1} Pb, 3.7, 5.4, and 5.2 mg l^{-1} Cd, and 87,185, and 57 mg l^{-1} Zn, respectively (Machiraju 2011). High loading of HMs was recorded in Roca watershed of Nebraska, the United States, due to greater runoff flow during high rainfall period (Elrashidi et al. 2015). Negative effect of untreated sewage sludge or biosolids on HM content, pathogenic microbial population, and toxicity on other life forms has been reported in many studies (Natal-Da-Luz et al. 2009; Artuso et al. 2011). Therefore, there is a need to find more ecofriendly and sustainable ways to handle such organic manures for agriculture.

3.3 Impact of Contaminants on Human Life

Contaminants from various sources of inorganic and organic sources pose serious threat to plant, animal, and human life on one hand, at the same time affect environment quality (Huang and Jin 2008; Srivastava et al. 2017). These HMs not only decrease crop yield but also affect the soil properties, thereby deteriorating soil health. Heavy metal accumulation poses potential health hazard to human because of their entry into food chain through agriculture production. Fruits and vegetables are the major source of human nutrition after cereals. Some of common dangerous HMs such as Cr, Cd, Pb, As, and Hg are taken through food and are deleterious at high concentrations. Most of these HMs are thermostable and nonbiodegradable and, therefore, accumulate to toxic level in air, water, and soil (Sharma et al. 2007; Lokeswari and Chandrappa 2006). Rai et al. (2019) in review paper highlighted that daily intake of metals (DIM) and health risk index (HRI) values for a study was <1 indicating low risk; however, associated risk due to interaction of HMs with skin and inhalation cannot be neglected in human, in particular to children being most vulnerable. Currently, many studies worldwide had grabbed attention on risk assessment based on HMs in edible plant tissues (Antoniadis et al. 2017; Shahid et al. 2012a, b; Xiong et al. 2014). One important pathway of heavy metal entry into food chain is through the soil and water, which is polluted by agrichemicals. These HMs are then taken up by crops through roots and foliage, which is consumed by human and animals. Long-term exposure to HMs such as As and Cr targets the pulmonary organs, nervous system respiratory disorders, cancer, etc. Cd, the most common HM, is accumulated in soil by phosphatic fertilizers. HMs like Cu, Fe, Zn, and Cr (III) are essential for human metabolic process and play important role in several enzyme activity. However, other metals like As, Hg, Pb, and Cd are nonessential for human beings and classified as the most dangerous elements as per USEPA (Rai et al. 2019). These HM accumulations in animals include a few short-term effects, including vomiting, abdominal pain, and nausea. Few serious health disorders like cancer, liver and kidney damage, endocrine disruption, developmental retardation, and other diseases from severe exposure to toxic compounds have been reported by Mahurpawar (2015). Some papers have described several parameters like hazard quotient (HQ), translocation factor (TrF), health risk index (HRI), estimated daily intake (EDI), life time cancer risk (ILTCR), and bioaccumulation potential (BAP) as important indicators for determining risk assessment of metals and their interaction with soil and plants (Antoniadis et al. 2017; Xiong et al. 2016).

Other major contaminant is radionuclides that enter food chain through crops grown with P fertilizers. Phosphate deposits are usually enriched with radioactive and HM contaminants. Despite the positive effect of crop yield improvement with RP application, accumulation of radioactive elements such as Ur, Th, and Ra and their decay products has been reported in many studies (Lema et al. 2014; Khan et al. 2018). Eisenbud and Gesell (1997) suggested that long-time application of P fertilizers increased radionuclide levels in the soil. Phosphogypsum, a common by-product from P fertilizers, also contains substantial amount of radionuclides to contaminant soil environment. This type of by-product without testing applied to soil can directly affect human and animal health. Foods grown on such sites are very harmful as these contaminants interfere with metabolic process of plants that are consumed by human. Nowak (2013) highlighted the potential threat to human and animal exposed to radioactivity internally and externally by accumulation in different body organs. Continuous and excessive exposure often results in harmful health risks to human. Contaminants like soil- and water-borne pathogens from organic manure are also a potential risk for human, especially to those who handle the manures for field applications. Moreover, such pathogens and pests that once enter into agricultural soils are difficult to manage as they start their life cycle and get associated with crops and other host plants like weeds. Water is the reservoir for a number of contaminants in environment. Both inorganic and organic fertilizers when used indiscriminately by farmers are lost via runoff and leaching, thus contaminating water bodies, and are harmful for human and animals. This nutrient-enriched water is home for a number of pathogenic microbes and animals that spread disease epidemics affecting human health and are difficult to control. Pretreatment of such manures is an important step before land application.

The behavior and bioavailability of various contaminants present in inorganic and organic fertilizers vary depending upon the source, climatic conditions, application rate, method, agricultural practices, etc. Though various research studies have highlighted the critical levels for various contaminants like HMs and radionuclides, there is still a need for creating awareness among both producers and consumers associated with farming. It is, therefore, recommended that there should be proper monitoring of HMs, radionuclide, pathogens, and other contaminant forms in the fertilizers and manures, before application on fields in order to combat entry of pollutants in human food chain. Site-specific information which includes detailed knowledge of soil type, crops to be grown, and awareness level of farmers followed by the potential for transfer of contaminants to soil, water, air and reaching human should be know. These parameters would help in setting guide before loading pollutants in cultivable lands. The role of these contaminants is well understood with respect to human health risk in many studies. Therefore, environment friendly and economically feasible techniques could be utilized for remediation from the contaminants before its entry into food chain. Modern technologies like nanotechnology and creating awareness among farmers about contaminants associated with inorganic and organic fertilizers would not only reduce adverse effect of contaminants in food crops but could also improve their livelihood security.

4 Conclusion

This chapter explains the role of inorganic and organic fertilizers in agriculture and their adverse impact on soil and water quality. Fertilizers improve soil fertility by enriching it with essential nutrients for crops; however, they also contribute various contaminants to environment. Inorganic fertilizer is no doubt a backbone for agriculture, but sometimes, it is equally harmful when used indiscriminately by producers. Organic manures like compost and manures are known as soil conditioners as they supply plant nutrients but improve the physical and chemical properties of soil and improve the carrying capacity of soil health. Organic fertilizers such as MSW and biosolids generated from human waste are sources of nutrients but contain various toxic contaminants such as HMs, radionuclide, pathogen, and organic pollutants, which deteriorate soil and water quality. Compared to inorganic fertilizers, organic fertilizers are more beneficial for soil system, and therefore, measure should be taken to reduce concentration of contaminants in such materials before its transfer to crops.

Long-term and excessive application of inorganic fertilizers has adverse effect on soil properties and environment system. Continuos application of urea decrease soil pH, leaching losses into underground water, and volatilization loss. Another macronutrient P which is considered as major constraint in crop production is critical for water contamination. High P buildup often leads to surface runoff and leaching loss to water bodies causing eutrophication. Phosphatic fertilizers also accumulate HMs like Cd and radionuclide in soil system, which is considered to have adverse effect on human health. Similarly, repeated application of animal manure especially in many Western countries had resulted in contamination of water bodies and soil system, causing huge losses of nutrients. The contaminants from fertilizers and manures not only degrade water quality but also enter into food chain with direct effect on human health.

The use of manures and fertilizers free from contaminants, pathogen, and pollutants with minimum loss should be studied. Contamination via inorganic fertilizers and organic manures should be checked by various technologies like phytoremediation, applying lower or optimum dose of fertilizers, proper waste water treatment, use of models for analyzing and monitoring pollutant in soil, air, and water, public awareness programs, etc. However, agricultural emissions from diffuse sources are of even greater importance and could be analyzed using models. Integration of agricultural best management practices (BMPs) with other mitigation techniques for tackling different contaminants could be more economical and environment friendly so as to lower the risk of environmental pollution to a greater extent. Thus, the movement of active contaminants from soil to crops is to be minimized and suppressed for human health, welfare and environment security.

References

Allison S, Martiny J (2008) Resistance, resilience, and redundancy in microbial communities. Proc Nat Aca Sci USA 105(Suppl 1):11512–11519. https://doi.org/10.1073/pnas.0801925105

Almasri MN, Kaluarachchi JJ (2004) Assessment and management of long-term nitrate pollution of ground water in agriculture-dominated watersheds. J Hydro 295:225–245

Antoniadis V, Levizou E, Shaheen SM, Ok YS, Sebastian A, Baum C, Prasad MN, Wenzel WW, Rinklebe J (2017) Trace elements in the soil-plant interface: phytoavailability, translocation, and phytoremediation a review. Earth Sci Rev 171:621–645

Antonkiewicz J, Pełka R (2014) Fractions of heavy metals in soil after the application of municipal sewage sludge, peat, and furnace ash. Soil Sci Ann 65:118–125

Arovuori K, Karikallio H (2009) Consumption patterns and competition in the world fertilizer markets. In: Paper prepared in ' symposium of the international food and agribusiness management association'. Budapest, Hungary

Artuso N, Kenedy TF, Connery J, Grant J, Schmidt O (2011) Effects of biosolids at varying rates on earthworms (*Eisenia foetida*) and springtails (*Folsomia candida*). App Environ Soil Sci:1–10

Atafar Z, Mesdaghinia A, Nouri J, Homaee M, Yunesian M, Ahmadimoghaddam M, Mahvi AH (2010) Effect of fertilizer application on soil heavy metal concentration. Environ Mon Assess 160:83

Bai Y et al (2010) Tradeoffs and thresholds in the effects of nitrogen addition on biodiversity and ecosystem functioning: evidence from inner Mongolia grasslands. Glob Change Biol 16:358–372

Baldwin KR, Shelton JE (1999) Availability of heavy metals in compost-amended soil. Bioresour Technol 69:1–14

Banerjee D, Bairagi H, Mukhopadhyay S, Pal A, Bera D, Ray L (2010) Heavy metal contamination in fruits and vegetables in two districts of West Bengal, India. Elec J Environ Agri Food Chem 9(9):1423–1432

Bartrons M, Penuelas J (2017) Pharmaceuticals and personal-care products in plants. Trends Plant Sci 22(3):194–203

Battaglin WA, Thurman EM, Kalkhoff SJ, Porter SD (2003) Herbicides and transformation products in surface waters of the Midwestern United States. J Am Water Resourc Assoc 39:743–756

Bennett EM, Carpenter SR, Caraco NF (2001) Human impact on erodable phosphorus and eutrophication: a global perspective. BioSci 51:227–234

Bhatnagar A, Sillanpaa M (2011) A review of emerging adsorbents for nitrate removal from water. Chemi Engg J 168:493–504

Bittman B, Forge TA, Kowalenko CG (2005) Response of the bacterial and fungal biomass in a grassland soil to multiyear applications of dairy manure slurry and fertilizer. Soil Bio Biochem 37:613–623

Boers P (1996) Nutrient emissions from agriculture in the Netherlands, causes and remedies. Water Sci Technol 33:183–189

Bose S, Bhattacharyya AK (2008) Heavy metal accumulation in wheat plant grown in soil amended with industrial sludge. Chemosphere 70:1264–1272

Bruun S, Hansen TL, Christensen TH, Magid J, Jensena LS (2006) Application of processed organic municipal solid waste on agricultural land—a scenario analysis. Environ Model Assess 11:251–265

Bureau of Indian Standard (BIS) (2012) Indian standard drinking water specification, New Delhi. Report 1–11

Carpenter SR (2005) Eutrophication of aquatic ecosystems: biostability and soil phosphorus. PNAS 102:10002–10005

Carpenter SR, Caraco NF, Correll DL, Howarth RW, Sharpley AN, Smith VH (1998) Nonpoint pollution of surface waters with phosphorus and nitrogen. Eco Appl 8:559–568

Chabukdhara M, Munjal A, Nema AK, Gupta SK, Kaushal RK (2016) Heavy metal contamination in vegetables grown around peri-urban and urban-industrial clusters in Ghaziabad, India. Hum Ecol Risk Assess 22(3):736–752

Chapman S (1930) On ozone and atomic oxygen in the upper atmosphere. Philos Mag 10:369–383

Chen D, Lan Z, Bai X, Grace JB, Bai Y (2013) Evidence that acidification-induced declines in plant diversity and productivity are mediated by changes in below-ground communities and soil properties in a semi-arid steppe. J. Ecology 101:1322–1334

Chen C, Gao M, Xie D, Ni J (2016) Spatial and temporal variations in non-point source losses of nitrogen and phosphorus in a small agricultural catchment in the Three Gorges. Region Environ Mon Assess 188:257. https://doi.org/10.1007/s10661-016-5260-0

Cheraghi M, Lorestani B, Merrikhpour H (2012) Investigation of the effects of phosphate fertilizer application on the heavy metal content in agricultural soils with different cultivation patterns. Bio Trace Elem Res 145:87–92

Chhonkar PK (2003) Organic farming: science and belief. J Soil Sci Soc India 51(4):365–377

Chien SH, Collamer DJ, Gearhart MM (2008) The effect of different ammonical nitrogen sources on soil acidification. Soil Sci 173:544–551

Correll DL (1998) The role of phosphorus in the eutrophication of receiving waters: a review. J Environ Qual 27:261–266

Czarnecki S, During RA (2015) Influence of long-term mineral fertilization on metal contents and properties of soil samples taken from different locations in Hesse, Germany. Soil 1:23–33. https://doi.org/10.5194/soil-1-23-2015

Dai J, Chen L, Zhao J, Ma N (2006) Characteristics of sewage sludge and distribution of heavy metal in plants with amendment of sewage sludge. J Environ Sci 18:1094–1100

Daniel H, Perinaz BT (2012) What a waste: a global review of solid waste management. World Bank

Darch T, Blackwell MSA, Hawkins JMB, Haygarth PM, Chadwick D (2014) A meta-analysis of organic and inorganic phosphorus in organic fertilizers, soils, and water: implications for water quality. Crit Rev Environ Sci Tech 44:2172–2202

Darusman LRS, Whitney DA, Janssen KA, Long JH (1991) Soil properties after twenty years of fertilization with different nitrogen sources. Soil Sci Soc Am J 55:1097–1100

De Vries W, Romkens PFAM, Van Leeuwen T, Bronswijk JJB (2002) Heavy metals. In: Haygarth PM, Jarvis SC (eds) Agriculture, hydrology and water quality. CABI, Wallingford, pp 107–132

Deka G, Bhattacharyya KG (2009) Enrichment of Cr, Mn, Ni and Zn in surface soil. In: Proceedings of international conference on energy and environment, March 19–21, pp 301–303. ISSN: 2070-3740

Eghball B, Gilley JE (2001) Phosphorus risk assessment index evaluation using runoff measurements. J Soil Water Conserv 56:202–206

Egiarte G, Corti G, Pinto M, Arostegui J, Macías F, Ruiz-Romero E, Camps Arbestain M (2009) Fractionation of Cu, Pb, Cr, and Zn in a soil column amended with an anaerobic municipal sewage sludge. Water Air Soil Pollu 198:133–148

Eisenbud M, Gesell T (1997) Natural radioactivity. In: Eisenbud M, Gesell T (eds) Environmental radioactivity. Academic Press, London, pp 134–200

Elloumi N, Belhaj D, Jerbi B, Zouari M, Kallel M (2016) Effects of sewage sludge on bio-accumulation of heavy metals in tomato seedlings. Span J Agri Res 14:1–13

Elrashidi MA, Seybold CA, Wysocki DA (2015) Effects of annual precipitation on heavy metals in runoff from soils in the US Great Plains. Water, Air Soil Pollu 226(12):417. https://doi.org/10.1007/s11270-015-2684-z

EPA (Environmental Protection Agency) (2010) Methane and nitrous oxide emissions from natural sources. EPA 430-R-10-001

EPAR (2001) United States Environmental Protection Agency. Chicago, pp 1–7

Erisman JW, Sutton MA, Galloway J, Klimont Z, Winiwarter W (2008) How a century of ammonia synthesis changed the world. Nat Geosci 1:636–639

Fagodiya RK, Pathak H, Kumar A, Bhatia A, Jain N (2017) Global temperature change potential of nitrogen use in agriculture: a 50-year assessment. Sci Rep 7:44928. (1–13)

FAO (2009) Resource STAT-fertilizer. Food and Agriculture Organization of the United Nations. [Online]. Available: http://faostat.fao.org/site/575/DesktopDefault.aspx?
FAO (2016) Summary report. World fertilizer trends and outlook to 2019
FAO (2017) Summary report. World fertilizer outlook 2020
FAOSTAT (Food and Agriculture Organization Corporate Statistical Database) (2015) FAO online database
Gall JE, Boyd RS, Rajakaruna N (2015) Transfer of heavy metals through terrestrial food webs: a review. Environ Monit Assess 187:201
Gallego SM, Benavy des MP, Tomaro M (1996) Effect of heavy metal ion excess on sunflower leaves: evidence for involvement of oxidative stress. Plt Sci 121:151–159
Garcia-Gil JC, Paza C, Soler-Rovira P, Polo A (2000) Long-term effects of municipal solid waste compost application on soil enzyme activities and microbial biomass. Soil Bio Biochem 32:1907–1913
Garg AK, Aulakh MS (2010) Effect of long-term fertilizer management and crop rotations on accumulation and downward movement of phosphorus in semi-arid subtropical irrigated soils. Comm Soil Sci Plant Anal 41:848–864
Garg VK et al (2014) Heavy metals bioconcentration from soil to vegetables and assessment of health risk caused by their ingestion. Biol Trace Elem Res 157(3):256–265
Geiseller D, Scow KM (2014) Long-term effects of mineral fertilizers on soil microorganisms—a review. Soil Bio Biochem 75:54–63
Ghosh AK, Bhatt MA, Agrawal HP (2012) Effect of long-term application of treated sewage water on heavy metal accumulation in vegetables grown in northern India. Environ Monit Assess 184:1025–1036
Gong W, Yan X, Wang J, Hu T, Gong Y (2009) Long-term manuring and fertilization effects on soil organic carbon pools under a wheat–maize cropping system in North China plain. Plant Soil 314:67–76
Gove L, Cooke CM, Nicholson FA, Beck AJ (2001) Movement of water and heavy metals (Zn, Cu, Pb and Ni) through sand and sandy loam amended with biosolids under steady state hydrological conditions. Bioresour Technol 78:171–179
Grant CA, Bailey LD, Therrien MC (1996) The effect of N, P and KCl fertilizers on grain yield and Cd concentration of malting barley. Fert Res 45:153–161
Guo JH, Liu XJ, Zhang Y, Shen JL, Han WX, Zhang WF, Christie P, Goulding KWT, Vitousek PM, Zhang FS (2010) Significant acidification in major Chinese croplands. Science 327:1008–1010
Gupta DK, Sandallo LM (2011) Metal toxicity in plants: perception, signaling and remediation. Springer, London
Hanson CR (2002) Nitrate concentrations in Canterbury groundwater: a review of existing data Environment Canterbury. Report No R02/17
Hargreaves JC, Adl MS, Warman PR (2008) A review of the use of composted municipal solid waste in agriculture. Agric. Ecosyst Environ 123:1–14
Hassan MJ, Wang F, Ali S, Zhang GP (2005) Toxic effect of cadmium on rice as affected by nitrogen fertilizer form. Plant Soil 277:359–365
Hassan NM, Mansour NA, Fayez-Hassan M, Sedqy E (2016) Assessment of natural radioactivity in fertilizers and phosphate ores in Egypt. J Taibah Uni Sci 10:296–306
Hill DD, Owens WE, Tchounwou PB (2005) Impact of animal waste application on runoff water quality in field experimental plots. Int J Environ Res Public Health 2:314–321
Huang SW, Jin JY (2008) Status of heavy metals in agricultural soils as affected by different patterns of land use. Environ Monit Assess 139:317–327
Huang B, Kuo S, Bembenek R (2004) Availability of cadmium in some phosphorus fertilizers to field-grown lettuce. Water Air Soil Pollu 158:37–51
Huang S, Peng X, Huang Q, Zhang W (2010) Soil aggregation and organic carbon fractions affected by long-term fertilization in a red soil of subtropical China. Geoderma 154:364–369
ICPDR (International Commission for the Protection of the Danube River) (2013) Reducing nutrient pollution, challenges in agriculture. www.icpdr.org

International fertilizer association (IFA) (2017) Fertilizer outlook 2017–2021

Interweck A, Stone LR, Ellis R, Whitney DA (1982) Influence of fertilizer nitrogen source on soil physical and chemical properties. Soil Sci Soc Am J 46:832–836

Jackie LS, Zhang H, Girma K, Raun WR, Penn CJ, Payton ME (2011) Soil acidification from long-term use of nitrogen fertilizers on winter wheat. Soil Sci Soc Am J 75:957–964

Jara SJ, Moral R, Pearez MMD, Paredes C, Galvez SL, Gavilanes TI et al (2015) Urban waste management and potential agricultural use in south American developing countries: a case study of Chimborazo region (Ecuador). Comm Soil Sci Plant Anal 46:157–169

Jenkins M, Truman CC, Siragusa GR, Line JE et al (2008) Rainfall and tillage effects on transport of fecal bacteria and sex hormones. Sci Total Environ 403:154–163

Jiao J, Du P, Lang C (2015) Nutrient concentrations and fluxes in the upper catchment of the Miyun reservoir, China, and potential nutrient reduction strategies. Environ Mon Assess 187:110

Ju XT, Xing GX, Chen XP, Zhang SL, Zhang LJ, Liu XJ, Cui ZL, Yin B, Christie P, Zhu ZL, Zhang FS (2009) Reducing environmental risk by improving N management in intensive Chinese agricultural systems. Proc Natl Acad Sci U S A 106:3041–3046

Juhler RK, Sorensen SR, Larsen L (2001) Analysing transformation products of herbicide residues in environmental samples. Water Res 35:1371–1378

Kabata-Pendias A, Pendias H (1992) Trace elements in soils and plants, 2nd edn. CRC Press, Boca Rotan

Kelepertzis E (2014) Accumulation of heavy metals in agricultural soils of Mediterranean: insights from Argolida basin, Peloponnese, Greece. Geoderma 221:82–90

Khaliq SJA, Ahmed M, Al-Wardy M, Al-Busaidi A, Choudri BS (2017) Wastewater and sludge management and research in Oman: an overview. J Air Waste Manag Assoc 67(3):267–278

Khan SA, Mulvaney RL, Ellsworth TR, Boast CW (2007) The myth of nitrogen fertilization for soil carbon sequestration. J Environ Qual 36:1821–1832

Khan S, Cao Q, Zheng YM, Huang YZ, Zhu YG (2008) Health risks of heavy metals in contaminated soils and food crops irrigated with wastewater in Beijing. China Environ Pollu 152:686–692

Khan SA, Mulvaney RL, Ellsworth TR (2013) The potassium paradox: implications for soil fertility, crop production and human health. Renew Agric Food Syst 29(1):1–25

Khan MN, Mobin M, Abbas ZK, Alamri SA (2018) Fertilizers and their contaminants in soils, surface and groundwater. In: DellaSala DA, Goldstein MI (eds) The encyclopedia of the anthropocene, vol 5. Elsevier, Oxford, pp 225–240

Khani A, Mirzaei M (2008) Comparative study of nitrate removal from aqueous solution using powder activated carbon and carbon nanotubes. In: 2nd international IUPAC conference on green chemistry, Russia, vol 2. pp 14–19

Kolpin DW, Furlong ET, Meyer MT, Thurman EM, Zaugg SD, Barber LB, Buxton HT (2002) Pharmaceuticals, hormones, and other organic wastewater contaminants in US streams, 1999–2000: a national reconnaissance. Environ Sci Technol 36:1202–1211

Korschens M, Albert E, Armbruster M, Barkusky D, Baumecker M et al (2013) Effect of mineral and organic fertilization on crop yield, nitrogen uptake, carbon and nitrogen balances, as well as soil organic carbon content and dynamics: results from 20 European long-term field experiments of the twenty-first century. Arch Agron Soil Sci 59:1017–1040

Kritee K, Drishya N, Daniel ZA, Jeremy P, Joseph R et al. (2018) High nitrous oxide fluxes from rice indicate the need to manage water for both long- and short-term climate impacts. Proc Nat Acad Sci (USA):1–6. https://doi.org/10.1073/pnas.1809276115

Kronvang B, Græsboll P, Larsen SE, Svendsen LM, Andersen HE (1996) Diffuse nutrient losses in Denmark. Water Sci Technol 33:81–88

Kumazawa K (2002) Nitrogen fertilization and nitrate pollution in groundwater in Japan: present status and measures for sustainable agriculture. Nutr Cycl Agroecosyst 63:129–137

Lakhdar A, Achiba W, Montemurro F, Jedidi N, Abdelly C (2009) Effect of municipal solid waste compost and farmyard manure application on heavy-metal uptake in wheat. Comm Soil Sci Plant Anal 40(21–22):3524–3538

Lary D (1997) Catalytic destruction of stratospheric ozone. J Geophys Res 102(21):515–526
Latare AM, Kumar O, Singh SK, Gupta A (2014) Direct and residual effect of sewage sludge on yield, heavy metals content and soil fertility under rice–wheat system. Ecol Eng 69:17–24
Lavado RS, Rodriguez MB, Taboada MA (2005) Treatment with biosolids affects soil availability and plant uptake of potentially toxic elements. Agri Ecosyst Environ 109:360–364
Lema MW, Jasper NI, Karoli NN, Patrick AN (2014) Environmental contamination by radionuclides and heavy metals through the application of phosphate rocks during farming and mathematical modeling of their impacts to the ecosystem. Int J Eng Res Gen Sci 2(4):852–863
Li SP (2002) Environmental biology (In Chinese). Chinese Agricultural Press, Beijing, pp 426–442
Lin CG (1996) Soil pollution and its control (In Chinese). Chinese Agricultural Press, Beijing, pp 44–58
Line DE, McLaughlin RA, Osmond DL, Jennings GD, Harman WA, Tweedy KL et al (1998) Nonpoint sources. Water Environ Res 70:895–912
Lipoth SL, Schoenau JJ (2007) Copper, zinc, and cadmium accumulation in two prairie soils and crops as influenced by repeated applications of manure. J Plant Nutr Soil Sci 170:378–386
Liu L, Gundersen P, Zhang T, Mo J (2012) Effects of phosphorus addition on soil microbial biomass and community composition in three forest types in tropical China. Soil Biol Biochem 44:31–38
Lokeswari H, Chandrappa GT (2006) Impact of heavy metal contamination of. Bellandur Lake on soil and cultivated vegetation Current Science 91(5):622–627
Lopes C, Herva M, Franco-Uria A, Roca E (2011) Inventory of heavy metal content in organic waste applied as fertilizer in agriculture: evaluating the risk of transfer into the food chain. Environ Sci Pollu Res Int 18(6):918–939
Lopez RP, Jose MN, Israel LC, Juan LA, Juan PB, Maria S (2010) Dynamics of contaminants in phosphogypsum of the fertilizer industry of Huelva (SW Spain): from phosphate rock ore to the environment. Appl Geochem 25:705–715
Lu C, Tian H (2017) Global nitrogen and phosphorus fertilizer use for agriculture production in the past half century: shifted hot spots and nutrient imbalance. Earth Syst Sci Data 9:181–192
Lucas RW, Klaminder J, Futter MN, Bishop KH, Egnell G, Laudon H, Hogberg P (2011) A meta-analysis of the effects of nitrogen additions on base cations: implications for plants, soils, and streams. Forest Ecol Manag 262:95–104
Luczkiewicz A (2006) Soil and groundwater contamination as a result of sewage sludge land application. Polish J Environ Stud 15:869
Lun F, Junguo L, Philippe C, Thomas N, Jinfeng C, Rong W, Daniel G, Jordi S, Josep P, Michael O (2018) Global and regional phosphorus budgets in agricultural systems and their implications for phosphorus-use efficiency. Earth Syst Sci Data 10:1–18
Machiraju PVS (2011) Industrial sludge as a fertilizer substitute. Int. J Pharma Bio Sci 2(1):193–199
Madrid L, Diaz-Barrientos E, Madrid F (2002) Distribution of heavy metal contents of urban soils in parks of Seville. Chemosphere 49:1301–1308
Mahurpawar M (2015) Effects of heavy metals on human health. Int J Res Granthaalayah:1–7
Marathe RA, Bharambe RR (2005) Physical characteristics of vertisol as influenced by integrated nutrient management system under sweet orange. PKV Res J 29(2):179–182
McGrath SP, Zhao FJ, Dunham SJ, Crosland AR, Coleman K (2000) Long-term changes in the extractability and bioavailability of zinc and cadmium after sludge application. J Environ Qual 29:875–883
Miettinen IT, Vartiainen T, Martikainen PJ (1997) Phosphorus and bacterial growth in drinking water. Appl Environ Microbio 63:3242–3245
Minnesota Department of Health (MDH) (2008) Screening evaluation of heavy metals in inorganic fertilizers. Minnesota Deparment of Agriculture, USA
Mirlean M, Roisenberg A (2007) Fluoride distribution in the environment along the gradient of a phosphate-fertilizer production emission (Southern Brazil). Environ Geochem Health 29:179–187

Mondal S, Singh RD, Patra AK, Dwivedi BS (2015) ICAR Changes in soil quality in response to short-term application of municipal sewage sludge in a typic haplustept under cowpea-wheat cropping system. Environ Nanotech Monit Manag 4:37–41
Mondini C, Sequi P (2008) Implication of soil C sequestration on sustainable agriculture and environment. Waste Manag 28(4):678–684
Mulvaney RL, Khan SA, Ellsworth TR (2009) Synthetic nitrogen fertilizers deplete soil nitrogen: a global dilemma for sustainable cereal production. J Environ Qual 38:2295–2314
Nagajyoti PC, Lee KD, Sreekanth VM (2010) Heavy metals, occurrence and toxicity for plants: a review. Environ Chem Lett 8:199–216
Natal-Da-Luz T, Tidona S, Jesus B, Morais PV, Sousa JP (2009) The use of sewage sludge as soil amendment- the need for an ecotoxicological evaluation. J Soils Sediments 3:246–260
National Oceanic and Atmospheric Administration (2009) Nitrous oxide now top ozone-depleting emission. Science Daily
Niassy S, Diarra K (2012) Effect of organic inputs in urban agriculture and their optimization for poverty alleviation in Senegal, West Africa. In: Singh RP (ed) Organic fertilizers: types, production and environmental impact. Nova Science Publisher, Hauppauge, pp 1–22
Nicholson FA, Chambers BJ, Williams JR, Unwin RJ (1999) Heavy metal contents of livestock feeds and animal manures in England and Wales. Bioresour Technol 70:23–31
Nicholson FA, Smith SR, Alloway BJ, Carlton-Smith C, Chambers BJ (2003) An inventory of heavy metal inputs to agricultural soils in England and Wales. Sci Total Environ 311:205–219
Nowak K (2013) Radionuclides content in selected mineral fertilizers available in Poland. In: ECO pole'12 conference, Zakopane. https://doi.org/10.2429/proc.2013.7(2)065
Pagliari PH, Laboski CAM (2012) Investigation of the inorganic and organic phosphorus forms in animal manure. J Environ Qual 41:901–910
Parris K (2011) Impact of agriculture on water Pollution in OECD Countries: recent trends and future prospects. Int J Water Res Develop 27(1):33–52
Patra R (2012) Eutrophication of important lakes in India. https://www.indiawilds.com/diary/the-death-of-a-waterbody-2/
Perianez R, Martinez-Aguirre A, Garcia-Leon M (1996) Uranium and Th-isotopes in an estuarine system in southwest Spain: tidal and seasonal variations. Appl Radiat Isotopes 47:1121–1125
Petersen SO, Henriksen K, Mortensen GK, Krogh PH, Brandt KK, Sorensen J, Madsen T, Petersen J, Gron C (2003) Recycling of sewage sludge and household compost to arable land: fate and effects of organic contaminants, and impact on soil fertility. Soil Tillage Res 72:139–152
Pinamonti F, Stringari G, Gasperi F, Zorzi G (1997) The use of compost: its effects on heavy metal levels in soil and plants. Resour Conserv Recycl 21:129–143
Powlson DS (1998) Phosphorus, agriculture and water quality. Soil Use Manag 14:123
Rai SN (2003) Groundwater pollution in India- an overview. Groundwater pollution. Allied Publishers Pvt. Ltd., New Delhi, pp 419–436
Rai PK, Lee J, Kailasa SK et al (2018) A critical review of ferrate (VI)-based remediation of soil and groundwater. J Environ Res 160:420–448
Rai PK, Leeb SS, Zhangc M, Tsangd YF, Kime K-H (2019) Heavy metals in food crops: health risks, fate, mechanisms, and management. Environ Int 125:365–385
Ravishankara AR, Daniel JS, Portmann RW (2009) Nitrous oxide (N_2O): the dominant ozone depleting substance emitted in the 21st century. Science 326:123–125
Rees HW, Chow TL, Zebarth BJ, Xing Z, Toner P, Lavoie J, Daigle JL (2011) Effects of supplemental poultry manure applications on soil erosion and runoff water quality from a loam soil under potato production in northwestern New Brunswick. Can J Soil Sci 91:595–613
Rehman S, Imtiaz N, Faheem M, Matiullah AA (2006) Determination of ^{238}U contents in ore samples using CR-39 based radon dosimeter disequilibrium case. Radiat Meas 4:471–476
Righi S, Luciallib P, Bruzzi L (2005) Health and environmental impacts of a fertilizer plant—Part I: assessment of radioactive pollution. J Environ Radioact 82:167–182
Robinson S (2014) How the ozone hole affects weather pattern. World Economic Forum

Ryan MH, Small DR, Ash JE (2000) Phosphorus controls the level of colonisation by arbuscular mycorrhizal fungi in conventional and biodynamic irrigated dairy pastures. Aust J Exp Agric 40:663–670

Sachan S (2007) Buildup of heavy metals in soil – water – plant continuum as influenced by irrigation with contaminated effluent. J Environ Sci Eng 49:293–296

Saggar S, Hedley CB, Giddens KM, Salt GJ (2000) Influence of soil phosphorus status and nitrogen addition on carbon mineralization from ^{14}C-labelled glucose in pasture soils. Biol Fertil Soils 32:209–216

Saha S et al (2015) Assessment of heavy metals contamination in different crops grown in long-term sewage-irrigated areas of Kolkata, West Bengal, India. Environ Monit Assess 187:4087

Saleh IH, Hafez AF, Elanany NH, Motaweh HA, Naim MA (2007) Radiological study on soils, foodstuff and fertilizers in the Alexandria Region, Egypt. Turk J Eng Environ Sci 31:9–17

Saueia CHR, Mazzilli BP (2006) Distribution of natural radionuclides in the production and use of phosphate fertilizers in Brazil. J Environ Radioact 89:229–239

Savci S (2012) An agricultural pollutant: chemical fertilizer. Int J Environ Sci Develop 3(1):77–80

Sebilo M, Mayer B, Nicolardot B, Pinay G, Mariotti A (2013) Long-term fate of nitrate fertilizer in agricultural soils. Proc Nat Acad Sci 110(45):18185–18189

Selvi D, Santhy P, Dhakshinamoorthy M (2005) Effect of inorganics alone and in combination with farmyard manure on physical properties and productivity of vertic haplustepts under long-term fertilization. J Indian Soc Soil Sci 53(3):302–307

Shahid M, Dumat C, Aslam M, Pinelli E (2012a) Assessment of lead speciation by organic ligands using speciation models. Chem Speciat Bioavailab 24:248–252

Shahid M, Dumat C, Silvestre J, Pinelli E (2012b) Effect of fulvic acids on lead induced oxidative stress to metal sensitive *Vicia faba* L. plant. Biol Fertil Soils 48:689–697

Shanker AK (2005) Chromium toxicity in plants. Environ Int 31:739–753

Sharma RK, Agarwal M, Marshall F (2007) Heavy metal Contamination of soil & vegetables in suburban areas of Varanasi, India. Ecotoxicol Environ Saf 66:258–266

Sharma B, Sarkar A, Singh P, Singh RP (2017) Agricultural utilization of biosolids: a review on potential effects on soil and plant grown. Waste Manag 64:117–132

Sharpley AN, Charpra SC, Wedepohl R, Sims JT, Daniel TC, Reddy KR (1994) Managing agricultural phosphorus for protection of surface waters: issues and options. J Environ Qual 23:437–451

Shayler H, McBride M, Ellen H (2009) Sources and impacts of contaminants in soils. Cornell Waste Management. http://cwmi.css.cornell.edu

Singh RP, Agrawal M (2007) Effect of sewage sludge amendment on heavy metal accumulation and consequent responses of Beta vulgaris plant. Chemosphere 67:2229–2240

Singh RP, Agrawal M (2008) Potential benefits and risks of land application of sewage sludge. Waste Manage 28:347–358

Singh RP, Agrawal M (2010) Effect of different sewage sludge applications on growth and yield of Vigna radiata L. field crop: metal uptake by plant. Eco Eng 36:969–972

Singh RP, Singh P, Ibrahim MH, Hashim R (2012) Land application of sewage sludge: physicochemical and microbial response. In: Whitacre DM (ed) Reviews of environmental contamination and toxicology. Springer, New York, pp 41–61. https://doi.org/10.1007/978-1-4614-0668-6_3

Skiba UM, Rees RM (2014) Nitrous oxide, climate change and agriculture. CAB Rev 9:1–7

Smith SR (2009) Critical review of the bioavailability and impacts of heavy metals in municipal solid waste composts compared to sewage sludge. Environ Int 35:142–156

Soliva M, Paulet S (2001) Composting of organic wastes and agricultural application. In: Boixadeira J, Teira MR (eds) Agricultural application of organic wastes. University of Lleida, Spain, pp 63–78. (in Spanish)

Sonmez KM, Sonmez S (2007) An investigation of seasonal changes in nitrate contents of soils and irrigation waters in greenhouses located in antalya-demre region. Asian J Chem 19:5639–5646

Speir TW, Horswell J, van Schaik AP, McLaren RG, Fietje G (2004) Composted biosolids enhance fertility of a sandy loam soil under dairy pasture. Biol Fertil Soils 40:349–358

Sposito G (1989) The chemistry of soils. Oxford University Press, New York, pp 127–146
Srivastava V, Ismail SA, Singh P, Singh RP (2015) Urban solid waste management in the developing world with emphasis on India: challenges and opportunities. Rev Environ Sci Bio 14:317–337
Srivastava V, de Araujo ASF, Vaish B, Bartelt-Hunt S, Singh P, Singh RP (2016) Biological response of using municipal solid waste compost in agriculture as fertilizer supplement. Rev Environ Sci Biol 15:677–696
Srivastava V, Abhijit S, Sonu S, Pooja S, Ademir SFA, Rajeev PS (2017) Agroecological responses of heavy metal pollution with special emphasis on soil health and plant performances. Front Environ Sci 5:64. https://doi.org/10.3389/fenvs.2017.00064
Sun C, Liu J, Wang Y, Sun L, Yu H (2013) Multivariate and geostatistical analyses of the spatial distribution and sources of heavy metals in agricultural soil in Dehui, Northeast China. Chemosphere 92:517–523
Sutton MA, Oenema O, Erisman JW, Leip A, van Grinsven H, Winiwarter W (2011) Too much of a good thing. Nature 472:159–161
Tao R, Yongchao L, Steven A, Wakelinc GC (2015) Supplementing chemical fertilizer with an organic component increases soil biological function and quality. Appl Soil Eco 96:42–51
Thein SJ (1976) Stabilizing soil aggregates with phosphoric acid. Soil Sci Soc Am Proc 40:105–108
Thurston-Enriquez JA, Gilley JE, Eghball B (2005) Microbial quality of runoff following land application of cattle manure and swine slurry. J Water Health 3:157–171
Tirado R, Allsopp M (2012) Phosphorus in agriculture: problems and solutions. Greenpeace Research Laboratories Technical Report (Review)
Toth G, Hermann T, Da Silva MR, Montanarella L (2016) Heavy metals in agricultural soils of the European Union with implications for food safety. Environ Pollut 88:299–309
Triberti L, Nastri A, Giordani G, Comellini F, Baldoni G, Toderi G (2008) Can mineral and organic fertilization help sequestrate carbon dioxide in cropland. Eur J Agron 29:13–20
U.S. EPA (1977) National Interim primary drinking water regulations. EPA 570/9–76-003. Washington, DC: U.S. Environmental Protection Agency
Varma VS, Kalamdhad AS (2015) Evolution of chemical and biological characterization during thermophilic composting of vegetable waste using rotary drum composter. Int. J Environ Sci Tech 12(6):2015–2024
Venglovsky J, Sasakova N, Placha I (2009) Pathogens and antibiotic residues in animal manures and hygienic and ecological risks related to subsequent land application. Biores Tech 100(22):5386–5391
Vijayvirgia RP (2008) Eutrophication: a case study of highly eutrophicated Lake Udaisagar, Udaipur (Raj.), India with regards to its nutrient enrichment and emerging consequences. In: Proceedings of Taal 2007: the 12th World Lake conference, vol. 1557, 1560p
Wang T, Camps-Arbestain M, Hedley M, Bishop P (2012) Predicting phosphorus bioavailability from high-ash biochars. Plant Soil. https://doi.org/10.1007/s11104-012-1131-9
Wang Z, Zheng P, Bao MS, Bai LY (2013) Study on passivation sewage sludge with coal fly ash. Environ Sci Tech 36:343–347
Wang H, Gao J, Li X, Zhang S, Wang H (2015) Nitrate accumulation and leaching in surface and ground water based on simulated rainfall experiments. PLoS One 10:e0136274. https://doi.org/10.1371/journal.pone.0136274
Weber J, Karczewska A, Drozd J, Licznar M, Licznar S, Jamroz E, Kocowicz A (2007) Agricultural and ecological aspects of a sandy soil as affected by the application of municipal solid waste composts. Soil Bio Biochem 39:294–1302
WHO (1958) International Standards for Drinking-Water. World Health Organization, Geneva
WHO (1985) Health hazards from nitrates in drinking water. Report on a WHO Meeting. Copenhagen, 5–9 March 1984. Geneva: World Health Organization
Xie J, Zhang X, Xu Z, Yuan G, Tang X, Sun X, Ballantine DJ (2014) Total phosphorus concentrations in surface water of typical agro- and forest ecosystems in China, 2004–2010. Front Environ Sci Engg 8:561–569

Xiong X, Yanxia L, Wei L, Chunye L, Wei H, Ming Y (2010) Copper content in animal manures and potential risk of soil copper pollution with animal manure use in agriculture. Resour Conserv Recycl 54:985–990
Xiong T, Leveque T, Shahid M, Foucault Y, Mombo S, Dumat C (2014) Lead and cadmium phytoavailability and human bioaccessibility for vegetables exposed to soil or atmospheric pollution by process ultrafine particles. J Environ Qual 43:1593–1600
Xiong T, Dumat C, Pierart A, Shahid M, Kang Y, Li N, Bertoni G, Laplanche C (2016) Measurement of metal bioaccessibility in vegetables to improve human exposure assessments: field study of soil-plant-atmosphere transfers in urban areas, South China. Environ Geochem Health 38:1283–1301
Xu Y, Yu W, Ma Q, Zhou H (2013) Accumulation of copper and zinc in soil and plant within ten-year application of different pig manure rates. Plant Soil Environ 11:492–499
Yeoh NS, Oades JM (1981a) Properties of soils and clays after acid treatment. I. Clay minerals. Aust J Soil Res 19:147–158
Yeoh NS, Oades JM (1981b) Properties of soils and clays after acid treatment: II. Urrbrae fine sandy loam. Aust. J. Soil Res 19:159–166
Zhang MK, He ZL, Stoffella PJ, Calvert DV, Yang XE, Xia YP, Wilson SB (2004) Solubility of phosphorus and heavy metals in potting media amended with yard waste biosolids compost. J Environ Qual 33:373–379
Zhang F, Li Y, Yang M, Wei L, Yan W (2011) Copper residue in animal manures and the potential pollution risk in Northeast China. J Resour Ecol 2(1):91–96
Zhao XL, Mu ZJ, Cao CM, Wang DY (2012) Growth and heavy metal uptake by lettuce grown in soils applied with sewage sludge compost. Commu Soil Sci Plant Anal 43:1532–1541
Zhuang P, McBride MB, Xia H, Li N, Li Z (2009) Health risk from heavy metals via consumption of food crops in the vicinity of Dabaoshan mine, South China. Sci Total Environ 407:1551–1561

Role of Nutrients in Plant Growth and Development

Preksha Shrivastav, Mrinalini Prasad, Teg Bahadur Singh, Arti Yadav, Deepika Goyal, Akbar Ali, and Prem Kumar Dantu

Contents

1 Introduction

The study of activities by which plants obtain their nutrition is called mineral nutrition. In recent decades, this area has become central to climate change, specifically environmental protection and modern agriculture. Crop yield is linearly related to fertilizer applied and its absorption. To meet the increasing food demand, the world consumption of primary elements, mainly N and P, has increased during the last few decades. However, crop plants use less than half of the fertilizer applied (Loomis and Connor 1992); remaining nutrients leach into surface water or groundwater. Some nutrients become attached to soil particles and contribute to air pollution. As a consequence of leaching, many water wells in the USA no longer meet the federal

P. Shrivastav · M. Prasad · T. B. Singh · A. Yadav · D. Goyal · A. Ali · P. K. Dantu (✉)
Department of Botany, Dayalbagh Educational Institute, (Deemed University), Agra, India

M. Naeem et al. (eds.), *Contaminants in Agriculture*,
https://doi.org/10.1007/978-3-030-41552-5_2

standards for nitrate concentrations in drinking water (Nolan and Stoner 2000). It can be stated that plants can help in recycling of animal waste and prove to be beneficial for detoxification of waste dumps (Macek et al. 2000).

Light and water must be supplied to plants for efficient utilization of nutrients. Plants are specific for their nutrient requirements and have certain optimum range for each nutrient, below which plants show nutrient deficiency symptoms. Excessive nutrient can also cause poor growth because of toxicity. Hence, adequate amount of nutrient supply is required for healthy crop production. Various tests have been developed to assess nutrient availability in soil as well as in subsequent crop plants. These data help plant scientists to determine nutrient need for a given plant in a specific soil. Availability of soil nutrients depends upon soil pH. Most micronutrients are either present in lesser concentration in soil or found as their respective salts, depending upon the pH. It has been observed that soil and plants are deficient in these essential elements; hence, there is a need to focus on this issue. Biofortification is becoming widespread, which depends on the soil–plant interactions (Olsen and Palmgren 2014; Patto et al. 2015; Tan et al. 2015). This chapter briefly describes availability of some essential nutrients, forms in which they are available to plants, their function in plants, and their availability in soil.

2 Nutrients That Limit Plant Growth and Development

Mineral nutrients have specific and essential functions in plant metabolism. Nitrogen, phosphorus, and potassium (macronutrient) are most important nutrients for plant growth and development. Nitrogen is a major component of proteins and is therefore required in higher concentration. It also catalyzes enzymatic reactions. Phosphorus is the main component of ATP and NADPH. Nucleic acids have nitrogen and phosphorus; therefore, these two mineral nutrients are the necessary requirement for any biological body (Marschner 1995).

2.1 Nitrogen

Green plants are unique in their ability to reduce atmospheric carbon via photosynthesis, and while doing this, they provide the energy source for all life. Nitrogen plays an essential role for protein and nucleic acid synthesis, which forms the living materials. Proteins are required as enzyme catalyst, while nucleic acids are required for translation of genetic material (Novoa and Loomis 1981). Nitrogen plays a role in C_3 plant photosynthesis; proteins of Calvin cycle and thylakoid represent leaf nitrogen. One study suggests that thylakoid nitrogen is proportional to chlorophyll content. Since nitrogen is a major element of leaves, in the absence of nitrogen, leaves show chlorosis. In the absence of nitrogen, leaves become yellow and this color appears in older leaves, while younger ones have mobilized nitrogen from older ones. Even in the absence of nitrogen, plants exhibit slender stem due to excess production of carbohydrates and lesser production of proteins. In this regard, the plant begins to synthesize anthocyanin and subsequent purple coloration of leaves, petioles, and stems.

2.2 *Phosphorus*

Phosphorus is an integral element of all membrane proteins, lipids, and nucleic acids. It plays a vital role in respiration and photosynthesis, that is, as sugar–phosphate intermediates, and is an important in energy storage or structural integrity. Its deficiency leads to stunted growth of young plants and dark green coloration of leaves. Absence of phosphorus causes small spots of dead tissues often called necrotic spots. A slight purple color may appear in leaves due to excess production of anthocyanin, but it is not associated with chlorosis. Phosphorus deficiency leads to delay in plant maturation and appearance of slender stem similar to nitrogen deficiency. To increase the yield of agricultural crops, it is required that phosphorus should be present in sufficient concentration (Tisdale and Nelson 1975; Denison and Kiers 2005). To maintain productive soils for agricultural crops, it is necessary to apply available forms of phosphorus (Sims 2000; Fixen 2005; White and Brown 2010).

2.3 *Potassium*

Potassium plays an important role in osmotic potential regulation of plant cells. It plays a role in enzyme activation involved in respiration and photosynthesis. Its deficiency causes mottled or marginal chlorosis and subsequently, necrosis effects mostly at the tips of leaves. Potassium is mobilized to the younger leaves; therefore, deficiency symptoms appear mainly in mature and older leaves. In case of monocots, necrotic lesions initially appear at tips and margins of leaves and then at base. Plants may also show slender and weak stems with abnormally short intermodal regions. Potassium-deficient plant roots exhibit increased susceptibility for root-rotting fungi present in the soil. It is extremely dynamic in its ionic form in plants but moderate inside soil (Ranade–Malvi 2011).

2.4 *Sulfur*

Two amino acids containing sulfur are constituents of vitamin complexes and coenzymes that are essential for metabolism. Symptoms of sulfur deficiency are mostly similar to nitrogen deficiency, including chlorosis, stunting of growth, and anthocyanin accumulation, since sulfur and nitrogen both form the basic structure of proteins. Nitrogen deficiency causes chlorosis in older leaves, while sulfur deficiency occurs initially in mature and older leaves due to the fact that in most species, sulfur is not easily remobilized to the younger leaves.

2.5 *Calcium*

In cell wall synthesis, calcium plays a crucial role, especially in case of middle lamellae. It plays a significant role in mitotic spindle formation required for proper functioning of membranes and also serves as the second messenger for signaling processes. It forms the calcium–calmodulin complex, a protein found in cytosol, which regulates many metabolic process. Its deficiency leads to necrosis at tips of roots and young leaves. Since it is involved in cell wall synthesis, its deficiency leads to death of young meristematic regions.

2.6 *Magnesium*

Magnesium ions have a specific role in activation of some enzymes that are involved in synthesis of nucleic acid, respiration, and photosynthesis. It is a part of the chlorophyll molecule. Its deficiency causes chlorosis; in case of extensive deficiency, leaves may become white or yellowish. Premature leaf abscission may also occur due to its deficiency.

3 Micronutrients

Micronutrients are those elements that are required in lesser quantities and necessary for plant metabolic activities, specifically enzyme activation for reaction catalysis (Epstein 1965).

3.1 *Boron*

Till now, the precise function of Boron is not clear; it is suggested that it plays a vital role in cell elongation nucleic acid synthesis. This micronutrient is essential for the normal growth of plants and certain diatom species. Its deficiency causes anatomical changes with corresponding change in physiology and biochemistry of cell. But it is difficult to determine the primary role of boron; probably it is involved at membrane-level functions. Other possible roles of boron are sugar transport, integrity of cell wall structure, lignification, respiration, IAA metabolism, and phenol metabolism. The available concentration of Boron may vary from soil to soil, while it is reported in range of 20–200 mg B/Kg (Ahmad et al. 2012).

3.2 *Zinc*

Zinc in its ionic form is required by plants for enzyme activation involved in many metabolic activities such as DNA replication, for activation of DNA polymerases, and for chlorophyll biosynthesis in some plants, hydrogenase and carbonic anhydrase stabilization of ribosomal fractions, and synthesis of cytochrome (Tisdale et al. 1984). Deficiency of zinc appears as reduction in intermodal growth resulting in growth. The leaves may appear small. Chlorosis of plant leaves infers requirement of zinc for chlorophyll biosynthesis. Plants activated by zinc are also involved in carbohydrate metabolism and pollen formation (Marschner 1995). Zinc is required for tryptophan biosynthesis, which is a precursor of auxin, hence required for hormone biosynthesis (Alloway 2004). Interaction of zinc with sulfhydryl group of membrane proteins and phospholipids helps in membrane maintenance (Kabata-Pendias and Pendias 2001; Dang et al. 2010; Alloway 2004).

Zinc deficiency is widespread and crops respond positively to application of zinc (Welch 2002). Zinc is present in soil primarily due to geochemical and pedochemical weathering process from rocks. The amount of zinc in soil depends on the type, intensity of weathering, and other climatic factors that affect soil genesis (Saeed and Fox 1977).

Availability of zinc in soil reduces due to high pH, high $CaCO_{3,}$ clay, and phosphate, as these factors fix available zinc in soil (Imtiaz 1999). Zinc is generally found at a lower concentration in acidic and sandy soil. About 30% of cultivable land soil of the world contains low levels of Zinc (Sillanpaa 1990).

The change in pH affects the availability of zinc in soil because of formation of insoluble complexes. It usually forms complexes with Mn and Fe hydroxides (Sajwan and Lindsay 1988) Microorganisms play a key role in availability of nutrients; among the nutrients, Zinc is the cofactor and mineral activator of many enzymes (Venkatakrishnan et al. 2003). At a higher level, it might limit the cell and bacterial growth (Baath 1992).

3.3 *Manganese*

Manganese in its ionic form is required by plants for the activation of enzymes, specifically those that are involved in TCA and ETS. It helps in the assimilation of carbon dioxide during photosynthesis and evolution of oxygen from water-splitting complex (Marschner 1995) and chlorophyll biosynthesis. Its activity is required in the formation of ascorbic acid, riboflavin, and carotene. It is a necessary element required by plant in lesser concentration; hence, it may become toxic for plant when available in excess and interfere with utilization of other minerals such as Ca, Mg, Fe, and P via some inhibitory effects on absorption and translocation (Clark 1982). High concentration of Mn affects enzymatic activities and hormonal balance in plants; hence, Mn catalyzed reaction becomes less active or sometime nonfunctional (Horst 1988). Intervenous chlorosis (chlorophyll deficiency) with consequent development of necrotic spot is a major symptom of Mn deficiency. Depending upon plant species, it may occur in younger or older leaves. Deficiency of micronutrients in soil is widespread; many millions of hectares of arable land in the world are deficient in one or more micronutrients (Rengel 2015).

Availability of Mn in soil depends on the oxidation state of this element; it has been observed that Mn^{4+} is unavailable for plants, while it is available as Mn^{2+} (reduced form). Reduction may be biological or chemical in nature (Rengel 2000). At alkaline pH, the availability of Mn may decrease; however, the chemistry of that Mn is not clear (Clark and Baligar 2000; Pan et al. 2014). It has been observed that the concentration of Mn^{2+} in soil decreases 100 fold with every unit increase in pH (Barber 1995). Supply of Mn is a complex variable that is dependent not only on soil chemistry but also on responses of plants and microorganisms. The mechanism for mobilization of Mn surrounding the root zone via root exudate is not clear (Gherardi and Rengel 2004; Mora et al. 2009; George et al. 2014). Nutrient deficiency symptoms in plants occur when the amount of nutrient required is below that permissible or optimum range in the soil that cannot be taken up by plants. This may occur due to low solubility of nutrients, or poor soil–microbe–plant interactions (Marschner et al. 2011).

3.4 Molybdenum

Molybdenum is a transition metal required by plants for the activation of enzymatic reactions including nitrogen assimilation, purine degradation, hormone synthesis, and sulfite detoxification. It is actually inactive in its native state and needs to be complexed by specific organic pterin, which serves as a prosthetic group, molybdenum cofactor. Recent studies reveal that the concentration of molybdate is controlled by molybdate transporters (Bittner 2014). Molybdenum and iron have a close connection, as most molybdo enzymes need iron containing redox groups. These ions are components of enzymes as nitrate reductase and nitrogenase. Deficiency of this element indicates chlorosis between veins and necrosis of older leaves. It may prevent flower formation and also nitrogen deficiency, if the plant depends on symbiotic nitrogen fixation. Plants require molybdenum in very low concentration; hence, in molybdenum-deficient soil, supply of molybdenum in small quantity may increase crop production.

In soil, availability of molybdate is favored above pH 5.5 and lesser pH impairs the availability absorption by soil oxides. Under lower pH conditions, its assimilation is limited leading to molybdenum deficiency and subsequent reduction in yield and growth of plants. It can be overcome by fertilization. Excess molybdenum characterized by yellowish leaves (Kaiser et al. 2005) and reduction in anthocyanin and seedling growth (Kumchai et al. 2013).

Molybdenum concentration in agricultural soil ranges from 0.2 to 5.0 mg/kg (Scheffer and Schachtschabel 2002). Soil solutions have molybdate ions, which are available to plants. The content of Fe, Mn, Al oxides, clay minerals, and organic carbon influences availability of Mo. Soil pH has a major role on the release of ions into the soil solution. It is observed that at pH range from 4 to 5, maximum adsorption of molybdenum occurs on positively charged metal oxides (Riley et al. 1987; Xie et al. 1993; Gupta 1978; Xu et al. 2013).

In acidic conditions, anions of molybdate are adsorbed on Fe, Mn, and Al oxides, on clay minerals and organic colloids. Its availability increases with pH through decreased adsorption of metal oxides (Jiang et al. 2015; Smith et al. 1997). Well-drained sandy soils have a lesser concentration of molybdenum due to leaching, while wet soil tends to accumulate higher levels (Riley et al. 1987).

In one study, the concentration of molybdenum in the soil solutions was determined and it was observed that it ranges from 0.002 to 0.100 μmol/L. It was also differentiated depending on different properties of soil. In one study, some soil parameters have been analyzed; among them, soil pH has been suggested to be the most important factor that affects the concentration of Mo in soil solution. It has been observed that in acid sandy soils, the Mo concentration in the soil solution is too low to sustain the nutritional need of the plants. Regular liming of soils and phosphorus supply can improve the availability of molybdenum to plants (Rutkowska et al. 2017).

3.5 Iron

Iron plays an important role as enzyme component, which is involved in electrons transfer reactions (redox reactions). It is reversibly oxidized from Fe^{2+} to Fe^{3+} during electron transfer. Intervenous chlorosis is a characteristic symptom of iron deficiency. In cases of prolonged deficiency, the veins may also appear chlorotic, turning the whole leaf to white. As iron is required for chlorophyll–protein complex synthesis, leaves may become chlorotic. Due to its precipitation in the older leaves, low mobility of iron as insoluble oxides or phosphates is observed. Complexes with phytoferritin, an iron-binding protein, are also observed in the leaf and other plant parts (Oh et al. 1996).

3.6 Copper

Similar to iron, copper is an element associated with enzymes that are involved in redox reactions. Plastocyanin, an enzyme involved in electron transfer during light reactions of photosynthesis is one example (Haehnel 1984). Dark green leaves, which may contain necrotic spots, are an initial symptom of copper deficiency. Leaves may abscise prematurely under extreme copper deficiency.

3.7 *Nickel*

Nickel, the 22nd most abundant element in the earth's crust, is found in natural soils in trace concentrations (Hussain et al. 2013). It is an essential element for metabolic activities of plants and many bacteria (Brown 2007). Ni is present in several enzymes in prokaryotes (e.g., glyoxalase-I, hydrogenases, some superoxide dismutases, carbon monoxide dehydrogenase, and methyl-coenzyme M reductase (Ragsdale 1998)), while urease is the only known nickel-containing enzyme in higher plants (Polacco et al. 2013). Nickel plays an important role in nitrogen fixation; nitrogen-fixing microbes require nickel for the enzymes that reprocess hydrogen gas liberated during fixation. Nickel deficiency appears in plants as leaf tip necrosis and urea accumulation. However, these symptoms occur rarely in plants.

3.8 *Chlorine*

Chlorine in its ionic form is required by plants during photosynthesis in water-splitting complex. It plays a role in cell division in leaves and roots (Harling et al. 1997; Clarke and Eaton-Rye 2000). Bronze like color appears in plant leaves due to chlorine deficiency; it may show stunted and thickened root tips. Some plants absorbed higher concentration of chlorine than required by plants for normal metabolic activities.

4 Availability of Mineral Nutrients in Soil

One of the most important components of organic material is nitrogen, next to carbon. Both these are essential for fertility of soil. The biogeochemical cycle of C and N plays an important role in global warming (Yang et al. 2010). The ratio of these two regulates the mineralization process in soil, specifically organic matter, which eventually releases soil nitrogen (Deng et al. 2013). Mineralization occurs via decomposition process. Significant decline in carbon storage has been observed due to change in C and N ratio (Aitkenhead and McDowell 2000). There are many factors that influence the biogeochemical cycle, namely climate, topography, and some basic soil properties, which eventually change the C and N storage. Land use is the most significant factor among all (Yang et al. 2010). Organic matter is the main source of carbon in soil and C:N represents its degradation. Since soil mechanism is governed by climate factors, soil organic carbon is the main factor that determines some important component of terrestrial ecosystem (Sakin et al. 2010; Garcia and Alcantara 2013; Zhang et al. 2007). Regarding N and P, their cycling shows many differences. The main source of N is atmosphere, while P is derived from rock weathering; due to this fact, the former one is usually absent in newly formed soil, thus not involved in net primary productivity (Tilman 1986; Berendse 1990; Vitousek et al. 1987).

Due to the mobile nature of nitrogen in soil, it is leached away; it can easily move from the ecosystem in a gaseous form as in cases of frequent fires and denitrification. Therefore, on the extent of nitrogen losses, soil may remain N limited for a long period of time. Nitrogen is carbon-bonded, while phosphorus is ester-bonded and often soluble, hence easily available for plants to absorb (Hunt et al. 1983; Howarth 1988), while carbon-bonded nitrogen is immobilized for a long time and thus promotes nitrogen limitation. Biochemistry is not the only feature that is responsible for this difference, but the external environment also affects the nitrogen and phosphorus availability. Since these are essential nutrients required by plants in excess, there is a need to determine the limitation of these elements in plants and soil (Boeye et al. 1997). Factorial fertilizer experiment can be used for macronutrients estimation, but these are time consuming, laborious, and impart some disturbances. Interpretation of such results causes difficulty due to disturbances at specific sites (Bobbink 1992). Plant responses for nutrient addition are affected by chemical adsorption and microbial immobilization.

One group of researcher has suggested that N:P mass ratios in plants indicate the limitation of certain nutrients (Koerselman and Meuleman 1996), but it is difficult to assess at community and species level; further, the N:P ratio is itself a limiting factor for plant growth and development (DiTomasso and Aarssen 1989). Plants grown in soil with lower fertility have high capacity to uptake mobile ions (Veerkamp and Kuiper 1982) and a comparatively lower capacity to absorb immobile ions (Chapin et al. 1986; Raab et al. 1998). Nitrogen found in the soil in the form of nitrate, ammonium, and as organic nitrogen, so plants absorb any form of nitrogen (soluble form), depending upon their preferences on the basis of different carrier proteins (Atkin 1996). It is reported that in Arctic plants, where a high concentration of amino acid occurs, plant growth preferentially depends on amino acids (Keilland 1994), while spruce grows on acidic soil, absorbs ammonium instead of nitrate (Kronzucker et al. 1997).

pH is a relevant property of soil which can even determine the yield of certain crops (Moody et al. 1998). It is a dynamic feature with significant differences (Behera and Shukla 2015; Kariuki et al. 2010). These differences are due to seasonal variations. During rainfall when evapotranspiration exceeds precipitation, salt concentration increases, which forces H^+ ions in soil, thereby decrease in pH, whereas in wet seasons, soil salts are removed, and hence, pH increases (Rengel 2002). These fluctuations are seasonal and not to be confused with changes in pH over centuries (Tang and Rengel 2003). Soil pH is an important factor, which has a dominant effect on the solubility and availability of ions (Clark and Baligar 2000). Iron toxicity occurs in soil with pH (<3.2), that is, acidic and anaerobic conditions

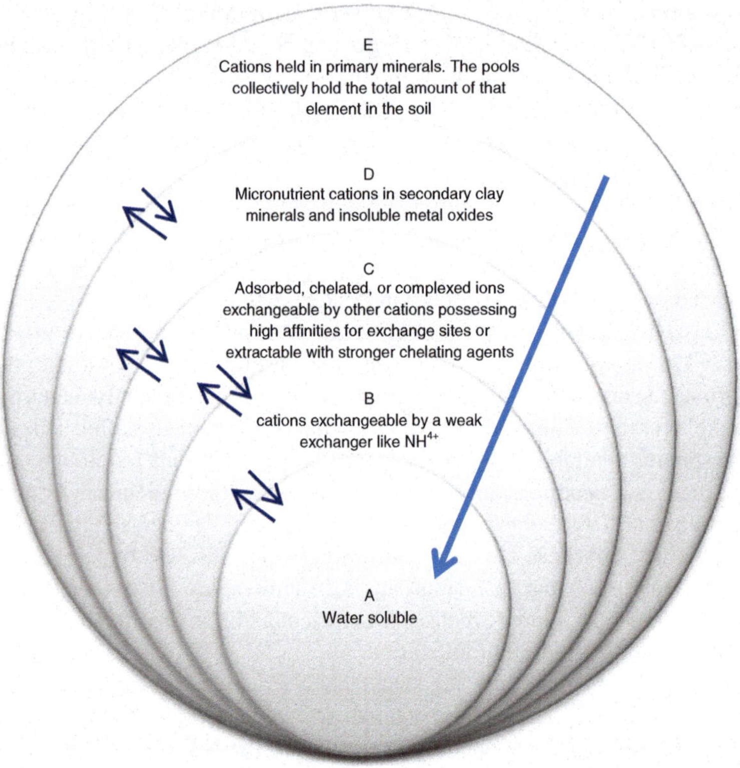

Fig. 1 Five major cation pools of micronutrients

(Khabaz-Saberi et al. 2010). Sometimes, manganese becomes toxic in poorly drained soil when reducing conditions dominate.

Micronutrient cations occur mainly in five pool types on the basis of their availability and solubility. These are postulated as A, B, C, D, and E (Fig. 1).

Pool A consists of non-adsorbed ions and ions adsorbed on colloids. Soil pH, redox potential, and concentration of other ions affect this zone in terms of ions present in this pool. In this pool, small or lesser concentration of Zn and Cu is observed, while Mn and Fe may be present in very smaller concentration (< 1 ppm). Low redox potential and low pH can increase the pool size for Mn and Fe, but has negligible effect on Zn and Cu.

Pool B includes water-soluble pool A and is larger than A. However, it is smaller for Zn and Cu except for some in which fertilization has been done for these elements. To predict the adequacy of Mn, exchangeable Mn of pool B is frequently used (Sherman 1957).

Pool C contains those cations which can be exchanged by the mass action of cations with affinities for the absorbent or by extraction through chelating agents. It has been suggested that this pool contains cations absorbed with great affinity by clay and humus of the soil.

It has been found that pool A, B, and C are in reversible equilibrium shown in Fig. 1 (designated by double arrows). The availability of micronutrients in these pools is greater than that of others, especially in pool C.

Pool D and pool E consist of secondary minerals around these three pools. These two cannot be separated by chemical methods due to precipitation of secondary minerals being highly resistant to weathering (Viets 1962).

Temperature, after pH, is the major factor, which can regulate biogeochemical processes, for example, soil respiration (Raich and Schlesinger 1992), N mineralization and nitrification (MacDonald et al. 1995), litter decomposition (Meentemeyer 1978; Jansson and Berg 1985; Hobbie 1996), denitrification (Malhi et al. 1990), CH_4 emission (Crill et al. 1988; Crill 1991; Johnson et al. 1996), fine root dynamics (Boone et al. 1998; Pregitzer et al. 2000; Gill and Jackson 2000), plant productivity (WarrenWilson 1957), and plant nutrient uptake (BassiriRad 2000). Anthropogenic activities have an impact on increased concentration of green house gases (Intergovernmental Panel on Climate Change (IPCC) 1996). Green house gases have a potential ability to capture heat energy and thus increased global mean temperature by 0.3–0.6 °C over the last century (IPCC 1996; Rind 1999; Karl et al. 2000). Global warming affects most of the processes on earth; however, it is not clear which processes will be most affected by warming. One researcher has reported that there are some factors that affect ecosystem response, such as stocks and initial turnover rates of labile soil C and N, relative size of the plant and soil C pools, dominant form of available N in the soil, soil water and precipitation regimes, the chemical composition and turnover rates of plant residues, and the longevity of individuals and population turnover rates of dominant species (Shaver et al. 2000) and availability of minerals in soil. Soil respiration rates generally increase with warmer temperatures (Peterjohn et al. 1993, 1994; McHale et al. 1998; Rustad and Fernandez 1998). Plant productivity have all been shown to be affected by climate warming (Van Cleve et al. 1990; Joslin and Wolfe 1993; Peterjohn et al. 1993, 1994; Harte and Shaw 1995; Hantschel et al. 1995; Robinson et al. 1995; Hobbie 1996; Lukewille and Wright 1997; Ineson et al. 1998; Jamieson et al. 1998).

5 Conclusion

Mineral nutrients are essential for plant growth and development. They are present in soil in the form of either cation or anion, depending upon their oxidation and reduction reactions. Availability of these elements is much affected by pH of the soil. It has been observed that alkaline pH is not favorable for soil health because these soils are either micronutrients-deficient or have lesser concentration of the same. Besides pH, temperature is another aspect that affects nutrient availability and other more parameters of soil and surrounding environment. Researches have proved that increase in temperature has a significant effect on ecosystem responses, including biogeochemical cycling. Due to disturbances in this cycle, the whole criteria that play governing role for ecosystem functioning have changed. Since nutrients are the major source for growth and development of plants, and the reservoir of these nutrients is soil, it is a necessary field of study with a broad scope. As India is the country of farmers and most of the population depends on agriculture, a great attention is a need of today. In this regard, necessary steps should be taken by keeping in mind the basic criteria of fertilizer supply and this can be done only when one has a knowledge of appropriate concentration of nutrients and factors that govern availability of the same.

References

Ahmad W, Zia MH, Malhi SS, Niaz A, Saifullah (2012) Boron deficiency in soils and crops: a review, crop plant. In: Goyal A (ed.), InTech. ISBN: 978-953-51-0527-5. Available from: http://www.intechopen.com/books/crop-plant/boron-deficiency-in-soils-and-crops-a-review

Aitkenhead JA, McDowell WH (2000) Soil C:N ratio as a predictor of annual riverine DOC flux at local and global scales. Glob Biogeochem Cycles 14(1):127–138

Alloway BJ (2004) In zinc in soil and crop nutrition, vol 2. International Zinc Association, Brussels, pp 32–34

Atkin OK (1996) Reassessing the nitrogen relations of arctic plants: a mini-review. Plant Cell Environ 19:695–704

Baath E (1992) Measurement of heavy metal tolerance of soil bacteria using thymidine incorporation into bacteria extracted after homogenization-centrifugation. Soil Biol Biochem 24:1167–1172

Barber SA (1995) Soil nutrient bioavailability. A mechanistic approach, 2nd edn. Wiley, New York

BassiriRad H (2000) Kinetics of nutrient uptake by roots: responses to global change. New Phytol 147:155–169

Behera SK, Shukla AK (2015) Spatial distribution of surface soil acidity, electrical conductivity, soil organic carbon content and exchangeable potassium, calcium and magnesium in some cropped acid soils of India. Land Degrad Dev 26:71–79

Berendse F (1990) Organic matter accumulation and nitrogen mineralization during secondary succession in heathland ecosystems. J Ecol 78:413–427

Bittner F (2014) Molybdenum metabolism in plants and crosstalk to iron. Department of Plant Biology, Braunschweig University of Technology, Braunschweig, Germany, Mini Review Article

Bobbink R (1992) Critical loads for nitrogen eutrophication of terrestrial and wetland ecosystems based upon changes in vegetation and fauna. In: Grennfelt P, Thornelof R (eds) Critical loads for nitrogen. Nord Miljrapport 41, Nordic Council of Ministers, Copenhagen, pp 111–159

Boeye D, Verhagen B, Van HV, Verheyen RF (1997) Nutrient limitation in species-rich lowland fens. J Veg Sci 8:415–424

Boone RD, Naddelhoffer KJ, Canary JD, Kaye JP (1998) Roots exert a strong influence on the temperature sensitivity of soil respiration. Nature 396:570–572

Brown PH (2007) Nickel. In: Barkerand AV, Pilbean DJ (eds) Handbook of plant nutrition. CRC Taylor and Francis, NewYork, pp 395–402

Chapin FS III, Shaver GR, Kedrowski RA (1986) Environmental controls over carbon, nitrogen and phosphorus chemical fractions in *Eriophorum vaginatum* L. in Alaskan tussock tundra. J Ecol 74:167–195

Clark RB (1982) Plant response to mineral element toxicity and deficiency. In: Christiansen MN, Lewis CF (eds) Breeding plants for less favorable environments. Wiley, New York, pp 71–142

Clark RB, Baligar VC (2000) Acidic and alkaline soil constraints on plant mineral nutrition. In: Wilkinson RE (ed) Plant-environment interactions. Marcel Dekker Inc, New York, pp 133–177

Clarke SM, Eaton-Rye JJ (2000) Amino acid deletions in loop C of the chlorophyll a-binding protein CP47 alter the chloride requirement and/or prevent the assembly of photosystem II. Plant Mol Biol 44:591–601

Crill PM (1991) Seasonal patterns of methane uptake and carbon dioxide release by a temperate woodland soil. Glob Biogeochem Cycles 5:319–334

Crill PM, Bartlett KB, Hariss RC, Gorham E, Verry ES, Sebacher DI, Madzar L, Sanner W (1988) Methane flux from Minnesota peatlands. Global Biogeochem Cycl 2:371–384

Dang HR, Li Y, Sun X, Zhang LY (2010) Absorption, accumulation and distribution of zinc in highly-yielding winter wheat. Agric Sci China 9(7):965–973

Deng Q, Cheng X, Zhou G, Liu J, Liu S, Zhang Q, Zhang D (2013) Seasonal responses of soil respiration to elevated CO_2 and N addition in young subtropical forest ecosystems in young subtropical forest ecosystems in southern China. Ecol Eng 61:65–73

Denison RF, Kiers ET (2005) Sustainable crop nutrition: constraints and opportunities. In: Broadley MR, White PJ (eds) Plant nutritional genomics. Blackwell, Oxford, pp 242–286

DiTomasso A, Aarssen LW (1989) Resource manipulations in natural vegetation: a review. Vegetatio 84:9–29

Epstein E (1965) Mineral metabolism. In: Bonner J, Varner JE (eds) Plant biochemistry. Academic Press, London, pp 438–466

Fixen PE (2005) Decision support systems in integrated crop nutrient management. In: Proceedings no. 569. International Fertilizer Society, New York

George TS, French AS, Brown LK, Karley AJ, White PJ, Ramsay L, Daniell TJ (2014) Genotypic variation in the ability of landraces and commercial cereal varieties to avoid manganese deficiency in soils with limited manganese availability: is there a role for root-exuded phytases ? Physiol Plant 151:243–256

Gherardi MJ, Rengel Z (2004) The effect of manganese supply on exudation of carboxylates by roots of lucerne (*Medicago sativa*). Plant Soil 260:271–282

Gill RA, Jackson RB (2000) Global patterns of root turnover for terrestrial ecosystems. New Phytol 147:13–31

Gupta UC (1978) Soil and plant factors affecting molybdenum uptake by plants. In: Gupta UC (ed) Molybdenum in agriculture. Cambridge University Press, Cambridge, pp 71–91

Haehnel W (1984) Photosynthetic electron transport in higher plants. Annu Rev Plant Physiol 35:659–693

Hantschel RE, Kamp T, Beese F (1995) Increasing soil temperature to study global warming effects on the soil nitrogen cycle in agroecosystems. J Biogeogr 22:375–380

Harling H, Czaja I, Schell J, Walden R (1997) A plant cation chloride co-transporter promoting auxin-independent tobacco protoplast division. EMBO J 16:5855–5866

Harte J, Shaw MR (1995) Shifting dominance within a montane vegetation community: results from a climate-warming experiment. Science 267:876–880

Hobbie SE (1996) Temperature and plant species control over litter decomposition in Alaskan tundra. Ecol Monogr 66:503–522

Horst WJ (1988) The physiology of Mn toxicity. In: Graham RD, Hannam RJ, Uren NC (eds) Manganese in soils and plants. Kluwer Academic Publishers, Dordrecht, pp 175–188

Howarth RW (1988) Nutrient limitation of net primary production in marine ecosystems. Annu Rev Ecol Syst 19:89–110

Hunt HW, Stewart JB, Cole CV (1983) A conceptual model for interactions of carbon, nitrogen, phosphorus, and sulphur in grasslands. In: Bolin B, Cook RB (eds) The major biogeochemical cycles and their interactions. Wiley, New York, pp 303–326

Hussain MB, Azam AS, Hina S, Farooq MA, Ali B (2013) Morphological, physiological and biochemical responses of plants to nickel stress: a review. Afr J Agric Res 8:1596–1602. https://doi.org/10.5897/AJAR12.407

Imtiaz M (1999) Zn deficiency in cereals. PhD thesis. Reading University, U.K

Ineson P, Benham DG, Poskitt J, Harrison AF, Taylor K, Woods C (1998) Effects of climate change on nitrogen dynamics in upland soils. 2. A soil warming study. Glob Chang Biol 4:153–161
Intergovernmental Panel on Climate Change (IPCC) (1996) Climate change 1995: the science of climate change. In: Houghton JT, Meira Filho LG, Callander BA, Harris N, Kattenberg A, Maskell K (eds) Contribution of working group 1 to the second assessment report of the intergovernmental panel on climate change. Cambridge University Press, Cambridge
Jamieson N, Barraclough D, Unkovich M, Monaghan R (1998) Soil N dynamics in a natural calcareous grassland under a changing climate. Biol Fertil Soils 27:267–273
Jansson PE, Berg B (1985) Temporal variation of litter decomposition in relation to simulated soil climate. Long-term decomposition in a scots pine forest. V. Can J Bot 63:1008–1016
Jiang W, Yang ZF, Yu T, Hou QY, Zhong C, Zheng GD, Yang ZQ, Li J (2015) Evaluation of the potential effects of soil properties on molybdenum availability in soil and its risk estimation in paddy rice. J Soils Sediments 15:1520–1530
Johnson LC, Shaver GR, Giblin AE, Nadelhoffer KJ, Rastetter ER, Laundre JA, Murray GL (1996) Effects of drainage and temperature on carbon balance of tussock tundra microcosms. Oecologia 108:737–774
Joslin JD, Wolfe MH (1993) Temperature increase accelerates nitrate release from high elevation red spruce soils. Can J For Res 23:756–759
Kabata-Pendias A, Pendias H (2001) Trace elements in soils and plants. CRC Press, Boca Raton/London/New York/Washington D.C
Kaiser BN, Gridley KL, Ngaire BJ, Phillips T, Tyerman SD (2005) The role of molybdenum in agricultural plant production. Ann Bot 96:745–754. https://doi.org/10.1093/aob/mci226
Kariuki SK, Schroder JL, Zhang HL, Hanks T, McGrath JM, Payton ME (2010) Temporal variability of soil property dynamics in a grazed pasture. Commun Soil Sci Plant Anal 41:27442754
Karl TR, Knight RW, Baker B (2000) The record breaking global temperature of 1997 and 1998: evidence for an increase in the rate of global warming? Geophys Res Lett 27:719
Keilland K (1994) Amino acid absorption by arctic plants: implications for plant nutrition and nitrogen cycling. Ecology 75:2373–2383
Khabaz-Saberi H, Rengel Z, Wilson R, Setter TL (2010) Variation for tolerance to high concentration of ferrous iron (Fe^{2+}) in Australian hexaploid wheat. Euphytica 172:275–283
Koerselman W, Meuleman AFM (1996) The vegetation N: P ratio: a new tool to detect the nature of nutrient limitation. J Appl Ecol 33:1441–1450
Kronzucker HJ, Siddiqi MY, Glass AM (1997) Conifer root discrimination against soil nitrate and the ecology of forest succession. Nature 385:59–61
Kumchai J, Huang JZ, Lee CY, Chen FC, Chin SW (2013) Proline partially overcomes excess molybdenum toxicity in cabbage seedlings grown in vitro. Genet Mol Res 12:5589–5601. https://doi.org/10.4238/2013.November.18.8
Loomis RS, Connor DJ (1992) Crop ecology: productivity and Management in Agricultural Systems. Cambridge University Press, Cambridge
Lozano-Garcia B, Parras-Alcantara L (2013) Land use and management effects on carbon and nitrogen in Mediterranean Cambisols. Agric Ecosyst Environ 179:208–214
Lukewille A, Wright RF (1997) Experimentally increased soil temperature causes release of nitrogen at a boreal forest catchment in southern Norway. Glob Chang Biol 3:13–21
MacDonald NW, Zak DR, Pregitzer KS (1995) Temperature effects on kinetics of microbial respiration and net nitrogen and sulfur mineralization. Soil Sci Soc Am J 59:233–240
Macek T, Mackova M, Kas J (2000) Exploitation of plants for the removal of organics in environmental remediation. Biotechnol Adv 18:23–34
Malhi SS, McGrill WB, Nyborg N (1990) Nitrate losses in soils: effects of temperature, moisture, and substrate concentration. Soil Biol Biochem 22:917–927
Marschner H (1995) Mineral nutrition of higher plants. Academic Press, London
Marschner P, Crowley D, Rengel Z (2011) Rhizosphere interactions between microorganisms and plants govern iron and phosphorus acquisition along the root axis - model and research methods. Soil Biol Biochem 43:883–894

McHale PJ, Mitchell MJ, Bowles FP (1998) Soil warming in a northern hardwood forest: trace gas fluxes and leaf litter decomposition. Can J For Res 28:1365–1372

Meentemeyer V (1978) Macroclimate and lignin control of litter decomposition rates. Ecology 59:465–472

Moody PW, Aitken RL, Dickson T (1998) Field amelioration of acidic soils in south-east Queensland: III. Relationships of maize yield response to lime and unamended soil properties. Aust J Agric Res 49:649–656

Mora ML, Rosas A, Ribera A, Rengel Z (2009) Differential tolerance to Mn toxicity in perennial ryegrass genotypes: involvement of antioxidative enzymes and root exudation of carboxylates. Plant Soil 320:79–89

Nolan BT, Stoner JD (2000) Nutrients in groundwater of the center conterminous United States 1992–1995. Environ Sci Technol 34:1156–1165

Novoa R, Loomis RS (1981) Nitrogen and plant production. Plant Soil 58:177–204

Oh SH, Cho SW, Kwon TH, Yang MS (1996) Purification and characterization of phytoferritin. J Biochem Mol Biol 29:540–544

Olsen LI, Palmgren MG (2014) Many rivers to cross: the journey of zinc from soil to seed. Front Plant Sci 5:30

Pan Y, Koopmans GF, Bonten LTC, Song J, Luo Y, Temminghoff EJM, Comans RNJ (2014) Influence of pH on the redox chemistry of metal hydroxides and organic matter in paddy soils. J Soils Sediments 14:1713–1726

Peterjohn WT, Melillo JM, Bowles ST (1993) Soil warming and trace gas fluxes: experimental design and preliminary flux results. Oecologia 93:18–24

Peterjohn WT, Melillo JM, Steudler PA, Newkirk KM, Bowles ST, Aber JD (1994) Responses of trace gas fluxes and N availability to experimentally elevated soil temperatures. Ecol Appl 4:617–625

Polacco JC, Mazzafera P, Tezotto T (2013) Nickel and urease in plants: still many knowledge gaps. Plant Sci 19:79–90. https://doi.org/10.1016/j.plantsci.2012.10.010

Pregitzer KS, King JS, Burton AJ, Brown SE (2000) Responses of tree fine roots to temperature. New Phytol 147:10–11

Raab TK, Lipson DA, Monson RK (1998) Soil amino acid utilization among species of the Cyperaceae: plant and soil processes. Ecology 80(7):2408–2419

Ragsdale SW (1998) Nickel biochemistry. Curr Opin Chem Biol 2:208–215. https://doi.org/10.1016/S1367-5931(98)80062-8

Raich JW, Schlesinger WH (1992) The global carbon dioxide flux in soil respiration and its relationship to vegetation and climate. Tellus 44:81–89

Ranade-Malvi U (2011) Interaction of micronutrients with major nutrients with special reference to potassium. Institute for Micronutrient Technology, Pune. Karnataka J Agric Sci 24(1):106–109

Rengel Z (2000) Uptake and transport of manganese in plants. In: Sigel A, Sigel H (eds) Metal ions in biological systems. Marcel Dekker, New York, pp 57–87

Rengel Z (2002) Role of pH in availability of ions in soil. In: Rengel Z (ed) Handbook of plant Growth. pH as a master variable in plant growth. Marcel Dekker, New York, pp 323–350

Rengel Z (2015) Availability of Mn, Zn and Fe in the rhizosphere. Soil science and plant nutrition. J Soil Sci Plant Nutr 2015 15(2):397–409

Riley MM, Robson AD, Gartrell JW, Jeffery RC (1987) The absence of leaching of molybdenum in acidic soils from Western Australia. Aust J Soil Res 25:179–184

Rind D (1999) Complexity and climate. Science 284:105–107

Robinson CH, Wookey PA, Parsons AN, Potter JA, Callaghan TV, Lee JA, Press MC, Welker JM (1995) Responses of plant litter decomposition and nitrogen mineralization to simulated environmental change in a high arctic polar semi-desert and a subarctic dwarf shrub heath. Oikos 74:503–512

Rustad LE, Fernandez IJ (1998) Experimental soil warming effects on CO_2 and CH_4 flux from a low elevation spruce fir forest soil in Maine, USA. Glob Chang Biol 4:597–607

Rutkowska B, Szulc W, Spychaj-Fabisiak E, Pior N (2017) Prediction of molybdenum availability to plants in differentiated soil conditions. Plant Soil Environ 63:491–497
Saeed M, Fox RL (1977) Relation between suspension pH and Zn solubility in acid and calcareous soils. Soil Sci 124:199–204
Sajwan KS, Lindsay WL (1988) Effect of redox, zinc fertilisation and incubation time on DTPA-extractable zinc, iron and manganese. Commun Soil Sci Plant Anal 19:1–11
Sakin E, Deliboran A, Sakin ED, Aslan H (2010) Carbon and nitrogen stocks and C:N ratio of Harran plain soils. Not Sci Biol 2(4):104–110
Scheffer F, Schachtschabel P (2002) Textbook of soil science, vol 15. Auflage Spektrum Akademischer Verlag, Heidelberg, p 593
Shaver GR, Canadell J, Chapin FS III, Gurevitch J, Harte J, Henry G, Ineson I, Jonasson S, Melillo J, Pitelka L, Rustad L (2000) Global warming and terrestrial ecosystems: a conceptual framework for analysis. Bioscience 50:871–882
Sherman GD (1957) Yearbook of agricultural, U. S. Department of Agriculture, pp 135–139
Sillanpaa M (1990) Micronutrients assessment at the country level. An international study FAO. Soils Bull 63. Food and agriculture Organization of the United Nations
Sims T (2000) Soil fertility evaluation. In: Sumner ME (ed) Hand book of soil science. CRC Press, Boca Raton/London/New York/Washington DC
Smith KS, Balistrieri LS, Smith SS, Severson RC (1997) Distribution and mobility of molybdenum in the terrestrial environment. In: Gupta UC (ed) Molybdenum in agriculture. Cambridge University Press, Cambridge, pp 23–46
Tan S, Han R, Li P, Yang G, Li S, Zhang P, Wang W, Zhao W, Yin L (2015) Overexpression of the MxIRT1 gene increases iron and zinc content in rice seeds. Transgenic Res 24:109–122
Tang C, Rengel Z (2003) Role of plant cation/anion uptake ratio in soil acidification. In: Rengel Z (ed) Handbook of soil acidity. Marcel Dekker, New York, pp 57–81
Tilman D (1986) Nitrogen-limited growth in plants from different successional stages. Ecology 67:555–563
Tisdale SL, Nelson WL (1975) Soil fertility and fertilizers, 3rd edn. Collier Macmillan Publication, London/New York
Tisdale SL, Nelson WL, Beaten JD (1984) Zinc in soil fertility and fertilizers, 4th edn. Macmillan Publishing Company, New York, pp 382–391
University of Western Australia, 35 Stirling Highway, Perth WA 6000, Australia
Van Cleve K, Oechel WC, Hom JL (1990) Response of black spruce (Picea mariana) ecosystems to soil temperature modifications in interior Alaska. Can J For Res 20:1530–1535
Vankatakrishnan SS, Sudlayandy RS, Savariappan AR (2003) Assessing in vitro solubilization potential of different zinc solubilizing Bacteria (ZSB) isolates. Braz J Microbiol 34:121–125
Vaz Patto MC, Amarowicz R, Aryee ANA, Boye JI, Chung H, Martin-Cabrejas MA, Domoney C (2015) Achievements and challenges in improving the nutritional quality of food legumes. Crit Rev Plant Sci 34:105–143
Veerkamp MT, Kuiper PJC (1982) The uptake of potassium by Carex species from swamp habitats varying from oligotrophic to eutrophic. Physiol Plant 55:237–241
Viets FG Jr (1962) U. S. Department of Agriculture, Fort Collins, Colo. Chem Availab Micronutr Soils 10(3)
Vitousek PM, Walker LR, Whiteaker LD, Mueller-Dombois D, Matson PA (1987) Biological invasion by Myrica faya alters ecosystem development in Hawaii. Science 238:802404
Warren-Wilson J (1957) Arctic plant growth. Adv Sci 13:383–388
Welch RM (2002) The impact of mineral nutrients in food crops on global human health. Plant Soil 247:83–90
White PJ, Brown PH (2010) Plant nutrition for sustainable development and global health. Ann Bot 105:1073–1080
Xie RJ, MacKenzie AF, Lou ZJ (1993) Causal modeling pH and phosphate effects on molybdate sorption in three temperate soils. Soil Sci 155:385–397

Xu N, Braida W, Christodoulatos C, Chen JP (2013) A review of molybdenum adsorption in soils/bed sediments: speciation, mechanism, and model applications. Soil Sediment Contam Int J 22:912–929

Yang YH, Fang JY, Guo DL, Ji CJ (2010) Vertical patterns of soil carbon, nitrogen and carbon: nitrogen stoichiometry in Tibetan grasslands. Biogeosci Discuss 7:1–24

Zhang HB, Lou YM, Wong MH, Zhao QG, Zhang GL (2007) Soil organic carbon storage and change with reduction in agricultural activities in Hong Kong. Geoderma 139:412–419

Role of Organic Fertilizers in Improving Soil Fertility

Teg Bahadur Singh, Akbar Ali, Mrinalini Prasad, Arti Yadav, Preksha Shrivastav, Deepika Goyal, and Prem Kumar Dantu

Contents

1 Introduction

Fertilizer Fertilizer is a chemical mixture of various macro- and micronutrients (in a proper ratio) required for sustainable growth and development of the plants. Organic fertilizer is used to maintain soil fertility. Organic fertilizers are biodegradable material, which makes better nutrient sources. Organic carbon content of organic fertilizers has equivalent or higher value of nitrogen and phosphorus contents. An organic fertilizer enhances potential growth of heterotrophic bacteria in soil of root zone and stimulates primary and secondary productivity in plants (Anderson 1987; Qin et al. 1995; Bokhtiar and Sakurai 2005). Quality and health of the soil could be improved by the application of organic manures at an optimum level (Yanan et al. 1997). A diagrammatic picture of different types of organic fertilizer is presented in Fig. 1.

Soil Fertility The inherent capacity of the soil to supply all essential macro- and micronutrients for the survival of plants in available forms and in a suitable proportion is known as soil fertility (Boulaine 1989). The soil fertility mainly depends on the mineralogical composition of the parent material, topography, and biological activities in the soil and local climatic conditions as temperature, solar radiation and rainfall required for pedogenesis. It is the outcome of the interactions between the

T. B. Singh · A. Ali · M. Prasad · A. Yadav · P. Shrivastav · D. Goyal · P. K. Dantu (✉)
Department of Botany, Dayalbagh Educational Institute (Deemed University), Agra, India

M. Naeem et al. (eds.), *Contaminants in Agriculture*,
https://doi.org/10.1007/978-3-030-41552-5_3

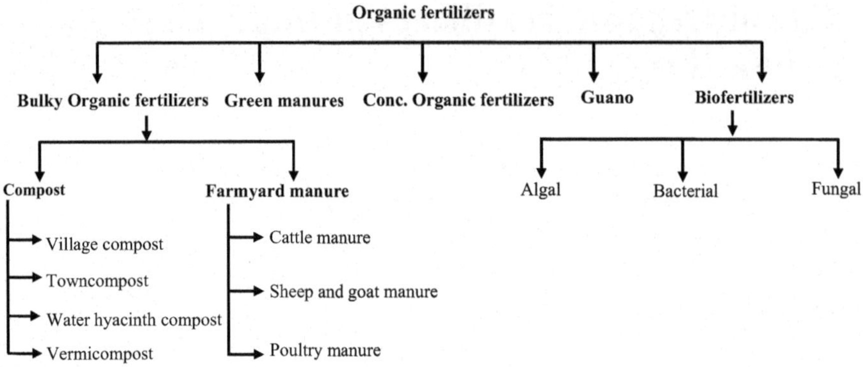

Fig. 1 Types of organic fertilizers

biological, chemical and physical properties of soil due to soil nature and the effects of climate (Liu et al. 2010).

Maintenance of Soil Fertility Soil fertility maintenance is the retaining, cycling and supplying of mineral nutrients required for the growth of plant over several years. It can be maintained by organic amendments (plant material, animal residues and sewage sludge) to soil, which are rapidly decomposed by enzymatic actions of rhizospheric microbes such as cellulase, pectinase and protease, releasing mineral nutrients such as nitrogen (N), phosphorus (P), potassium (K) and sulphur (S) (Gianfreda and Bollag 1996; Zhang et al. 2012; Dong et al. 2012; Patil et al. 2014; Liu et al. 2014). Soil microbial biomass carbon (C) and nitrogen (N) represent the microbial size and soil fertility status and maintain nutrient pool in soil (Nair and Ngouajio 2012). There is an immense role of microbial flora to increase soil fertility, which is stimulated by many physical and chemical parameters of soil such as soil organic matter (SOM), acidity, alkalinity, and clay content (Nautiyal et al. 2010; Xun et al. 2015).

The activity of soil organisms (macroorganism and microorganisms), mineral types, soil–air exchange rates and other biological, chemical or physical processes are related to soil fertility (Das and Mukherjee 1990; Diacono and Montemurro 2010). Soil microorganisms play an important role in agro-ecosystem by degrading soil organic matter, nutrient cycling and bioremediation of pollutant of soil (Shaheen et al. 2007; Dong et al. 2014; Luo et al. 2015; Matulich and Martiny 2015). The microbial population of soil is influenced by soil organic matter that helps in improving soil fertility (Ge et al. 2010). Soil organic matter (SOM) plays a very important role in soil fertility. SOM is a very important factor that can improve soil structure stability, resistance to rainfall impact, rate of infiltration and faunal activities (Roose and Barthes 2001).

2 Types of Soil Fertility

(a) **Permanent Soil Fertility:** This type of fertility depends on the inherent nature of the soil itself. It can be improved and sustained by soil management practices.
(b) **Temporary Soil Fertility:** This type of soil fertility is acquired by appropriate soil management. This soil fertility is extremely dependent on the permanent fertility that is already available.

Organic Fertilizer Organic fertilizers are naturally occurring fertilizers produced by both plants and animals.

3 Types of Organic Fertilizer

1. Bulky organic fertilizer
   (A) Compost
   (B) Farm yard manure
2. Green manures.
3. Concentrated organic fertilizers:
   (A) Oil cakes: richest sources of plant nutrient of all organic manures
   (B) Other concentrated organic fertilizers
4. Guano
5. Biofertilizers
   (A) Algal biofertilizer
   (B) Fungal biofertilizer
   (C) Bacterial biofertilizer

1. Bulky Organic Fertilizers: These Types of Fertilizers Contain Low Amount of Plant Nutrient

   (A) Compost

Compost is the rotten organic matter that has been decomposed by the process of composting. The process of composting can be enhanced by providing controlled optimum conditions for the detritus eating microorganisms. Microorganisms play a vital role in converting heterogeneous organic matters into humus substance by the composting process. The main decomposers of organic matter are bacteria (including actinomycetes), fungi and protozoa. Protozoa, mites, nematodes and springtails feed on microbes that are involved in decomposing of organic matter (Neher 1999). Compost can be prepared from a range of waste materials like crop stubbles, straw, crop residue such as sugar cane trash and rice husks, litter, weeds, leaves, and kitchen waste. This type of compost contains 0.7–1.0% K_2O, 0.3–0.6% P_2O_5 and 0.4

0.8% N (Verma and Verma 2012). Degradation of organic matter is dominantly by fungi in early stages due to the high ratio of lignin and nitrogen and later bacteria are the secondary colonizers (Beare et al. 1992). The microbial populations in soil affects the rate of decomposition of organic matter and the successive release of available plant nutrients in the soil (Mukherjee et al. 1991; Debnath et al. 1994).

(i) **Village Compost:** The compost made from farm refuses such as crop residue, weeds, stubble and leaves of tree and vegetable plants. The farm refuses are collected and stored in the pit and mixed with cow dung slurry and water or soil and water. Microorganism inoculants are used to decompose this residue. After turning and storage for about 6 months it is suitable for application in fields. The average nutrient content of village compost is 0.4–0.8% N, 0.3–0.6%P_2O_5 and 0.7–1.0% K_2O (Verma and Verma 2012).

(ii) **Town Compost:** The compost made from town refuses like night soil, street sweepings, and dustbin refuse are called town compost. It contains 1–2% N, 1% P_2O_5 and 1.5% K_2O (Verma and Verma 2012). The biodegradable organic material is transformed into humic substances collected from various sources of the urban area (Golueke 1977; Wiles 1978; Bewick 1980). The chemical composition of solid urban waste is heterogeneous as organic (biodegradable) volatile matter, protein, lipids, sugar, cellulose, starch, lignin, phosphorus, potassium, crude fibre, etc. Mineralization and partial humification of these substances are carried out through composting (de Bertoldi et al. 1983).

(iii) **Water Hyacinth Compost:** Water hyacinth–derived manures are called water hyacinth compost. For preparing water hyacinth compost, the plants are chopped into small pieces of about 5–10 cm in length to increase the surface area for microbial action, dried in sunlight for 3 days and filled in composting boxes. Water hyacinth decomposes in only 30 days as compared to other crop plants that require up to 2–3 months. This compost could be used as surface mulch or as compost. The crude powder obtained from roots of water hyacinth has successfully been used to support vegetable crop production. As a fertilizer, water hyacinth compost contains 2% nitrogen or the equivalent to 10.5 kg of ammonium sulphate; 1.1% P_2O_5 equivalent to 6.9 kg single superphosphate; 2.5% K_20 equivalent to 5.0 kg muriate of potash (Sharma 1971). Water hyacinth has high concentrations of nutrients and faster growth rate, so it has a great potential as fertilizer for the nutrient-deficient soil. Compost of water hyacinth is a potential source of available form of phosphorus and exchangeable potassium for higher crop production. Higher application of hyacinth compost increases the soil pH, which is favourable for better growth of soil microorganisms. Hyacinth compost can act as a soil stabilizer and hence increase soil productivity.

(iv) **Vermicompost:** The use of earthworms to convert biodegradable organic waste materials into compost is known as vermicomposting. Vermicompost is prepared by a variety of organic solid wastes of domestic animals, agro-industries, human wastes, etc. The nutrient content of vermicompost is 2–3% N, 1.55–2.25% P_2O_5 and 1.85–2.25% K_2O (Sinha et al. 2009). Vermicomposting differs from other composting, as this process is active between 10 °C and 32 °C, and is faster in comparison to composting and involve earthworms rather than fungi or bacteria. Several species of earthworms such as *Eiseniafoetida*, *Perionyx excavatus*, *Eudrilus eugeniae* and *Lumbricus rubellus* are used to convert the organic wastes into high-quality compost. The gut of earthworms produces humic acid, which helps in stabilizing pore space distribution of soil. The soil cast of earthworms has the available form of plant nutrients (nitrogen as nitrate, exchangeable Ca, Mg, K and P) and organic matter. Microbial growth is enhanced due to the nutrient-enriched casting in the gut of earthworm (Lee 1985). The microbial community in the gut of earthworms helps in the decomposition of organic waste (Fischer et al. 1995; Karsten and Drake 1997). Edward (1998) reported that vermicompost is rich in microbial populations and diversity, particularly actinomycetes bacteria and fungi. In the vermiculture (culture of earthworms), young earthworms grow and reproduce in the pit having organic waste. The faecal waste and dead tissues from a large number of earthworms could be used as manure. The use of vermicompost for higher crop production is increasing due to awareness of the adverse effects of agrochemicals (Follet et al. 1981).

Impacts of Various Composts on Soil Fertility

Compost has the same effect on soil fertility as other organic fertilizers. The physical, chemical and biological properties of soil improve by the application of compost. The application of compost makes compacted to sandy soils and clay soils loose. Compost improves the permeability of soils by increasing in the water-holding and heat-absorbing capacity. Plant growth improves in alkaline and saline soils to which compost has been added and also makes this soil less deleterious. Compost is made of plant residues having all the nutrients that are needed by the plants. Mineral nutrients are released slowly from organic matter by the activity of microbes, which improves soil fertility, resulting in higher crop yield production (Stelly 1977; Loehr et al. 1979). Utilization of organic manure as compost has many advantages such as recycling of farm waste, improved soil fertility and reduced loss of nutrients through runoff (Gandolfi et al. 2010).

(B) **Farm Yard Manure (FYM)**

Farmyard manure is the traditional and most common manure used by farmers to maintain fertility of the soil. It is easily available and has all the essential nutrients that are required by the plants. This is a heterogeneous organic material consisting of dung and urine of farm animals, crop residue that left part of fodder fed to the cattle, and household waste sweeping at various stages of decomposition. It is an important component of sustainable agricultural production. Application of FYM improves the chemical, physical and biological properties of the soil (Lee and Wani 1989). It maintains soil fertility by recycling of the plant nutrients by applying FYM to the field (Parker 1990). FYM can reduce the dependency on costly chemical fertilizer (King 1990). FYM amendments to the soil improve soil quality and crop yield (Nambiar and Abrol 1989). FYM is the most important component of integrating nutrient management. Insoluble and undigested residues of fodder are passed out in the excreta that have nitrogen and potassium, mostly in liquid form in urine, while phosphorus is present in the solid dung (Thorneby et al. 1999).

(i) **Cattle Manures:** The farm yard manure is a heterogeneous mixture of solid and liquid excreta of farm animals along with the crops residue that are left after cattle feeding. The farm yard manure is the most popular and oldest bulky organic manure that is utilized by farmers. This type of manure has about 0.5–1.5% of N, 0.4–0.8% of P_2O_5 and 0.5–1.9% K_2O (Verma and Verma 2012).
(ii) **Sheep and Goat Manure:** Sheep and goat's droppings are a good source of organic manure. It has higher nutrients as compared to the farmyard manure and are easily available to crop plants. This type of manure has about 3.0% of N, 1% of P_2O_5 and 2% K_2O (Verma and Verma 2012).
(iii) **Poultry Manures**: Poultry manure is an extremely rich source of nitrogen and organic matter. This type of manure has 1–1.8% of N, 1.4–1.8% of P_2O_5 and 0.8–0.9% of K_2O (Verma and Verma 2012). Poultry manure is a suitable manure for all crops and soils. The production of poultry manure is increasing due to the rapid growth in the poultry industry. Poultry litter has all essential plant nutrients (N, P, K, S, Ca, Mg, B, Cu, Fe, Mn, Mo and Zn) and is an outstanding fertilizer (Subramanian and Gupta 2006). Poultry manure has the ability to modify the soil environment and enhance plant growth (Chalker-Scott 2007).

Impacts of Farm Yard Manures on Soil Fertility

FYM is the traditional organic manure used by the farmer. It releases nutrients slowly as it decomposes and enhances organic matter of the soil. The nutrients of FYM are not entirely available to the crop in the year of application. Generally, 30% of N, 60–70% of P_2O_5 and 75% of K_2O become available to the first crop and rest of nutrients become available to the subsequent crops. This phenomenon of availability of plant nutrients to the subsequent crop is known as the residual effect (Gaur and Singh 1995). FYM is very effective in increasing the organic and nitrogen content of the soils.

2. Green Manures

Green manure is the un-decomposed green plant material that could be obtained in two ways: (i) by collecting green leaves and twigs of trees, shrubs and herbs that grow up in field forest and wastelands and (ii) by cultivating some legume crops and subsequently ploughing or turning them into the soil to enhance physical structure as well as soil fertility.

The important plant species useful for green-leaf manure are neem (*Azadirachta indica*, mahua (*Madhuca longifolia*), wild indigo (*Baptisia australis*), Mexican *Lilac* (*Gliricidia sepium*), Karanji (*Pongamia glabra*), Mudar (*Calotropis procera*), Agati (*Sesbania grandiflora*), subabul (*Leucaena leucocephala*) and other shrubs. The most important green manure crops are sunn hemp (*Crotalaria juncea*), dhaincha (*Sesbania bispinosa*), Ranmoong (*Phaseolus trilobus*), Guar (*Cyamopsis tetragonoloba*) and Sesbania (*Sesbania rostrata*).

Characteristics of Green Manure Crops

- Green manure crops should be fast growing, non-woody and of shorter duration so that they could be fitted in a cropping system.
- Green manure crops should produce high biomass and should be succulent for rapid decomposition.
- Green manure crops should have the ability to grow on low fertility soils.
- Green manure crops should be mixed in the soil after attaining vegetative growth because they are grown for their green leafy materials that are high in nutrients. The main purpose of practice of ploughing and mixing into the soil of un-decomposed green plant tissue is to increase soil organic carbon that helps in maintaining soil fertility.
- The green manure of legumes crop adds mainly nitrogen to the soil.
- There are improvements in physical, chemical and biological properties of the soil by using green manuring and it also enhances mobilization of minerals, leaching and percolation. The effect of green manure on soils is similar to that of farmyard manures. It is cheap and is the best method to increase soil fertility, as it can supplement farmyard and other organic manures without involving much cost. The legumes crops used as green manuring crops provide nitrogen, phosphorus, potassium as well as soil organic matter, while non-leguminous crops provide only organic matter to the soil.
- Impacts of Green Manures on Soil Fertility

Green manuring adds organic matter and nutrients to the soil macro (N, P, K), secondary and micronutrients that help in the maintenance of soil fertility needed for optimum plant growth and higher yield. Humus formed from green manures enhances water-holding capacity of soil, promotes aeration and drainage, decreases soil loss by erosion, conserves moisture and prevents nutrient leaching and granulation, which help the plant growth. Deep-rooted green-manure crops help in the turning of nutrient from deep soil layer to topsoil layer. The organic matter of soil increases through green manuring that stimulates the activity of soil microorganism. These microorganisms enhance the rate of decomposition of un-decomposed green plant material and change the biochemical properties of soil. All legume crops have the ability to fix free nitrogen from the atmosphere due to root nodule bacteria (*Rhizobium* sp.) and improve the nitrogen status of the soil that can minimize the use of nitrogenous fertilizers. Weed proliferation and weed growth can be reduced by cultivating green manure crops in the off-season. The alkaline problem of soil could be improved through application of green manure.

3. Concentrated Organic Manures

   (A) **Oil Cakes:** Oil cakes are by-products of oilseed crops and are known as concentrated organic manures. These are a good source of organic nitrogen and also have a small amount of phosphorus and potassium. In organic farming, oil cakes are used as manure for increasing crop production. This manure adds organic carbon to the soil and enhances growth of beneficial microbes of soil. Oil cakes are divided into two groups.

(i) **Edible Oil Cakes**: These oil cakes are safely fed to livestock as concentrates, for example, coconut cake, linseed cake, groundnut cake, mustard oil cakes, til cake, etc.
(ii) **Non-Edible Oilcakes**: Oil cakes that are not suitable for feeding livestock and hence mainly used for manuring crops, for example, castor cake, cotton seed cake, mahua cake, neem cake, etc.

Both these types of oil cakes can be used as manure; however, non-edible oil cakes are used as manures principally for horticultural crops. After application of oil cake in the agriculture fields, the nutrient is available for crop plant in 7–10 days. Oil cakes should be ground very fine before application for even distribution in the soil (Daji 1955).

(B) **Other Concentrated Organic Manures**

Blood-meal, meat-meal, fish-meal, horn and hoof meal, raw bone-meal and steamed bone meal are dried and powdered and after that can be used as manure. These concentrated organic manures are a good source of nitrogen, phosphorus and potassium.

Impacts of Concentrated Organic Manures on Soil Fertility

The non-edible oil seed cakes such as castor cake (*Ricinus communis*), jatropha cake (*Jatropha curcas*) and neem cake (*Azadirachta indica*), along with other manure as poultry and grass trimmings, are used to improve soil fertility. Addition of cellulolytic fungi enhances the decomposition and improves the soil quality (Gaur et al. 1982; Gaind and Nain 2007).

4. Guano

Guano is a natural organic fertilizer that is obtained from the excreta and dead bodies of birds. It has a high nutrient content (7–8% nitrogen, 11–14% phosphorus and 2.3–3% potassium). It has been reported that the nitrogen of organic fertilizers is water-insoluble and is gradually released with the decomposition of fertilizers (Cooke 1972). However, the rate of nitrification in guano is more rapid in comparison to other organic fertilizers (Owen et al. 1950). Guano was the most popular fertilizer before the development of fertilizer through free nitrogen of the atmosphere (Hadas and Rosenberg 1992). Guano has a higher nitrogen content and is rapidly decomposable. It is a costly fertilizer (Hadas and Kautsky 1994). Feather meal can be substituted for guano, as it is considerably cheaper and also has a high nitrogen content. It is obtained as a by-product from poultry processing plant (Hadas and Kautsky 1994).

Impacts of Guano on Soil Fertility

This type of organic fertilizer is nutrient-rich and has the ability to provide nutrient quickly. It is suitable for all crops and can apply before or at sowing time. The guano of seabirds have the highest nutrient value (10–16% nitrogen, 8–12% phosphorus and 2–3% potassium), while bats and other birds have a lower nutrient content. Guano stimulates soil microorganism activity by introducing an enzyme that enhances the process of decomposition of organic matter and improves the soil texture. Beneficial microbes are found in guano that increases soil fertility and controls fungi and nematodes. It works in similar way to compost and helps in increasing soil fertility. Guano acts as a soil binder by binding soil particles together. It does not leach out easily by natural weathering and helps to build up better aeration in the soil. Small oceanic islands are the common nesting places of seabirds where excreta of these birds alter physical, chemical and biological properties of soil and plant communities (McColl and Burger 1976; Nelson 1979). Phosphate, nitrate, and ammonium content are extremely high in ornithogenic soils that enhance growth of plants (Hutchinson 1950; Wainright et al. 1998; Anderson and Polis 1999). Soil moisture increases with deposition of guano; this may be due to increasing soil organic matter, while pH of soil decreases with its increment in the soil (Wait et al. 2005).

5. Biofertilizer

Biofertilizers are products that contain living microorganisms and enhance plant growth by increasing the availability of primary nutrient in the rhizosphere of the host plant. These are generally applied to the surface of plants and seeds. Biofertilizers increase soil fertility by adding nutrients through the natural process as nitrogen fixation, phosphorus solubilization or nutrient mobilization. Biofertilizers could reduce the dependency on chemical fertilizer. These biofertilizers are cost-effective and renewable source of plant nutrients to supplement the chemical fertilizers for sustainable agriculture. The organic matter of the soil can be increased by application of biofertilizers (Zaccaro et al. 1999; Maqubela et al. 2009), thus improving the soil structure (De Cano et al. 2002; Maqubela et al. 2009; Saadatnia and Riahi 2009).

Biofertilizers have been classified into three classes based on microorganisms used.

(A) **Algal Biofertilizer**

Algae are a natural organic source of biofertilizers that can be thought as one of the best substitutes to the chemical fertilizers. Algal biofertilizer is formed by algae as azolla and blue-green algae (BGA). Azolla (*Azolla pinnata*) is a water fern (Pteridophyte). It is also known as the aquatic weed that is commonly found floating in pond, lakes, shallow trenches and channels. Azolla is commonly found in the rice fields. Blue-green algae (BGA), also known as cyanobacteria, are commonly found in the rice fields and are capable of photosynthesis. BGA also produce plant growth regulators, stimulate the transport of nutrients from soil to plants, cause cluster of soil and improve the physical and chemical properties of the soil (Singh et al. 2016; Naser et al. 2017). These BGA are also able to fix free nitrogen and enhance the level of phosphorus by converting insoluble phosphorus into a soluble form (Irisarri et al. 2001). Some BGA live in symbiotic association as *Anabaena Azolla* live in the epidermal cavity of the leaf of *Azolla pinnata* where it fixes atmospheric nitrogen. This association is termed as *Azolla Anabaena* complex. Fresh *Azolla* has about 90–95% water. Decomposed *Azolla* contains 4–6% nitrogen, 0.5–0.9% phosphorus, 2–6% potassium, 0.4–1.0% calcium, 0.5% magnesium, 0.11–0.16% manganese, 0.06–0.16% iron and 9–10% total ash. BGA biofertilizer enriches nitrogen and decreases the stress affecting growth and yield of plants (Alam et al. 2014; Singh et al. 2014). Some microalgae as *Tetraselmis* sp. are used in the production of biofertilizers that are eco-friendly and good for maintaining soil quality.

Algal biofertilizer plays a significant role in conservation and build-up of soil fertility, therefore increasing the growth and yield of plant as a natural biofertilizer (Song et al. 2005). It can be cultivated in barren areas (Saadaoui et al. 2016) and is a good source of majority of micro- and macronutrients that are necessary for plant growth. It enhances nutrient transport from soil to the plant and helps in the reduction of soil salinity (Saadatnia and Riahi 2009). Algal biofertilizers enhance the level of phosphorus in the soil by the production of organic acids (Wilson 2006). Application of algae biofertilizer increases the population of beneficial microorganisms in the soil (Mishra et al. 2013). It secretes some growth-promoting substance like hormones (auxin, cytokinins gibberellin and abscisic acids), vitamins, amino acids (Roger and Reynaud 1982; Rodriguez et al. 2006). Algae biofertilizer can fix CO_2 through photosynthesis that decreases the level of CO_2 in the atmosphere and finally reduces the effect of global warming. It increases the pore size of the soil through the filamentous structure and production of adhesive substances, enhances the water-holding capacity of the soil through production of viscous substance (Roger and Reynaud 1982) and enhances the soil organic matter after death and decomposition (Saadatnia and Riahi 2009).

(B) **Fungal Biofertilizer**

Fungal biofertilizers include plant growth stimulating fungi e.g. *Aspergillus, Fusarium, Penicillium, Piriformospora, Phoma* and *Trichoderma,* mycorrhizal fungi (ectomycorrhiza, e.g., *Pisolithus tinctonus* and arbuscular mycorrhizae, for example, *Glomus mosseae* or *Glomus intraradices,* which form mutualistic relationship with plants, enzymatic producing fungi as *Myriococcumthermophilum, Thermoascus aurantiacus*, and *Thermomyces lanuginosus* for compost production and P-solubilizing fungi (*Penicillium spp.* and *Fusarium spp.*) and K-solubilizing fungi (*Aspergillus niger, Aspergillus terreus*) Fungal biofertilizers play a significant role in stimulating plant growth, productivity and improving soil fertility. Mycorrhizae are fungi which form mutualistic relationships with roots of 90% of plants (Das et al. 2007; Rinaldi et al. 2008). Mycorrhizae promote absorption of nutrients and water, control plant diseases, and improve soil structure (Chandanie et al. 2006; Rinaldi et al. 2008). The use of bioinoculants of phosphorus solubilizing fungi (*Penicillium* and *Aspergillus*) to soil for improving phosphorus uptake is becoming popular.

(C) **Bacterial Biofertilizer**

Plant growth promoting rhizobacteria (PGPR) are used in the production of biofertilizer. Bacteria that colonize roots of plant are known as PGPR (Kloepper and Schroth 1978). PGPR are mostly free-living and soil-born bacteria that are isolated from soil for the production of biofertilizers. Biofertilizers are applied to the seeds and crops to enhance growth of the plant. PGPRs are used in the production of biofertilizer and have at least one characteristics as suppression of plant disease, improved nutrient acquirement and phytohormone production (Kloepper et al. 1980). The direct mode of action involves phosphorus solubilization and its uptake by roots of plants, free nitrogen fixation, production of siderophores, production of phytohormones like auxins, cytokinins and gibberellins, and depressing level ethylene in plants. In last few decades, a large array of bacteria including species of *Azospirillum, Azotobacter, Alcaligenes, Arthrobacter, Burkholderia, Bacillus, Enterobacter, Klebsiella, Pseudomonas* and *Serratia* have been reported to enhance the plant growth (Kloepper et al. 1989; Glick 1995). Presently several biofertilizers are available in the market for increasing the uptake of nitrogen through nitrogen-fixing bacteria associated with the root (*Azotobacter* and *Azospirillum*), iron uptake from siderophores-producing bacteria (*Pseudomonas, Bacillus*), sulphur uptake from sulphur-oxidizing bacteria (*Thiobacillus*) and phosphorus uptake from phosphate mineral-solubilizing bacteria (*Bacillus*, *Pseudomonas* and *Azotobacter*). *Rhizobium* are gram-negative soil bacteria that fix free nitrogen from atmosphere in the root nodules of legumes crop plant, and these bacteria were the first biofertilizer identified and applied in legumes crops for over 100 years ago (Kannaiyan 2002).

Impacts of Biofertilizer on Soil Fertility

Biofertilizers enhance soil fertility by adding organic matter to the soil that acts as binder for the soil particles together, preventing soil erosion, eructing, and desertification. It also increases the water-holding capacity of the soil (Swathi 2010). Biofertilizers are an alternative source of chemical fertilizer to increase soil fertility. Biofertilizers play an important role in the increment of crop productivity and sustainability of the soil (Obana et al. 2007; Malamissa et al. 2007; Pandey et al. 2008; Khosro and Yousef 2012). Application of biofertilizers increases the biodiversity of beneficial microorganisms as algae, bacteria (plant growth-promoting rhizobacteria, (PGPR) and nitrogen fixers) and fungi including the arbuscular mycorrhiza fungi (AMF). Biofertilizers release the nutrients slowly. Soil fertility increases by the long-term application of biofertilizers, which leads to the buildup of nutrients in fields.

4 Advantages of Using Organic Fertilizers

1. Organic fertilizers are suitable because they supply balanced nutrients that help to keep plants healthy.
2. These fertilizers enhance soil biological activity that improves nutrient mobilization from organic and chemical sources through the process of decomposition.
3. Enhancement in root system has been found by using organic fertilizers due to better soil structure.
4. Organic fertilizers increase the organic matter content of soil and improve soil texture, water retention and resistance to erosion. Therefore, it helps in the improvement of soil physical and physiological structure.
5. Organic fertilizers have the ability to release nutrients slowly and contribute to the residual pool of organic N and P in the soil, reducing N-leaching loss and P fixation and also supply micronutrients.

5 Disadvantages of Using Organic Fertilizers

1. Organic fertilizers are comparatively low in nutrient content, so larger volumes are needed to provide enough nutrients for crop growth and yield.
2. The nutrient release rate is too slow to meet crop requirements in a short time, hence some nutrient deficiency may occur.
3. The sufficient quantity of nutrients does not exist in organic fertilizers to sustain maximum crop growth.
4. The cost of compost production is high as compared to chemical fertilizers.
5. Long-term or heavy application to agricultural soils may result in salt, nutrient or heavy metal accumulation and may adversely affect plant growth, soil organisms, water quality and animal and human health.

6 Conclusions and Future Perspectives

In agriculture, intensive use of various kinds of chemical fertilizers has reduced soil fertility and made soil unsuitable for crop plants. This huge application of chemical fertilizers has also led to severe health and environmental threats such as soil erosion, water pollution, pesticide harming, water logging and reduction of biodiversity. Crop production is increased by the intensive use of inorganic fertilizer, but it causes soil fertility depletion. To minimize this adverse effect of chemical fertilizers, a new approach has been developed, called organic agriculture. Organic agriculture includes the use of organic fertilizers. Organic fertilizers are gaining familiarity in many countries, these being eco-friendly and cost-effective. The best practice is organic farming to conserve soil fertility and the environment. In this farming system, different types of organic material are used as compost (village compost, town compost, water hyacinth compost, vermicompost), farmyard manure (cattle manures, sheep penning, poultry manures), green manures (leguminous plant and non-leguminous plant), biofertilizers (algal biofertilizers, fungal biofertilizers, bacterial biofertilizers or plant growth-promoting rhizobacteria, (PGPR)). These fertilizers are able to increase crop yields and minimize the evil effect of chemical fertilizers, pesticide and herbicides. Therefore, there is an urgent need for the involvement of governmental and international policies for the development of eco-friendly production technologies to reduce the adverse effect of intensive farming to discontinue use of various kinds of chemical fertilizers, pesticides and insecticides.

References

Alam S, Seth RK, Shukla DN (2014) Role of blue green algae in Paddy crop. Euro J Exp Bio 4:24–28

Anderson JPE (1987) Handling and storage of soils for pesticide experiments. In: Sommerville L, Greaves MP (eds) Pesticide effects on soil microflora. Taylor & Francis, London, pp 45–60

Anderson WB, Polis GA (1999) Nutrient fluxes from water to land: seabirds affect plant nutrient status on gulf of California Islands. Oecologia 118:324–332

Beare MH, Parmelee RW, Hendrix PF, Cheng W, Coleman D, Crossley JDA (1992) Microbial and faunal interactions and effects on litter nitrogen and decomposition in agro ecosystems. Ecol Monogr 62(569):591

Bewick MWM (1980) Handbook of organic waste conversion. Van Nostrand Reinhold Co, New York

Bokhtiar SM, Sakurai K (2005) Effects of organic manure and chemical fertilizer on soil fertility and productivity of plant and ratoon crops of sugarcane. Arch Agron Soil Sci 51:325–334

Boulaine J (1989) History of soil scientists and soil science. INRA Editions, Paris

Chalker-Scott L (2007) Impact of Mulches on landscape plants and the environment — a review. J Environ Hort 25:239–249

Chandanie WA, Kubota M, Hyakumachi M (2006) Interactions between plant growth promoting fungi and arbuscular mycorrhizal fungus *Glomus mosseae* and induction of systemic resistance to anthracnose disease in cucumber. Plant Soil 286:209–217

Cooke GW (1972) Fertilizing for maximum yield. Crosby Lockwood & Son Ltd, London, pp 13–27

Daji JA (1955) Manures and manuring. Root and tuber manuring. Farm Bull 7:24–25. lCAR, New Delhi

Das AC, Mukherjee D (1990) Microbiological changes during decomposition of wheat straw and neem cake in soil. Environ Ecol 8:1012–1015

Das A, Prasad R, Srivastava A, Giang HP, Bhatnagar K, Varma A (2007) Fungal siderophores: structure, functions and regulation. In: Varma A, Chincholkar SB (eds) Soil biology microbial siderophores, Springer, Berlin/Heidelberg, pp 1–42

De Bertoldi MD, Vallini G, Pera A (1983) The biology of composting: a review. Waste Manag Res 1:157–176

De Cano MMS, De Caire GZ, De Mulé MCZ, Palma RM (2002) Effect of Tolypothrix tenuis and Microchaetetenera on biochemical soil properties and maize growth. J Plant Nutr 25:2421–2431

Debnath A, Das AC, Mukherjee D (1994) Studies on the decomposition of organic wastes in soil. Microbiol Res 149:195–201

Diacono M, Montemurro F (2010) Long-term effects of organic amendments on soil fertility. a review. Agron Sustain Dev 30:401–422

Dong W, Zhang X, Wang H, Dai X, Sun X (2012) Effect of different fertilizer application on the soil fertility of Paddy soils in red soil region of southern China. PLoS One 7:e44504
Dong W, Zhang XY, Dai XQ, Fu XL, Yang FT, Liu XY, Sun XM, Wen XF, Schaeffer S (2014) Changes in soil microbial community composition in response to fertilization of paddy soils in subtropical China. Appl Soil Ecol 84:140–147
Edward CA (1998) The use of earthworms in the break down and management of organic wastes. In: Earthworm ecology. CRC Press, Boca Raton, pp 327–354
Fischer K, Hahn D, Amann RI, Daniel O, Zeyer J (1995) In situ analysis of the bacterial community in the gut of the earthworm *Lumbricus terrestris* L. by whole-cell hybridization. Can J Microbiol 4:666–673
Follet R, Donahue R, Murphy L (1981) Soil and soil amendments. Prentice Hall Inc., New Jersey
Gaind S, Nain L (2007) Chemical and biological properties of wheat soil in response to paddy straw incorporation and its biodegradation by fungal inoculants. Biodegradation 18:495–503
Gandolfi I, Sicolo M, Franzetti A, Fontanarosa E, Santagostino A, Bestetti G (2010) Influence of compost amendment on microbial community and ecotoxicity of hydrocarbon contaminated soils. Bioresour Technol 101:568–575
Gaur AC, Singh G (1995) Recycling of rural and urban wastes through conventional and vermicomposting. In: Tandon HLS (ed) Recycling of crop, animal, human and industrial wastes in agriculture. FDCO, New Delhi, pp 31–35
Gaur AC, Sadasivam KV, Mathur RS, Magu SP (1982) Role of mesophilic fungi in composting. Agric Wastes 4:453–460
Ge G, Li Z, Fan F, Chu G, Hou Z, Liang Y (2010) Soil biological activity and their seasonal variations in response to long-term application of organic and inorganic fertilizers. Plant Soil 326:31–44
Gianfreda L, Bollag JM (1996) Influence of natural and anthropogenic factors on enzyme activity in soil. In: Stotzky G, Bollag JM (eds) Soil biochemistry, vol 9. Marcel Dekker, New York, pp 123–193
Glick BR (1995) The enhancement of plant growth by free-living bacteria. Can J Microbiol 41(109):117
Golueke CG (1977) Biological reclamation of solid wastes. Rodale Press, Emmaus
Hadas A, Kautsky L (1994) Feather meal, a semi slow-release nitrogen fertilizer for organic farming. Fert Res 38:165–170
Hadas A, Rosenberg R (1992) Guano as a nitrogen source for fertigation in organic farming. Fert Res 32:209–214
Hutchinson GE (1950) Survey of existing knowledge of biogeochemistry: 3, the biogeochemistry of vertebrate excretion. Bull Am Mus Nat Hist 96:1–554
Irisarri P, Gonnet S, Monza J (2001) Cyanobacteria in Uruguayan rice fields: diversity, nitrogen fixing ability and tolerance to herbicides and combined nitrogen. J Biotechnol 191:95–103
Kannaiyan S (2002) Biofertilizers for sustainable crop production. Biotechnology of biofertilizers. Narosa Publishing House, New Delhi, p 377
Karsten GR, Drake HL (1997) Denitrifying bacteria in the earthworm gastrointestinal tract and in vivo emission of nitrous oxide (N_2O) by earthworms. Appl Environ Microbiol 63:1878–1882
Khosro M, Yousef S (2012) Bacterial bio-fertilizers for sustainable crop production: a review. APRN J Agric Biol Sci 7:237–308
King LD (1990) Soil nutrient management in the United States. In: Edwards CA et al (eds) Sustainable agricultural systems. Soil and Water Conservation society, Ankeny, pp 89–104
Kloepper JW, Schroth MN (1978) Plant growth-promoting rhizobacteria on radishes. In: Proceedings of the 4th international conference on plant pathogenic Bacteria, veterinary pathology and phytobacteriology station. INRA, Angers, pp 879–882
Kloepper JW, Leong J, Teintza M, Schorth MN (1980) Enhanced plant growth by siderophores produced plant growth promoting rhizobacteria. Nature 286:885–886
Kloepper JW, Lifshitz R, Zablotowicz RM (1989) Free-living bacterial inocula for enhancing crop productivity. Trends Biotechnol 7:9–44

Lee KE (1985) Earthworms, their ecology and relationships with land use. Academic Press, Sydney, p 411
Lee KK, Wani SP (1989) Significance of biological nitrogen fixation and organic manures in soil fertility management. In: Special publication of the international fertilizer development center. pp 89–108I
Liu E, Yan CY, Mei XR, He WQ, Bing SH, Ding LP, Liu Q, Liu S, Fan TL (2010) Long-term effect of chemical fertilizer, straw, and manure on soil chemical and biological properties in Northwest China. Geoderma 158:173–180
Liu SL, Huang DY, Chen AL, Wei WX, Brookes PC, Li Y, Wu JS (2014) Differential responses of crop yields and soil organic carbon stock to fertilization and rice straw incorporation in three cropping systems in the subtropics. Agric Ecosyst Environ 184:51–58
Loehr RC, Jewell WJ, Novak JD, Clarkson WW, Friedman GS (1979) Land application of wastes. Van Nostrand Reinhold Co., New York
Luo P, Han X, Wang Y, Han M, Shi H, Liu N, Bai H (2015) Influence of long-term fertilization on soil microbial biomass, dehydrogenase activity, and bacterial and fungal community structure in a brown soil of northeast China. Ann Microbiol 65:533
MalamIssa O, Defarge C, Bissonnais YL, Marin BD, Bruand A, D'Acqui LP, Nordenberg S, Annerman M (2007) Effects of the inoculation of cyanobacteria on the microstructure and the structural stability of a tropical soil. Plant Soil 290:209–219
Maqubela MP, Mnkeni PNS, Malamissa O, Pardo MT, D'Acqui LP (2009) Nostoccyanobacterial inoculation in south African agricultural soils enhances soil structure, fertility and maize growth. Plant Soil 315:79–92
Matulich KL, Martiny JBH (2015) Microbial composition alters the response of litter decomposition to environmental change. Ecology 96:154–163
McColl JG, Burger J (1976) Chemical inputs by a colony of Franklin's gulls nesting in cattails. Am Midl Nat 96:270–280
Mishra DJ, Singh R, Mishra UK, Kumar SS (2013) Role of bio-fertilizer in organic agriculture: a review. Res J Recent Sci 2:39–41
Mukherjee D, Mitra S, Das AC (1991) Effect of oi1 cakes on changes in carbon, nitrogen and microbial population in soil. J Indian Soc Soil Sci 39:457–462
Nair A, Ngouajio M (2012) Soil microbial biomass, functional microbial diversity, and nematode community structure as affected by cover crops and compost in an organic vegetable production system. Appl Soil Ecol 92:45–55
Nambiar KKM, Abrol IP (1989) Long-term fertilizer experiments in India (an overview). Fertiliser News 34:11–20
Naser FK, Alshekhmoham KSN, Altaher QMA, Makani H, Salunke A (2017) Improving soil health and plant growth with combination of BGA and micronutrients. Imp J Interdiscip Res 3:315–319
Nautiyal C, Chauhan P, Bhatia CR (2010) Changes in soil physico-chemical properties and microbial functional diversity due to 14 years of conversion of grassland to organic agriculture in semi-arid agroecosystem. Soil Till Res 109:55–60
Neher D (1999) Nematode communities in organically and conventionally managed agricultural soils. J Nematol 31:142–154
Nelson B (1979) Seabirds: their biology and ecology. A&W Publishers/Nova Science Publishers Inc, New York, pp 218–233
Obana S, Miyamoto K, Morita S, Ohmori MI (2007) Effect of Nostoc sp. on soil characteristics, plant growth and nutrient uptake. J Appl Phycol 16:641–646
Owen O, Rogers DW, Winsor GW (1950) The nitrogen status of soils. Part I The nitrification of some nitrogenous fertilizers. J Agric Sci 40:185–190
Pandey SK, Singh JP, Gopal J (2008) Potato varieties and cropping systems in India. Potato J 35:103–110
Parker CF (1990) Role of animals in sustainable agriculture. In: Edwards CA et al (eds) Sustainable agricultural systems. Soil and Water Conservation Society, Ankeny, pp 438–450

Patil M, Bheemappa A, Angadi JG, Guledgudda SS (2014) A critical analysis on economics and constraints in adoption of organic vegetable cultivation in Belgaum district. Karnataka J Agric Sci 27:539–541

Qin J, Culver DA, Yu N (1995) Effect of organic fertilizer on heterotrophs and autotrophs: implications for water quality management. Aquac Res 26:911–920

Rinaldi AC, Comandini O, Kuyper TW (2008) Ectomycorrhizal fungal diversity: separating the wheat from the chaff. Fungal Divers 33:1–45

Rodriguez AA, Stella AM, Storni MM, Zulpa G, Zaccaro MC (2006) Effects of cyanobacterial extracellular products and gibberellic acid on salinity tolerance in *Oryza sativa* L. Saline Syst 2:7

Roger PA, Reynaud PA (1982) Free living blue green algae in tropical soils. In: Dommer-Gues Y, Diem H (eds) Microbiology of tropical soils and plant productivity. Martinus Nijhoff Publisher, La Hague, pp 147–168

Roose E, Barthes B (2001) Organic matter management for soil conservation and productivity restoration in Africa: a contribution from francophone research. Nutr Cycl Agroecosys 61:59–170

Saadaoui I, Emadi MA, Bounnit T, Schipper K, Jabri HA (2016) Cryopreservation of microalgae from desert environments of Qatar. J Appl Phycol 28:2233–2240

Saadatnia H, Riahi H (2009) Cyanobacteria from paddy fields in Iran as a biofertilizer in rice plants. Plant Soil Environ 55:207–212

Shaheen A, Fatma M, Rizk A, Singer SM (2007) Growing onion plants without chemical fertilization. Res J Agr Biol Sci 3:95–104

Sharma A (1971) Eradication and utilization of water hyacinth-a review. Curr Sci 40:51–55

Singh S, Singh BK, Yadav SM, Gupta AK (2014) Potential of biofertilizers in crop production in Indian agriculture. Am J Plant Nutr Fert Technol 4:33–40

Singh AK, Singh AP, Gaurav N, Srivastava A, Gariya HS (2016) Growth of BGA on different types of soil, effect of BGA on physical and chemical properties of soil for paddy plants. J Med Plants Stud 4:111–114

Sinha R, Herat S, Valani D, Chauhan K (2009) Earthworms vermicompost: a powerful crop nutrient over the conventional compost & protective soil conditioner against the destructive chemical fertilizers for food safety and security am-Euras. J Agric Environ Sci 5:01–55

Song T, Martensson L, Eriksson T, Zheng W, Rasmessen U (2005) Biodiversity and seasonal variation of the cyanobacterial assemblage in rice paddy field in Fujian, China. FEMS Microbiol Ecol 54:131–140

Stelly M (1977) Soils for management of organic wastes and waste waters. Soil Sciences Society of America, American Society of Agronomy and the Crop Science Society of America, Madison

Subramanian B, Gupta G (2006) Adsorption of trace elements from poultry litter by montmorillonite clay. J Hazard Mater 128:80–83

Swathi V (2010) The use and benefits of bio-fertilizer and biochar on agricultural soils. B.Sc. thesis, Department of Chemical and Biological Engineering, Chalmers University of Technology, Goteborg Sweden, pp 20–24

Thorneby L, Persson K, Tragardh G (1999) Treatment of liquid effluents from dairy cattle and pigs using reverse osmosis. J Agric Eng Res 73:159–170

Verma JP, Verma R (2012) Organic fertilizers and their impact on agricultural production system: In organic fertilizers: types, production and environmental impact edited by Singh RP. Nova Science Publishers, Inc. New York , pp 218–232

Wainright SC, Haney JC, Kerr C, Golovkin AN, Flint MV (1998) Utilization of nitrogen derived from seabird guano by terrestrial and marine plants at St. Paul, Pribilof Islands, Bering Sea, Alaska. Mar Biol 131:63–71

Wait DA, Aubrey DP, Anderson WB (2005) Seabird guano influences on desert islands: soil chemistry and hernaceous species richness and productivity. J Arid Environ 60:681–695

Wiles CC (1978) Composting of refuse. In: Composting of municipal residues and sludges. Information Transfer Inc. and Hazardous Material Control Research Institute, Rockville, p 20

Wilson LT (2006) Cyanobacteria: a potential nitrogen source in Rice fields. Texas Rice 6:9–10

Xun W, Huang T, Zhao J, Ran W, Wang B, Shen Q (2015) Environmental conditions rather than microbial inoculum composition determine the bacterial composition, microbial biomass and enzymatic activity of reconstructed soil microbial communities. Soil Biol Biochem 90:10–18

Yanan T, Emteryd O, Dianqing L, Grip H (1997) Effect of organic manure and chemical fertilizer on nitrogen uptake and nitrate leaching in a Eumorthic anthrosols profile. Nutr Cycl Agroecosys 48:225–229

Zaccaro MC, De Caire GZ, De Cano MS, Palma RM, Colombo K (1999) Effect of cyanobacterial inoculation and fertilizers on rice seedlings and postharvest soil structure. Comm Soil Sci Plant Anal 30:97–107

Zhang FS, Cui ZL, Chen XP, Ju XT, Shen JB, Chen Q (2012) Integrated nutrient management for food security and environmental quality in China. AdvAgron 116:1

Effect of Heavy Metals on Plant Growth: An Overview

Deepika Goyal, Arti Yadav, Mrinalini Prasad, Teg Bahadur Singh, Preksha Shrivastav, Akbar Ali, Prem Kumar Dantu, and Sushma Mishra

Contents

1 Introduction

The term "heavy metal" refers to a metal or metalloid with atomic density at least five times greater than that of water (Hawkes 1997; Tchounwou et al. 2012). The common heavy metals found in environment include cadmium (Cd), lead (Pb), nickel (Ni), silver (Ag), zinc (Zn), cobalt (Co), chromium (Cr), iron (Fe), arsenic (As), and mercury (Hg). The presence of high concentration of heavy metals in water, air, and soil poses a threat to all forms of life (Oliveira and Pampulha 2006).

D. Goyal · A. Yadav · M. Prasad · T. B. Singh · P. Shrivastav · A. Ali
P. K. Dantu · S. Mishra (✉)
Department of Botany, Dayalbagh Educational Institute (Deemed University), Agra, India

M. Naeem et al. (eds.), *Contaminants in Agriculture*,
https://doi.org/10.1007/978-3-030-41552-5_4

This chapter, however, focuses on the effects of heavy metals on metabolism of plants and their overall growth. An outline of the source of heavy metals, their techniques of remediation, and various defense mechanisms present in heavy metal-tolerant plants, has also been discussed.

2 Source of Heavy Metals

Gathering of heavy metals in biosphere may take place by both natural and human activities (Fig. 1). While, weathering of rocks is the chief natural source of heavy metal contamination in environment, the anthropogenic sources include mining, smelting operations, and agricultural activities (Herawati et al. 2000).

2.1 *Natural Sources of Heavy Metal Contamination*

2.1.1 Weathering of Rocks

Heavy metals derived from rock materials represent the "lithogenic" component. The type of parent rock is the factor which determines the concentration and composition of heavy metals formed in soil. The principal heavy metal pollutants contributed by parent rock include Co, Cr, Fe, Mn, Ni, and Zn. Weathering of igneous rocks, such as Augite, Olivine, and Hornblende, gives rise to considerable amounts of heavy elements, while sedimentary rocks contribute only a small fraction (Nagajyoti et al. 2010).

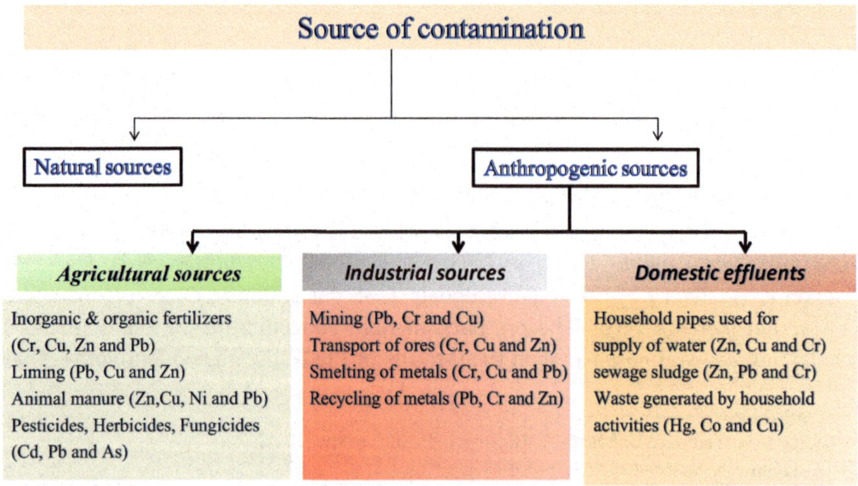

Fig. 1 Natural and anthropogenic sources of heavy metals and their composition

2.1.2 Other Natural Sources

Apart from rocks, volcanoes, wind-blown dusts and storms, natural fires, sea sprays, and aerosols (in coastal areas) are other natural sources of heavy metals (Seaward and Richardson 1990). Geothermal sources, likewise volcanic eruptions, have raised noteworthy atmospheric toxic wastes and contaminants (Eshleman et al. 1971).

2.2 *Anthropogenic Sources*

2.2.1 Agricultural Sources

The inorganic and organic fertilizers, manure, limes, pesticides, etc., used in agriculture contain variable amounts of Cr, Cd, Ni, Zn, Pb, and other heavy metals (Nriagu 1989; Yanqun et al. 2005). Similarly, most of the commonly used chemical pesticides like Bordeaux mixture and lead arsenate contain Cu, Hg, Mn, Pb, or Zn. Apart from these, use of municipal and industrial wastewater for irrigation is also a predominant source of heavy metals in soil.

2.2.2 Industrial Sources

Industrial processes like mining, smelting, and metal processing mainly generate Cr and Ni, while Vanadium (V), Titanium (Ti), and Mn are mainly derived from oil and coal-related activities (Guan et al. 2018). Coal mining also liberates significant levels of As, Cd, and Fe, while gold mining increases the level of Hg in the environment (Lacerda 1997). The heavy metals are generally generated in particulate and vapor forms, which, on combining with water present in the atmosphere, form aerosols. Aerosols may be either spread by wind (dry deposition) or precipitated in rainfall (wet deposition), and cause contamination of soil and water bodies (Nagajyoti et al. 2010). Also, coal and petroleum combustion and nuclear power stations give rise to heavy metals such as Se, Cd, B, Cu, Cs, Zn, and Ni to the atmosphere (Verkleji 1993). Processing of plastics, microelectronics, wood preservation, textiles and paper processing also cause heavy metal toxicity to the environment (Tchounwou et al. 2012).

2.2.3 Domestic Effluents

Domestic effluents constitute the major source of heavy metal contamination in water bodies. Domestic effluents may include untreated wastewater substances and passed through the filters in biological treatment plants and waste substances are passed through sewage outfalls. Most of the widely used enzymatic detergents contain trace amounts of elements like Fe, Cr, Mn, Zn, Co, Sr, and B, thereby contributing toward heavy metal pollution (Angino et al. 1970).

3 Impacts of Heavy Metal Toxicity on Plants

While, in aquatic systems, complete plant body is accessible to heavy metals and ions, terrestrial plants uptake heavy metals from the soil mainly through their roots. The uptake process is facilitated by some transport proteins, chelating agents produced in the rhizosphere, and plant-induced pH changes (Tangahu et al. 2011). Other means of entry of heavy metals are stomata, lenticels, wounds, etc. (Shahid et al. 2017). Heavy metals are also absorbed directly through the leaves because of particles accumulated on the foliar surfaces of leaves.

Heavy metals basically cause formation of reactive oxygen species (ROS) and free radicals that lead to uncontrolled oxidation and radical chain reactions, ultimately damaging the cellular biomolecules like nucleic acids, lipids, and proteins (Phaniendra et al. 2015). Plants, being primary producers, form the base of ecological pyramid; thus, the heavy metals entering the plant body make their way through successive trophic levels of the food chain. This problem gets exacerbated for the heavy metals that are bioaccumulative, that is, they are neither degraded in the environment nor easily metabolized by plants. In fact, some of the heavy metal-tolerant plant species like *B. napus*, *B. Juncea* have intrinsic ability to accumulate heavy metals in their body, thereby, threatening the contamination of food webs (Gall et al. 2015; Mourato et al. 2015).

3.1 Heavy Metals That Function As Micronutrients

Some heavy metals, like Cu, Fe, Mn, Mo, and Zn, function as micronutrients; that is, these are required in small quantities (less than 1 pound per acre) for important physiological functions of plants (Misra and Mani 1991). The optimum concentration of these heavy metals and their role(s) in plant development have been mentioned in Table 1. Just as lack of these nutrients results in deficiency symptoms, their elevated levels in soil also lead to toxicity effects, which have been described below.

3.1.1 Iron

Iron is an important constituent of several plant proteins and enzymes like leghemoglobin, cytochromes, ferrodoxin, catalase, peroxidase, aconitase, and superoxide dismutase (Marschner 1995). However, elevated levels of iron cause production of ROS, that is, free radicals, which alter membrane permeability and damage membrane structure (De Dorlodot et al. 2005).

Table 1 Range and functional roles of a few environmentally important heavy metals in plants

Elements	Land plants (μg/g dry wt)	Roles in plants
Fe	140	Component of hemoglobin, myoglobin, and cytochromes
Cu	4.15	Required for photosynthesis, acts as cofactor of superoxide dismutase, ascorbate oxidase
Zn	8–100	Main player of replication and transcription, cofactor of carbonic anhydrase, alcohol dehydrogenase
Mn	15–100	Required for photosynthesis during splitting of water, cofactor of malic dehydrogenase, oxalosuccinic dehydrogenase
Co	0.05–0.5	Found in the form of vitamin B_{12}
Ni	1	Fixes nitrogen in legumes, main component of urease enzyme

3.1.2 Copper

Copper (Cu) is a vital co-factor for plastocyanin and cytochrome oxidase, which are involved in key physiological processes of plants like photosynthesis and respiration. However, just like other micronutrients, excessive amount of Cu has been reported to adversely affect plant growth in *Brassica juncea* (Singh and Tewari 2003), *Solanum melongena* (Neelima and Reddy 2002), *Alyssum montanum* (Ouzounidou 1994), among others. A conspicuous impact of copper toxicity is thickening of root apices in *Pinus* seedlings, inhibition of production of root hair in *Betula papyrifera,* and seedlings of *Lonicera tatarica* (Arduini et al. 1995; Patterson and Olson 1983). Excessive Cu causes production of some new roots and thickening of taproot in *Citrus paradisi* (Zhu and Alva 1993). Excessive Cu causes chlorosis in *Banksia ericifolia* (heath banksia), *Casuarina distyla* (she-oak) and *Eucalyptus eximia* (yellow bloodwood) (Mitchell et al. 1988).

3.1.3 Zinc

Like other micronutrients, zinc (Zn) is an essential micronutrient for many metabolic processes of plants. However, beyond its optimum range (15–60 ppm), it adversely affects roots, shoots, seed germination, and flowering responses in French marigold (Choi et al. 1996). Elevated levels of Zn in soil cause decline in the level of chlorophyll pigments leading to chlorosis in younger leaves. Zn toxicity promotes senescence and causes reduction in plant biomass (Mirshekali et al. 2012). Moreover, it also acts as genotoxic pollutant by causing structural and numerical aberrations in chromosome of plants, thus affecting cell division (Sharma and Talukdar 1987).

3.1.4 Manganese

Manganese (Mn) is an important micronutrient and cofactor of enzymes required in photosynthesis. However, excess amount of this heavy metal causes reduction in plant growth and visible symptoms like leaf bronzing and shortening of internodes (Crawford et al. 1989). While Mn concentration in the range of 500 μM caused reduced shoot growth in rice, soybean showed chlorosis at a concentration of 200 μM (Lidon and Teixeira 2000; Lavres et al. 2009).

3.1.5 Cobalt

Cobalt (Co) occurs naturally in earth's crust in the form of erythrite [$Co_3(AsO_4)_2$], cobaltite [CoAsS] and smaltite [$CoAs_2$] (Nagajyoti et al. 2010). The effects of Co toxicity have been studied in various crops like barley, tomato, and oilseed rape, where it has been found to inhibit biomass and shoot growth (Li et al. 2009). Excess Co causes leaf distortion, giving an appearance of a structure like hook with rudimentary leaflets. Exogenous application of Co to tomato leaves leads to greatest accumulation of this heavy metal in roots and older leaves, while, lowest in stem. The plant showed altered enzymatic activity of peroxidase, catalase, ribonuclease, and acid phosphatase (Gopal et al. 2003). Co toxicity also causes decreased transpiration rate and water potential, an effect that is not found in Cu or Cr toxicity.

3.1.6 Nickel

Nickel (Ni) acts as a cofactor of urease, the enzyme which metabolizes urea into ammonia (usable form of nitrogen) within plants. Deficiency of Ni causes accumulation of toxic urea within the tissue and formation of necrotic legions on the leaf tips (Bhalerao et al. 2015).This enzyme helps in nitrogen fixation in many plant species. Ni also helps in disease tolerance (Sengar et al. 2008), but the mechanismis not very clear. Deficiency of this mineral nutrient causes reduction in size of leaflets with small rounded tips; this condition is called mouse-ear. At higher concentration (>50 μg/g dry weight), Ni acts as a phytotoxic metal and adversely affects growth inmany plant species (Crooke 1956). Ni is reported to inhibit gas exchange and photosynthesis in plants such as maize and sunflower (Lo and Chen 1994; Mishra et al. 1973). Seregin and Kozhevnikova (2006) reported effect of Ni toxicity on wheat and found 1 mM NiSO4 solution causes decrease in the mesophyll thickness, reduction in size of vascular bundles, change in vessels diameter etc.

3.2 Heavy Metals That Are Not Vital for Plant Growth

Other heavy metals like lead, arsenic, manganese, and cadmium are highly deleterious to plants. Table 2 shows a comparative toxicity effect of different heavy metals on plants, which varies with different species.

3.2.1 Lead

Lead (Pb) is a widespread toxic element in the soil, ranging from about 400–800 mg/kg in most soils, to up to 1000 mg/kg in soil in industrialized areas (Sharma and Dubey 2005). At present, there is no report on the role of Pb in plant growth and development. The problem of Pb toxicity is aggravated by the fact that it is hard to degrade and is extremely persistent in both water and soil. In fact, according to Saxena et al. (1999), Pb has been found to be present in the environment for 150–5000 years. Pb exposure causes inhibition of plant root growth and cell division in root tips of *Lemna minor* (Eun et al. 2002). Pb toxicity causes reduction in the root length and root dry mass in *Pisum sativum, Triticum aestivum, Zea mays,* and *Sedum alfredii* (Samardakiewicz and Wozny 2005). Lead causes inhibition of stem and root elongation, and leaf expansion in *Allium* species (Gruenhage and Jager 1985), *Hordeum vulgare* (Juwarkar and Shende 1986), and *Raphanus sativus*. In *Zea mays*, Pb accumulation interferes with microtubule organization and disruption in microtubules, resulting in altered cell division (Eun et al. 2002). It is found through numerous studies that Pb leads to increased production of ROS in plant cells, resulting in lipid peroxidation, membrane damage, and oxidative stress (Reddy et al. 2005; Liu et al. 2010). Pb toxicity hinders photosynthesis by obstructing electron transport, inhibition of Calvin cycle, and stomatal closure. It also causes reduced uptake of magnesium and iron (the constituents of chlorophyll), resulting in inhibition of chlorophyll synthesis (Burzynski 1987), and increase in chlorophyllase enzyme activity (Drazkiewicz 1994) and alterations in lipid composition of

Table 2 The range of metal toxicity in several plant species

Plant species	Range of metal toxicity	References
Hordeum vulgare	Hg > Pb > Cu > Cd > Cr > Ni > Zn	
Lolium perenne	Cu > Ni > Mn > Pb > Cd > Zn > Al > Hg > Cr > Fe	
Triticum aestivum	Cu > Cr > Ni > Zn > Pb ≈ Cd > Al > Fe	
Vicia faba	Cd > Ni > Zn ≈ Co	
Zea mays	Cu ≈ Tl > Ag > Cd > Hg > Co > Zn > Pb or Tl3+ > Cu2+ > Ag+ > Hg^{2+} ≈ Cd^{2+} > Ni^{2+} > Zn^{2+} ≈ Pb^{2+} ≈ Co^{2+} > Sr^{2+}	Ivanov et al. (2003)
Phaseolus sp.	Hg^2 ≈ Cr	Parmar et al. (2002)

thylakoid membrane (Stefanov et al. 1995). It has also been reported that Pb alters nitrate assimilation, shoot nitrate content, free amino acids, and growth in *Brassica pekinensis* (Xiong et al. 2006).

3.2.2 Arsenic

Arsenate (Ar) is an analog of phosphate and a toxic metalloid. It is widely distributed in the environment due to various natural and anthropogenic activities such as mining and fossil fuel combustion (Bhattacharyya et al. 2003). High concentration of As causes detrimental effects on plant growth by causing cell necrosis, chlorosis, and electrolyte leakage from cell membranes (Singh et al. 2006). The toxicity of As also leads to formation of ROS that can damage nucleic acids, proteins, and cause peroxidation of lipids present in membrane (Moller et al. 2007). Chinese brake fern (*Pteris vitatta*) was found to be the first hyperaccumulator of this heavy metal, and today, this species is currently being used as a potent As tolerant and the potential of tolerance is up to 1500 ppm in soil (Shen et al. 2014).

3.2.3 Mercury

Mercury (Hg) is found to occur in different forms like metallic mercury (HgS and methyl-Hg), inorganic mercury, and organic mercury (Hg^{2+}). Excessive level of Hg^{2+} is toxic to plant cells and induces evident injuries and physiological disorders in plants (Zhou et al. 2007). Hg^{2+} affects mitochondria, leaf stomata, binds to water channel proteins, and acts as a barrier of water flow in plants (Zhang and Tyerman 1999; Messer et al. 2005). Photosynthesis, transpiration rate, and water uptake and chlorophyll synthesis are also adversely affected by exposure to mercury. Hg is found to cause loss of magnesium, manganese, potassium, and accumulation of iron (Boening 2000). Hg affects the antioxidant defense system, by causing interference with the nonenzymatic antioxidants like glutathione, thiols, and enzymatic antioxidants like ascorbate peroxidase, superoxide dismutase, and glutathione reductase (Israr et al. 2006). Other harmful effects of Hg include interference with the transport of vital micronutrients, inactivation or denaturation of proteins, and disruption of cell organelles and cell membranes.

3.2.4 Cadmium

Cadmium (Cd), like Hg, Co, Pb, and As, is a nonessential element for plants. The impacts of Cd toxicity can be seen on the activity of enzymes involved in photosynthesis and nitrogen metabolism (Alcántara et al. 1994; Mathys 1975). Cd also adversely affects stomatal opening, mineral uptake, and water balance in plants. Cd accumulation reduces ATPase activity of plasma membrane and alters membrane permeability and metal ion homeostasis (Fodor et al. 1995). High Cd levels have been reported to reduce the level of osmoprotectants, mainly proline, and alter the genetic stability in *Solanum nigrum* (Al Khateeb and Al-Qwasemeh 2014).

4 Plant Defense Mechanisms Adopted by Plants Against Heavy Metal Stress

As mentioned earlier, heavy metals interact with biomolecules likewise nuclear proteins and DNA and cause production of ROS. This causes morphological, metabolic, and physiological abnormalities in plants (Manara 2012). Hence, plants have evolved some defense mechanisms for heavy metal tolerance (Table 3). These include two broad strategies: either to *avoid*, or *tolerate* heavy metal toxicity (Fig. 2). Physical barriers like thick cuticle, trichomes, cell wall, plasma membrane, and mycorrhiza constitute the first line of protection in plants to counteract heavy metals (Hall 2002; Wong et al. 2004; Harada et al. 2010). If, however, the heavy metals manage to cross these biophysical barriers, plants synthesize and secrete certain chemicals to mitigate the harmful effects of heavy metals. These chemicals include metallochaperones or chelators like nicotianamine, spermine, putrescine, mugineic acids, organic acids, phytochelatins, glutathione, etc., and metallothioneins or cellular exudates such as phenolic compounds and flavonoids, protons, heat shock proteins, some amino acids (proline and histidine), and hormones (jasmonic acid, salicylic acid, and ethylene) (Viehweger 2014; Dalvi and Bhalerao 2013; Sharma and Dietz 2006).

Table 3 Potential plant defense mechanisms involved in the detoxification of heavy metals

Mechanism of heavy metal tolerance	Metal	References
Trichomes	Zn, Cu, Cd, Ni	Emamverdian et al. (2015)
Mycorrhizas	Zn, Cu, Cd	Jentschke and Godbold (2000)
Cell wall, exudates	Ni, Al	Ma et al. (1997)
Plasma membrane		
Reduced uptake	Arsenate, Ni	Meharg and Macnair (1992)
Active efflux	Zn	
Phytochelatins	Cd	Cobbett (2000)
Metallothioneins	Cu	Murphy and Taiz (1995)
Organic acids, amino acids	Cd, Al, Cu, Ni, Al	Rauser (1999)
Heat shock proteins	Cd	Neumann et al. (1994)
Vacuolar compartmentation	Zn	

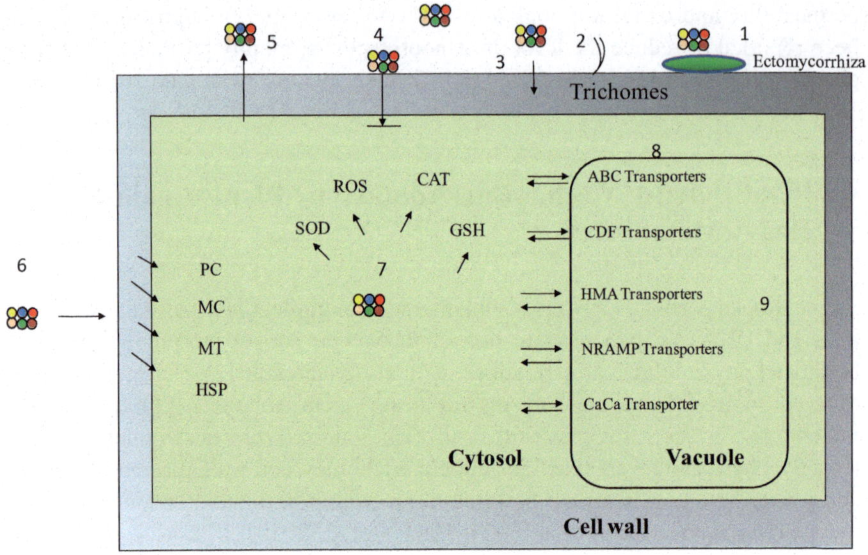

Fig. 2 Detoxification strategies of heavy metals. (1) Restriction of metal translocation to roots by ectomycorrhiza. (2) Deposition of heavy metals into trichomes. (3) Metal ion binding to cell wall and root exudates. (4) Restriction of metal ions movement to cytosol. (5) Effluxing of metal ions by the cell membrane to apoplast region. (6) Metal chelation in the cytosol by phytochelatins, metallothioneins, and heat shock proteins. (7) Induction of oxidative stress and defense mechanisms. (8) Transport of metal ions in cytosol by various transporters. (8) Sequestration of metal ions into vacuole. heavy metals, MC metal chelating protein, MT metallothionein, HSP heat shock protein, SOD superoxide dismutase, CAT catalase, ROS reactive oxygen species, GSH glutathiones, PC plastocyanin

4.1 Trichomes

Trichomes, the hair-like structures on plant surface, which are commonly associated with their role in protecting the plants against herbivorous insects, also function in heavy metal tolerance. Trichomes either serve as a depository site for heavy metals or secrete various secondary metabolites that lead to detoxification of toxic heavy metals (Emamverdian et al. 2015). The first mechanism of heavy metal tolerance has been reported for trichomes of *Alyssum lesbiacum*, which store significant amount of Ni (Krämer et al. 1997); and trichomes of *Arabidopsis halleri* that store higher concentrations of Zn (Sarret et al. 2002). The trichome number was found to increase in tobacco seedlings treated with Cd (Choi et al. 2001).

4.2 Cell Wall and Root Exudates

The role of cell wall in defense response has not been very well documented. Bringezu et al. reported that *Silene vulgaris*, a heavy metal-tolerant plant, accumulated toxic heavy metals in epidermal cell walls, either bound to a protein or as silicates (Bringezu et al. 1999). Root exudates act as metal chelators, which enhance the uptake of certain metals from soil. For instance, Ni-chelating exudates from roots were found to chelate histidine, while citrate accumulated in the roots of tolerant plant *Thalspi* sp. helped to reduce Ni uptake, thus playing a role in Ni-detoxification strategy (Hall 2002). Roots of *Fagopyrum esculentum* (Buckwheat) secrete oxalic acid under conditions of Al stress, and accumulate nontoxic Al-oxalate in the leaves and detoxify Al both externally and internally (Ma et al. 1997).

4.3 Plasma Membrane

As mentioned above in the previous sections, heavy metals alter plasma membrane permeability, causing increased ion leakage from cells and hence disrupting ion homeostasis. For example, high levels of Cu and Cd have been reported to cause increased potassium efflux from roots by altering the H^+ ATPase pump and changing the lipid composition (Wainwright and Woolhouse 1977; Ros et al. 1990; Fodor et al. 1995; Demidchik et al. 1997). Therefore, to overcome this, some heavy metal-tolerant plants modify their plasma membrane to reduce heavy metal uptake and facilitate more efflux of crucial ions like potassium. *Holcus lanatus* shows arsenic tolerance by suppressing high-affinity arsenate transport system and synthesizing phytochelatins and oligomers of gluthathiones (Meharg and Macnair 1992). In tobacco, plasma membrane transporters bind calmodulin and extend Ni tolerance and Pb^{2+} hypersensitivity (Arazi et al. 1999). In *Arabidopsis thaliana,* cadmium metal ions are effluxed from the plasma membrane by ABC transporters (Kim et al. 2007).

4.4 Vacuolar Compartmentalization

Apart from efflux of ions through plasma membrane (discussed above), sequestration into vacuole is another protective mechanism adopted by plants to reduce the levels of toxic metals in cytosol (Ernst et al. 1992). Meristematic cells of *Festuca rubra* and *Hordeum vulgare* store zinc in vacuole and help in the detoxification of heavy metals (Davies et al. 1991; Brune et al. 1994). Vesicles from the roots of Zn-tolerant and -sensitive plant of *Silene vulgaris* were isolated, and Zn transport was found to be 2.5 times higher in vesicles of the tolerant lines in comparison to the sensitive ones, suggesting the role of tonoplast in zinc tolerance.

4.5 *Mycorrhizae*

Mycorrhizae, particularly ectomycorrhizae, are found to be effective in mitigating the impacts of metal toxicity on host plant (Marschner 1995; Huttermann et al. 1999; Jentschke and Godbold 2000). Mycorrhizae have been reported to protect plants against heavy metal stress by a number of mechanisms, such as depositing heavy metals within cortical cells (Turnau 1998), chemical binding to cell wall or mycelium, sequestration into vacuole or other storage organelles (Hall 2002), and alteration in the pH of soil, resulting in immobilization of metals (Bano and Ashfaq 2013). *Glomus mosseae,* an arbuscular mycorrhiza of maize, binds some heavy metals like Zn, Cu, and Pb, and lowers its uptake by plants (Huang et al. 2005). In a study, inoculation of *Glomus etunicatum* in *Lactuca sativa* has been reported to increase the sequestration of zinc as compared to noninoculated plants (Farshian et al. 2007).

4.6 *Phytochelatins*

Phytochelatins (PCs) are short-chain sulfur-rich repetitions of peptides synthesized by the enzyme phytochelatin synthase (PCS). These metal binding proteins are synthesized nontranslationally from reduced glutathione. PCs are reported to occur in plants, microbes, animals, yeasts, and maintain metals homeostasis (Silva 2012; Bian et al. 2013). PCs are used as biomarkers for the initial detection of heavy metal toxicity in plants. PCs are found to be transported from roots to shoots, and this translocation is mediated by phloem sap and xylem in *Brassica napus* (Yadav 2010). A high concentration of Cu and As stimulated synthesis of PCs in *Solanum nigrum* and *Oryza sativa,* which resulted in the immobilization of these metals by complex formation (Fidalgo et al. 2013; Lemos Batista et al. 2014). It is reported that when PCs genes are transferred in *Arabidopsis thaliana*, it shows an enhanced tolerance to Cd and As (Guo et al. 2012).

4.7 *Metallothioneins*

Metallothioneins (MTs) are cysteine-rich, cytoplasmic metal-binding and low molecular weight proteins found in both prokaryotes and eukaryotes. These proteins are classified into four classes according to the arrangement of cysteine residues. In soybean, MT1, MT2, and MT3 detoxify Cd and MT4 detoxifies Zn (Pagani et al. 2012). In plants, MT nullifies heavy metals toxicity through homeostasis of intracellular metal ions, cellular sequestration, and metal transport adjustment (Prasad 2004; Kohler et al. 2004).

5 Remediation of Heavy Metals

Heavy metal toxicity imposed on plants is the most challenging problem in most countries. Removal of heavy metals from a contaminated site is known as remediation (Khan et al. 2000). Some of the conventional methods for remediation include dredging (physical removal of the contaminated sediment layers), capping (covering the contaminated sediment surface with clean material, thus isolating the sediments), and incineration (waste treatment technology that involves the combustion of organic substances contained in waste materials) (Azubuike et al. 2016).

Bioremediation, that is, the use of living organisms (mainly microbes) for the removal of pollutants from contaminated site, has gained immense popularity in recent years (Table 4). Some microbes are unique in producing enzymes that degrade organic contaminants into nontoxic compounds. The remediation techniques of heavy metals could be classified into four major types depending upon the type of microorganism used: cyanoremediation, bioremediation, mycoremediation, and phytoremediation; these remediation strategies have been discussed below.

5.1 *Cyanoremediation*

Cyanoremediation is the use of algae for the removal of organic and inorganic pollutants from the contaminated site. Some algae such as *Chlorella, Spirulina, Spirogyra, Oedogonium* have been used for phycoremediation purposes. Deng et al. (2007) reported the use of *Cladophora fascicularis* as an effective material, which absorbs Pb (II). In another study, observed the biosorption capacity of *Cladophora and Spirogyra,* for copper (Cu^{2+}) and lead (Pb^{2+}), respectively. In recent years, the focus on algae for remediation has increased due to their wide occurrence, central role in carbon dioxide fixation, and potential source of biofuel (Chekroun et al. 2013).

5.2 *Bioremediation*

In bioremediation, living organisms are used to degrade environmental contaminants into less toxic forms (Vidali 2001). In bioremediation, microbes enzymatically attack the pollutants and convert them to harmless products. Some of the widely used microorganisms in bioremediation are *Pseudomonas putida, Dechloromonas aromatica, Deinococcus radiodurans, Methylibium petroleiphilum,* and *Alcanivorax borkumensis* (Mani and Kumar 2014). Apart from being a cost-effective process, bioremediation involves minimal site disruption, and the possibility of simultaneous treatments. Although this approach is commonly used, it does have some drawbacks. Bacteria are not capable of working on a broad category of

Table 4 Some of the widely used heavy metal remediation techniques using living organisms

Living organism	Species	Metals	References
Plants	*Pteris vittata*	Cu, Ni, Zn, As	Ma et al. (2001)
	Brassica juncea	Se, Cd	Banuelos et al. (2005)
	Helianthus annuus	Cd	Mani and Kumar (2014)
	Populus sp.	Hg	Lyyra et al. (2007)
	Brassica napus	Cd	Selvam and Wong (2008)
	Typha latifolia	Pb	Tiwari et al. (2008)
	Nelumbo nucifera	Zn, Cu, Pb, Ni	Kumar et al. (2008)
	Amaranthus viridis	Cr	Liu et al. (2008)
	Helianthus annuus	Cu, Zn, Pb, Hg, As, Cd, Ni	Mani et al. (2012)
	Trifoliumpretense	Cd	Wu et al. (2009)
	Spinacea oleracea	Pb, Zn	Mani et al. (2012)
	Vetiveriazizanioides	Cd, Pb	Danh et al. (2009)
	Nicotiana tabacum	Cd	Wojas et al. (2009)
	Brassica juncea	Pb	Zarei et al. (2010)
	Pistia stratiotes	Cd, Pb, Zn	Veselý et al. (2012)
	Populustremula	Zn, Cd, Cu	Ruiz et al. (2011)
	Gmelina arborea	Al	Dudhane et al. (2012)
Bacteria	*Pseudomonas veronii*	Cd, Zn, Cu	Vullo et al. (2008)
	Burkholderia species	Cd, Pb	Jiang et al. (2008)
	Bacillus sp.	Cd, Pb, Cu	Guo et al. (2010)
	Kocuria flava	Cu	Achal et al. (2011)
	Serratia marcescens	U	Kumar et al. (2011)
	Pseudomonas aeruginosa	U	Choudhary and Sar (2011)
	Bacillus cereus	Cd, Zn	Hrynkiewicz and Baum (2012)
	Halomonas sp.	Sr	Achal et al. (2012a)
	Sporosarcinaginsengisoli	As	Achal et al. (2012b)
Fungi	*Penicillium canescens*	Cr	Say et al. (2003)
	Ganoderma lucidum	Ar	Loukidou et al. (2003)
	Aspergillus fumigates	Pb	Ramasamy et al. (2011)
Lichen	*Cladoniarangiformis*	Pb	Ekmekyapar et al. (2012)
Algae	*Chlorella pyrendoidosa*	U	Singhal et al. (2004)
	Chlorella fusca	Pb, Zn, Cd, Cr, Cu, Ni	Ahluwalia and Goyal (2007)
	Spirogyra sp.	Pb, Cu	Lee and Chang (2011)
	*Spirullina*sp.	Cr, Cu, Fe, Mn, Se, Zn	Mane and Bhosle (2012)
	*Hydrodictylon*sp.	V, As	Saunders et al. (2012)
	*Oedogonium*sp.	V, As	Saunders et al. (2012)

organic compounds; some are not capable of degrading contaminants such as chlorinated organic or high aromatic hydrocarbons.

5.3 Mycoremediation

Besides the use of bacteria, fungal species such as *Aspergillus niger*, *Aureobasidium pullulans*, *Ganoderma lucidum*, and *Cladosporium resinae* are found to be capable of mycoremediation (Mani and Kumar 2014). Fungi secrete more potent enzymes even in nutrient-deficient conditions, which act on a broad category of natural chemicals. Remediation through fungus may proceed faster than bacterial degradation because of their large filament surface area (Gadd 2010). Some fungal species are reported to metabolize hydrocarbons and hence, used in mycoremediation of oil-polluted regions. These are many fungi being used for mycoremediation purpose such as *Acremonium, Aspergillus, Aureobasidium, Candida, Cephalosporium, Cladosporium, Cunninghamella, Fusarium, Geotrichum, Gliocladium, Graphium, Hansenula,* and *Mortierella*. Some fungi, such as *Trichoderma*, act as biocontrol agents as well as remediate agricultural waste. *Lentinus edodes*, the gourmet mushroom, has the potential of remediating more than 60% of pentachlorophenol from soil (Pletsch et al. 1999). Such a potent fungus is being used as a boon in oil industries and refineries. *Phanerochaete chrysosporium* and other white rot fungi degrade some xenobiotics like DDT and lindane (Kirk et al. 1992). Mycoremediation is a very efficient method of remediation that produces fewer toxic chemicals as by-products (Gadd 2009).

5.4 Phytoremediation

Phytoremediation is the application of plants and their associated microflora to clean up environment in a cost-effective and noninvasive manner. This could be brought about by a number of mechanisms like phytoextraction, phytostabilization, phytoevaporation, and rhizodegradation.

Phytoextraction, also known as phytoabsorption, is a remediation process by which heavy metals and other contaminated soil pollutants are absorbed by plant roots and transported to shoots, forming the harvestable plant biomass. Till date, more than 400 hyperaccumulator plants, mainly belonging to *Poaceae, Brassicaceae, Asteraceae, Caryophyllaceae, Lamiaceae, Fabaceae, Violaceae*, and *Euphorbiaceae,* have been reported. Our understanding of hyperaccumulation of heavy metals has increased due to recent researches up to molecular level. Heavy metal storing plants are characterized by greater abundance of roots in comparison to shoot biomass, faster transpiration rate, and slow growth as compared to that of excluders.

Phytostabilization, also known as phytodeposition, phytosequestration, and phytoaccumulation, refers to the decrease in the availability and mobility of heavy metals from soil. The mobility of contaminants is lowered by absorption and accumulation into plant roots or immobilization within the rhizosphere, thereby reducing off-site contamination. The main aim is to prevent the migration of contaminants by wind and water erosion and leaching. Phytostabilization can be enhanced by using soil amendments like phosphates, organic matter and alkalizing agents etc. that decrease solubility of metals in soil and minimize its leaching to groundwater. Some plant species like *Agrostis* sp. and *Festuca* sp. are the most common species used in the phytostabilization of Zn, Cu, and Pb. Willows (*Salix sp.*) facilitate both phytostabilization and phytoextraction.

Phytoevaporation is another method of phytoremediation by which plants take up heavy metals like Se and Hg from soil and convert them into volatile form, thereby releasing it into the atmosphere. Species of the Brassicaceae family like *Brassica juncea* can remediate up to 40 g Se/ha.

Rhizodegradation is the degradation of heavy metals by soil microflora, where the enzymatic activity of soil microbes is enhanced by plant root exudates. *Carex pendula* accumulates Pb in the roots under in situ conditions.

6 Future Perspective and Conclusions

Heavy metal-contamination of agricultural land is one of the abiotic stresses that limit crop productivity. Genetic engineering and the recent genome editing approaches have been used to confer heavy metal resistance in plants (Mishra 2019; Sedeek et al. 2019). One of the strategies of tackling heavy metal toxicity in plants is to target the initial step of uptake of heavy metals by plants. Research on membrane proteins like ion channels and pumps should be promoted to understand the molecular mechanism involved in the transport of heavy metals across cell membranes and within the cells. Heavy metal accumulators, such as *B. napus* and *B. juncea*, are widely used for phytoremediation due to their large biomass. A crucial, yet often neglected, aspect of phytoremediation involves recovery or disposal of heavy metals (accumulated in plants) in such a manner that the plant biomass can be properly handled and the associated environmental risks could be reduced. Further, investigations should be focused on rhizosphere and soil microbial diversity, which affect heavy metal solubility. Work could also be directed toward endophytes (non-pathogenic microbes inside the plant organs) that provide resistance against heavy metals. The role of government and environment protection agencies is imperative in initiating awareness among people and formulating stringent laws in order to check the anthropogenic production of these heavy metals. The knowledge about harmful impacts of heavy metals on plant growth is very important not only for improved plant growth and yield, but also to achieve pollution-free environment and ecological harmony.

Acknowledgments The authors would like to thank Director, DEI, for his continuous support and encouragement. SM is grateful to Dayalbagh Educational Institute, Deemed University, Agra, for sanctioning the Research Project, DEI/Minor Project/2017-18 (iv), as a startup grant. DG is thankful to DST-INSPIRE for providing fellowship.

The authors declare no conflict of interest.

References

Achal V, Pan X, Zhang D (2011) Remediation of copper-contaminated soil by *Kocuriaflava* CR1, based on microbially induced calcite precipitation. Ecol Eng 37(10):1601–1605

Achal V, Pan X, Zhang D (2012a) Bioremediation of strontium (Sr) contaminated aquifer quartz sand based on carbonate precipitation induced by Sr resistant *Halomonas* sp. Chemosphere 89(6):764–768

Achal V, Pan X, Fu Q, Zhang D (2012b) Biomineralization based remediation of As (III) contaminated soil by *Sporosarcina ginsengisoli*. J Hazard Mater 201:178–184

Ahluwalia SS, Goyal D (2007) Microbial and plant derived biomass for removal of heavy metals from wastewater. Bioresour Technol 98(12):2243–2257

Al Khateeb W, Al-Qwasemeh H (2014) Cadmium, copper and zinc toxicity effects on growth, proline content and genetic stability of Solanum nigrum L., a crop wild relative for tomato; comparative study. Physiol Mol Biol Plants 20:31–39

Alcántara E, Romera FJ, Cañete M, De la Guardia MD (1994) Effects of heavy metals on both induction and function of root Fe (lll) reductase in Fe-deficient cucumber (*Cucumissativus* L.) plants. J Exp Bot 45(12):1893–1898

Angino EE, Magnuson LM, Waugh TC, Galle OK, Bredfeldt J (1970) Arsenic in detergents-possible danger and pollution hazard. Science 168:389–392

Arazi T, Sunkar R, Kaplan B, Fromm H (1999) A tobacco plasma membrane calmodulin-binding transporter confers Ni^{2+} tolerance and Pb^{2+} hypersensitivity in transgenic plants. Plant J 20:171–182

Arduini I, Godbold DL, Onnis A (1995) Influence of copper on root growth and morphology of *Pinus pinea* L. and *Pinus pinaster*Ait. seedlings. Tree Physiol 15:411–415

Azubuike CC, Chikere CB, Okpokwasili GC (2016) Bioremediation techniques–classification based on site of application: principles, advantages, limitations and prospects. World J Microbiol Biotechnol 32(11):180

Bano SA, Ashfaq D (2013) Role of mycorrhiza to reduce heavy metal stress. Nat Sci 5:16–20

Banuelos G, Terry N, LeDuc DL, Pilon SEA, Mackey B (2005) Field trial of transgenic Indian mustard plants shows enhanced phytoremediation of selenium-contaminated sediment. Environ Sci Technol 39(6):1771–1777

Bhalerao SA, Sharma AS, Poojari AC (2015) Toxicity of nickel in plants. Int J Pure and Appl Biosci 3(2):345–355

Bhattacharyya R, Chatterjee D, Nath B, Jana J, Jacks G, Vahter M (2003) High arsenic groundwater: mobilization, metabolism and mitigation – an overview in the Bengal Delta Plain. Mol Cell Biochem 253:347–355

Bian M, Zho M, Sun D, Li C (2013) Molecular approaches unravel the mechanism of acid soil tolerance in plants. Crop J 1(2):91–104

Boening DW (2000) Ecological effects, transport, and fate of mercury: a general review. Chemosphere 40:1335–1351

Bringezu K, Lichtenberger O, Leopold I, Neumann D (1999) Heavy metal tolerance of Silene vulgaris. J Plant Physiol 154:536–546

Brune A, Urbach W, Dietz KJ (1994) Compartmentation and transport of zinc in barley primary leaves as basic mechanisms involved in zinc tolerance. Plant Cell Environ 17(2):153–162

Burzynski M (1987) The uptake and transpiration of water and the accumulation of lead by plants growing on lead chloride solutions. Acta Soc Bot Pol 56:271–280

Chekroun KB, Baghour M (2013) The role of algae in phytoremediation of heavy metals: a review. J Mater Environ Sci 4(6):873–880
Choi JM, Pak CH, Lee CW (1996) Micronutrient toxicity in French marigold. J Plant Nutr 19:901–916
Choi YE, Harada E, Wada M, Tsuboi H, Morita Y, Kusano T, Sano H (2001) Detoxification of cadmium in tobacco plants: formation and active excretion of crystals containing cadmium and calcium through trichomes. Planta 213(1):45–50
Choudhary S, Sar P (2011) Uranium biomineralization by a metal resistant *Pseudomonas aeruginosa* strain isolated from contaminated mine waste. J Hazard Mater 186(1):336–343
Cobbett CS (2000) Phytochelatins and their roles in heavy metal detoxification. Plant Physiol 123(3):825–832
Crawford TW, Stroehlein JL, Kuehl RO (1989) Manganese and rates of growth and mineral accumulation in cucumber. J Am Soc Hort Sci 114:300–306
Crooke WM (1956) Effect of soil reaction on uptake of nickel from a serpentine soil. Soil Sci 81(4):269–276
Dalvi A, Bhalerao SA (2013) Response of plants towards heavy metal toxicity: an overview of avoidance, tolerance and uptake mechanism. Ann Plant Sci 2:362–368
Danh LT, Truong P, Mammucari R, Tran T, Foster N (2009) Vetiver grass, Vetiveria zizanioides: a choice plant for phytoremediation of heavy metals and organic wastes. Int J Phytoremediation 11(8):664–691
Davies KL, Davies MS, Francis D (1991) Zinc-induced vacuolation in root meristematic cells of Festuca rubra L. Plant Cell Environ 14:399–406
De Dorlodot S, Lutts S, Bertin P (2005) Effects of ferrous iron toxicity on the growth and mineral composition of an inter specific rice. J Plant Nutr 28:1–20
Demidchik V, Sokolik A, Yurin V (1997) The effect of $Cu2^{+}$on ion transport systems of the plant cell plasmalemma. Plant Physiol 114:1313–1325
Deng L, Su Y, Su H, Wang X, Zhu X (2007) Sorption and desorption of lead (II) from wastewater by green algae Cladophora fascicularis. J Hazard Mater 143(1–2):220–225
Drazkiewicz M (1994) Chlorophyll occurrence, functions, mechanism of action, effects of internal and external factors. Photosynthetica 30:321–331
Dudhane M, Borde M, Jite PK (2012) Effect of aluminium toxicity on growth responses and antioxidant activities in *Gmelina arborea* Roxb. inoculated with AM fungi. Int J Phytoremediation 14(7):643–655
Ekmekyapar F, Aslan A, Bayhan YK, Cakici A (2012) Biosorption of Pb (II) by nonliving lichen biomass of *Cladoniarangi formis* Hoffm. Int J Environ Res 6(2):417–424
Emamverdian A, Ding Y, Mokhberdoran F, Xie Y (2015) Heavy metal stress and some mechanisms of plant defense response. Sci World J 2015:756120. https://doi.org/10.1155/2015/756120
Ernst WHO, Verkleij JAC, Schat H (1992) Metal tolerance in plants. Acta Bot Neerl 41:229–248
Eshleman A, Siegel SM, Siegel BZ (1971) Is mercury from Hawaiian volcanoes a natural source of pollution? Nature 223:471–475
Eun SO, Youn HS, Lee Y (2002) Lead disturbs microtubule organization in the root meristem of *Zea mays*. Physiol Plant 110:357–365
Farshian S, Khara J, Malekzadeh P (2007) Influence of arbuscular mycorrhizal fungus (*Glomus etunicatum*) with lettuce plants under zinc toxicity in nutrient solution. Pak J Biol Sci 10:2363–2367
Fidalgo F, Azenha M, Silva AF, de Sousa A, Santiago A, Ferraz P, Teixeira J (2013) Copper-induced stress in Solanum nigrum L. and antioxidant defense system responses. Food Energy Secur 2(1):70–80
Fodor E, Szabo NA, Erdei L (1995) The effects of cadmium on the fluidity and H^{+}-ATPase activity of plasma membrane from sunflower and wheat roots. J Plant Physiol 147:87–92
Gadd G (2009) Heavy metal pollutants: environmental and biotechnological aspects. In: Encyclopedia of microbiology Elsevier, pp 321–324
Gadd GM (2010) Metals, minerals and microbes: geomicrobiology and bioremediation. Microbiology 156(3):609–643

Gall JE, Boyd RS, Rajakaruna N (2015) Transfer of heavy metals through terrestrial food webs: a review. Environ Monit Assess 187(4):201

Gopal R, Dube BK, Sinha P, Chatterjee C (2003) Cobalt toxicity effects on growth and metabolism of tomato. Commun Soil Sci Plant Anal 34:619–628

Gruenhage L, Jager IIJ (1985) Effect of heavy metals on growth and heavy metals content of *Allium Porrum* and Pisum sativum. Angew Bot 59:11–28

Guan Q, Wang F, Xu C, Pan N, Lin J, Zhao R, Luo H (2018) Source apportionment of heavy metals in agricultural soil based on PMF: a case study in Hexi Corridor, northwest China. Chemosphere 193:189–197

Guo H, Luo S, Chen L, Xiao X, Xi Q, Wei W, He Y (2010) Bioremediation of heavy metals by growing hyperaccumulaor endophytic bacterium *Bacillus* sp. L14. Bioresour Technol 101(22):8599–8605

Guo J, Xu W, Ma M (2012) The assembly of metals chelation by thiols and vacuolar compartmentalization conferred increased tolerance to and accumulation of cadmium and arsenic in transgenic Arabidopsis thaliana. J Hazard Mater 199:309–313

Hall JL (2002) Cellular mechanisms for heavy metal detoxification and tolerance. J Exp Bot 53:1–11

Harada E, Kim JA, Meyer AJ, Hell R, Clemens S, Choi YE (2010) Expression profiling of tobacco leaf trichomes identifies genes for biotic and abiotic stresses. Plant Cell Physiol 51:1627–1637

Hawkes JS (1997) Heavy metals. J Chem Educ 74:1369–1374

Herawati N, Suzuki S, Hayashi K, Rivai IF, Koyoma H (2000) Cadmium, copper and zinc levels in rice and soil of Japan, Indonesia and China by soil type. Bull Environ Contam Toxicol 64:33–39

Hrynkiewicz K, Baum C (2012) The potential of rhizosphere microorganisms to promote the plant growth in disturbed soils. In: Environmental protection strategies for sustainable development. Springer, Dordrecht, pp 35–64

Huang Y, Tao S, Chen YJ (2005) The role of arbuscular mycorrhiza on change of heavy metal speciation in rhizosphere of maize in wastewater irrigated agriculture soil. J Environ Sci 17:276–280

Huttermann A, Arduini I, Godbold DL (1999) Metal pollution and forest decline. In: Prasad NMV, Hagemeyer J (eds) Heavy metal stress in plants: from molecules to ecosystems. Springer-Verlag, Berlin, pp 253–272

Israr M, Sahi S, Datta R, Sarkar D (2006) Bioaccumulation and physiological effects of mercury in *Sesbania drummonii*. Chemosphere 65:591–598

Ivanov VB, Bystrova EI, Seregin IV (2003) Comparative impacts of heavy metals on root growth as related to their specificity and selectivity. Russian J Plant Physiol 50(3):398–406

Jentschke G, Godbold DL (2000) Metal toxicity and ectomycorrhizas. Physiol Plant 109:107–116

Jiang CY, Sheng XF, Qian M, Wang QY (2008) Isolation and characterization of a heavy metal-resistant Burkholderia sp. from heavy metal-contaminated paddy field soil and its potential in promoting plant growth and heavy metal accumulation in metal-polluted soil. Chemosphere 72(2):157–164

Juwarkar AS, Shende GB (1986) Interaction of Cd-Pb effect on growth yield and content of Cd, Pb in barley. Ind J Environ Health 28:235–243

Khan AG, Kuek C, Chaudhry TM, Khoo CS, Hayes WJ (2000) Role of plants, mycorrhizae and phytochelators in heavy metal contaminated land remediation. Chemosphere 41(1–2):197–207

Kim DY, Bovet L, Maeshima M, Martinoia E, Lee Y (2007) The ABC transporter AtPDR8 is a cadmium extrusion pump conferring heavy metal resistance. Plant J 50:207–218

Kirk TK, Lamar RT, Glaser JA (1992) The potential of white-rot fungi in bioremediation. In Biotechnology and Environmental Science. Springer, Boston MA, pp 131–138

Kohler A, Blaudez D, Chalot M, Martin F (2004) Cloning and expression of multiple metallothioneins from hybrid poplar. New Phytol 164:83–93

Krämer U, Grime GW, Smith JAC, Hawes CR, Baker AJM (1997) Micro-PIXE as a technique for studying nickel localization in leaves of the hyperaccumulator plant Alyssum lesbiacum. Nucl Instrum Methods Phys Res, Sect B 130(1–4):346–350

Kumar JN, Soni H, Kumar RN, Bhatt I (2008) Macrophytes in phytoremediation of heavy metal contaminated water and sediments in Pariyej Community Reserve, Gujarat, India. Turkish J of Fisheries and Aquatic Sci 8(2):193–200

Kumar R, Acharya C, Joshi SR (2011) Isolation and analyses of uranium tolerant *Serratiamarcescens* strains and their utilization for aerobic uranium U (VI) bioadsorption. J Microb 49(4):568

Lacerda LD (1997) Global mercury emissions from gold and silver mining. Water Air Soil Pollut 97:209–221

Lavres J, Malavolta E, Nogueira NL, Moraes MF, Rodrigues A, Lanzoni M, Pereira C (2009) Changes in anatomy and root cell ultrastructure of soybean genotypes under manganese stress. Rev Bras Cienc Solo 33:395–403

Lee YC, Chang SP (2011) The biosorption of heavy metals from aqueous solution by *Spirogyra* and *Cladophora* filamentous macroalgae. Bioresour Technol 102(9):5297–5304

Lemos Batista B, Nigar M, Mestrot A, Alves Rocha B, Barbosa Júnior F, Price AH, Feldmann J (2014) Identification and quantification of phytochelatins in roots of rice to long-term exposure: evidence of individual role on arsenic accumulation and translocation. J Exp Bot 65(6):1467–1479

Li HF, Gray C, Mico C, Zhao FJ, McGrath SP (2009) Phytotoxicity and bioavailability of cobalt to plants in a range of soils. Chemosphere 75:979–986

Lidon FC, Teixeira MG (2000) Rice tolerance to excess Mn implications in the chloroplast lamellae and synthesis of a novel Mn protein. Plant Physiol Biochem 38:969–978

Liu D, Zou J, Wang M, Jiang W (2008) Hexavalent chromium uptake and its effects on mineral uptake, antioxidant defence system and photosynthesis in *Amaranthus viridis* L. Bioresour Technol 99(7):2628–2636

Liu N, Lin ZF, Lin GZ, Song LY, Chen SW, Mo H (2010) Lead and cadmium induced alterations of cellular functions in leaves of Alocasia macrorrhiza L. Schott. Ecotoxicol Environ Saf 73:1238–1245

Lo TY, Chen SM (1994) The effect of chemical treatment on the carotene and ascorbic acid contents of tomato. J Chin Chem Soc 11:95–98

Loukidou MX, Matis KA, Zouboulis AI, Liakopoulou-Kyriakidou M (2003) Removal of As(V) from wastewaters by chemically modified fungal biomass. Water Res 37(18):4544–4552

Lyyra S, Meagher RB, Kim T, Heaton A, Montello P, Balish RS, Merkle SA (2007) Coupling two mercury resistance genes in Eastern cottonwood enhances the processing of organomercury. Plant Biotechnol J 5(2):254–262

Ma JF, Zheng SJ, Matsumoto H (1997) Detoxifying aluminium with buckwheat. Nature 390:569–570

Ma LQ, Komar KM, Tu C, Zhang W, Cai Y, Kennelley ED (2001) A fern that hyperaccumulates arsenic. Nature 409(6820):579

Manara A (2012) Plant responses to heavy metal toxicity. In: Plants and heavy metals. Springer, Dordrecht, pp 27–53

Mane PC, Bhosle AB (2012) Bioremoval of some metals by living algae Spirogyra sp. and *Spirullina* sp. from aqueous solution. Int J Environ Res 6(2):571–576

Mani D, Kumar C (2014) Biotechnological advances in bioremediation of heavy metals contaminated ecosystems: an overview with special reference to phytoremediation. Int J Environ Sci Technol 11(3):843–872

Mani D, Sharma B, Kumar C, Balak S (2012) Cadmium and lead bioaccumulation during growth stages alters sugar and vitamin C content in dietary vegetables. Proc Natl Acad Sci India Sect B Biol Sci 82(4):477–488

Marschner H (1995) Mineral nutrition of higher plants, 2nd edn. Academic Press, Toronto

Mathys W (1975) Enzymes of heavy-metal-resistant and non-resistant populations of *Silenecucubalus* and their interaction with some heavy metals in vitro and in vivo. Physiol Plant 33(2):161–165

Meharg AA, Macnair MR (1992) Genetic correlation between arsenate tolerance and the rate of influx of arsenate and phosphate in Holcus lanatus. Heredity 69:336–341

Messer RL, Lockwood PE, Tseng WY, Edwards K, Shaw M, Caughman GB, Lewis JB, Wataha JC (2005) Mercury (II) alters mitochondrial activity of monocytes at sublethal doses via oxidative stress mechanisms. J Biomed Mater Res B 75:257–263

Mirshekali H, Hadi H, Amirnia R, Khodaverdiloo H (2012) Effect of zinc toxicity on plant productivity, chlorophyll and Zn contents of sorghum (Sorghum bicolor) and common lambsquarter (Chenopodium album). Int J Agric Res Rev 2:247–254

Mishra D, Kar M, Pradhan PK (1973) Chemical regulation of acid inorganic pyrophosphates activity during senescence of detached leaves. Exp Gerantol 8:165–167

Mishra S (2019) Targeted genome editing tools in plants. In: Innovations in life science research. NOVA Science Publisher, New York. ISBN: 978-1-53615-868-7

Misra SG, Mani D (1991) Soil pollution. Ashish Publishing House, New Delhi

Mitchell RL, Burchett MD, Pulkownik A, McCluskey L (1988) Effects of environmentally hazardous chemicals on the emergence and early growth of selected Australian native plants. Plant Soil 112:195–199

Moller IM, Jensen PE, Hansson A (2007) Oxidative modifications to cellular components in plants. Annu Rev Plant Biol 58:459–481

Mourato MP, Moreira IN, Leitão I, Pinto FR, Sales JR, Martins LL (2015) Effect of heavy metals in plants of the genus Brassica. Int J Mol Sci 16(8):17975–17998

Muhammad D, Chen F, Zhao J, Zhang G, Wu F (2009) Comparison of EDTA- and citric acidenhanced phytoextraction of heavy metals in artificially metal contaminated soil by Typha angustifolia. Int J Phytoremediation 11:558–574

Murphy A, Taiz L (1995) Comparison of metallothionein gene expression and nonproteinthiols in ten *Arabidopsis* ecotypes (correlation with copper tolerance). Plant Physiol 109(3):945–954

Nagajyoti PC, Lee KD, Sreekanth TVM (2010) Heavy metals, occurrence and toxicity for plants: a review. Environ Chem Lett. 8:199–216

Neelima P, Reddy KJ (2002) Interaction of copper and cadmium with seedlings growth and biochemical responses in Solanum melongena. Nat Environ Pollut Technol 1:285–290

Neumann D, Lichtenberger O, Günther D, Tschiersch K, Nover L (1994) Heat-shock proteins induce heavy-metal tolerance in higher plants. Planta 194(3):360–367

Nriagu JO (1989) A global assessment of natural sources of atmospheric trace metals. Nature 338:47–49

Oliveira A, Pampulha ME (2006) Effects of long-term heavy metal contamination on soil microbial characteristics. J Biosci Bioeng 102(3):157–161

Ouzounidou G (1994) Change in chlorophyll fluorescence as a result of copper treatment: dose response relations in Silene and Thlaspi. Photosynthetica 29:455–462

Pagani MA, Tomas M, Carrillo J (2012) The response of the different soybean metallothionein isoforms to cadmium intoxication. J Inorg Biochem 117:306–315

Parmar NG, Vithalani SD, Chanda SV (2002) Alteration in growth and peroxidase activity by heavy metals in Phaseolus seedlings. Acta Physiol Plant 24(1):89–95

Patterson WA, Olson JJ (1983) Effects of heavy metals on radicle growth of selected woody species germinated on filter paper, mineral and organic soil substrates. Can J For Res 13(2):233–238

Phaniendra A, Jestadi DB, Periyasamy L (2015) Free radicals: properties, sources, targets, and their implication in various diseases. Indian J Clin Biochem 30(1):11–26

Pletsch M, de Araujo BS, Charlwood BV (1999) Novel biotechnological approaches in environmental remediation research. Biotechnol Adv 17(8):679–687

Prasad MNV (2004) Phytoremediation of metals in the environment for sustainable development. Proc Indian Natl Sci Acad 70:91–98

Ramasamy P, Vino AB, Saravanan R, Subhapradha N, Shanmugam V, Shanmugam A (2011) Screening of antimicrobial potential of polysaccharide from cuttlebone and methanolic extract from body tissue of Sepia prashadi Winkworth, 1936. Asian Pac J Trop Biomed 1(2):244–248

Rauser WE (1999) Structure and function of metal chelators produced by plants. Cell Biochem Biophys 31(1):19–48

Reddy AM, Kumar SG, Jyothsnakumari G, Thimmanaik S, Sudhakar C (2005) Lead induced changes in antioxidant metabolism of horsegram (*Macrotyloma uniflorum* (Lam.) Verdc.) and bengalgram (*Cicer arietinum* L.). Chemosphere 60:97–104

Ros ROC, Cooke DT, Burden RS, James CS (1990) Effects of the herbicide MCPA and the heavy metals, cadmium and nickel on the lipid composition, Mg^{2+}-ATPase activity and fluidity of plasma membranes from rice, *Oryza sativa* (cv. Bahia) shoots. J Exp Bot 41:457–462

Ruiz ON, Alvarez D, Torres C, Roman L, Daniell H (2011) Metallothionein expression in chloroplasts enhances mercury accumulation and phytoremediation capability. Plant Biotechnol J 9(5):609–617

Samardakiewicz S, Wozny A (2005) Cell division in Lemna minor roots treated with lead. Aquat Bot 83:289–295

Sarret G, Saumitou-Laprade P, Bert V, Proux O, Hazemann JL, Traverse A, Manceau A (2002) Forms of zinc accumulated in the hyperaccumulator Arabidopsis halleri. Plant Physiol 130(4):1815–1826

Saunders RJ, Paul NA, Hu Y, de Nys R (2012) Sustainable sources of biomass for bioremediation of heavy metals in waste water derived from coal-fired power generation. PLoS One 7(5):e36470

Saxena PK, Krishna Raj S, Dan T, Perras MR, Vettakkorumakankav NN (1999) Phytoremediation of heavy metal contaminated and polluted soils. In: Heavy metal stress in plants. Springer, Berlin, pp 305–329

Say R, Yilmaz N, Denizli A (2003) Removal of chromium (VI) ions from synthetic solutions by the fungus *Penicillium canescens*. Eur J Miner Process Environ Prot 3(1):36–41

Seaward MRD, Richardson DHS (1990) Atmospheric sources of metal pollution and effects on vegetation. In: Shaw AJ (ed) Heavy metal tolerance in plants evolutionary aspects. CRC Press, Boca Raton, pp 75–94

Sedeek KE, Mahas A, Mahfouz M (2019) Plant genome engineering for targeted improvement of crop traits. Front Plant Sci 10:114. https://doi.org/10.3389/fpls

Selvam A, Wong JWC (2008) Phytochelatin systhesis and cadmium uptake of *Brassica napus*. Environ Technol 29(7):765–773

Sengar RS, Gupta S, Gautam M, Sharma A, Sengar K (2008) Occurrence, uptake, accumulation and physiological responses of nickel in plants and its effects on environment. Res J Phytochem 2(2):44–60

Seregin IV, Kozhevnikova AD (2006) Physiological role of nickel and its toxic effects on higher plants. Russ J Plant Physiol 53:257–277

Shahid M, Dumat C, Khalid S, Schreck E, Xiong T, Niazi NK (2017) Foliar heavy metal uptake, toxicity and detoxification in plants: a comparison of foliar and root metal uptake. J Hazard Mater 325:36–58

Sharma SS, Dietz KJ (2006) The significance of amino acids and amino acid-derived molecules in plant responses and adaptation to heavy metal stress. J Exp Bot 57:711–726

Sharma P, Dubey RS (2005) Lead toxicity in plants. Braz J Plant Physiol 17:35–52

Sharma A, Talukdar G (1987) Effects of metals on chromosomes of higher organisms. Environ Mutagen 9:191–226

Shen H, He Z, Yan H, Xing Z, Chen Y, Xu W, Wenzhong X, Ma M (2014) The fronds tonoplast quantitative proteomic analysis in arsenic hyperaccumulator *Pteris vittata* L. J Proteome 105:46–57

Silva S (2012) Aluminium toxicity targets in plants. J Bot 2012:219462

Singh PK, Tewari SK (2003) Cadmium toxicity induced changes in plant water relations and oxidative metabolism of *Brassica juncea* L. plants. J Environ Biol 24:107–117

Singh N, Ma LQ, Srivastava M, Rathinasabapathi B (2006) Metabolic adaptations to arsenic induced oxidative stress in *Pteris vittata* L. and Pteris ensiformis L. Plant Sci 170:274–282

Singhal R, Joshi S, Tirumalesh K, Gurg R (2004) Reduction of uranium concentration in well water by Chlorella (*Chlorella pyrendoidosa*) a fresh water algae immobilized in calcium alginate. J Radioanal Nucl Chem 261(1):73–78

Stefanov K, Seizova K, Popova I, Petkov VL, Kimenov G, Popov S (1995) Effects of lead ions on the phospholipid composition in leaves of *Zea mays* and Phaseolus vulgaris. J Plant Physiol 147:243–246

Tangahu BV, Abdullah S, Rozaimah S, Basri H, Idris M, Anuar N, Mukhlisin M (2011) A review on heavy metals (As, Pb, and Hg) uptake by plants through phytoremediation. Int J Chem Eng 2011:939161: 1–31. https://doi.org/10.1155/2011/939161

Tchounwou PB, Yedjou CG, Patlolla AK, Sutton DJ (2012) Heavy metal toxicity and the environment. In: Molecular, clinical and environmental toxicology. Springer, Basel, pp 133–164

Tiwari S, Kumari B, Singh SN (2008) Evaluation of metal mobility/immobility in fly ash induced by bacterial strains isolated from the rhizospheric zone of *Typha latifolia* growing on fly ash dumps. Bioresour Technol 99(5):1305–1310

Turnau K (1998) Heavy metal content and localization in mycorrhizal Euphorbia cyparissias zinc wastes in Southern Poland. Acta Soc Bot Pol 67:105–113

Vidali M (2001) Bioremediation: an overview. Pure Appl chem 73(7):1163–1172

Verkleji JAS (1993) The effects of heavy metals stress on higher plants and their use as bio monitors. In: Markert B (ed) Plant as bioindicators: indicators of heavy metals in the terrestrial environment. VCH, New York, pp 415–424

Veselý T, Tlustos P, Száková J (2012) Organic acid enhanced soil risk element (Cd, Pb and Zn) leaching and secondary bioconcentration in water lettuce (*Pistia stratiotes* L.) in the rhizofiltration process. Int J Phytoremediation 14(4):335–349

Viehweger K (2014) How plants cope with heavy metals. Bot Stud 55:1–12

Vullo DL, Ceretti HM, Daniel MA, Ramírez SA, Zalts A (2008) Cadmium, zinc and copper biosorption mediated by *Pseudomonas veronii* 2E. Bioresour Technol 99(13):5574–5581

Wainwright SJ, Woolhouse HW (1977) Some physiological aspects of copper and zinc tolerance in Agrostis tenuis Sibth.: cell elongation and membrane damage. J Exp Bot 28:1029–1036

Wojas S, Hennig J, Plaza S, Geisler M, Siemianowski O, Skłodowska A, Antosiewicz DM (2009) Ectopic expression of *Arabidopsis* ABC transporter MRP7 modifies cadmium root-to-shoot transport and accumulation. Environ Pollut 157(10):2781–2789

Wong HL, Sakamoto T, Kawasaki T, Umemura K, Shimamoto K (2004) Down-regulation of metallothionein, a reactive oxygen scavenger, by the small GTPase OsRac1 in rice. Plant Physiol 135:1447–1456

Wu H, Tang S, Zhang X, Guo J, Song Z, Tian S, Smith DL (2009) Using elevated CO_2 to increase the biomass of a *Sorghum vulgare*× Sorghum vulgare var. sudanense hybrid and *Trifoliumpratense* L. and to trigger hyperaccumulation of cesium. J Hazard Mater 170(2–3):861–870

Xiong ZT, Zhao F, Li MJ (2006) Lead toxicity in Brassica pekinensis Rupr.: effect on nitrate assimilation and growth. Environl Toxicol: An Int J 21(2):147–153

Yadav SK (2010) Heavy metals toxicity in plants: an overview on the role of glutathione and phytochelatins in heavy metal stress tolerance of plants. S Afr J Bot 76(2):167–179

Yanqun Z, Yuan L, Jianjun C, Haiyan C, Li Q, Schratz C (2005) Hyper accumulation of Pb, Zn and Cd in herbaceous grown on lead-zinc mining area in Yunnan, China. Environ Int 31:755–762

Zarei GH, Shamsi HO, Dehghani SM (2010) The effect of drought stress on yield, yield components and seed oil content of three autumnal rapeseed cultivars (*Brassica napus* L.). J Res Agric Sci 6(1):29–37

Zhang WH, Tyerman SD (1999) Inhibition of water channels by HgCl2 in intact wheat root cells. Plant Physiol 120:849–857

Zhou ZS, Huang SQ, Guo K, Mehta SK, Zhang PC, Yang ZM (2007) Metabolic adaptations to mercury-induced oxidative stress in roots of Medicago sativa L. J Inorg Biochem 101:1–9

Zhu B, Alva AK (1993) Distribution of trace metals in some sandy soils under citrus production. Soil Sci Soc Am J 57:350–355

Effect of Acid Rain on Plant Growth and Development: Physiological and Molecular Interventions

Biswojit Debnath and Golam Jalal Ahammed

Contents

1 Introduction

1.1 Basic Concept of Acid Rain

Since the beginning of civilization, progressive societies have been using numerous natural means. The earth's energy has been used to produce different facilities that make our life easier. Conversely, it results in environmental pollution due to the release of unsafe substances to the environment. Burning of fossil fuel for transportation, industrial development, and urbanization has significantly increased the smoky and particulate impurities in atmosphere, leading to massive air pollution in many cities of the world (Dwivedi and Tripathi 2007). Acid rain is formed from air

B. Debnath
College of Horticulture, Fujian Agriculture and Forestry University, Fuzhou, Fujian, China

Department of Horticulture, Sylhet Agricultural University, Sylhet, Bangladesh

G. J. Ahammed (✉)
College of Forestry, Henan University of Science and Technology, Luoyang, Henan, China
e-mail: ahammed@haust.edu.cn

M. Naeem et al. (eds.), *Contaminants in Agriculture*,
https://doi.org/10.1007/978-3-030-41552-5_5

pollution and it is a major environmental hazard worldwide, especially in Europe, East Asia, North America including Canada (Bouwman et al. 2002). In China, acid rain is documented as one of the major environmental issues in current eras because of prompt commercial expansion (Larssen et al. 1999; Feng 2000b; Chen et al. 2010). The dissemination of acid rain is predominantly identified in Southern parts of China. The occurrence of acid rain has been growing in China ever since 1970s. It was estimated that the damage of ecological benefits due to acidic rain is more than 16 billion Yuan (2.4 billion US$) per year only in 11 provinces of South China (Feng 2000a). The soils of this vast region have been acidified and major ecosystems are also in severe risk (Chen et al. 2010).

In 1852, Robert Angus Smith, a pharmacist of Manchester, England, discovered acid rain for the first time (Fairfax and Lepp 1975). He observed high acidity levels in rainwater over industrial regions of England. On the other hand, he also detected a lesser acidity in less polluted parts, especially near the coastal area (Seinfeld and Pandis 2012). Until 1950s, very little attention was paid to his work. But, when biologists identified a shocking drop of fish numbers in the southern Norway lakes and linked the problem to acid rain, a significant awareness was aroused among global communities (Seinfeld and Pandis 2012). Simultaneously, the influence of acid rain on vegetation was put forward in the forefront. After that, researchers focused on the studies on acid rain to know the origin of the acid rain and its impacts.

2 Causes and Formation of Acid Rain

The major reasons for acid rain formation entail the occurrence of sulfur dioxide (SO_2), nitrogen oxide (NO_x), and ozone in air. These pollutants profoundly originate due to human anthropogenic activities, such as burning of flammable waste and fossil petroleum in thermal power plants, and gas emission from vehicles (Singh and Agrawal 2007). The manmade causes of air pollution by sulfur dioxides (SO_2) are the fiery coal as well as petroleum and several industrial processes. The manufacturing of iron, zinc, copper, sulfuric acid, and petroleum industry are the other sources of sulfur dioxides (Cullis and Hirschler 1980). Although the contribution of NO_x to acid rain formation is lesser compared to SO_2, its volume is increasing day by day. The natural sources of NO_x are ignition, volcanic eruption, biological process, etc., and the artificial causes are means of transportation exhausts and manufacturing emission.

Normally, the rainwater is somewhat acidic due to the formation of carbonic acid through the reaction with atmospheric carbon dioxide (CO_2) and rainwater in nature:

$$CO_2 + H_2O \rightarrow H_2CO_3$$

In addition, insignificant amounts of sulfuric acid and nitric acid are formed in normal rainwater:

$$SO_2 + H_2O \rightarrow H_2SO_3$$

$$2N_2 + 5O_2 + 2H_2O \rightarrow 4\,HNO_3$$

The absorption of H+ ion higher than 2.5 eq^{-1} and pH value lower than 5.6 in rain water is measured to be acid rain (Evans 1984). Air pollutants such as SO_2 and NO_x are emitted into atmosphere, and sulfuric acid and nitric acid mists are formed due to interaction with gases in the prevailing winds. These acids persist in vapor state beneath the widespread high temperature at surroundings. The aerosol droplets formed at a lower temperature remain black color and acidic due to unburnt carbon elements in nature. This substance is known as "acid smut." The incidence of oxidizing means and the features of the reaction upsets the proportion of acid generation (Calvert et al. 1985).

2.1 *Acid Formation in the Presence of Sulfur*

The amount of sulfur is high in coal and oxidization occurs when coal is burned

$$S + O \rightarrow 2SO_2$$

In addition, SO_2 is formed directly in the flame and discharged into the atmosphere commencing from the burned stacks. SO_2 is gradually oxidized to SO_3^{2-} at normal temperature in the normal wind:

$$2SO_2 + O2 \rightarrow 2\,SO_3^{2-}$$

$$SO_3^{2-} + H_2O \rightarrow H_2SO_4$$

$$SO_2 + H_2O \rightarrow H_2SO_3$$

Furthermore, SO_3^{2-} is converted to SO_4 due to its oxidant property in atmosphere. The plenteous amount of ammonia and O_3 in severely air-polluted areas may cause sulfur dioxide oxidation in clouds. These catalysts aid the conversion of more SO_2 to sulfuric acid.

$$H_2O_2 + HSO_3^{-} \rightarrow HSO_4^{-} + H_2O$$

2.2 *Acid Formation in the Presence of Nitrogen*

$$N_2 + O_2 \rightarrow 2NO$$

$$2NO + O_2 \rightarrow 2NO_2$$

$$4NO_2 + O_2 + 2H_2O \rightarrow 4HNO_3$$

$$O_3 + NO_2 \rightarrow NO_3 + O_2$$

$$NO_3 + NO_2 \rightarrow N_2O_5$$

$$N_2O_5 + H_2O \rightarrow HNO_3$$

2.3 *Acid Formation in the Presence of Ozone (O_3)*

$$O_3 \rightarrow O_2 + O$$

$$O + H_2O \rightarrow OH^{\bullet}\ (\text{hydroxy radical})$$

$$OH^{\bullet} + SO_2 \rightarrow HSO_3$$

$$HSO_3^{-} + OH^{\bullet} \rightarrow H_2SO_4$$

$$OH^{\bullet} + NO \rightarrow 2HNO_3$$

$$HSO_3 + O_2 \rightarrow SO_3^{2-} + HO^{\bullet}_2\ (\text{peroxy radical})$$

In addition to the above-mentioned reactions, formic acid, acetic acid, and other organic acids are formed from the reaction of peroxy radical with formaldehyde and acetaldehyde, which contribute 5–20% of acidity of acid rain.

3 Problems Related to Acid Rain on Plants

As shown in Fig. 1, acid precipitations as rain, snow, mist, or dew adversely impact the state of natural ecosystems as well as negatively affect plants either directly by damaging the photosynthetic organs or indirectly through the soil and root system, promoting higher solubility of toxic metals such as cadmium, mercury, and lead and their release from sediments and soils (Likens and Bormann 1974). Acidity, as well as the injurious occurrence of toxic elements, harms vegetation while susceptible microbial species are abolished from the soil due to decay and breakdown of organic debris, which also affect the capacity of a stable regulation of nutrients. During last few decades, acid rain caused extensive dieback in the upper canopy of Masson pine and a huge forest areas declined in Southern China (Hogan 1998).

At present, acid rain is considered as one of the serious environmental problems for agriculture and forestry. Researchers piloted their research by means of

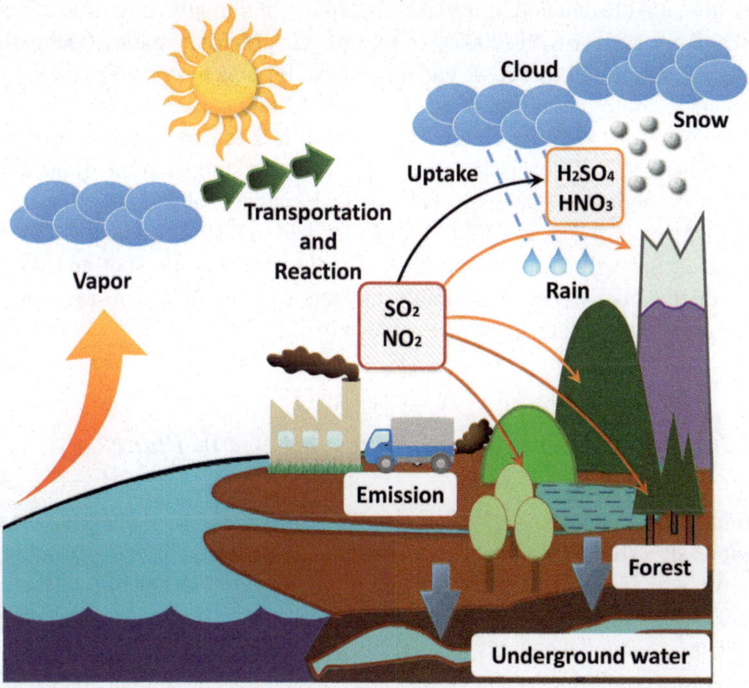

Fig. 1 A schematic representation showing water cycle and acid rain formation. All substances remaining in the air are transported from the atmosphere to the ground and water bodies through water circulation. Natural and anthropogenic activities such as fossil fuel combustion increase air pollution. Air pollutants such as sulfur oxide and nitrogen oxide are mixed with rainwater as acid and fall onto the ground. The phenomenon is generally termed "acid rain" (Seto and Shepherd 2009; Xalxo and Sahu 2017)

simulated acid rain (SAR) that changes the growth, development, physiological and molecular activities of plants (Ren et al. 2018; Debnath et al. 2018b). As vegetation and soils are the primary receptors of acid precipitation (Fig. 1), our discussion is confined to the impact of acid rain on plants in this literature.

3.1 *Impact of Acid Rain on Plant Growth*

Acid rain is a relatively new and overriding abiotic stress factor that impacts a wide range of physiological and metabolic processes, leading to a significant reduction in normal growth and development of plants. For instance, simulated acid rain (SAR) hampers the key plant growth parameters such as plant height, leaf number, stem diameter, shoot and root fresh weight, suggesting that SAR-induced stress has significant negative impacts on plant growth (Liu et al. 2007, 2011b; Dolatabadian et al. 2013). This might be partially because of lesser water latent in cells which causes stomatal closure and limits CO_2 absorption and finally inhibition of cell division (Dwivedi and Tripathi 2007). In one of our previous studies (Debnath et al. 2018c), we also found that SAR had an adverse influence on both plant height and stem diameter in tomato. It was revealed that acid rain caused marked symptoms of phytotoxicity, for example, chlorosis, necrosis, less leaf production, leaf curling, leaves withering, leaf abscission, and plant become stubby, stunted and brittle in severe stress conditions (Debnath et al. 2018c; Eguagie et al. 2016; Odiyi and Eniola 2015). Likewise, Silva et al. (2005) found that acid rain (lower than pH 3.0) showed stunted plant growth commonly with leaf chlorosis, necrotic spot and occasional wilting of the plants. The undesirable impacts of simulated acid rain on growth features in different plant species are shown in Table 1.

3.2 *Impact of Acid Rain on Photosynthesis in Plants*

The components of photosynthetic apparatus and its membrane integrity are altered by various stressful environments, such as drought, salinity, high temperature, cold, and so on, which finally lead to reduction of photosynthetic capability (Sheng et al. 2008). The photosynthetic capability of plants largely relies on the chlorophyll content and environmental hazards remarkably decline the photosynthetic capability of plants (Amirjani 2011). As shown in Table 1, the chlorophyll contents of leaves in different plant species decreased noticeably due to acid rain deposition. Likewise, acid rain hampers the photosynthetic activity; nonetheless, the effect of acid rain on plant photosynthetic activities varies with plant species (Chen et al. 2013; Kováčik et al. 2011). Arti et al. (2010) reported that the old leaves of *Lycopersicon esculentum* Mill. (Tomato), *Solanum melongea* L. (Brinjal) and *Capsicum annuum* L. (Chilly) plants were readily exposed and abscission of leaves started on the first week of exposure to SAR. Additionally, the osmotic consequence due to stress may cause disturbance to the water stability in plants and prevent growth in addition to stomatal closure and thus declining photosynthesis (Hernández and Almansa 2002). Our previous study (Debnath et al. 2018c) also showed that simulated acid rain causes leaf chlorosis and necrosis in the leaves of tomato genotypes. Moreover, the experiment revealed that the chlorophyll a, b, and total carotenoids contents in tomato leaves declined considerably following exposure to simulated acid rain. Eventually, ultrastructures of chloroplast and plasma membrane of leaves were affected by acid rain deposition alongside the chlorophyll degradation and lower photosynthetic activity (Debnath et al. 2018a; Wen et al. 2011). The Calvin cycle, the Krebs cycle, glycolysis, and the Pentose phosphate pathway were affected under acid rain condition, indicating that primary respiratory pathways and photosynthesis were inhibited by simulated acid rain treatment (Liu et al. 2011a; Wang et al. 2013). Irregular white-to-tan lesions were seen on both upper and lower surfaces of tomato leaves, which were related to attenuated chlorophyll content due to simulated acid rain (Shaukat and Khan 2008). Shan (1998) also found that the efficiency of chlorophyll use in photosynthesis markedly reduced due to acid rain, which

Table 1 Effects of acid rain on different growth, biochemical, and physiological parameters

Crop species	Family	Effects	Reference
Solanum lycopersicum L. cv. Micro-Tom	Solanaceae	Reduced growth and photosynthesis, altered antioxidant activity and reduced yield	Debnath et al. (2018a, b)
Solanum lycopersicum L. cv. Red Rain	Solanaceae	Reduced growth and photosynthesis, altered antioxidant activities	Debnath et al. (2018c)
Solanum esculentum	Solanaceae	Reduced growth, chlorophyll, soluble sugar, phenol	Arti et al. (2010) and Shaukat and Khan (2008)
Capsicum annuum L.	Solanaceae	Reduced growth, chlorophyll content and yield	Bamidele and Eguagie (2015) and Arti et al. (2010)
Solanum melongea L.	Solanaceae	Reduced growth, chlorophyll content	Arti et al. (2010)
Oryza sativa	Poaceae	Altered antioxidant activity	Ren et al. (2018)
Triticum sp	Poaceae	Altered photosynthesis and antioxidant activity	Dolatabadian et al. (2013)
Cunninghamia lanceolata (Lamb.) Hook	Cupressaceae	Attenuated growth and photosynthesis	Liu et al. (2018)
Pisum sativum L.	Fabaceae	Damage to photosynthetic apparatus	Polishchuk et al. (2016)
Vigna unguiculata L. Walps	Fabaceae	Reduced growth and development, harvest index	Odiyi and Eniola (2015)
Glycine max	Fabaceae	Inhibited growth and photosynthesis, antioxidant enzyme system	Wen et al. (2011), Hu et al. (2014) and Zhang et al. (2015)
Dimocarpus longana Lour. cv. Wulongling	Sapindaceae	Changes in protein expression	Pan et al. (2015)
Cucumis sativus L.	Cucurbitaceae	Reduced photosynthesis and antioxidant enzymes	Yu et al. (2002)
Ricinus communis	Euphorbiaceae	Reduced growth	Madiha et al. (2015)
Manihot esculenta (Crantz)	Euphorbiaceae	Reduced growth and development, harvest index	Odiyi and Bamidele (2013)
Abelmoschus caillei	Malvaceae	Reduced growth, chlorophyll content and yield	Eguagie et al. (2016)
Liquidambar formosana Hance	Altingiaceae	Altered photosynthesis and antioxidant responses	Chen et al. (2013)
Schima superba Gardn. et Champ.	Theaceae	Altered photosynthesis and antioxidant responses	Chen et al. (2013)
Carica papaya	Caricaceae	Reduced growth	Madiha et al. (2015)
Tillandsia albida	Bromeliaceae	Altered photosynthesis, amino acid and antioxidant compounds	Kováčik et al. (2011)
Hypogymnia physodes	Parmeliaceae	Changes in photosynthesis, amino acid and antioxidant compounds	Kováčik et al. (2011)

(continued)

Table 1 (continued)

Crop species	Family	Effects	Reference
Xanthoria parietina	Teloschistaceae	Altered photosynthesis, amino acid and antioxidant compounds	Kováčik et al. (2011)
Brassica napus	Brassicaceae	Reduced growth and yield component	Cao et al. (2010)

might be associated with the rise in the chlorophyll degradation rate to pheophytin "a" that results in fall of net photosynthetic rate. The maximal photochemical efficiency of photosystem II (Fv/Fm) is considered a reliable indicator of stress-cased damage (Qin et al. 2011). A number of previous studies showed a decline in Fv/Fm and Φ_{PSII} as reliable indicators of photoinhibition in response to acid rain treatments (Liu et al. 2018; Chen et al. 2013).

3.3 Impact of Acid Rain on Antioxidant System in Plants

Exposure of plants to stresses results in the generation of large quantities of reactive oxygen species (ROS) (Schutzendubel and Polle 2002), which oxidize proteins, lipids and nucleic acids consequential to the anomalies in the cell (Di Toppi and Gabbrielli 1999). Plants use different enzymatic and non-enzymatic antioxidant compounds to minimize ROS levels when challenged with any abiotic or biotic stressor (Kholová et al. 2009; Ruiz-Lozano 2003). Moreover, both enzymatic and non-enzymatic antioxidant compounds participate in the comprehensive antioxidant defense system to balance ROS and to maintain cell membrane stability under stress conditions (Fontenele et al. 2017).

Similarly, the disproportionate accumulation of ROS such as hydrogen peroxide and superoxide anion, and malondialdehyde (MDA) contents is a common consequence detected in plants following exposure to acid rain (Ren et al. 2018; Chen et al. 2013; Kováčik et al. 2011). In different crops, the enzymatic antioxidant activity, for instance, superoxide dismutase (SOD), peroxidase (POD), catalase (CAT), ascorbate peroxide (APx), glutathione transferase (GST), etc., and the non-enzymatic antioxidant compounds such as phenolic, flavonoids, ascorbic acid, and proline content increase in response to moderate acid rain stress, which potentially minimizes ROS and membrane damage (Debnath et al. 2018c; Ren et al. 2018). Nonetheless, plant antioxidant systems collapse under severe acid rain stress; as a result, the enzymatic and non-enzymatic antioxidant activities decline probably because of the changed metabolic status or else their biosynthesis (Debnath et al. 2018c; Ren et al. 2018).

3.4 Impact of Acid Rain on Gene Expression in Plants

Genes that play a significant role in fine categorized biochemical pathways of photosynthesis, reactive oxygen species (ROS) production, antioxidant defense, and expression of transcriptional factor are remarkably modulated in responses to acid rain (Liu et al. 2013). A proteomics study showed that the ribulose-1,5-bis-phosphate (RuBP) carboxylase/oxygenase (RuBisCO) expression was markedly dropped under acid rain conditions in *Arabidopsis thaliana* and the carbonic anhydrase gene expression pattern altered noticeably, signifying that photosynthesis is somewhat vulnerable to acid rain (Liu et al. 2011a). Various genes that are involved in light reaction of photosynthesis such as photosynthetic electron transport chain-related genes (At2g01590 and At4g27880) and light harvesting complex in PSI and PSII constituent protein-related genes (At2g34430, At2g05070, At3g08940, At3g27690, At1g15820, At5g54270, and At1g03130) were repressed under simulated acid rain treatment (Liu et al. 2013). The superoxide anion and hydrogen peroxide production increased due to acid rain treatments which triggered malondialdehyde (MDA) content by affecting membrane permeability. In contrast, the expression of ROS scavenging antioxidant enzymes and non-enzymatic antioxidant-related genes such as Cu/Zn superoxide dismutase (SOD), L-ascorbate peroxidase (APx) peroxidase (POD1), catalase (CAT1), glutathione S transferase (GST), and ascorbic acid (AA) were changed under acid rain condition in different plants (Ren et al. 2018; Chen et al. 2013). In addition, ROS scavenging-related pathway genes such as two superoxide dismutase (At1g08830 and At4g25100), glutathione peroxidase (At4g11600 and At2g25080), a class III peroxidase (At3g49120), a monodehydroascorbate reductase (At5g03630), a peroxisomal catalase (At4g35090), and three thioredoxins (At5g16400, At1g07700, and At1g08570) were induced under simulated acid rain conditions (Liu et al. 2013). Moreover, different abiotic and biotic stress-related transcription factors such as WRKY transcription factors, zinc finger proteins, MYB transcription factor, and calcium signal pathway associated genes were induced after simulated acid rain treatment (Liu et al. 2013).

4 Conclusions

Recently, acid rain has received important consideration as a major abiotic factor that causes severe damage to plants and ecosystem. Air pollutants (SO_2 and NO_x) responsible for the formation of acid rain are increasing day by day due to fast economic and industrial growth. Acid rain causes primary as well as secondary injuries to the plants including growth reduction, decline of photosynthetic activity, oxidative stress, and membrane damage, variation of enzymatic and non-enzymatic antioxidant activity, alteration of gene expression and yield penalty. Despite limited research reports on the effect of acid rain on plants, the present review would assist the researchers to get a better understanding of the impact of acid rain on plants at physiological and molecular levels, and to adopt strategies to overcome the hazardous effects of acid rain on natural ecosystem and crop production.

References

Amirjani MR (2011) Effect of salinity stress on growth, sugar content, pigments and enzyme activity of rice. Int J Bot 7(1):73–81

Arti V, Ashish T, Abdullah A (2010) Impact of simulated acid rain of different pH-levels on some major vegetable plants in India. Rep Opin 2:38–40

Bamidele J, Eguagie M (2015) Ecophysiological response of *Capsicum annuum* L. exposed to simulated acid rain. Nigerian J Biotechnol 30(1):48–52

Bouwman AF, Vuuren DPV, Derwent RG, Posch M (2002) A global analysis of acidification and eutrophication of terrestrial ecosystems. Water Air Soil Pollut 141(1–4):349–382

Calvert JG, Lazrus A, Kok GL, Heikes BG, Walega JG, Lind J, Cantrell CA (1985) Chemical mechanisms of acid generation in the troposphere. Nature 317(6032):27

Cao C-x, Zhou Q, Han L-l, Zhang P, Jiang H-d (2010) Effects of simulated acid rain on oilseed rape (*Brassica napus*) physiological characteristics at flowering stage and yield. Ying Yong Sheng Tai Xue Bao 21(8):2057

Chen J, Li W, Gao F (2010) Biogeochemical effects of forest vegetation on acid precipitation-related water chemistry: a case study in southwest China. J Environ Monit 12(10):1799–1806

Chen J, Wang W-H, Liu T-W, Wu F-H, Zheng H-L (2013) Photosynthetic and antioxidant responses of *Liquidambar formosana* and *Schima superba* seedlings to sulfuric-rich and nitric-rich simulated acid rain. Plant Physiol Biochem 64:41–51

Cullis C, Hirschler M (1980) Atmospheric sulphur: natural and man-made sources. Atmos Environ (1967) 14(11):1263–1278

Debnath B, Hussain M, Irshad M, Mitra S, Li M, Liu S, Qiu D (2018a) Exogenous melatonin mitigates acid rain stress to tomato plants through modulation of leaf ultrastructure, photosynthesis and antioxidant potential. Molecules 23(2):388

Debnath B, Hussain M, Li M, Lu X, Sun Y, Qiu D (2018b) Exogenous melatonin improves fruit quality features, health promoting antioxidant compounds and yield traits in tomato fruits under acid rain stress. Molecules 23(8):1868

Debnath B, Irshad M, Mitra S, Li M, Liu S, Rizwan HM, Pan T, Qiu D (2018c) Acid rain deposition modulates photosynthesis, enzymatic and non-enzymatic antioxidant activities in tomato. Int J Environ Res 12(2):203–214

Di Toppi LS, Gabbrielli R (1999) Response to cadmium in higher plants. Environ Exp Bot 41(2):105–130

Dolatabadian A, Sanavy SAMM, Gholamhoseini M, Joghan AK, Majdi M, Kashkooli AB (2013) The role of calcium in improving photosynthesis and related physiological and biochemical attributes of spring wheat subjected to simulated acid rain. Physiol Mol Biol Plants 19(2):189–198

Dwivedi A, Tripathi B (2007) Pollution tolerance and distribution pattern of plants in surrounding area of coal-fired industries. J Environ Biol 28(2):257–263

Eguagie M, Aiwansoba R, Omofomwan K, Oyanoghafo O (2016) Impact of simulated acid rain on the growth, yield and plant component of *Abelmoschus caillei*. J Adv Biol Biotechnol 6(1):1–6

Evans LS (1984) Botanical aspects of acidic precipitation. Bot Rev 50(4):449–490

Fairfax J, Lepp N (1975) Effect of simulated 'acid rain'on cation loss from leaves. Nature 255(5506):324

Feng Z (2000a) Ecological effects and control strategies of acid deposition on ecosystems in China. Yunnan Environmental Science S 1.

Feng Z (2000b) Impacts and control strategies of acid deposition on terrestrial ecosystems in China. Eng Sci 9:5–11

Fontenele NMB, Otoch MLO, Gomes-Rochette NF, de Menezes Sobreira AC, Barreto AAGC, de Oliveira FDB, Costa JH, Borges SSS, do Nascimento RF, de Melo DF (2017) Effect of lead on physiological and antioxidant responses in two *Vigna unguiculata* cultivars differing in Pb-accumulation. Chemosphere 176:397–404

Hernández JA, Almansa MS (2002) Short-term effects of salt stress on antioxidant systems and leaf water relations of pea leaves. Physiol Plant 115(2):251–257

Hogan GD (1998) Effect of simulated acid rain on physiology, growth and foliar nutrient concentrations of sugar maple. Chemosphere 36(4–5):633–638

Hu H, Wang L, Liao C, Fan C, Zhou Q, Huang X (2014) Combined effects of lead and acid rain on photosynthesis in soybean seedlings. Biol Trace Elem Res 161(1):136–142

Kholová J, Sairam R, Meena R, Srivastava G (2009) Response of maize genotypes to salinity stress in relation to osmolytes and metal-ions contents, oxidative stress and antioxidant enzymes activity. Biol Plant 53(2):249–256

Kováčik J, Klejdus B, Bačkor M, Stork F, Hedbavny J (2011) Physiological responses of root-less epiphytic plants to acid rain. Ecotoxicology 20(2):348–357

Larssen T, Seip HM, Semb A, Mulder J, Muniz IP, Vogt RD, Lydersen E, Angell V, Tang D, Eilertsen O (1999) Acid deposition and its effects in China: an overview. Environ Sci Pol 2(1):9–24

Likens GE, Bormann FH (1974) Acid rain: a serious regional environmental problem. Science 184(4142):1176–1179

Liu J, Zhou G, Yang C, Ou Z, Peng C (2007) Responses of chlorophyll fluorescence and xanthophyll cycle in leaves of *Schima superba* Gardn. & Champ. and *Pinus massoniana* Lamb. to simulated acid rain at Dinghushan Biosphere Reserve, China. Acta Physiol Plant 29(1):33–38

Liu T-W, Fu B, Niu L, Chen J, Wang W-H, He J-X, Pei Z-M, Zheng H-L (2011a) Comparative proteomic analysis of proteins in response to simulated acid rain in Arabidopsis. J Proteome Res 10(5):2579–2589

Liu T-W, Wu F-H, Wang W-H, Chen J, Li Z-J, Dong X-J, Patton J, Pei Z-M, Zheng H-L (2011b) Effects of calcium on seed germination, seedling growth and photosynthesis of six forest tree species under simulated acid rain. Tree Physiol 31(4):402–413

Liu TW, Niu L, Fu B, Chen J, Wu FH, Chen J, Wang WH, Hu WJ, He JX, Zheng HL (2013) A transcriptomic study reveals differentially expressed genes and pathways respond to simulated acid rain in *Arabidopsis thaliana*. Genome 56(1):49–60

Liu X, Fu Z, Zhang B, Zhai L, Meng M, Lin J, Zhuang J, Wang GG, Zhang J (2018) Effects of sulfuric, nitric, and mixed acid rain on Chinese fir sapling growth in Southern China. Ecotoxicol Environ Saf 160:154–161

Madiha Y, Khan AA, Darma ZU (2015) Effect of acid rain on growth of Papaya (*Carica papaya*) and Castor (*Ricinus communis*) plants. e-JST 9(1): 43–47.

Odiyi B, Bamidele J (2013) Effects of simulated acid rain on growth and yield of Cassava *Manihot esculenta* (Crantz). J Agric Sci 6(1):96

Odiyi BO, Eniola AO (2015) The effect of simulated acid rain on plant growth component of cowpea (*Vigna unguiculata*) L. Walps Jordan J Biol Sci 8(1):51–54

Pan T, Li Y, Ma C, Qiu D (2015) Calcium affecting protein expression in longan under simulated acid rain stress. Environ Sci Pollut Res Int 22(16):12215–12223

Polishchuk AV, Vodka MV, Belyavskaya NA, Khomochkin AP, Zolotareva EK (2016) The effect of acid rain on ultrastructure and functional parameters of photosynthetic apparatus in pea leaves. Cell Tissue Biol 10(3):250–257

Qin L-Q, Li L, Bi C, Zhang Y-L, Wan S-B, Meng J-J, Meng Q-W, Li X-G (2011) Damaging mechanisms of chilling-and salt stress to *Arachis hypogaea* L. leaves. Photosynthetica 49(1):37–42

Ren X, Zhu J, Liu H, Xu X, Liang C (2018) Response of antioxidative system in rice (*Oryza sativa*) leaves to simulated acid rain stress. Ecotoxicol Environ Saf 148:851–856

Ruiz-Lozano JM (2003) Arbuscular mycorrhizal symbiosis and alleviation of osmotic stress. New perspectives for molecular studies. Mycorrhiza 13(6):309–317

Schutzendubel A, Polle A (2002) Plant responses to abiotic stresses: heavy metal-induced oxidative stress and protection by mycorrhization. J Exp Bot 53(372):1351–1365

Seinfeld JH, Pandis SN (2012) Atmospheric chemistry and physics: from air pollution to climate change. Wiley, Hoboken

Seto KC, Shepherd JM (2009) Global urban land-use trends and climate impacts. Curr Opin Environ Sustain 1:89–95

Shan Y (1998) Effects of simulated acid rain on Pinus densiflora: inhibition of net photosynthesis by the pheophytization of chlorophyll. Water Air Soil Pollut 103(1):121–127

Shaukat SS, Khan MA (2008) Growth and physiological responses of tomato (*Lycopersicon esculentum* Mill.) to simulated acid rain. Pak J Bot 40(6):2427–2435

Sheng M, Tang M, Chen H, Yang B, Zhang F, Huang Y (2008) Influence of arbuscular mycorrhizae on photosynthesis and water status of maize plants under salt stress. Mycorrhiza 18(6–7):287–296

Silva L, Azevedo A, Silva E, Oliva M (2005) Effects of simulated acid rain on the growth of five Brazilian tree species and anatomy of the most sensitive species (*Joannesia princeps*). Aust J Bot 53:789–796

Singh A, Agrawal M (2007) Acid rain and its ecological consequences. J Environ Biol 29(1):15

Wang X, Liu Z, Niu L, Fu B (2013) Long-term effects of simulated acid rain stress on a staple forest plant, *Pinus massoniana* Lamb: a proteomic analysis. Trees 27(1):297–309

Wen K, Liang C, Wang L, Hu G, Zhou Q (2011) Combined effects of lanthanumion and acid rain on growth, photosynthesis and chloroplast ultrastructure in soybean seedlings. Chemosphere 84(5):601–608

Xalxo R, Sahu K (2017) Acid rain-induced oxidative stress regulated metabolic interventions and their amelioration mechanisms in plants. Biologia 72:1382

Yu JQ, Ye SF, Huang LF (2002) Effects of simulated acid precipitation on photosynthesis, chlorophyll fluorescence, and antioxidative enzymes in *Cucumis sativus* L. Photosynthetica 40(3):331–335

Zhang X, Du Y, Wang L, Zhou Q, Huang X, Sun Z (2015) Combined effects of lanthanum (III) and acid rain on antioxidant enzyme system in soybean roots. PLoS One 10(7):e0134546

Part II
Sources of Contaminants and Their Impacts on Agriculture

Heavy Metals: Source, Toxicity Mechanisms, Health Effects, Nanotoxicology and Their Bioremediation

Ayesha Siddiqa and Muhammad Faisal

Contents

1 Introduction

Heavy metals contribute to the large volume of constituents of the Earth's crust. They along with metalloids are derived from lithogenic sources naturally. They always stayed undisturbed until the unmethodical human activities started disturbing the natural balance of geochemical cycles. Modern industrialization has caused the accumulation of heavy metals to the toxicity level. Anthropogenic sources of heavy metals are industrial waste, leaded gasoline and paints, disposal of high metal waste, mine tailing, petrochemical spillage, waste water irrigation, unregulated use of pesticides, coal combustion residues, tanning of leather and atmospheric deposition.

A heavy metal is defined as 'any chemical element that possesses comparatively high density and due to which is potentially toxic to the all biological systems'.

Almost 90 elements fall under the category of heavy metals and they all have highest bioavailability and deposition equally in land and aquatic biomes (Mi et al. 2019; Nabuyanda et al. 2019; Redan et al. 2019); nevertheless, they have lesser bioavailability in atmosphere in the form of aerosols (Wan et al. 2016). Toxicity of

A. Siddiqa · M. Faisal (✉)
Department of Microbiology and Molecular Genetics, University of the Punjab, Lahore, Pakistan
e-mail: faisal.mmg@pu.edu.pk

M. Naeem et al. (eds.), *Contaminants in Agriculture*,
https://doi.org/10.1007/978-3-030-41552-5_6

heavy metals varies from one element to another (Zhang et al. 2019a; Liu et al. 2017). It is also dependent on several parameters: their bioavailable forms (Honda et al. 2015), their concentration in environment (Hu and Long 2019), physical condition of the environment and their available chemical forms (Men et al. 2018; Acat et al. 2017; Cao et al. 2018).

One important thing that cannot be ignored is that most of these notorious heavy metals are also essential nutrients, i.e. Se, Mg, Cu, Co and Zn (Doron et al. 2018; Knop et al. 2017). Copper and zinc are cofactors of enzymes in important life processes (Knop et al. 2017; Fasken et al. 2019). Some of them act as the prosthetic group in metalloprotein (Chaudhuri et al. 2018). Many of these elements take part in redox bioreactions (Su et al. 2019), impart structural function in macromolecules (DNA) metabolism and help in electron transport chain (ETC) (Sankaralingam et al. 2018). Some of them are the part of amino acids like Se (Meng et al. 2019; Addinsall et al. 2018), but these benefits can become their toxicity potential and can thus make heavy metals major environmental contaminants.

Metals are classified according to their characteristics i.e. coordination chemistry, toxicity and essentiality. Thus, class B heavy metals contain all toxic elements, for instance, Ni, Ag, Pb and Hg (Ahmad and Mahmood 2019; Alho et al. 2019). Heavy metals are considered as toxic because they are persistent in the environment, and their bioaccumulation and bioavailability are very high. They are not transformed easily into lesser toxic and non-reactive forms (Ahamed et al. 2019). These class B heavy metals assimilate into biological systems, for example, roots of plants (rhizobium deposition) (Hammami et al. 2016) and leaves of plants (foliar deposition) (Goudarzi et al. 2018; Franca et al. 2017; Fumagalli et al. 2010). Heavy metals enter the human body either by inhalation of ingestion (Honda et al. 2015; Guo et al. 2017, 2018; Huang et al. 2018). Whatever the route of entrance may be, heavy metal toxicity is always due to the interaction of their reactive forms at molecular and cellular levels (Jadoon et al. 2018; Chen et al. 2019; Queiroz et al. 2018; Zong et al. 2018). They may disrupt the protein structure by binding to sulphydryl groups of protein (Paschoalini et al. 2019), or they may cause oxidative stress in the cellular environment by generating reactive oxygen species (Nishio et al. 2019).

2 Occurrence of Heavy Metals in Biosphere

Heavy metals when found in excessive quantities can cause adverse effects on human health. Agricultural drainage water, mining and industrial effluents in general can cause increased level of toxic heavy metals in landfills and aquatic bodies (Yin et al. 2018). These metals exist in elemental forms or in conjugation with other elements (compounds), i.e. sodium selenite, ferrous Oxide (Nour-Eldein et al. 2018). Depending on the types of heavy metals and their interaction with biomolecules, their toxicity may differ, i.e. Se in its oxyanions (Selenate and Selenite) is highly reactive and toxic (Nour-Eldein et al. 2018); however, in its elemental form, it is non-reactive and consequently non-toxic. Similarly, Cris is lethal in chromates and trioxide form but trivalent chromium is an essential dietary mineral in low quantities (Zhang et al. 2019b).

Accumulation of excessive metals and metalloids in landfills may lead to damaging ecological changes. Out of all their bio-geochemical changes, the most adverse effect is the bioaccumulation and absorption of toxic heavy metals in biological bodies i.e. organisms (Ahamed et al. 2019; Jadoon et al. 2018). The point of entry of these heavy metals into food web is the 'producers'. There are a number of reported cases that determined several small plants, which are hyperaccumulators and can store high concentrations of metals into them (Lin et al. 2017; Akram et al. 2015; Xia et al. 2018). Thus, these plants open the portal for toxic heavy metals in the food web. The direct consumption of these plants will lead to harmful effects on their consumers (animals and humans) (Alho et al. 2019; Chen et al. 2019; Zhang et al. 2019b; Jaiswal et al. 2018; Mwakalapa et al. 2019). The areas of high metal contamination may cause deleterious effects on human health even during harvesting (Pokorska-Niewiada et al. 2018). However, this happens only in highly contaminated soils (Khan et al. 2018).

One point is of utter importance that humans are also prone to the metal contamination from soil particles lingering on the plants or the parts of plants. This happens in leafy vegetables i.e. spinach, lettuce (Franca et al. 2017; Talukdar and Talukdar 2013; Boostani et al. 2019; Younis et al. 2016; Wang et al. 2018a) and root crops i.e. potatoes, carrots and the parts of the plants that have a direct contact with soil (Peng et al. 2018). This is because it becomes difficult to wash off all soil particles while cutting and cooking.

Humans are at a greater risk of metal toxicity because being omnivores, they have diverse food sources. If plants and animals bioaccumulate, the toxic metals they may or may not biotransform or detoxify the toxic forms of heavy metals and when consumed by humans as their daily food source, they can cause metal intolerance (Boostani et al. 2019). Animals do not secrete metal; instead, they store them in their bodies, which is a natural way. However, humans, upon meat consumption, become exposed to these metals (Yin et al. 2018). The following are the examples of metal bioaccumulation in plants and animals. Shaheen et al. (2016) reported higher levels of As, Cd, Pb, Cr, Mn, Ni, Cu and Zn in Banana (*Musa paradisiaca*), Mango (*Mangifera indica*), Brinjal (*Solanum melongena*), Carrot (*Daucus carota*), Jackfruit (*Artocarpus heterophyllus*), Bean (*Dolichos lablab*), Green Chilies (*Capsicum frutescens*), Potato (*Solanum tuberosum*), Onion (*Allium cepa*) and Tomato (*Lycopersicon esculentum*) (Shaheen et al. 2016); moreover, many others also reported high concentrations of heavy metals in crops (Onakpa et al. 2018; Sharma et al. 2018). Another study in North East of Varanasi estimated the accumulation of Cd, Pb, Cu, Zn, Cr and Ni in all widely planted vegetables and cereals and milk samples (Singh et al. 2010). It was estimated in various studies that leafy vegetables had a higher concentration of metals than root crops and this might be due to atmospheric pollution and foliar deposition of metals. Several studies have shown that the crops produced in the soil, polluted with irrigation water, showed higher concentrations of Cu, Cd, Cr, Pb, Ni, Zn, etc. (Genthe et al. 2018). A study conducted in different districts of Tianjin, China, by Wang et al. (2005) showed high deposition of Cd, Zn, Pb, Ni, Cu and Cr in locally produced vegetables and fish (Wang et al. 2005). The possible cause of these high concentrations was assumed to be the use of sludge and industrial wastewater in agricultural purposes and irrigated wastewater consumption in fisheries. The deposition of heavy metals i.e. Pb, Hg and Cd has been reported and analysed in sea foods in various studies. However, a comparative study was conducted among 35 country samples (Petroczi and Naughton 2009) that brought the bioaccumulation of heavy metals in sea foods in the calculated data. Meanwhile, high concentrations of Al, Zn, Cd, Cr, As, Cu, Mn, Pb, Se and Hg were reported in the food samples (cabbage, beef, bean and fish) taken from two cities of Africa (Nuapia et al. 2018). Two species of fish taken from Lake Taihu, China, showed metal deposition (Cr, Pb, Cd) in several organs and bioaccessibility of these to their consumers (Rajeshkumar and Li 2018).

3 Toxicity of Heavy Metals

Although most of the metals found in soil have biological importance and result from bio-geochemical cycle, their unnecessary abundance in the soil and water bodies has deleterious effects on the animals and humans. These metals, in their reactive forms, may chemically coordinate with the host macromolecules at the cellular levels and cause severe cellular damage (Rajeshkumar and Li 2018). Their redox properties give them a way out from host control mechanisms i.e. transport, binding to cellular receptors, intra-cellular compartmentalization and homeostasis (Zong et al. 2018; Polykretis et al. 2019; Ben Massoud et al. 2018). The resultant damage causes the malfunctioning of cells, which may lead to the irreversible organ damage (Lash 2019; Khushboo et al. 2018). Table 1 shows an overview of sources, ways of exposures and toxicity of heavy metals at cellular and molecular levels. Till now, most of the metal oxidative damages reported are biomacromolecule oxidations, i.e. DNA damage and protein disruption (Jain et al. 2018). Here is the review of the extent of toxicity mechanisms of few metals and their effects on humans, plants and animals.

3.1 Lead

Lead is known to be toxic to human health for long and reportedly has caused excessive damage to the environment and its indigenous species. The main sources of lead deposition in landfills and aquatic environment are fertilizers (Pourrut et al. 2011), pesticides, smoke from factories (Graeme and Pollack 1998), smelting of ores, metal plating and battery-making industries and smoke from gasoline industries and vehicles. However, the exposure to lead in human population is either by drinking polluted water (Pourrut et al. 2011) or by consuming contaminated food

Table 1 A brief toxicity review of non-essential heavy metals, their distribution and sources or contamination

Sr. no.	Non-essential heavy metals	Sources of deposition in biosphere	Ways of chronic exposure	Toxicity at molecular and structural levels	Reference
1.	Lead (Pb)	Fertilizers, smelting of ores, battery making, gasoline industries, pesticides, metal plating	By drinking contaminated water, lead-containing aerosols	Damage to the structural integrity of bio-molecules;disrupt the balance of ROS and antioxidants; lipid peroxidation due to the available free radicals;interruption in apoptosis, protein folding and maturation, cell adhesion and cell recognition	(Sai Siva Ram et al. (2018), Li et al. (2016b), Butt et al. (2018), Shi et al. (2019), Muszynska et al. (2018), Minigalieva et al. (2017)
2.	Arsenic (As)	Arsenical pesticides, natural deposition by weathering and erosion of arsenic-containing rocks	by contaminated water, dusty air, due to ecological and anthropogenic deposition	Oxidative stress leading to cellular damage; molecular damage (protein instability, DNA damage, lipid peroxidation) Keratosis; hypertension; non-cancerous hypo- and hyper-pigmentation	Hossain et al. (2018), Kumar Dubey et al. (2018), Rana et al. (2018), Saifullah et al. (2018), Vineetha et al. (2018).
3.	Chromium (Cr)	Electroplating, metallurgy, agricultural wastes, pigment factories, tanning and paper production	By contaminated water and crops	Uncontrolled generation of ROS (superoxide ions, hydroxyl radicals and hydrogen peroxide) leading to molecular damage and carcinogenesis; unregulated production of antioxidants leading to imbalance	Antoniadis et al. (2018), Hao et al. (2017), Junaid et al. (2016), Sihag et al. (2019), Zhong et al. (2017), Stambulska et al. (2018).
4.	Mercury (Hg)	Industrial effluents, mining, agricultural waste, municipal and household waste water and incineration; smelting of Zinc and copper ores, bio-geochemical cycles and microbial methylation, volcanic eruption and mine tailings	By contaminated drinking water, inhalation of polluted air, consuming contaminated sea foods	Increased ROS generation causing molecular damage; non-functional DNA leading to reduced protein generation, reduced antioxidant production; retarded cell adhesion.	Ahmad and Mahmood (2019), Cabrita et al. (2019), Rana et al. (2018), Aslanturk et al. (2014), Leon-Canedo et al. (2019).

(continued)

Table 1 (continued)

Sr. no.	Non-essential heavy metals	Sources of deposition in biosphere	Ways of chronic exposure	Toxicity at molecular and structural levels	Reference
5.	Cadmium (Cd)	Ore outcropping, metal smelting, household waste and sewage sludge, agricultural waste and pesticides.	By dinking contaminated water and consuming contaminated food, i.e. crops, sea food	Abnormalities in calcium, phosphorous and bone metabolism; Iincreased ROS production leading to apoptosis; acute toxicity leads to multiple renal failure	Irvine et al. (2016), Venter et al. (2017), Bakti et al. (2018), Amna et al. (2015)
6.	Silver (Ag)	Industrial effluents from food manufacturing and packaging, jewellery, pharmaceuticals' production, ore leaching and smelting	Ingestion or inhalation of silver nanoparticles	Silver nanoparticles induce ROS generation, imbalance of cytokines' production; molecular damage, i.e. DNA damage; structural damage, i.e. mitochondrial damage; induces necrosis; discoloration of skin; disrupted liver enzymes' activity, interrupted neurotransmitter levels, deformed organs	Hadrup et al. (2018), Tao et al. (2016), Thummabancha et al. (2016), Lee et al. (2018b), Konop et al. (2019)
7.	Aluminium (Al)	Natural deposition, mining, acidification of the soil due to ore leaching, acid rain	By contaminated soil, water and air	Inhibits important enzymes, i.e. phosphodiesterase, hexokinases and phosphatases, thus interrupting metabolic function; causes instability of plasma membrane, thus interfering several cellular functions, i.e. cellular interaction and signalling, cellular communication and secretory functions	Maleki et al. (2019), Gouda et al. (2018), Kura et al. (2015)

mostly by plants and aquatic food (Zhang et al. 2019c). Unlike other heavy metals, e.g. Zn, Cu, Mg and Se, lead does not play any biologically important role (Rana et al. 2018). Nevertheless, many toxicological aspects of lead have been unleashed during the decades of research. In some cases, lead exposure resulted in less biomass and reduced the quality of crop by changing its components (Malar et al. 2016). In plants, lead has reportedly increased the ROS release (Malar et al. 2016) that usually interferes with the biochemical reactions involved in lipid peroxidation (Ashraf and Tang 2017), which ultimately leads to the disruption of photosynthetic process by damaging the chlorophyll (Malar et al. 2016; Ommati et al. 2019). In other words, through the disruption of bio-chemical processes, lead accumulation leads to the retarded growth of plants. With the disruption of intracellular compartments, lead is also known to cause the ion instability in cells that also contributes to the defaulted photosynthetic machinery (Kupper 2017).

3.1.1 Toxicity Mechanism of Lead

Lead toxicity to the biological systems works in two ways. First, lead upon contact imbalances the reduced and oxidized forms of glutathione reductase, consequently increases the level of ROS species in cells by decreasing the antioxidants (Dewanjee et al. 2015). Second, lead causes the lipid peroxidation due to the production of high levels of free radicals (Malar et al. 2016). At higher concentration, lead can cause structural damage to the cells as the structural integrity of the bio-macromolecules is not maintained (Nariya et al. 2018). Another way of ionic imbalance that lead can do is to replace the important bivalents i.e. Fe^{2+}, Mg^{2+} and Ca^{2+} and monovalent Na^{+} in essential biochemical reactions. Through this, lead causes the disruption of cellular mechanisms like apoptosis, ionic transportation, maturation, protein folding, intra- and inter-cellular interaction, cellular recognition and cell adhesion, enzymatic reactions and an interplay of neurotransmitters (Ashraf and Tang 2017; Tian et al. 2014; Agnihotri et al. 2018; Okesola et al. 2019; Toz and Deger 2018; Aouini et al. 2018; Liu et al. 2018).

3.2 *Arsenic*

Arsenic is a semi-metalloid which is of great health concern due to its ecological and anthropogenic deposition (Debure et al. 2018). Despite its utter toxicity to all life forms and no biological importance (Sai Siva Ram et al. 2018), human activities have greatly increased its environmental contamination (Zwolak 2019), for instance the contamination of drinking water with arsenical pesticides (Bera et al. 2010). Moreover, it already occurs naturally and has continuous deposition (Norwood et al. 2007). Arsenic is one of the many heavy metals whose inorganic forms are extremely toxic to human health; its inorganic forms are arsenate and arsenite, which are dispensed in the ecosystem by industrial sources and natural resources (Saifullah et al. 2018).

3.2.1 Mechanism of Arsenic Toxicity

Many Algae, Fungi and Bacterial species, which are resistant to arsenic, biotransform the toxic organic form of arsenic into inorganic methylated non-toxic form enzymatically (Mateos et al. 2017). The inorganic methylated forms are monomethylarsonic acid (MMA) and dimethylarsinic acid (DMA V) (Bilinsky et al. 2019) and these are the biomarkers of arsenic exposure and chronic toxicity. Hence, this biomethylation is the detoxification method that many microbes employ (Pantoja Munoz et al. 2016). However, in human this biomethylation is done at the cellular level and monomethylarsonic acid (MMA V) and dimethylarsinic acid (DMA V) are released through urine, but MMA (III) is not excreted and stays inside the cells (Xue et al. 2017). This MMA (III) is the intermediate of biomethylation and potentially toxic to cells when compared to other arsenicals (Wang et al. 2018b). MMA (III) is reported to be the cause of cancer, upon arsenic exposure. During this biomethylation and arsenic V to III bioconversion, the reactive oxygen species are produced, these both mechanisms lead to the high oxidative stress leading to cellular damage (Wang et al. 2007; Orihuela et al. 2013). This oxidative stress interferes with biomolecules, for instance, lipid peroxidation, protein instability and DNA damage (Urrialde et al. 2017; Acharyya et al. 2015). Arsenic (III) formed also interacts with phytochelatins, glutathione and sulphydryl groups of protein, which results in increasing the ROS release (Maheshwari et al. 2018; Xu et al. 2019).

In plants, the uptake and biosorption of arsenic is dependent on the species of arsenic; for instance, arsenic V is taken up by the Pi transporters and inorganic arsenic III is taken up by the protein channels of three aquaporin subfamilies i.e. membrane protein (PIP), nodulin 26-like intrinsic protein (NIP) (Ji et al. 2017) and tonoplast intrinsic protein (Yang et al. 2015; Shen et al. 2014; Rosen 1999). Upon exposure, plants tend to promote the synthesis of antioxidants (Thangapandiyan et al. 2019) i.e. catalase, superoxide dismutase SOD, glutathione reductase, ascorbate peroxidase (APX), glutathione peroxidase and glutathione S-transferase (Raza et al. 2018; Praveen and Gupta 2018). Moreover, plants start producing γ-Glu-Cys-Gly-tripeptide glutathione upon exposure, which helps in tolerating the high arsenicals by producing phytochelatins (Shri et al. 2019; Das et al. 2018; Kumari et al. 2018; Hossain et al. 2018).

3.3 Chromium

Chromium is one of the most abundant heavy metals on the Earth's crust and comes at number seven with relevance to its occurrence, abundance, and it exists in seven interchangeable oxidation states (Cr^{+2} to Cr^{+6}) (Debure et al. 2018), (Onakpa et al. 2018; Lecoanet et al. 2019; Jaiswar et al. 2018). It is an essential nutrient in smaller quantities for the lipid and carbohydrate metabolism (Maret 2019; Hoffman et al. 2014), but its toxicity is dependent on the oxidation states it exists (Levina and Lay 2008; Yadav and Khandegar 2019). Among all these oxidation states, trivalent Chromium Cr^{+3} and hexavalent Chromium Cr^{+6} are toxic to all biological systems

(Masood et al. 2019; Filice et al. 2019; Zhan et al. 2019). Natural sources of chromium deposition are coal and oil burning (Hua et al. 2018), pigments (Lee et al. 2018a), chromium steel (Kamerud et al. 2013; OuYang et al. 2019), catalyst (Fernandez et al. 2018; Marinho et al. 2019), mining and oil drilling and tanneries (Fontaine et al. 2019; Miao et al. 2019). Chromium in its oxidation state III is not harmful when present in the environment; nevertheless, in its oxidative form IV, it is highly toxic to humans; this is possibly due to its high mobility and solubility when present in the oxidized form IV (Tamindzija et al. 2019). Chromium III does not require any transport channels or proteins and simply enters the cell through diffusion; thus, it poses a potential threat to all biological systems. However, chromium VI enters the cells more readily into the Cr III and considered to be number one carcinogen (Gang et al. 2019; Zeng et al. 2014).

Environmental pollution with hexavalent chromium is of great concern and there is an outrage to lower the use of chromium in industries or to find a substitute. The biggest source of chromium pollution is anthropogenic activities, i.e. agricultural wastes, dyes and pigment factories, tanning, wood preservation, electroplating, metallurgy, paper production and chemical industries (Tseng et al. 2019; Slejko et al. 2019; Hausladen et al. 2018; Lacerda et al. 2019). These activities have led to the extremely high quantities of the oxidized chromium in environment (Hausladen et al. 2018). However, a lot of research is required to understand the phytotoxicity of chromium species; however, continuous accumulation of chromium affects crop production and its quality (Onakpa et al. 2018; Sharma et al. 2018), meanwhile posing threats to the human lives. Chromium in excess retards the plant growth (chlorosis, lesser biomass, retarded growth and seed germination) (Amin et al. 2019; Stambulska et al. 2018), and upon consumption, it enters the food chain.

3.3.1 Mechanism of Toxicity

In the latest studies, it is reported that higher quantities (100 mg/kg) of hexavalent chromium retarded the shoot and root growth in lemon grass (Patra et al. 2018). It greatly affected the nutrient uptake efficiency of roots, thus eliciting poor growth. Chromium toxicity also decreased the root numbers and root hair formation. However, quite interestingly, lower quantities of chromium promoted the growth of the plant possibly due to the improvement in chlorophyll ultrastructure (Patra et al. 2018). Nevertheless, with an increasing concentration, the chlorophyll content decreases due to the inhibition of Co_2 absorption (Stambulska et al. 2018). High concentrations of chromium also increased the lignification and changed the morphology by disrupting the endodermis and hypodermis (Stambulska et al. 2018). When the plants grow in Cr stress, they tend to accumulate the proline in cells, to maintain the osmotic conditions; this is their tolerance mechanism (Kundu et al. 2018). Moreover, Cr toxicity results in reduced biomass due to the degradation of proteins (Cortes-Eslava et al. 2018), delayed seed germination and sometimes seed death (Stambulska et al. 2018). This adverse effect on seed health is possibly due to the inhibition of amylases and production of proteases (Stambulska et al. 2018).

Hexavalent chromium is absorbed by the plants through sulphate and phosphate protein channels; it is very oxidizing and is converted to pentavalent or tetravalent chromium; these chromium states have short lives and are ultimately converted to trivalent chromium. Trivalent Cr is less soluble and consequently less toxic than hexavalent Cr (Fuentes-Gandara et al. 2018). These reduction reactions result in the production of ROS i.e. superoxide ions, hydroxyl radicals and hydrogen peroxide and cause protein and DNA damage (Ni et al. 2014). These ROS induce the overproduction of antioxidant enzymes, e.g. catalase, superoxide dismutase and peroxide, which imbalance the growth mechanisms.

In humans, chromium is also taken up as hexavalent VI through respiratory and digestive systems more rapidly than trivalent III (Fuentes-Gandara et al. 2018). Human exposure to chromate is usually occupational exposure, i.e. in cement, rubber, pigment, paint, paper, metal plating and tanning industries (Jaiswar et al. 2018). Wounds when come in contact with chromium can transform into ulcers and take months to heal. However, this is not the sole exposure to chromium; our drinking water is also contaminated with Cr (hexavalent) already crossing the critical values. In vivo studies in rodents have reported the accumulation of heavy metals including Cr in placenta (Banu et al. 2018). Chromium high concentrations inhibit the glutathione reductase in red blood cells and reduce the conversion to methemoglobin to haemoglobin. However, the ultimate effect of Cr on human is genotoxicity or mutagenesis (Marat et al. 2018), due to the formation of ROS during the reduction of hexavalent to trivalent, as discussed above.

3.4 Mercury

Mercury is a shiny, liquid, odourless and naturally occurring metal that upon heating is converted into an odourless and colourless gas (Bailon et al. 2018). Mercury is highly toxic to all biological forms as being extremely bioaccumulative (Hashemi and Tabibian 2018), and thus exists only in smaller quantities. However, water bodies, especially marine water, are contaminated by mercury through anthropogenic activities (Schneider et al. 2019; Budnik and Casteleyn 2019). Some major sources of mercury contamination are industrial effluents, agricultural waste, mining, municipal and household waste water and incineration (Budnik and Casteleyn 2019; Janiga and Haas 2019; Palacios-Torres et al. 2018). Due to the increased mercury pollution and its extreme toxicity, marine water is highly polluted and aquatic life is in grave danger. Nevertheless, its toxicity depends on the type of its oxidation state (Rua-Ibarz et al. 2019). Mercury exists in three forms: inorganic salts, metallic elements and compounds and organic compounds, and each has its own severity of toxicity and bioavailability (Li et al. 2016a). Pollution of the environment with mercury is primarily caused by industrial and domestic use, i.e. mercury arc lamps, fluorescent lamps, barometers, hydrometers, thermometers. It is widely used industries as a catalyst, as an amalgam in dental clinics and batteries making, in electrical appliances and relays, in coal-powered plants, in chlor-alkali production, in PVC production, in paper and pulp making and in mining (Emmerton et al. 2018; Arvay et al. 2017). Mercury has been reported to be used in old medicinal remedy Zuotai, an old Tibetan regime (Li et al. 2018).

Mercury exists in the Earth's crust as a result of two procedures; first due to geochemical reactions and second is the result of microbial methylation and demethylation biochemical reactions (Wang et al. 2018c; Mondal et al. 2018). So, there is an intricate balance between them. Smelting of zinc and copper ores also release mercury in the environment, volcanic eruption and mine tailings are reported to add mercury on the crust (Liao et al. 2019; Kerfoot et al. 2018).

3.4.1 Toxicity Mechanism of Mercury

The major exposure of toxic mercury to humans is through aquatic animal i.e. by eating contaminated sea food. Aquatic plants can accumulate inorganic mercury (Beauvais-Fluck et al. 2018) (mercuric chlorides and methyl, butyl mercury compounds) in roots (Ritger et al. 2018; Hang et al. 2018), which is more toxic than organic mercuric compounds. Earlier research showed noticeable growth retarding effects of mercury on various plants i.e. (*Pistia stratiotes, Elodea Canadensis, Lemna minor*), *Pistia stratiotes* and Marine diatom, e.g. *Chaetoceros costatum* (Leon-Canedo et al. 2019; Li et al. 2017). In the exposed plants, they detected the reduced protein content, degraded chlorophyll, lower level of RNA, denatured and less functional DNA, lower levels of functionally active antioxidant enzymes and proteases activity (Lima et al. 2019a). Mercury accumulation in aquatic plants is associated with the mercury concentrations provided and mercury deposition is linked to the accumulation of mercury in the proximity of cell surface or in cell wall (Lima et al. 2019b; Cabrita et al. 2019).

Microorganisms show less adversity when it comes to mercury toxicity as they possess various resistance mechanisms as they can bind mercury to their cell surfaces (Campos et al. 2018), they contain enzymatic polymers for chelation i.e. metallothionein, many microbes have efflux pumps to expel the mercuric ions out of the cell; microbes can transform toxic forms of mercury into less damaging form enzymatically (Jan et al. 2012). Many of them form mercuric complexes with sulphides and oxides and precipitate them on the cell wall to lower the toxicity and lastly microbes can methylate it and transport the mercury out of cell through the cell membrane. Some plants are also tolerant to the toxic effect of mercuric compounds by adopting several tolerance mechanisms, i.e. increased glutathione production (Kim et al. 2017; Cozzolino et al. 2016).

4 Role of Nanoparticles in Heavy Metal Accumulation and Toxicity

With the advent of nanobiotechnology and its worldwide possible application in various aspects of life, their increased use leads to their disposal in landfills and water bodies. As nanoparticles are readily available to the biological systems due to their small size, they interfere with the normal physiological functioning of cells (Liu et al. 2019; Fajardo et al. 2019; Feng et al. 2017). A study reported the lung injuries and inflammation resulted by the metal nanoparticles oxidative stress. This was due to the inhalation of metal oxide nanoparticulates in an industry dealing with nanoparticle production (Assadian et al. 2018). A group of scientists studied the relationship of nanoparticles of twenty-four metal oxides with their cellular toxicity in vivo (Dankers et al. 2018). Moreover, other studies also reported the cytotoxic effects of magnesium oxide, cobalt oxide and copper oxide on bronchial epithelial cell lines, which led to the secretion of cytokines (Mangalampalli et al. 2017, 2018; Chen et al. 2018; Mansano et al. 2018). Dispensing the nanoparticles of CuO in rhizosphere may lead to the deteriorated plant roots (Gao et al. 2018). Nanoparticles are being used in developing and improving existing drug delivery methods and to adopt anticancer strategies as they work by inducing the Trojan-horse-type mechanism at cellular level (Li et al. 2019; Zhang et al. 2018). This mechanism elicits the induced apoptosis, autophagy or necrosis (Souza et al. 2018). However, this mode of action can become deleterious when healthy cells remain in contact with nanoparticles for a prolonged time (Souza et al. 2018). The increased use of nanoparticles of heavy metals i.e. Silver may cause hindrance to biotic environment by interfering at molecular level. A study showed the increased toxicity of Silver nanoparticles to Zebrafish embryos even upon storage in stabilized conditions. This happened due to the possible release of silver ions (Lee et al. 2018b). Another study compared the inhibitory effects of $AgNo_3$ and silver nanoparticles on photosynthesis in *Chlamydomonas reinhardtii* and reported that due to the availability of Silver ions and binding to cysteine, these nanoparticles abolished photosynthesis in Algae (Navarro et al. 2008). Bondarenko et al. (2013) estimated the toxic effects of copper oxide, silver and zinc oxide nanoparticles, which were released from household effluents, on crustaceans, fish algae, bacteria, yeast, protozoa, nematodes and mammalian cell lines considering them as the non-targeted organisms. They all were reported to be biocidal to natural biome (Bondarenko et al. 2013). We can estimate the toxicity of nanoparticles by considering their solubility, availability and interaction with macromolecules (Boran and Saffak 2018), for instance, zinc oxide nanoparticles and silver nanoparticles are more toxic as compared to copper oxide nanoparticles as they are solubilized well and readily available. However, silver nanoparticles are more toxic than zinc oxides due to their interaction with ligands (Bondarenko et al. 2013; Kwak et al. 2016). Increased applications of gold nanoparticles in medicine have raised a concern regarding their biodistribution and absorption, pharmacokinetics, nanotoxicology/genotoxicity and elimination of gold NP from the body. Few studies reported the toxicity of gold ultra-small (1.5 nm) to the spleen and liver. Moreover, the stabilizer used in the nanotubes of gold is CTAB, which is toxic to biological systems (Khlebtsov and Dykman 2011). However, the nanosize of these gold NPs and their toxicity has a critical relationship (Glazer et al. 2011), which should be evaluated since the beginning of their bulk uses, everyone is doing independent research with no complete submission of facts and reported data vary tremendously. Copper oxide nanoparticles have been used in medicine as antimicrobial agents for the infections of *E. coli* and MRSA and in technology as catalysts, sensors, ceramic pigments and semiconductors (Assadian et al. 2019; Bugata et al. 2019). Their toxicity to human lymphocytes was estimated by Assadian et al. (2018) and reported to be cytotoxic. They reduced the number of viable cells in the blood in a concentration-dependent manner. They increased the production of reactive oxygen species and caused the lysosomal and mitochondrial leaking and oxidative stress (Assadian et al. 2018, 2019).

It is advised to assure the implementation of general guidelines for the usage of nanoparticles and research of nanotoxicology on non-targeted species. We can assume by studying that bacteria are the most targeted organisms, but as compared to the range of targeted to non-targeted organisms that have been tested, they are proven to be the least sensitive. Moreover, the use of NPs in medicine should be tested before implementation and their biodistribution, bioavailability and biocompatibility should be estimated.

References

Acat M, Aydemir Y, Yazici O et al (2017) Comparison of the fractional exhaled nitric oxide levels in adolescents at three schools located three different distances from a large steel mill. Can Respir J 2017:6231309. https://doi.org/10.1155/2017/6231309

Acharyya N, Deb B, Chattopadhyay S, Maiti S (2015) Arsenic-induced antioxidant depletion, oxidative DNA breakage, and tissue damages are prevented by the combined action of folate and vitamin B12. Biol Trace Elem Res 168(1):122–132. https://doi.org/10.1007/s12011-015-0324-5

Addinsall AB, Wright CR, Shaw CS et al (2018) Deficiency of selenoprotein S, an endoplasmic reticulum resident oxidoreductase, impairs the contractile function of fast-twitch hindlimb muscles. Am J Physiol Regul Integr Comp Physiol 315(2):R380–R396. https://doi.org/10.1152/ajpregu.00244.2017

Agnihotri A, Gupta P, Dwivedi A, Seth CS (2018) Counteractive mechanism (s) of salicylic acid in response to lead toxicity in Brassica juncea (L.) Czern. cv. Varuna. Planta 248(1):49–68. https://doi.org/10.1007/s00425-018-2867-0

Ahamed M, Akhtar MJ, Alhadlaq HA (2019) Preventive effect of TiO2 nanoparticles on heavy metal Pb-induced toxicity in human lung epithelial (A549) cells. Toxicol In Vitro 57:18–27. https://doi.org/10.1016/j.tiv.2019.02.004

Ahmad S, Mahmood R (2019) Mercury chloride toxicity in human erythrocytes: enhanced generation of ROS and RNS, hemoglobin oxidation, impaired antioxidant power, and inhibition of plasma membrane redox system. Environ Sci Pollut Res Int 26:5645–5657. https://doi.org/10.1007/s11356-018-04062-5

Akram S, Najam R, Rizwani GH, Abbas SA (2015) Determination of heavy metal contents by atomic absorption spectroscopy (AAS) in some medicinal plants from Pakistani and Malaysian origin. Pak J Pharm Sci 28(5):1781–1787

Alho LOG, Gebara RC, Paina KA, Sarmento H, Melao M (2019) Responses of Raphidocelis subcapitata exposed to Cd and Pb: mechanisms of toxicity assessed by multiple endpoints. Ecotoxicol Environ Saf 169:950–959. https://doi.org/10.1016/j.ecoenv.2018.11.087

Amin H, Ahmed Arain B, Abbasi MS, Amin F, Jahangir TM, Soomro NU (2019) Evaluation of chromium phyto-toxicity, phyto-tolerance, and phyto-accumulation using biofuel plants for effective phytoremediation. Int J Phytoremediation:1–12. https://doi.org/10.1080/15226514.2018.1524837

Amna AN, Masood S et al (2015) Differential effects of cadmium and chromium on growth, photosynthetic activity, and metal uptake of Linum usitatissimum in association with Glomus intraradices. Environ Monit Assess 187(6):311. https://doi.org/10.1007/s10661-015-4557-8

Antoniadis V, Zanni AA, Levizou E et al (2018) Modulation of hexavalent chromium toxicity on Omicronriganum vulgare in an acidic soil amended with peat, lime, and zeolite. Chemosphere 195:291–300. https://doi.org/10.1016/j.chemosphere.2017.12.069

Aouini F, Trombini C, Volland M, Elcafsi M, Blasco J (2018) Assessing lead toxicity in the clam Ruditapes philippinarum: bioaccumulation and biochemical responses. Ecotoxicol Environ Saf 158:193–203. https://doi.org/10.1016/j.ecoenv.2018.04.033

Arvay J, Demkova L, Hauptvogl M et al (2017) Assessment of environmental and health risks in former polymetallic ore mining and smelting area, Slovakia: spatial distribution and accumulation of mercury in four different ecosystems. Ecotoxicol Environ Saf 144:236–244. https://doi.org/10.1016/j.ecoenv.2017.06.020

Ashraf U, Tang X (2017) Yield and quality responses, plant metabolism and metal distribution pattern in aromatic rice under lead (Pb) toxicity. Chemosphere 176:141–155. https://doi.org/10.1016/j.chemosphere.2017.02.103

Aslanturk A, Uzunhisarcikli M, Kalender S, Demir F (2014) Sodium selenite and vitamin E in preventing mercuric chloride induced renal toxicity in rats. Food Chem Toxicol 70:185–190. https://doi.org/10.1016/j.fct.2014.05.010

Assadian E, Zarei MH, Gilani AG, Farshin M, Degampanah H, Pourahmad J (2018) Toxicity of copper oxide (CuO) nanoparticles on human blood lymphocytes. Biol Trace Elem Res 184(2):350–357. https://doi.org/10.1007/s12011-017-1170-4

Assadian E, Zarei MH, Gilani AG, Farshin M, Dezhampanah H, Pourahmad J (2019) Correction to: toxicity of copper oxide (CuO) nanoparticles on human blood lymphocytes. Biol Trace Elem Res 192:345. https://doi.org/10.1007/s12011-019-01678-7

Bailon MX, David AS, Park Y, Kim E, Hong Y (2018) Total mercury, methyl mercury, and heavy metal concentrations in Hyeongsan River and its tributaries in Pohang city, South Korea. Environ Monit Assess 190(5):274. https://doi.org/10.1007/s10661-018-6624-4

Bakti F, Sasse C, Heinekamp T, Pocsi I, Braus GH (2018) Heavy metal-induced expression of PcaA provides cadmium tolerance to Aspergillus fumigatus and supports its virulence in the Galleria mellonella model. Front Microbiol 9:744. https://doi.org/10.3389/fmicb.2018.00744

Banu SK, Stanley JA, Taylor RJ et al (2018) Sexually dimorphic impact of chromium accumulation on human placental oxidative stress and apoptosis. Toxicol Sci 161(2):375–387. https://doi.org/10.1093/toxsci/kfx224

Beauvais-Fluck R, Slaveykova VI, Skyllberg U, Cosio C (2018) Molecular effects, speciation, and competition of inorganic and methyl mercury in the aquatic plant Elodea nuttallii. Environ Sci Technol 52(15):8876–8884. https://doi.org/10.1021/acs.est.8b02124

Ben Massoud M, Sakouhi L, Karmous I et al (2018) Protective role of exogenous phytohormones on redox status in pea seedlings under copper stress. J Plant Physiol 221:51–61. https://doi.org/10.1016/j.jplph.2017.11.014

Bera AK, Rana T, Das S et al (2010) Ground water arsenic contamination in West Bengal, India: a risk of sub-clinical toxicity in cattle as evident by correlation between arsenic exposure, excretion and deposition. Toxicol Ind Health 26(10):709–716. https://doi.org/10.1177/0748233710377775

Bilinsky LM, Thomas DJ, Fisher JW (2019) Using mathematical modeling to infer the valence state of arsenicals in tissues: a PBPK model for dimethylarsinic acid (DMA(V)) and dimethylarsinous acid (DMA(III)) in mice. J Theor Biol 461:215–229. https://doi.org/10.1016/j.jtbi.2018.10.051

Bondarenko O, Juganson K, Ivask A, Kasemets K, Mortimer M, Kahru A (2013) Toxicity of Ag, CuO and ZnO nanoparticles to selected environmentally relevant test organisms and mammalian cells in vitro: a critical review. Arch Toxicol 87(7):1181–1200. https://doi.org/10.1007/s00204-013-1079-4

Boostani HR, Najafi-Ghiri M, Mirsoleimani A (2019) The effect of biochars application on reducing the toxic effects of nickel and growth indices of spinach (Spinacia oleracea L.) in a calcareous soil. Environ Sci Pollut Res Int 26(2):1751–1760. https://doi.org/10.1007/s11356-018-3760-x

Boran H, Saffak S (2018) Comparison of dissolved nickel and nickel nanoparticles toxicity in larval zebrafish in terms of gene expression and DNA damage. Arch Environ Contam Toxicol 74(1):193–202. https://doi.org/10.1007/s00244-017-0468-8

Budnik LT, Casteleyn L (2019) Mercury pollution in modern times and its socio-medical consequences. Sci Total Environ 654:720–734. https://doi.org/10.1016/j.scitotenv.2018.10.408

Bugata LSP, Pitta Venkata P, Gundu AR et al (2019) Acute and subacute oral toxicity of copper oxide nanoparticles in female albino Wistar rats. J Appl Toxicol 39:702. https://doi.org/10.1002/jat.3760

Butt UJ, Shah SAA, Ahmed T, Zahid S (2018) Protective effects of Nigella sativa L. seed extract on lead induced neurotoxicity during development and early life in mouse models. Toxicol Res (Camb) 7(1):32–40. https://doi.org/10.1039/c7tx00201g

Cabrita MT, Duarte B, Cesario R et al (2019) Mercury mobility and effects in the salt-marsh plant Halimione portulacoides: uptake, transport, and toxicity and tolerance mechanisms. Sci Total Environ 650(Pt 1):111–120. https://doi.org/10.1016/j.scitotenv.2018.08.335

Campos JA, Esbri JM, Madrid MM et al (2018) Does mercury presence in soils promote their microbial activity? The Almadenejos case (Almaden mercury mining district, Spain). Chemosphere 201:799–806. https://doi.org/10.1016/j.chemosphere.2018.02.163

Cao Z, Wang M, Chen Q et al (2018) Preliminary assessment on exposure of four typical populations to potentially toxic metals by means of skin wipes under the influence of haze pollution. Sci Total Environ 613–614:886–893. https://doi.org/10.1016/j.scitotenv.2017.09.181

Chaudhuri P, Imam HT, Essig Y et al (2018) Molecular genetic and biochemical characterization of a putative family of zinc metalloproteins in Caenorhabditis elegans. Metallomics 10(12):1814–1823. https://doi.org/10.1039/c8mt00169c

Chen X, Zhang C, Tan L, Wang J (2018) Toxicity of Co nanoparticles on three species of marine microalgae. Environ Pollut 236:454–461. https://doi.org/10.1016/j.envpol.2018.01.081

Chen H, Li Y, Ma X et al (2019) Analysis of potential strategies for cadmium stress tolerance revealed by transcriptome analysis of upland cotton. Sci Rep 9(1):86. https://doi.org/10.1038/s41598-018-36228-z

Cortes-Eslava J, Gomez-Arroyo S, Risueno MC, Testillano PS (2018) The effects of organophosphorus insecticides and heavy metals on DNA damage and programmed cell death in two plant models. Environ Pollut 240:77–86. https://doi.org/10.1016/j.envpol.2018.04.119

Cozzolino V, De Martino A, Nebbioso A, Di Meo V, Salluzzo A, Piccolo A (2016) Plant tolerance to mercury in a contaminated soil is enhanced by the combined effects of humic matter addition and inoculation with arbuscular mycorrhizal fungi. Environ Sci Pollut Res Int 23(11):11312–11322. https://doi.org/10.1007/s11356-016-6337-6

Dankers ACA, Kuper CF, Boumeester AJ et al (2018) A practical approach to assess inhalation toxicity of metal oxide nanoparticles in vitro. J Appl Toxicol 38(2):160–171. https://doi.org/10.1002/jat.3518

Das S, Majumder B, Biswas AK (2018) Modulation of growth, ascorbate-glutathione cycle and thiol metabolism in rice (Oryza sativa L. cv. MTU-1010) seedlings by arsenic and silicon. Ecotoxicology 27(10):1387–1403. https://doi.org/10.1007/s10646-018-1994-5

Debure M, Tournassat C, Lerouge C et al (2018) Retention of arsenic, chromium and boron on an outcropping clay-rich rock formation (the Tegulines Clay, eastern France). Sci Total Environ 642:216–229. https://doi.org/10.1016/j.scitotenv.2018.06.037

Dewanjee S, Dua TK, Khanra R et al (2015) Water spinach, ipomoea aquatic (Convolvulaceae), ameliorates lead toxicity by inhibiting oxidative stress and apoptosis. PLoS One 10(10):e0139831. https://doi.org/10.1371/journal.pone.0139831

Doron L, Goloubinoff P, Shapira M (2018) ZnJ2 is a member of a large chaperone family in the chloroplast of photosynthetic organisms that features a DnaJ-like Zn-finger domain. Front Mol Biosci 5:2. https://doi.org/10.3389/fmolb.2018.00002

Emmerton CA, Cooke CA, Wentworth GR, Graydon JA, Ryjkov A, Dastoor A (2018) Total mercury and methylmercury in lake water of Canada's Oil Sands region. Environ Sci Technol 52(19):10946–10955. https://doi.org/10.1021/acs.est.8b01680

Fajardo C, Costa G, Nande M, Martin C, Martin M, Sanchez-Fortun S (2019) Heavy metals immobilization capability of two iron-based nanoparticles (nZVI and Fe3O4): soil and freshwater bioassays to assess ecotoxicological impact. Sci Total Environ 656:421–432. https://doi.org/10.1016/j.scitotenv.2018.11.323

Fasken MB, Corbett AH, Stewart M (2019) Structure-function relationships in the Nab2 polyadenosine-RNA binding Zn finger protein family. Protein Sci 28(3):513–523. https://doi.org/10.1002/pro.3565

Feng L, Yang X, Asweto CO et al (2017) Low-dose combined exposure of nanoparticles and heavy metal compared with PM2.5 in human myocardial AC16 cells. Environ Sci Pollut Res Int 24(36):27767–27777. https://doi.org/10.1007/s11356-017-0228-3

Fernandez CJ, Domini CE, Grunhut M, Lista AG (2018) A soft material for chromium speciation in water samples using a chemiluminescence automatic system. Chemosphere 196:361–367. https://doi.org/10.1016/j.chemosphere.2017.12.178

Filice FP, Henderson JD, Li MSM, Ding Z (2019) Correlating live cell viability with membrane permeability disruption induced by trivalent chromium. ACS Omega 4(1):2142–2151. https://doi.org/10.1021/acsomega.8b02113

Fontaine M, Clement Y, Blanc N, Demesmay C (2019) Hexavalent chromium release from leather over time natural ageing vs accelerated ageing according to a multivariate approach. J Hazard Mater 368:811–818. https://doi.org/10.1016/j.jhazmat.2018.12.112

Franca FC, Albuuerque AM, Almeida AC et al (2017) Heavy metals deposited in the culture of lettuce (Lactuca sativa L.) by the influence of vehicular traffic in Pernambuco, Brazil. Food Chem 215:171–176. https://doi.org/10.1016/j.foodchem.2016.07.168

Fuentes-Gandara F, Pinedo-Hernandez J, Marrugo-Negrete J, Diez S (2018) Human health impacts of exposure to metals through extreme consumption of fish from the Colombian Caribbean Sea. Environ Geochem Health 40(1):229–242. https://doi.org/10.1007/s10653-016-9896-z

Fumagalli A, Faggion B, Ronchini M et al (2010) Platinum, palladium, and rhodium deposition to the Prunus laurus cerasus leaf surface as an indicator of the vehicular traffic pollution in the city of Varese area: an easy and reliable method to detect PGEs released from automobile catalytic converters. Environ Sci Pollut Res Int 17(3):665–673. https://doi.org/10.1007/s11356-009-0166-9

Gang H, Xiao C, Xiao Y et al (2019) Proteomic analysis of the reduction and resistance mechanisms of Shewanella oneidensis MR-1 under long-term hexavalent chromium stress. Environ Int 127:94–102. https://doi.org/10.1016/j.envint.2019.03.016

Gao X, Avellan A, Laughton S et al (2018) CuO nanoparticle dissolution and toxicity to wheat (Triticum aestivum) in rhizosphere soil. Environ Sci Technol 52(5):2888–2897. https://doi.org/10.1021/acs.est.7b05816

Genthe B, Kapwata T, Le Roux W, Chamier J, Wright CY (2018) The reach of human health risks associated with metals/metalloids in water and vegetables along a contaminated river catchment: South Africa and Mozambique. Chemosphere 199:1–9. https://doi.org/10.1016/j.chemosphere.2018.01.160

Glazer ES, Zhu C, Hamir AN, Borne A, Thompson CS, Curley SA (2011) Biodistribution and acute toxicity of naked gold nanoparticles in a rabbit hepatic tumor model. Nanotoxicology 5(4):459–468. https://doi.org/10.3109/17435390.2010.516026

Gouda AS, El-Nabarawy NA, Ibrahim SF (2018) Moringa oleifera extract (Lam) attenuates Aluminium phosphide-induced acute cardiac toxicity in rats. Toxicol Rep 5:209–212. https://doi.org/10.1016/j.toxrep.2018.01.001

Goudarzi G, Alavi N, Geravandi S et al (2018) Health risk assessment on human exposed to heavy metals in the ambient air PM10 in Ahvaz, Southwest Iran. Int J Biometeorol 62(6):1075–1083. https://doi.org/10.1007/s00484-018-1510-x

Graeme KA, Pollack CV Jr (1998) Heavy metal toxicity, part II: lead and metal fume fever. J Emerg Med 16(2):171–177. https://doi.org/10.1016/S0736-4679(97)00283-7

Guo L, Lyu Y, Yang Y (2017) Concentrations and chemical forms of heavy metals in the bulk atmospheric deposition of Beijing, China. Environ Sci Pollut Res Int 24(35):27356–27365. https://doi.org/10.1007/s11356-017-0324-4

Guo ZX, Geng H, Zhang JH et al (2018) Ecological and health risks of trace heavy metals in atmospheric PM2.5 collected in Wuxiang town, Shanxi Province. Huan Jing Ke Xue 39(3):1004–1013. https://doi.org/10.13227/j.hjkx.201706011

Hadrup N, Sharma AK, Loeschner K (2018) Toxicity of silver ions, metallic silver, and silver nanoparticle materials after in vivo dermal and mucosal surface exposure: a review. Regul Toxicol Pharmacol 98:257–267. https://doi.org/10.1016/j.yrtph.2018.08.007

Hammami H, Parsa M, Mohassel MH, Rahimi S, Mijani S (2016) Weeds ability to phytoremediate cadmium-contaminated soil. Int J Phytoremediation 18(1):48–53. https://doi.org/10.1080/15226514.2015.1058336

Hang X, Gan F, Chen Y et al (2018) Evaluation of mercury uptake and distribution in rice (Oryza sativa L.). Bull Environ Contam Toxicol 100(3):451–456. https://doi.org/10.1007/s00128-017-2237-9

Hao P, Zhu Y, Wang S et al (2017) Selenium administration alleviates toxicity of chromium(VI) in the chicken brain. Biol Trace Elem Res 178(1):127–135. https://doi.org/10.1007/s12011-016-0915-9

Hashemi SA, Tabibian S (2018) Application of Mulberry nigra to absorb heavy metal, mercury, from the environment of green space city. Toxicol Rep 5:644–646. https://doi.org/10.1016/j.toxrep.2018.05.006

Hausladen DM, Alexander-Ozinskas A, McClain C, Fendorf S (2018) Hexavalent chromium sources and distribution in California groundwater. Environ Sci Technol 52(15):8242–8251. https://doi.org/10.1021/acs.est.7b06627

Hoffman NJ, Penque BA, Habegger KM, Sealls W, Tackett L, Elmendorf JS (2014) Chromium enhances insulin responsiveness via AMPK. J Nutr Biochem 25(5):565–572. https://doi.org/10.1016/j.jnutbio.2014.01.007

Honda A, Tsuji K, Matsuda Y et al (2015) Effects of air pollution-related heavy metals on the viability and inflammatory responses of human airway epithelial cells. Int J Toxicol 34(2):195–203. https://doi.org/10.1177/1091581815575757

Hossain MM, Khatun MA, Haque MN et al (2018) Silicon alleviates arsenic-induced toxicity in wheat through vacuolar sequestration and ROS scavenging. Int J Phytoremediation 20(8):796–804. https://doi.org/10.1080/15226514.2018.1425669

Hu L, Long Y (2019) Zinc leaching behavior in semi-aerobic landfill. Environ Technol 40(1):29–36. https://doi.org/10.1080/09593330.2017.1377293

Hua C, Zhou G, Yin X et al (2018) Assessment of heavy metal in coal gangue: distribution, leaching characteristic and potential ecological risk. Environ Sci Pollut Res Int 25(32):32321–32331. https://doi.org/10.1007/s11356-018-3118-4

Huang H, Jiang Y, Xu X, Cao X (2018) In vitro bioaccessibility and health risk assessment of heavy metals in atmospheric particulate matters from three different functional areas of Shanghai, China. Sci Total Environ 610–611:546–554. https://doi.org/10.1016/j.scitotenv.2017.08.074

Irvine GW, Pinter TB, Stillman MJ (2016) Defining the metal binding pathways of human metallothionein 1a: balancing zinc availability and cadmium seclusion. Metallomics 8(1):71–81. https://doi.org/10.1039/c5mt00225g

Jadoon WA, Khpalwak W, Chidya RCG et al (2018) Evaluation of levels, sources and health hazards of road-dust associated toxic metals in Jalalabad and Kabul cities, Afghanistan. Arch Environ Contam Toxicol 74(1):32–45. https://doi.org/10.1007/s00244-017-0475-9

Jain S, Muneer S, Guerriero G et al (2018) Tracing the role of plant proteins in the response to metal toxicity: a comprehensive review. Plant Signal Behav 13(9):e1507401. https://doi.org/10.1080/15592324.2018.1507401

Jaiswal A, Verma A, Jaiswal P (2018) Detrimental effects of heavy metals in soil, plants, and aquatic ecosystems and in humans. J Environ Pathol Toxicol Oncol 37(3):183–197. https://doi.org/10.1615/JEnvironPatholToxicolOncol.2018025348

Jaiswar A, Varshney D, Adholeya A, Prasad P (2018) Do environmentally induced DNA variations mediate adaptation in Aspergillus flavus exposed to chromium stress in tannery sludge? BMC Genomics 19(1):868. https://doi.org/10.1186/s12864-018-5244-2

Jan AT, Azam M, Ali A, Haq QM (2012) Molecular characterization of mercury resistant bacteria inhabiting polluted water bodies of different geographical locations in India. Curr Microbiol 65(1):14–21. https://doi.org/10.1007/s00284-012-0118-3

Janiga M, Haas M (2019) Alpine accentors as monitors of atmospheric long-range lead and mercury pollution in alpine environments. Environ Sci Pollut Res Int 26(3):2445–2454. https://doi.org/10.1007/s11356-018-3742-z

Ji R, Zhou L, Liu J et al (2017) Calcium-dependent protein kinase CPK31 interacts with arsenic transporter AtNIP1;1 and regulates arsenite uptake in Arabidopsis thaliana. PLoS One 12(3):e0173681. https://doi.org/10.1371/journal.pone.0173681

Junaid M, Hashmi MZ, Malik RN, Pei DS (2016) Toxicity and oxidative stress induced by chromium in workers exposed from different occupational settings around the globe: a review. Environ Sci Pollut Res Int 23(20):20151–20167. https://doi.org/10.1007/s11356-016-7463-x

Kamerud KL, Hobbie KA, Anderson KA (2013) Stainless steel leaches nickel and chromium into foods during cooking. J Agric Food Chem 61(39):9495–9501. https://doi.org/10.1021/jf402400v

Kerfoot WC, Urban NR, McDonald CP et al (2018) Mining legacy across a wetland landscape: high mercury in Upper Peninsula (Michigan) rivers, lakes, and fish. Environ Sci Process Impacts 20(4):708–733. https://doi.org/10.1039/c7em00521k

Khan ZI, Ugulu I, Ahmad K et al (2018) Assessment of trace metal and metalloid accumulation and human health risk from vegetables consumption through spinach and coriander specimens irrigated with wastewater. Bull Environ Contam Toxicol 101(6):787–795. https://doi.org/10.1007/s00128-018-2448-8

Khlebtsov N, Dykman L (2011) Biodistribution and toxicity of engineered gold nanoparticles: a review of in vitro and in vivo studies. Chem Soc Rev 40(3):1647–1671. https://doi.org/10.1039/c0cs00018c

Khushboo M, Murthy MK, Devi MS et al (2018) Testicular toxicity and sperm quality following copper exposure in Wistar albino rats: ameliorative potentials of L-carnitine. Environ Sci Pollut Res Int 25(2):1837–1862. https://doi.org/10.1007/s11356-017-0624-8

Kim YO, Bae HJ, Cho E, Kang H (2017) Exogenous glutathione enhances mercury tolerance by inhibiting mercury entry into plant cells. Front Plant Sci 8:683. https://doi.org/10.3389/fpls.2017.00683

Knop M, Dang TQ, Jeschke G, Seebeck FP (2017) Copper is a cofactor of the formylglycine-generating enzyme. Chembiochem 18(2):161–165. https://doi.org/10.1002/cbic.201600359

Konop M, Klodzinska E, Borowiec J et al (2019) Application of micellar electrokinetic chromatography for detection of silver nanoparticles released from wound dressing. Electrophoresis 40:1565–1572. https://doi.org/10.1002/elps.201900020

Kumar Dubey A, Kumar N, Ranjan R et al (2018) Application of glycine reduces arsenic accumulation and toxicity in Oryza sativa L. by reducing the expression of silicon transporter genes. Ecotoxicol Environ Saf 148:410–417. https://doi.org/10.1016/j.ecoenv.2017.10.047

Kumari P, Rastogi A, Shukla A, Srivastava S, Yadav S (2018) Prospects of genetic engineering utilizing potential genes for regulating arsenic accumulation in plants. Chemosphere 211:397–406. https://doi.org/10.1016/j.chemosphere.2018.07.152

Kundu D, Dey S, Raychaudhuri SS (2018) Chromium (VI) - induced stress response in the plant Plantago ovata Forsk in vitro. Genes Environ 40:21. https://doi.org/10.1186/s41021-018-0109-0

Kupper H (2017) Lead toxicity in plants. Met Ions Life Sci 17. https://doi.org/10.1515/9783110434330-015

Kura AU, Saifullah B, Cheah PS, Hussein MZ, Azmi N, Fakurazi S (2015) Acute oral toxicity and biodistribution study of zinc-aluminium-levodopa nanocomposite. Nanoscale Res Lett 10:105. https://doi.org/10.1186/s11671-015-0742-5

Kwak JI, Cui R, Nam SH, Kim SW, Chae Y, An YJ (2016) Multispecies toxicity test for silver nanoparticles to derive hazardous concentration based on species sensitivity distribution for the protection of aquatic ecosystems. Nanotoxicology 10(5):521–530. https://doi.org/10.3109/17435390.2015.1090028

Lacerda LM, Garcia SC, da Silva LB et al (2019) Evaluation of hematological, biochemical parameters and thiol enzyme activity in chrome plating workers. Environ Sci Pollut Res Int 26(2):1892–1901. https://doi.org/10.1007/s11356-018-3755-7

Lash LH (2019) Environmental and genetic factors influencing kidney toxicity. Semin Nephrol 39(2):132–140. https://doi.org/10.1016/j.semnephrol.2018.12.003

Lecoanet P, Blangis M, Garcia M, Legallois Y, Fabre T (2019) Chromium-cobalt intoxication with intense systemic complications following total hip revision after per-operative ceramic fracture. Case Rep Orthop 2019:4209796. https://doi.org/10.1155/2019/4209796

Lee PK, Chang HJ, Yu S et al (2018a) Characterization of Cr (VI) - containing solid phase particles in dry dust deposition in Daejeon, South Korea. Environ Pollut 243(Pt B):1637–1647. https://doi.org/10.1016/j.envpol.2018.09.127

Lee WS, Kim E, Cho HJ et al (2018b) The relationship between dissolution behavior and the toxicity of silver nanoparticles on zebrafish embryos in different ionic environments. Nanomaterials (Basel) 8(9):E652. https://doi.org/10.3390/nano8090652

Leon-Canedo JA, Alarcon-Silvas SG, Fierro-Sanudo JF et al (2019) Mercury and other trace metals in lettuce (Lactuca sativa) grown with two low-salinity shrimp effluents: accumulation and human health risk assessment. Sci Total Environ 650(Pt 2):2535–2544. https://doi.org/10.1016/j.scitotenv.2018.10.003

Levina A, Lay PA (2008) Chemical properties and toxicity of chromium(III) nutritional supplements. Chem Res Toxicol 21(3):563–571. https://doi.org/10.1021/tx700385t

Li Y, Zhao J, Li YF et al (2016a) Comparative metalloproteomic approaches for the investigation proteins involved in the toxicity of inorganic and organic forms of mercury in rice (Oryza sativa L.) roots. Metallomics 8(7):663–671. https://doi.org/10.1039/c5mt00264h

Li F, Song Z, Bo LC, Lyu L, Chen J (2016b) Protective effect of Nrf2 activation by curcumin against lead-induced toxicity and apoptosis in SH-SY5Y cells. Zhonghua Lao Dong Wei Sheng Zhi Ye Bing Za Zhi 34(6):401–405. https://doi.org/10.3760/cma.j.issn.1001-9391.2016.06.001

Li R, Wu H, Ding J, Fu W, Gan L, Li Y (2017) Mercury pollution in vegetables, grains and soils from areas surrounding coal-fired power plants. Sci Rep 7:46545. https://doi.org/10.1038/srep46545

Li C, Xu W, Chu S et al (2018) The chemical speciation, spatial distribution and toxicity of mercury from Tibetan medicine Zuotaibeta-HgS and HgCl2 in mouse kidney. J Trace Elem Med Biol 45:104–113. https://doi.org/10.1016/j.jtemb.2017.08.010

Li T, Yang J, Liu R et al (2019) Efficient fabrication of reversible pH-induced carboxymethyl chitosan nanoparticles for antitumor drug delivery under weakly acidic microenvironment. Int J Biol Macromol 126:68–73. https://doi.org/10.1016/j.ijbiomac.2018.12.178

Liao Y, Xu H, Liu W et al (2019) One step interface activation of ZnS using cupric ions for mercury recovery from nonferrous smelting flue gas. Environ Sci Technol 53:4511–4518. https://doi.org/10.1021/acs.est.9b01310

Lima FRD, Martins GC, Silva AO et al (2019a) Critical mercury concentration in tropical soils: impact on plants and soil biological attributes. Sci Total Environ 666:472–479. https://doi.org/10.1016/j.scitotenv.2019.02.216

Lima FRD, Engelhardt MM, Vasques ICF et al (2019b) Evaluation of mercury phytoavailability in Oxisols. Environ Sci Pollut Res Int 26(1):483–491. https://doi.org/10.1007/s11356-018-3623-5

Lin L, Zheng F, Zhou H, Li S (2017) Biomimetic gastrointestinal tract functions for metal absorption assessment in edible plants: comparison to in vivo absorption. J Agric Food Chem 65(30):6282–6287. https://doi.org/10.1021/acs.jafc.7b02054

Liu X, Liang Y, Guo J (2017) Heavy metal pollution in Nanchang City and its health implication on traffic policemen. Environ Sci Pollut Res Int 26:17885. https://doi.org/10.1007/s11356-017-0289-3

Liu X, Wei F, Cheng Y et al (2018) Auditory training reverses lead (Pb)-toxicity-induced changes in sound-azimuth selectivity of cortical neurons. Cereb Cortex 29:3294–3304. https://doi.org/10.1093/cercor/bhy199

Liu S, Cui M, Li X, Thuyet DQ, Fan W (2019) Effects of hydrophobicity of titanium dioxide nanoparticles and exposure scenarios on copper uptake and toxicity in Daphnia magna. Water Res 154:162–170. https://doi.org/10.1016/j.watres.2019.01.055

Maheshwari N, Khan FH, Mahmood R (2018) 3,4-Dihydroxybenzaldehyde lowers ROS generation and protects human red blood cells from arsenic(III) induced oxidative damage. Environ Toxicol 33:861–875. https://doi.org/10.1002/tox.22572

Malar S, Shivendra Vikram S, Jc Favas P, Perumal V (2016) Lead heavy metal toxicity induced changes on growth and antioxidative enzymes level in water hyacinths [Eichhornia crassipes (Mart.)]. Bot Stud 55(1):54. https://doi.org/10.1186/s40529-014-0054-6

Maleki A, Hosseini MJ, Rahimi N et al (2019) Adjuvant potential of selegiline in treating acute toxicity of aluminium phosphide in rats. Basic Clin Pharmacol Toxicol 125:62–74. https://doi.org/10.1111/bcpt.13207

Mangalampalli B, Dumala N, Grover P (2017) Acute oral toxicity study of magnesium oxide nanoparticles and microparticles in female albino Wistar rats. Regul Toxicol Pharmacol 90:170–184. https://doi.org/10.1016/j.yrtph.2017.09.005

Mangalampalli B, Dumala N, Grover P (2018) Allium cepa root tip assay in assessment of toxicity of magnesium oxide nanoparticles and microparticles. J Environ Sci (China) 66:125–137. https://doi.org/10.1016/j.jes.2017.05.012

Mansano AS, Souza JP, Cancino-Bernardi J, Venturini FP, Marangoni VS, Zucolotto V (2018) Toxicity of copper oxide nanoparticles to Neotropical species Ceriodaphnia silvestrii and Hyphessobrycon eques. Environ Pollut 243(Pt A):723–733. https://doi.org/10.1016/j.envpol.2018.09.020

Marat I, Arstan M, Galymzhan Y, Timur J, Yerbolat I, Almasbek Y (2018) Impact of chromium and boron compounds on the reproductive function in rats. Toxicol Ind Health 34(6):365–374. https://doi.org/10.1177/0748233718759162

Maret W (2019) Chromium supplementation in human health, metabolic syndrome, and diabetes. Met Ions Life Sci 19. https://doi.org/10.1515/9783110527872-015

Marinho BA, Cristovao RO, Boaventura RAR, Vilar VJP (2019) As(III) and Cr(VI) oxyanion removal from water by advanced oxidation/reduction processes-a review. Environ Sci Pollut Res Int 26(3):2203–2227. https://doi.org/10.1007/s11356-018-3595-5

Masood N, Farooqi A, Zafar MI (2019) Health risk assessment of arsenic and other potentially toxic elements in drinking water from an industrial zone of Gujrat, Pakistan: a case study. Environ Monit Assess 191(2):95. https://doi.org/10.1007/s10661-019-7223-8

Mateos LM, Villadangos AF, de la Rubia AG et al (2017) The arsenic detoxification system in corynebacteria: basis and application for bioremediation and redox control. Adv Appl Microbiol 99:103–137. https://doi.org/10.1016/bs.aambs.2017.01.001

Men C, Liu R, Xu F, Wang Q, Guo L, Shen Z (2018) Pollution characteristics, risk assessment, and source apportionment of heavy metals in road dust in Beijing, China. Sci Total Environ 612:138–147. https://doi.org/10.1016/j.scitotenv.2017.08.123

Meng XL, Chen CL, Liu YY et al (2019) Selenoprotein SELENOK enhances the migration and phagocytosis of microglial cells by increasing the cytosolic free ca(2+) level resulted from the up-regulation of IP3R. Neuroscience 406:38–49. https://doi.org/10.1016/j.neuroscience.2019.02.029

Mi Y, Tao X, Zhang X, Si Y (2019) Acute biotoxicity assessment of heavy metals in sewage sludge based on the glucose consumption of Escherichia coli. R Soc Open Sci 6(1):181769. https://doi.org/10.1098/rsos.181769

Miao X, Hao Y, Zhang F, Zou S, Ye S, Xie Z (2019) Spatial distribution of heavy metals and their potential sources in the soil of Yellow River Delta: a traditional oil field in China. Environ Geochem Health 42:7. https://doi.org/10.1007/s10653-018-0234-5

Minigalieva IA, Katsnelson BA, Panov VG et al (2017) In vivo toxicity of copper oxide, lead oxide and zinc oxide nanoparticles acting in different combinations and its attenuation with a complex of innocuous bio-protectors. Toxicology 380:72–93. https://doi.org/10.1016/j.tox.2017.02.007

Mondal P, de Alcantara MR, Jonathan MP, Biswas JK, Murugan K, Sarkar SK (2018) Seasonal assessment of trace element contamination in intertidal sediments of the meso-macrotidal Hooghly (Ganges) river estuary with a note on mercury speciation. Mar Pollut Bull 127:117–130. https://doi.org/10.1016/j.marpolbul.2017.11.041

Muszynska E, Hanus-Fajerska E, Ciarkowska K (2018) Studies on lead and cadmium toxicity in Dianthus carthusianorum calamine ecotype cultivated in vitro. Plant Biol (Stuttg) 20(3):474–482. https://doi.org/10.1111/plb.12712

Mwakalapa EB, Simukoko CK, Mmochi AJ et al (2019) Heavy metals in farmed and wild milkfish (Chanos chanos) and wild mullet (Mugil cephalus) along the coasts of Tanzania and associated health risk for humans and fish. Chemosphere 224:176–186. https://doi.org/10.1016/j.chemosphere.2019.02.063

Nabuyanda MM, van Bruggen J, Kelderman P, Irvine K (2019) Investigating Co, Cu, and Pb retention and remobilization after drying and rewetting treatments in greenhouse laboratory-scale constructed treatments with and without Typha angustifolia, and connected phytoremediation potential. J Environ Manag 236:510–518. https://doi.org/10.1016/j.jenvman.2019.02.016

Nariya A, Pathan A, Shah N et al (2018) Ameliorative effects of curcumin against lead induced toxicity in human peripheral blood lymphocytes culture. Drug Chem Toxicol 41(1):1–8. https://doi.org/10.3109/01480545.2015.1133637

Navarro E, Piccapietra F, Wagner B et al (2008) Toxicity of silver nanoparticles to Chlamydomonas reinhardtii. Environ Sci Technol 42(23):8959–8964

Ni W, Huang Y, Wang X, Zhang J, Wu K (2014) Associations of neonatal lead, cadmium, chromium and nickel co-exposure with DNA oxidative damage in an electronic waste recycling town. Sci Total Environ 472:354–362. https://doi.org/10.1016/j.scitotenv.2013.11.032

Nishio R, Tamano H, Morioka H, Takeuchi A, Takeda A (2019) Intake of heated leaf extract of Coriandrum sativum contributes to resistance to oxidative stress via decreases in heavy metal concentrations in the kidney. Plant Foods Hum Nutr 74:204–209. https://doi.org/10.1007/s11130-019-00720-2

Norwood WP, Borgmann U, Dixon DG (2007) Chronic toxicity of arsenic, cobalt, chromium and manganese to Hyalella azteca in relation to exposure and bioaccumulation. Environ Pollut 147(1):262–272. https://doi.org/10.1016/j.envpol.2006.07.017

Nour-Eldein NH, Hassanin EA, El-Sayed WM (2018) Mitigation of acute aluminum toxicity by sodium selenite and N-acetylcysteine in adult male rats. Biol Trace Elem Res 183(1):128–137. https://doi.org/10.1007/s12011-017-1126-8

Nuapia Y, Chimuka L, Cukrowska E (2018) Assessment of heavy metals in raw food samples from open markets in two African cities. Chemosphere 196:339–346. https://doi.org/10.1016/j.chemosphere.2017.12.134

Okesola MA, Ajiboye BO, Oyinloye BE, Ojo OA (2019) Effect of Zingiber officinale on some biochemical parameters and cytogenic analysis in lead-induced toxicity in experimental rats. Toxicol Mech Methods 29(4):255–262. https://doi.org/10.1080/15376516.2018.1558321

Ommati MM, Jamshidzadeh A, Heidari R et al (2019) Carnosine and histidine supplementation blunt lead-induced reproductive toxicity through antioxidative and mitochondria-dependent mechanisms. Biol Trace Elem Res 187(1):151–162. https://doi.org/10.1007/s12011-018-1358-2

Onakpa MM, Njan AA, Kalu OC (2018) A review of heavy metal contamination of food crops in Nigeria. Ann Glob Health 84(3):488–494. https://doi.org/10.29024/aogh.2314

Orihuela R, Kojima C, Tokar EJ et al (2013) Oxidative DNA damage after acute exposure to arsenite and monomethylarsonous acid in biomethylation-deficient human cells. Toxicol Mech Methods 23(6):389–395. https://doi.org/10.3109/15376516.2012.762570

OuYang S, Zhang Y, Chen Y et al (2019) Preparation of glass-ceramics using chromium-containing stainless steel slag: crystal structure and solidification of heavy metal chromium. Sci Rep 9(1):1964. https://doi.org/10.1038/s41598-018-37996-4

Palacios-Torres Y, Caballero-Gallardo K, Olivero-Verbel J (2018) Mercury pollution by gold mining in a global biodiversity hotspot, the Choco biogeographic region, Colombia. Chemosphere 193:421–430. https://doi.org/10.1016/j.chemosphere.2017.10.160

Pantoja Munoz L, Purchase D, Jones H et al (2016) The mechanisms of detoxification of As(III), dimethylarsinic acid (DMA) and As(V) in the microalga Chlorella vulgaris. Aquat Toxicol 175:56–72. https://doi.org/10.1016/j.aquatox.2016.02.020

Paschoalini AL, Savassi LA, Arantes FP, Rizzo E, Bazzoli N (2019) Heavy metals accumulation and endocrine disruption in Prochilodus argenteus from a polluted neotropical river. Ecotoxicol Environ Saf 169:539–550. https://doi.org/10.1016/j.ecoenv.2018.11.047

Patra DK, Pradhan C, Patra HK (2018) An in situ study of growth of lemongrass Cymbopogon flexuosus (Nees ex Steud.) W. Watson on varying concentration of chromium (Cr(+6)) on soil and its bioaccumulation: perspectives on phytoremediation potential and phytostabilisation of chromium toxicity. Chemosphere 193:793–799. https://doi.org/10.1016/j.chemosphere.2017.11.062

Peng Y, Yang R, Jin T, Chen J, Zhang J (2018) Risk assessment for potentially toxic metal(loid)s in potatoes in the indigenous zinc smelting area of northwestern Guizhou Province, China. Food Chem Toxicol 120:328–339. https://doi.org/10.1016/j.fct.2018.07.026

Petroczi A, Naughton DP (2009) Mercury, cadmium and lead contamination in seafood: a comparative study to evaluate the usefulness of target hazard quotients. Food Chem Toxicol 47(2):298–302. https://doi.org/10.1016/j.fct.2008.11.007

Pokorska-Niewiada K, Rajkowska-Mysliwiec M, Protasowicki M (2018) Acute lethal toxicity of heavy metals to the seeds of plants of high importance to humans. Bull Environ Contam Toxicol 101(2):222–228. https://doi.org/10.1007/s00128-018-2382-9

Polykretis P, Cencetti F, Donati C, Luchinat E, Banci L (2019) Cadmium effects on superoxide dismutase 1 in human cells revealed by NMR. Redox Biol 21:101102. https://doi.org/10.1016/j.redox.2019.101102

Pourrut B, Shahid M, Dumat C, Winterton P, Pinelli E (2011) Lead uptake, toxicity, and detoxification in plants. Rev Environ Contam Toxicol 213:113–136. https://doi.org/10.1007/978-1-4419-9860-6_4

Praveen A, Gupta M (2018) Nitric oxide confronts arsenic stimulated oxidative stress and root architecture through distinct gene expression of auxin transporters, nutrient related genes and modulates biochemical responses in Oryza sativa L. Environ Pollut 240:950–962. https://doi.org/10.1016/j.envpol.2018.04.096

Queiroz PS, Ruas FAD, Barboza NR, de Castro BW, Guerra-Sa R (2018) Alterations in the proteomic composition of Serratia marcescens in response to manganese (II). BMC Biotechnol 18(1):83. https://doi.org/10.1186/s12896-018-0493-3

Rajeshkumar S, Li X (2018) Bioaccumulation of heavy metals in fish species from the Meiliang Bay, Taihu Lake, China. Toxicol Rep 5:288–295. https://doi.org/10.1016/j.toxrep.2018.01.007

Rana MN, Tangpong J, Rahman MM (2018) Toxicodynamics of lead, cadmium, mercury and arsenic- induced kidney toxicity and treatment strategy: a mini review. Toxicol Rep 5:704–713. https://doi.org/10.1016/j.toxrep.2018.05.012

Raza M, Mahjabeen I, Fahim M et al (2018) Redox balance and DNA fragmentation in arsenic-exposed occupational workers from different industries of Pakistan. Environ Sci Pollut Res Int 25(33):33381–33390. https://doi.org/10.1007/s11356-018-3274-6

Redan BW, Jablonski JE, Halverson C, Jaganathan J, Mabud MA, Jackson LS (2019) Factors affecting transfer of the heavy metals arsenic, lead, and cadmium from diatomaceous-earth filter aids to alcoholic beverages during laboratory-scale filtration. J Agric Food Chem 67:2760. https://doi.org/10.1021/acs.jafc.8b06062

Ritger AL, Curtis AN, Chen CY (2018) Bioaccumulation of mercury and other metal contaminants in invasive lionfish (Pterois volitans/miles) from curacao. Mar Pollut Bull 131(Pt A):38–44. https://doi.org/10.1016/j.marpolbul.2018.03.035

Rosen BP (1999) Families of arsenic transporters. Trends Microbiol 7(5):207–212

Rua-Ibarz A, Bolea-Fernandez E, Maage A, Frantzen S, Sanden M, Vanhaecke F (2019) Tracing mercury pollution along the Norwegian coast via elemental, speciation, and isotopic analysis of liver and muscle tissue of deep-water marine fish (Brosme brosme). Environ Sci Technol 53(4):1776–1785. https://doi.org/10.1021/acs.est.8b04706

Sai Siva Ram AK, Pratap Reddy K, Girish BP, Supriya C, Sreenivasula Reddy P (2018) Arsenic aggravated reproductive toxicity in male rats exposed to lead during the perinatal period. Toxicol Res (Camb) 7(6):1191–1204. https://doi.org/10.1039/c8tx00146d

Saifullah DS, Naeem A et al (2018) Opportunities and challenges in the use of mineral nutrition for minimizing arsenic toxicity and accumulation in rice: a critical review. Chemosphere 194:171–188. https://doi.org/10.1016/j.chemosphere.2017.11.149

Sankaralingam M, Lee YM, Pineda-Galvan Y et al (2018) Redox reactivity of a mononuclear manganese-oxo complex binding calcium ion and other redox-inactive metal ions. J Am Chem Soc 141:1324. https://doi.org/10.1021/jacs.8b11492

Schneider L, Allen K, Walker M, Morgan C, Haberle S (2019) Using tree rings to track atmospheric mercury pollution in Australia: the legacy of mining in Tasmania. Environ Sci Technol 53:5697–5706. https://doi.org/10.1021/acs.est.8b06712

Shaheen N, Irfan NM, Khan IN, Islam S, Islam MS, Md. Kawser A (2016) Presence of heavy metals in fruits and vegetables: Health risk implications in Bangladesh. Chemosphere 152:431–438. https://doi.org/10.1016/j.chemosphere.2016.02.060

Sharma S, Nagpal AK, Kaur I (2018) Heavy metal contamination in soil, food crops and associated health risks for residents of Ropar wetland, Punjab, India and its environs. Food Chem 255:15–22. https://doi.org/10.1016/j.foodchem.2018.02.037

Shen H, He Z, Yan H et al (2014) The fronds tonoplast quantitative proteomic analysis in arsenic hyperaccumulator Pteris vittata L. J Proteome 105:46–57. https://doi.org/10.1016/j.jprot.2014.01.029

Shi WG, Liu W, Yu W et al (2019) Abscisic acid enhances lead translocation from the roots to the leaves and alleviates its toxicity in Populus x canescens. J Hazard Mater 362:275–285. https://doi.org/10.1016/j.jhazmat.2018.09.024

Shri M, Singh PK, Kidwai M et al (2019) Recent advances in arsenic metabolism in plants: current status, challenges and highlighted biotechnological intervention to reduce grain arsenic in rice. Metallomics 11(3):519–532. https://doi.org/10.1039/c8mt00320c

Sihag S, Brar B, Joshi UN (2019) Salicylic acid induces amelioration of chromium toxicity and affects antioxidant enzyme activity in Sorghum bicolor L. Int J Phytoremediation 21:1–12. https://doi.org/10.1080/15226514.2018.1524827

Singh A, Sharma RK, Agrawal M, Marshall FM (2010) Health risk assessment of heavy metals via dietary intake of foodstuffs from the wastewater irrigated site of a dry tropical area of India. Food Chem Toxicol 48(2):611–619. https://doi.org/10.1016/j.fct.2009.11.041

Slejko FF, Petrini R, Lutman A, Forte C, Ghezzi L (2019) Chromium isotopes tracking the resurgence of hexavalent chromium contamination in a past-contaminated area in the Friuli Venezia Giulia region, northern Italy. Isot Environ Health Stud 55(1):56–69. https://doi.org/10.1080/10256016.2018.1560278

Souza LRR, da Silva VS, Franchi LP, de Souza TAJ (2018) Toxic and beneficial potential of silver nanoparticles: the two sides of the same coin. Adv Exp Med Biol 1048:251–262. https://doi.org/10.1007/978-3-319-72041-8_15

Stambulska UY, Bayliak MM, Lushchak VI (2018) Chromium(VI) toxicity in legume plants: modulation effects of rhizobial symbiosis. Biomed Res Int 2018:8031213. https://doi.org/10.1155/2018/8031213

Su J, Hu TH, Murase R et al (2019) Redox activities of metal-organic frameworks incorporating rare-earth metal chains and tetrathiafulvalene linkers. Inorg Chem 58:3698–3706. https://doi.org/10.1021/acs.inorgchem.8b03299

Talukdar T, Talukdar D (2013) Response of antioxidative enzymes to arsenic-induced phytotoxicity in leaves of a medicinal daisy, Wedelia chinensis Merrill. J Nat Sci Biol Med 4(2):383–388. https://doi.org/10.4103/0976-9668.116989

Tamindzija D, Chromikova Z, Spaic A, Barak I, Bernier-Latmani R, Radnovic D (2019) Chromate tolerance and removal of bacterial strains isolated from uncontaminated and chromium-polluted environments. World J Microbiol Biotechnol 35(4):56. https://doi.org/10.1007/s11274-019-2638-5

Tao W, Chen G, Zeng G et al (2016) Influence of silver nanoparticles on heavy metals of pore water in contaminated river sediments. Chemosphere 162:117–124. https://doi.org/10.1016/j.chemosphere.2016.07.043

Thangapandiyan S, Ramesh M, Miltonprabu S, Hema T, Jothi GB, Nandhini V (2019) Sulforaphane potentially attenuates arsenic-induced nephrotoxicity via the PI3K/Akt/Nrf2 pathway in albino Wistar rats. Environ Sci Pollut Res Int 26:12247–12263. https://doi.org/10.1007/s11356-019-04502-w

Thummabancha K, Onparn N, Srisapoome P (2016) Analysis of hematologic alterations, immune responses and metallothionein gene expression in Nile tilapia (Oreochromis niloticus) exposed to silver nanoparticles. J Immunotoxicol 13(6):909–917. https://doi.org/10.1080/1547691X.2016.1242673

Tian T, Ali B, Qin Y et al (2014) Alleviation of lead toxicity by 5-aminolevulinic acid is related to elevated growth, photosynthesis, and suppressed ultrastructural damages in oilseed rape. Biomed Res Int 2014:530642. https://doi.org/10.1155/2014/530642

Toz H, Deger Y (2018) The effect of chitosan on the erythrocyte antioxidant potential of lead toxicity-induced rats. Biol Trace Elem Res 184(1):114–118. https://doi.org/10.1007/s12011-017-1164-2

Tseng CH, Lee IH, Chen YC (2019) Evaluation of hexavalent chromium concentration in water and its health risk with a system dynamics model. Sci Total Environ 669:103–111. https://doi.org/10.1016/j.scitotenv.2019.03.103

Urrialde V, Alburquerque B, Guirao-Abad JP, Pla J, Arguelles JC, Alonso-Monge R (2017) Arsenic inorganic compounds cause oxidative stress mediated by the transcription factor PHO4 in Candida albicans. Microbiol Res 203:10–18. https://doi.org/10.1016/j.micres.2017.06.004

Venter C, Oberholzer HM, Cummings FR, Bester MJ (2017) Effects of metals cadmium and chromium alone and in combination on the liver and kidney tissue of male Spraque-Dawley rats: An ultrastructural and electron-energy-loss spectroscopy investigation. Microsc Res Tech 80(8):878–888. https://doi.org/10.1002/jemt.22877

Vineetha RC, Binu P, Arathi P, Nair RH (2018) L-ascorbic acid and alpha-tocopherol attenuate arsenic trioxide-induced toxicity in H9c2 cardiomyocytes by the activation of Nrf2 and Bcl2 transcription factors. Toxicol Mech Methods 28(5):353–360. https://doi.org/10.1080/15376516.2017.1422578

Wan D, Han Z, Yang J, Yang G, Liu X (2016) Heavy metal pollution in settled dust associated with different urban functional areas in a heavily air-polluted city in North China. Int J Environ Res Public Health 13(11). https://doi.org/10.3390/ijerph13111119

Wang X, Sato T, Xing B, Tao S (2005) Health risks of heavy metals to the general public in Tianjin, China via consumption of vegetables and fish. Sci Total Environ 350(1–3):28–37. https://doi.org/10.1016/j.scitotenv.2004.09.044

Wang TC, Jan KY, Wang AS, Gurr JR (2007) Trivalent arsenicals induce lipid peroxidation, protein carbonylation, and oxidative DNA damage in human urothelial cells. Mutat Res 615(1–2):75–86. https://doi.org/10.1016/j.mrfmmm.2006.10.003

Wang YM, Zhang H, Xiong YT et al (2018a) Leaf aging effects on copper and cadmium transfer along the lettuce-snail food chain. Chemosphere 211:81–88. https://doi.org/10.1016/j.chemosphere.2018.07.141

Wang P, Yin N, Cai X et al (2018b) Nutritional status affects the bioaccessibility and speciation of arsenic from soils in a simulator of the human intestinal microbial ecosystem. Sci Total Environ 644:815–821. https://doi.org/10.1016/j.scitotenv.2018.07.003

Wang C, Pan D, Han H, Hu X (2018c) Vertical profile, contamination assessment of mercury and arsenic in sediment cores from typical intertidal zones of China. Environ Monit Assess 190(6):366. https://doi.org/10.1007/s10661-018-6732-1

Xia H, Liang D, Chen F et al (2018) Effects of mutual intercropping on cadmium accumulation by the accumulator plants Conyza canadensis, Cardamine hirsuta, and Cerastium glomeratum. Int J Phytoremediation 20(9):855–861. https://doi.org/10.1080/15226514.2018.1438356

Xu M, Niu Q, Hu Y, Feng G, Wang H, Li S (2019) Proanthocyanidins antagonize arsenic-induced oxidative damage and promote arsenic methylation through activation of the Nrf2 signaling pathway. Oxidative Med Cell Longev 2019:8549035. https://doi.org/10.1155/2019/8549035

Xue XM, Ye J, Raber G et al (2017) Arsenic methyltransferase is involved in arsenosugar biosynthesis by providing DMA. Environ Sci Technol 51(3):1224–1230. https://doi.org/10.1021/acs.est.6b04952

Yadav A, Khandegar V (2019) Dataset on statistical reduction of highly water-soluble Cr (VI) into Cr (III) using RSM. Data Brief 22:1074–1080. https://doi.org/10.1016/j.dib.2018.12.054

Yang Y, Wu S, Lilley RM, Zhang R (2015) The diversity of membrane transporters encoded in bacterial arsenic-resistance operons. Peer J 3:e943. https://doi.org/10.7717/peerj.943

Yin J, Wang AP, Li WF, Shi R, Jin HT, Wei JF (2018) Time-response characteristic and potential biomarker identification of heavy metal induced toxicity in zebrafish. Fish Shellfish Immunol 72:309–317. https://doi.org/10.1016/j.fsi.2017.10.047

Younis U, Malik SA, Rizwan M et al (2016) Biochar enhances the cadmium tolerance in spinach (Spinacia oleracea) through modification of Cd uptake and physiological and biochemical attributes. Environ Sci Pollut Res Int 23(21):21385–21394. https://doi.org/10.1007/s11356-016-7344-3

Zeng F, Wu X, Qiu B, Wu F, Jiang L, Zhang G (2014) Physiological and proteomic alterations in rice (Oryza sativa L.) seedlings under hexavalent chromium stress. Planta 240(2):291–308. https://doi.org/10.1007/s00425-014-2077-3

Zhan M, Yu H, Li L, Nguyen DT, Chen W (2019) Detection of hexavalent chromium by copper sulfide nanocomposites. Anal Chem 91(3):2058–2065. https://doi.org/10.1021/acs.analchem.8b04501

Zhang N, Zhang J, Wang P et al (2018) Investigation of an antitumor drug-delivery system based on anti-HER2 antibody-conjugated BSA nanoparticles. Anti-Cancer Drugs 29(4):307–322. https://doi.org/10.1097/CAD.0000000000000586

Zhang M, He P, Qiao G, Huang J, Yuan X, Li Q (2019a) Heavy metal contamination assessment of surface sediments of the Subei Shoal, China: spatial distribution, source apportionment and ecological risk. Chemosphere 223:211–222. https://doi.org/10.1016/j.chemosphere.2019.02.058

Zhang Q, Song Y, Amor K, Huang WE, Porcelli D, Thompson I (2019b) Monitoring Cr toxicity and remediation processes - combining a whole-cell bioreporter and Cr isotope techniques. Water Res 153:295–303. https://doi.org/10.1016/j.watres.2019.01.009

Zhang L, Belloc da Silva Muccillo V, CAM VG (2019c) A combined toxicokinetics and toxicodynamics approach to investigate delayed lead toxicity in the soil invertebrate Enchytraeus crypticus. Ecotoxicol Environ Saf 169:33–39. https://doi.org/10.1016/j.ecoenv.2018.10.084

Zhong X, Zeng M, Bian H, Zhong C, Xiao F (2017) An evaluation of the protective role of vitamin C in reactive oxygen species-induced hepatotoxicity due to hexavalent chromium in vitro and in vivo. J Occup Med Toxicol 12:15. https://doi.org/10.1186/s12995-017-0161-x

Zong L, Xing J, Liu S, Liu Z, Song F (2018) Cell metabolomics reveals the neurotoxicity mechanism of cadmium in PC12 cells. Ecotoxicol Environ Saf 147:26–33. https://doi.org/10.1016/j.ecoenv.2017.08.028

Zwolak I (2019) The role of selenium in arsenic and cadmium toxicity: an updated review of scientific literature. Biol Trace Elem Res 193:44–63. https://doi.org/10.1007/s12011-019-01691-w

Heavy Metal–Induced Gene Expression in Plants

Abdul Razaque Memon

Contents

1 Introduction

The increase in industrialization activities in the developed and developing countries has ultimately increased the global threat to the environment that can cause an unprecedented imbalances in the natural ecosystem. For example, industrial activities like mining, leather tanning industry, textile factories, petrol chemical industry, intensive farming, and other manufacturing industries are the main sources for polluting the environment (Bhargava et al. 2012). Each source of contamination has its own damaging effects to plants, animals, and ultimately human health, but those that add heavy metals to soils and waters are of serious concern due to their persistence in the environment and carcinogenicity to human beings.

A. R. Memon (✉)
Department of Molecular Biology and Genetics, Faculty of Science and Letters, Usak University, Usak, Turkey
e-mail: armemon@usak.edu.tr

M. Naeem et al. (eds.), *Contaminants in Agriculture*,
https://doi.org/10.1007/978-3-030-41552-5_7

The contamination of agricultural soil by heavy metals has become a critical environmental concern due to their potential adverse ecological effects. Such toxic elements are considered as soil pollutants due to their widespread occurrence and their acute and chronic toxic effect on plants grown of such soils (Chaney et al. 2018). Heavy metals are defined as metals with a density higher than 5 gcm^{-3}. Around 43 of 90 naturally occurring elements are heavy metals (Cevher-Keskin et al. 2019; Weast et al. 1988), but only some of them have biological importance (Marschner 2011). Based on their solubility under physiological conditions, 17 heavy metals can be absorbed by living cells and could be important for organisms and ecosystems (Weast et al. 1988). Among these metals, Fe, Mn, Zn, Cu, and Mo are important as micronutrients. V, Co, W, and Cr are toxic elements with high or low importance as trace elements. As, Hg, Ag, Sb, Cd, Pb, and U have no known function as nutrients and seem to be more or less toxic to plants and microorganisms (Weast et al. 1988; Memon et al. 2008). The metal contamination caused by the human activities is far greater than the natural process. For example, it was shown that around 13 times more Cu, 15 times more Cd, 21 times more Zn, and 100 times more Pb were released in the atmosphere by human activities than by natural processes (Campbell et al. 1983). Several laborious and expensive methods have been used to clean up these kinds of contaminants, but no optimal results were obtained. Presently, phytoremediation is being used as an alternative, cost-effective, and environmental friendly method for removing the metal pollutants from contaminated soil and water (van der Ent et al. 2018). Numerous accumulator plant species have been identified to absorb and remove toxic elements, such as cadmium, chromium, lead, arsenic, and variety of radionuclides, from the soils successfully. Phytoextraction, which is one of the category of phytoremediation technology, could be used to extract and remove majority of the toxic metals with unknown biological function (e.g., Cd, Cr, Pb, Co, Ag, Se, Hg) from the contaminated soils (Chaney and Baklanov 2017; Memon et al. 2001).

An interesting breakthrough that has emerged from the comparative physiological and molecular analyses of hyperaccumulators and related nonhyperaccumulators is that most key steps of hyperaccumulation rely on different regulations and gene expressions found in both kinds of plants. In particular, a determinant role in driving the uptake and translocation of heavy metals to leaves and finally sequestration in vacuoles or cell walls is one of the important part of detoxification mechanism in hyperaccumulator plants (Memon et al. 2001; Memon and Yatazawa 1982). The constitutive overexpression of genes encoding transmembrane transporters, such as members of ZIPs, HMAs, MATE, YSL, and MTPs families, has been reported in many of these accumulator plants (Memon 2016; Memon and Schröder 2009).

Recently, phytoremediation researchers have discovered that Indian mustard (*Brassica juncea* (L.) Czern, AABB genome) can accumulate high levels of metals, including Zn, Cd, Pb, and Se, and *B. nigra* (BB genome) has been shown a superb Cu accumulator (Cevher-Keskin et al. 2019; Dalyan et al. 2017; Memon and Zahirovic 2014). The metal-accumulating ability of these plants, coupled with the potential to rapidly produce large quantities of shoot biomass, makes these plants ideal for phytoextraction (Memon 2016), and these plant species have received huge attention in phytoremediation field.

2 Phytoremediation

Phytoremediation is an emerging cleanup technology defined as the use of plants to remove or contain or render harmless the contaminants such as toxic metals and organic and radioactive compounds from soils, sediments, and water (Chaney and Baklanov 2017). This technology is environmental friendly and potentially cost-effective. This green technology is unique in way that it takes the advantage of selective uptake capabilities of plant root systems, together with the translocation, bioaccumulation, and contaminant degradation abilities of the entire plant system (Lasat 2002).

However, the ability of plant to accumulate heavy metals varies significantly between species and among varieties within species, as different mechanisms of ion uptake and translocation are operative in each species, based on their genetic, morphological, physiological, and anatomical characteristics. Phytoremediation can be divided into several different categories, such as phytoextraction, phytofiltration, phytostabilization, phytovolatilization, and phytodegradation, depending on the remediation mechanisms (Ali et al. 2013; Memon et al. 2001).

Phytoextraction is also referred as phytoaccumulation, and it takes the approach to remove contaminants from soil without destroying the soil structure and other physical and chemical properties of the soil. The metal ion accumulated in the aerial parts of the plant can be removed to dispose or burnt to recover metals.

Phytofiltration is defined as the use of plants, both land and aquatic, to remove the contaminants from aqueous wastes.

Phytostabilization is generally used to remove the pollutants from the soil by absorbing the pollutants (e.g., toxic metals) through plant roots and keeping them in the rhizosphere, rendering them harmless by preventing them from leaching.

Phytovolatilization involves the use of plants to remove the contaminants from the soil, for example, mercury, selenium, and arsenic, by transforming them into volatile form and volatilize them into the atmosphere.

Phytodegradation is the breakdown of organics pollutants by the use of plants and associated microorganisms (Alkorta and Garbisu 2001). Most of the phytoremediation technologies could be used simultaneously, but the metal removal from the soil depends on its bioavailable form in the soil. The efficient use of phytotechnology depends on the plant species, for example, some plants may have one function, whereas others can have multiple functions of phytoremediation (Lone et al. 2008).

Phytoremediation is considered as environmentally friendly and cost-effective technology for cleaning up the contaminated soils. The phytoextraction is the most common and profitable technique mainly used for extracting heavy metals and radioactive elements from the soil (van der Ent et al. 2018).

There are several important components of phytoextraction which can play an important role in remediating the contaminated soils efficiently, which are as follows: appropriate accumulator plant species, metal availability in the soil and uptake by roots, metal translocation from roots to shoots, and plant tolerance to toxic metals. There are several plant species which are currently proposed for phytoextraction, such as *Arabidopsis halleri*, *Noccaea caerulescens,* and *Alyssum* sps (*A. murale, A. lesbiacum,* and *A. tenium*), which can absorb and accumulate high concentration of Zn and Cd in their shoots. However, the remediation capacity of these plant species is limited because of their slow growth and low biomass. Currently, a number of plant species as metal accumulators have been identified within the Brassicaceae family (Pollard et al. 2014). To understand the genomics of these metal accumulator plants, the vast genetic resources and bioinformatics data developed in model plant *A. thaliana* could be extended to the newly identified metal accumulator species that display traits absent in this model species (Reeves et al. 2017). For developing the new-generation phytoremediation technologies, the data of the multidisciplinary research including high-throughput sequencing, all available -omics data, genetic engineering, plant–microbe interaction, and agricultural and environmental engineering tools could be integrated and be applied for practical purpose in the contaminated areas (Chaney and Baklanov 2017; Memon and Schröder 2009).

3 Plants as Accumulators of Heavy Metals

Plant requires at least six major elements (N, P, K, Ca, Mg, and S) and eight minor elements (Cu, Zn, Mn, Fe, B, Cl, Ni, and Mo) for their growth and development. Plant roots take up the elements selectively, with some being preferentially acquired over others (Memon et al. 2008). The uptake and accumulation of heavy metals by plant roots depend on an interrelated network of physiological and molecular mechanisms, such as (i) binding of the metals to extracellular exudates and cell wall components; (ii) movement of the metals from cytoplasm to the vacuolar compartment; (iii) complexation of the metal with various compounds in the cell, such as several amino acids, organic acids, small molecular weight proteins like metallothioneins, and small metal-binding peptides like phytochelatins; (iv) heavy metal-induced antioxidative enzymes; and (v) modification of plant metabolism and quick repair and recover of damaged cell structures (Paunov et al. 2018).

Baker and co-workers (Baker et al. 2000), while working on the vegetation of metalliferous soils, have classified plants into three categories: (i) excluder plants, which prevent the uptake of toxic metals into root cells and keep the metal level in their shoots very low (De Vos and Schat 1991). Excluders can be used in the polluted soils for stabilization and to prevent further spread of contamination due to erosion (Lasat 2002). (ii) Accumulator plants can accumulate huge amount of metals in their aboveground parts when grown at either low or high metal soil concentrations. Accumulators have high metal uptake rate in the roots and can efficiently transport them in the shoots and have high metal accumulation capacity (Pollard et al. 2014). (iii) In indicator plants, internal metal concentration reflects the external levels (McGrath et al. 2002).

4 Hyperaccumulators

Plants with exceptional metal-accumulating capacity are known as hyperaccumulator plants. The hyperaccumulator plants, which are naturally growing in metal-rich habitats, can accumulate 100- to 1000-fold higher levels of metals than normal plants, since these plants take up two or three orders more of metals from the soil than plant species growing on uncontaminated soils (Lone et al. 2008). Accumulator species are naturally capable of accumulating heavy metals in their aboveground tissues, without developing any toxicity symptoms. They are often endemic to naturally mineralized soils and can accumulate metals at different level depending on metal species. For example, they can accumulate to a level of 0.1% (on leaf dry weight basis) for Ni, Co, Cr, Cu, Al, and Pb; at 1% level for Zn and Mn; and at 0.01% level for Cd and Se (Baker and Brooks 1989; Baker et al. 2000).

Considerable research work has been carried out in identifying the accumulator plant species and their mechanisms of metal uptake and hyperaccumulation. The field of phytotechnology has been revolutionized and got high momentum after the discovery of hyperaccumulator plant species since these plants have excellent capacity to absorb and accumulate metals at levels 50–500 times greater than average plants (Lasat 2000). Hyperaccumulator plants which are generally restricted to metalliferous soils and accumulate metals in their aboveground parts especially in the leaves are classified as "obligate" hyperaccumulators (Reeves et al. 2017). The other type of hyperaccumulator plants which can accumulate metal from both metalliferous and nonmetalliferous soils is classified as "facultative" hyperaccumulators (Pollard et al. 2014). This latter category includes many plant species which can accumulate metal not only from ultramafic soils but also from metal polluted soils. For example, *Biscutella laevigata* accumulates >1% thallium (Babst-Kostecka et al. 2016), *Pteris vittata* with up to 2.3% arsenic (Ma et al. 2001), and *Phytolacca americana* accumulates >1% manganese (Xu et al. 2009).

Around 450 plant species from 45 angiosperms families have been reported as metal hyperaccumulators, which include the members of the *Asteraceae, Brassicaceae, Caryophyllaceae, Cyperaceae, Cunoniaceae, Fabaceae, Flacourtiaceae, Lamiaceae, Poaceae, Violaceae,* and *Euphorbiaceae* (Padmavathiamma and Li 2007; Reeves et al. 2017). Interestingly, a large number of accumulator plant species are reported to be in *Brassicaceae* family especially in the genera *Alyssum* and *Noccaea*, wherein accumulation of more than one metal has been reported (Reeves and Baker 2000; Vamerali et al. 2010; Vara Prasad and de Oliveira Freitas 2003; Verbruggen et al. 2009). *Pteris vittata* (Chinese brake fern) is reported to accumulate up to 95% of the arsenic in its fronds (Ma et al. 2001; Zhang et al. 2002). *Noccaea caerulescens* (pennycress) is a well-known metal hyperaccumulator which can accumulate large amounts of Zn (39,600 mg/kg) and Cd

(1800 mg/kg) without any apparent damage (Basic et al. 2006; Hanikenne and Nouet 2011; Rascio and Navari-Izzo 2011). This diploid and self-pollinating plant can be easily grown under laboratory conditions and is an excellent experimental material for studying the mechanisms of metal uptake, accumulation, and tolerance. Apart from *N. caerulescens, Brassica juncea* and *B. nigra* have also been used as a model system to investigate the physiology and biochemistry of metal accumulation in plants especially for Zn, Cd, and Cu (Cevher-Keskin et al. 2019; Memon et al. 2001). These crop plants with high biomass production could be excellent candidates for phytoremediation in coming years.

Memon and co-workers reported several Mn accumulator plant species, which accumulated huge amount of Mn in their leaves. For example, *Acanthopanax sciadophylloides* Frach. & Sav. (Mn: 4600 ppm), *Ilex crenata* Thunb. var. paludosa (Mn:1155 ppm), and *Clethra barbinervis* Sieb. & Zucc. (Mn: 1374 ppm) accumulated very high amount of Mn in their leaves while naturally growing in the forest soil containing very low metal. These plants are possibly facultative Mn accumulators and can grow in both nonmetalliferous and metalliferous soils (Memon et al. 1979; Memon and Yatazawa 1982). Among these Mn accumulator plant species, *Acanthopanax sciadophylloides* was found to be superb Mn accumulator, achieving >1% Mn in their leaves when growing on soils with only background concentrations of this element (Memon et al. 1979; Memon and Yatazawa 1982). Similarly, *Noccaea caerulescens* and *Arabidopsis halleri* were also reported to be facultative Zn accumulators, achieving >1% foliar Zn when growing in soils containing very low Zn concentration (Reeves et al. 2001; Stein et al. 2017). Majority of the accumulator plants are reported to be grown in ultramafic soils and hyperaccumulate Ni, and some of them accumulate Ni, Co, and Mn (Reeves et al. 2018). Many obligate and facultative hyperaccumulator plants have reported to hyperaccumulate several toxic elements, for example, Cd, Cu, Co, Mn, Pb, Zn, Se, Tl, and some rare earth elements (Reeves et al. 2017).

The efficiency of the cleanup of metal contaminated soils by the use of accumulator plants depends on their biomass production as well as on their bioconcentration factor (BCF ratio of metal concentration in the shoot tissue to the soil) (McGrath and Zhao 2003). The BCF varies with plant species and is determined by the capacity of the roots to absorb, accumulate, store, and detoxify metals while maintaining metabolism, growth, and biomass production (Clemens et al. 2002; Gleba et al. 1999; Guerinot and Salt 2001). Hyperaccumulators have a bioconcentration factor greater than 1, sometimes reaching as high as 50–100 (Reeves et al. 2018). The metal bioconcentration factors in nonaccumulator plants is less than 1, which means that they are not able to reduce soil contamination by 50% for longer time (time longer than human lifespan) (Peuke and Rennenberg 2005). Multiple mechanisms are involved for metal tolerance in accumulator plants, such as high cell wall–binding capacity, active transport of metal ions into the vacuolar compartment, and formation of complexes with organic acids, metallothioneins, and/or chelation with phytochelatins (Memon et al. 2001; Memon and Schröder 2009).

There is general consensus that metal hyperaccumulation is an evolutionary adaptation by specific plants which can survive and live in naturally metal-rich habitats that confer on them high metal tolerance qualities, drought tolerance, and protection against herbivores or pathogens (Reeves et al. 2018). The most accepted hypothesis is that the hyperaccumulation character protects the plant against pathogens and herbivores (Boyd and Banzhaf 2007; Galeas et al. 2008; Huitson and Macnair 2003; Martens and Boyd 1994; Noret et al. 2007). However, the detailed mechanisms of metal uptake and tolerances need to be worked out in coming years.

Notable efforts were carried out to assemble a database for hyperaccumulator plants. For example, Environment Canada's PHYTOREM database and the METALS (metal-accumulating plants) database originally maintained by the Environmental Consultancy, University of Sheffield (ECUS Ltd., Sheffield, UK), were created. The problem in these databases is that not only accumulators were recorded but also other plant species grown in metalliferous soils were included. (Reeves et al. 2017). To circumvent this problem, a new database was established which was confined only with hyperaccumulator plants, and currently, around 700 pint species have been recorded and will eventually increase to 1000–1500 species. The Global Hyperaccumulator Database (www.hyperaccumulators.org) went online in 2015 under the administration of the Center for Mined Land Rehabilitation of the University of Queensland, Brisbane, Australia. This new database gives the information about all known metal and metalloid hyperaccumulator plant species and contains detailed information about the taxonomy, distribution, ecology, collection records, analytical data, and other useful studies related to these species (Reeves et al. 2017). This database is continuously updated and is freely available to all researchers.

5 Brassicaceae

The *Brassicaceae* (order Brassicales) is the largest family in Brassicales, commonly known as the mustard family, and is composed of approximately 338 genera and 3700 species. Genus Brassica contains around 100 species, including important oilseed crops (e.g., *B. napus, B. juncea*) and many common vegetable plants such as broccoli, cauliflower, cabbage, Brussels sprouts, radishes, turnip, and various gardening plants (wallflower, sweet alyssum, rock cress, etc.) (Ozturk et al. 2012; Warwick and Black 1991). The Brassica species are closely related to the model plant *Arabidopsis thaliana*, and their chromosome numbers vary from 2n = 10 to 2n = 38 (Lysak et al. 2005). Some of the Brassica species are diploid (e.g., *B. rapa, B. nigra, B. oleracea*), and some of them are allotetraploid (e.g., *B. napus, B. juncea, B. carinata*). The genome of *B. rapa* is the smallest, at ca. 529 Mb, and *B. napus* is the largest one, at ca 1132 Mb, in Brassica species represented in the "Triangle U" (Memon and Zahirovic 2014). The genome of both plants has been sequenced, and both sequences and sequence annotations are available in the public domain: http://www.brassicagenome.net/databases.php; https://www.ncbi.nlm.nih.gov/genome/?term=brassica%20napus (Memon 2016; Wang et al. 2016). The annotated Arabidopsis genome sequence can be exploited as a tool for carrying out the comparative analysis of *Arabidopsis* and *Brassica* genomes, for example, metal accumulator and tolerant species, *B. napus* and *B. juncea*, respectively. The biology of both *Arabidopsis* and *Brassica* is similar, and the comparative genetic mapping between species of *Brassicaceae* showed collinear blocks even though the species differed with respect to genome size, base chromosome number, and ploidy. Around 80–90% homology was found between the exons of putative orthologous genes in *Arabidopsis* and *Brassica*, and this clearly indicates that the knowledge from Arabidopsis is highly relevant for gene isolation and characterization in Brassica crops (Ozturk et al. 2012).

6 Brassica Species

The plant species in Brassica are the major source of vegetable oil in the world after palm and soybean oil. These plant species produce high biomass and also accumulate and tolerate high amount of metals in their tissues (Anjum et al. 2013; Kumar et al. 1995). There are six agronomical important species of Brassica which are commercially used for both oil and vegetable production, and among them, three are diploids (*B. nigra, B. oleracea,* and *B. campestris*) and the other three are amphidiploids (*B. juncea, B. napus,* and *B. carinata*). The genomic variation and relationships between these species have been described in the form of triangle, and in the literature, it is known as triangle of U (Nagaharu 1935; Ozturk et al. 2012). Furthermore, an extensive research work on metal tolerance and accumulation was carried out, and the differential metal accumulation pattern among Brassica species has been observed (Anjum et al. 2013; Diwan et al. 2010).

Several Brassica species have been reported to exhibit higher tolerance toward most of the toxic metals/metalloids, and majority of Brassica species are now known as good accumulators of toxic metals (including Cd, Cu, Ni, Pb, U, Zn) (Kumar et al. 1995; Ozturk et al. 2012), allocating large amounts of majority of these metals into aboveground parts, thus stand second to none in terms of their utility in toxic metal-remediation strategies. Moreover, as several species of Brassica are able to produce significant amounts of biomass (a required trait for phytoremediation) and are adaptable to a range of environmental conditions, there is the potential to develop superior genotypes of *Brassica* sps for phytoremediation through selection and breeding techniques (Anjum et al. 2012).

Brassica juncea needs special attention in the field of phytoremediation because it is not only a metal tolerant but also accumulates huge amount of several metals in its shoots, including Zn, Cd, and Pb. It is shown that *B. juncea* accumulates high amount of Cd in the shoots (1450 μg Cd/g dry wt), which is three times more than reported in *Brassica napus* (555 μg/g dry wt). In addition, this plant also extracts high amount of other metals from soil, such as Pb (28% reduction) and Se (reduced between 13% and 48%) (Szczygłowska et al. 2014). Interestingly, this plant is more effective at removing Zn from soil than *Noccaea caerulescens*, a known hyperaccumulator of zinc. This is due to the fact that *B. juncea* produces ten times more biomass than *N. caerulescens* (Anjum et al. 2013; Szczygłowska et al. 2014). In Southeastern Anatolia, several endemic metal accumulator was discovered, and among them, *Brassica nigra* Diyarbakir ecotype was found to be Cu accumulator (Cevher-Keskin et al. 2019). When this ecotype was regenerated from callus culture and grown in soil containing 500 uM Cu, the shoots accumulated around 20,000 ug g^{-1} DW Cu, which was around three times more than in the roots (Memon and Zahirovic 2014; Ozturk et al. 2012).

The high metal accumulation capacity of several plant species from the Brassicaceae family indicates that these accumulator plants especially *B. juncea* and *B. nigra* could be important candidates for phytoremediation of Zn, Cu, Cd, and Pd from the contaminated soils (Cevher-Keskin et al. 2019; Dalyan et al. 2017; Kumar et al. 2012; Memon and Zahirovic 2014).

7 Heavy Metal Uptake, Accumulation, and Subcellular Localization

Highly specific and very efficient mechanisms have been developed by plants in order to take up essential micronutrients from the soil, even when present at low quantities. Plant-induced pH changes, redox reactions, and plant-produced chelating agents in the rhizosphere help plants to absorb trace elements even from poor nutrient soils and translocate and store them in vacuole and other organelles. The same mechanism is also involved in the uptake, translocation, and storage of toxic elements, whose chemical properties simulate those of essential elements. Thus, the metal uptake, translocation, and accumulation mechanisms are of much interest in the area of phytoremediation (Memon and Schröder 2009; Tangahu et al. 2011).

B. nigra is shown to be a Cu accumulator (Memon et al. 2008), and our results with microarray analysis showed that many genes especially metal ATPases and other metal transporters were several hundred fold upregulated in the shoots of *B. nigra* Diyarbakir ecotype when plants were subjected to 500 uM Cu concentration (Memon and Zahirovic 2014). This Diyarbakır ecotype was named as facultative metallophyte because it can grow both at low metal and at very high metal concentration (e.g., 500 uM Cu) in the soil. We used Arabidopsis ATH1 genome array to analyze the gene expression pattern in our facultative accumulator *B. nigra* Diyarbakir ecotype. The gene expression pattern in the roots and shoots of *B. nigra* was compared when grown at 0 and 25 uM Cu. The differential response in terms of gene expression in roots and shoots was observed when plants were subjected to 25 uM Cu concentration for 72 h. The response to Cu was much stronger in roots (88 genes showed increased or decreased mRNA levels) than in leaf tissues (24 were up- or downregulated). These genes were identified as metal transporters, signal transduction and metabolism-related genes, and transport facilitation genes. Glutathione pathway-related genes (γ-ECS, PC, etc.) were also identified, and their mRNAs were differentially expressed in root and shoot tissues (Cevher-Keskin et al. 2019).

The range of known transport mechanisms or specialized proteins embedded in the plant cell plasma membrane involved in ion uptake and translocation include (1) proton pumps ATPases that consume energy and generate electrochemical gradients and (2) co- and anti-transporters (proteins that use the electrochemical gradients generated by H^+-ATPases to drive the active uptake of ions) and channels (proteins that facilitate the transport of ions into the cell). Each transport mechanism is likely to take up a range of ions. Plant uptake-translocation mechanisms are presumably closely regulated (Tangahu et al. 2011). Most of the crop plants require small amount of micronutrients for their metabolism and normal growth, which generally vary from 10 to 15 ppm. Some other plants which are labelled as "hyperaccumulators" can take up toxic metals at levels in the thousands of ppm without showing any toxicity symptoms. It is intriguing that how these accumulator plants cope with metal toxicity and how they avoid metal toxicity. What are the mechanisms responsible to make these metals in innocuous form? Multiple mechanisms are involved, especially the storage in the vacuole appears to be one of the major mechanism for metal detoxification (Memon et al. 2001; Reeves et al. 2018; Tangahu et al. 2011).

Clarkson and Lüttge (1989) reported that Cu and Zn, Ni and Cd compete for the same membrane carriers. Different chelators are reported to be involved in the translocation of metal cations through the xylem, such as organic acid chelators (e.g., malate, citrate, histidine (Salt et al. 1995; von Wirén et al. 1999), or nicotinamide (Stephan et al. 1996; von Wirén et al. 1999). Since the metal is complexed within a chelate, it can be translocated upward in the xylem without being adsorbed by the high cation exchange capacity of the xylem (von Wirén et al. 1999).

Several metal tolerant and accumulation mechanisms in plants growing in metal contaminated soils have been suggested (Memon 2016; Memon and Schröder 2009). Hyperaccumulation in plants is a part of metal homeostasis network in which metal is efficiently taken by the roots and then transported from the roots to shoots through xylem and then either complexed and sequestered in the subcellular compartments or secreted in the trichomes (Hanikenne and Nouet 2011; Memon and Schröder 2009; Ovečka and Takáč 2014). These accumulator plants have very unique eco-physiological character and have the ability to uptake, transport, and accumulate huge amount of metals in their shoots and leaves and compartmentalize them in the cell wall, vacuole, and other subcompartments in the cytosol in order to keep them away from metabolic activities in the cell (Memon and Schröder 2009; Memon and Yatazawa 1982). To determine the subcellular localization on Mn in the leaves of *Acanthopanax sciadophylloides,* electroprobe X-ray microprobe analysis was carried out with fresh leaves frozen down in liquid nitrogen. The microdistribution pattern of Mn at subcellular level showed that most of the Mn was deposited in the cell wall and vacuolar compartment of epidermal cells and it was kept away from metabolically active compartments, for example, cytosol, mitochondria, and chloroplast (Fig. 1; Memon and Yatazawa 1984). Interestingly, Mn was predominantly accumulated in the epidermal cells (E) and was almost absent from the cells of stomatal complex (Fig. 2, A. R. Memon, Unpublished data).

The metal distribution and accumulation pattern vary with plant species and the type of element. For example, *A. halleri* leaves accumulates Zn and Cd more in the

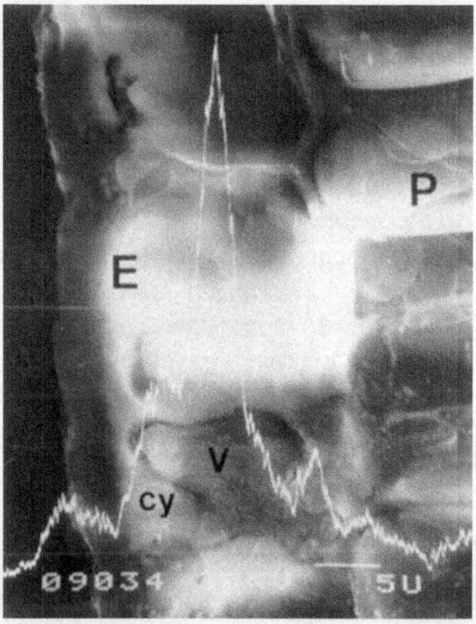

Fig. 1 Secondary electron image (SEI) showing a line scan profile of a leaf section of tea plant with Mn (Kα radiation) peaks. cy, cytosol; V, vacuole of epidermal cells (E); P, palisade parenchyma cells

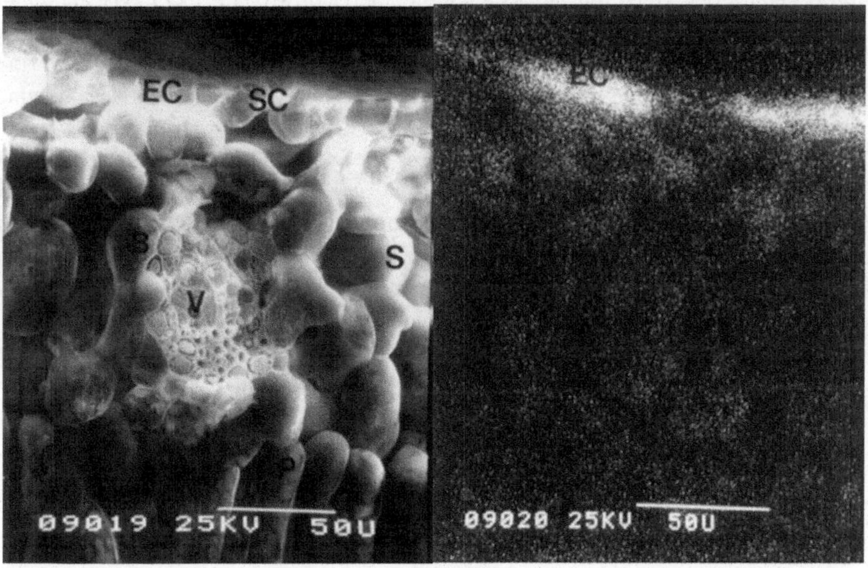

Fig. 2 Secondary electron (SEI) and Mn X-ray distribution images of the abaxial side of a tea leaf. A, SEI. B, Mn X-ray distribution image. EC, epidermal complex; SC, stomatal complex; S, spongy parenchyma; V, vascular bundle; P, parenchyma cells

mesophyll cells than in the epidermis (Küpper et al. 2000), whereas *N. caerulescens* accumulates six-fold higher Zn and Cd in epidermis cells than in mesophyll cells (Küpper et al. 1999). Interestingly, *B. juncea* which is reported to be metal tolerant and accumulator (Dalyan et al. 2017) and is a good candidate for phytoremediation is reported to accumulate 40-fold higher Cd in trichomes compared to the total leaf (Salt et al. 1995). Other hyperaccumulator species like *Alyssum lesbiacum* accumulate a significant portion of Zn and Ni in leaf trichomes (Reeves et al. 2018).

Cell fractionation analysis with *A. sciadophylloides* Frach & Sav leaves showed that most of the Mn was present in cell wall and in supernatant, and a very large amount of Mn in the supernatant fraction was found to be bound with organic acid having a molecular weight of approximately 145 g/mol (Memon and Yatazawa 1984). In order to identify the chemical forms of accumulated Mn in the cell, a high-performance liquid chromatography and high-voltage paper electrophoresis analysis were carried with the leaf extracts, and results showed that Mn was chelated with oxalic acid in vacuolar compartment (Memon and Yatazawa 1984). The detoxification mechanism of Mn proposed that Mn^{2+} is taken up at plasma membrane and binds with malate in cytoplasm, and this Mn-malate complex is transported through the tonoplast membrane into the vacuoles, where Mn dissociates from malate and complexes with oxalate. Here, malate functions as "transport vehicle" through the cytoplasm and oxalate as the "terminal acceptor" in the vacuole (Memon and Schröder 2009; Memon and Yatazawa 1984). Many other mechanisms have been proposed for detoxification and accumulation depending on the metal type and plant species (Isaure et al. 2015; Memon 2016; Rascio and Navari-Izzo 2011). Phytochelatins, metallothioneins, metalloenzymes, metal-activated enzymes, and various channel proteins have been reported to bind and sequester metals (Hanikenne and Nouet 2011; Isaure et al. 2015; Memon 2016).

Most of the current research on genetic, genomic, and transcriptome analysis of transporters and the participation of multiple gene families in response to metal stress have been obtained in a two-model accumulator plant species such as *Noccaea caerulescens* and *Arabidopsis halleri* (Verbruggen et al. 2009). The recent development of next-generation sequencing technologies, transcriptomics (microarray analysis, RNA-seq analysis), proteomics, and metabolomics opens up the new avenue to understand the function and regulation of genes, proteins, and metabolites in the cell when encountered high metal concentration in the environment (Verbruggen et al. 2013).

8 Heavy Metal Transporters in Plants

Recently, molecular biology and genetics studies with hyperaccumulators plants have identified several important gene families involved in metal transport and could play a key role in metal tolerance. So far, several classes of metal transporter proteins have been identified in plants, and they fulfil many important functions ranging from metal absorption, transport, sequestration, and storage in specific organelles. They also play an important role in metal homeostasis in plant cell (Memon 2016). Metal transporters are classified into six main groups, which include natural resistant-associated macrophage protein (NRAMP), ZRT/ RT-like protein (ZIP), cation diffusion facilitator (CDF) transporters, yellow stripe-like (YSL) proteins, and P_{1B}-type heavy metal ATPases (HMAs) (Memon 2016; Merlot et al. 2018). Plants grown in polluted or metalliferous soils have evolved several mechanisms in order to deal with toxic quantities of heavy metals, such as Cd, Ni, Cr, Pb, or Hg, or toxic excess of essential elements, such as Zn, Cu, Mn, B, and Co. A fine control of metal homeostasis is required to overcome the toxic and oxidative damage at subcellular level. Thus, it is not surprising that plant can activate several transporters which can function either in excluding metal at the root or in sequestering them at subcellular level in some compartments such as vacuole. *Arabidopsis thaliana* genome has uncovered several families of transporter genes, which include 15 members of zinc and iron transporters (ZIP), 8 members of cation diffusion facilitator (CDF), 6 members of copper transporters (CTR), 6 members of NRAMP homologues, and 8 members of Cu, Zn/Cd transporting ATPases (Mäser et al. 2001; Merlot et al. 2018) (http://www.cbs.umn.edu/arabidopsis/). In addition, some other transporter families, for example, vacuolar cation proton exchanger (CAX) and ABC transporters, are also involved in metal homeostasis in plant cells (Colangelo and Guerinot 2006; Hall and Williams 2003; Memon 2016; Memon and Schröder 2009; Sarma et al. 2018).

8.1 ZIP Gene Family

ZIP transporters are involved in the transport of four essential micronutrients: Zn, Fe, Mn, and Cu (Cohen et al. 2004; Eide et al. 1996; Grotz et al. 1998; Lin et al. 2009; Pedas et al. 2008; Pence et al. 2000; Wintz et al. 2003). This family is derived from the first defined members that include ZRT- (Zn-regulated carrier) and IRT- (Fe-regulated carrier) like proteins. It has been reported that ZIP family members can also transport heavy metals such as Cd (Zheng et al. 2018). Therefore, the ZIP family can play an important role in the transport of various heavy metals, both essential and nonessential (toxic) (Guerinot 2000; Pence et al. 2000; Rogers et al. 2000). They have been discovered in both dicots and monocots, including Arabidopsis, Medicago (Milner et al. 2013; Stephan et al. 1996; Stephens et al. 2011), rice, maize, and barley (Chen et al. 2008; Li et al. 2013; Stephens et al. 2011; Tiong et al. 2015). Arabidopsis has 15 members of ZIP transporters, but recently, 18 ZIPs from Arabidopsis and 16 ZIPs from rice has been annotated (Ivanov and Bauer 2017). To date, only three members of the ZIP family (AtIRT gene) have been isolated from Arabidopsis. The role and functions of Arabidopsis AtIRT1, AtIRT2, and AtIRT3 have been described in detail (Grotz et al. 1998; Guerinot 2000). Little is known about the functionality of the other 12 ZIP members in Arabidopsis. Recently, a distinct expression pattern of ZIP genes in Arabidopsis and rice was found in response to Cd stress (Zheng et al. 2018). ZIP genes in Arabidopsis were mainly upregulated in roots, while in rice, they were upregulated in shoots when subjected to Cd stress. Most of the characterized ZIP proteins are reported to be targeted to plasma membrane; however, some of them are found to be located in tonoplast or other endomembrane system (Milner et al. 2013; Ricachenevsky and Sperotto 2016). NcZNT1 is an important ZIP-like transporter involved in Zn and Cd hyper-accumulation and tolerance in *N. caerulescens* (Lin et al. 2016). It is orthologue of AtZIP4 from *A. thaliana* having 90% cDNA and 87% amino acid identity (Lin et al. 2014). NcZNt1 is differentially expressed between *A. thaliana* and *N. caerulescens* due to its differences in cis- and trans-regulatory elements. It is constitutively expressed in the stele of the roots of *N. caerulescens* and is responsible for xylem-mediated translocation of metals to the shoot (Lin et al. 2016).

8.2 CDF-Type Family

The cation diffusion facilitator (CDF) proteins play an important role in metal homeostasis and tolerance (Mäser et al. 2001). CDFs are membrane bound proteins and transport zinc and other heavy metal ions. They are involved in metal tolerance/resistance by efflux of ions and share a two-modular architecture consisting of a transmembrane domain (TMD) and C-terminal (Ricachenevsky et al. 2013) domain (CTD) that protrude into the cytoplasm (Kolaj-Robin et al. 2015). The first CDF gene in plants is zinc carrier ZAT1 characterized in *A. thaliana*. It was then renamed to metal tolerance protein 1 (MTP1) (Ricachenevsky et al. 2013). ZAT is expressed primarily throughout the plant, and its expression is enhanced by increase in Zn concentration. These carriers are known to play an important role in tolerating Zn^{2+}, Cd^{2+}, or Co^{2+} ions in plants, yeast, bacteria and animals (Memon 2016).

8.3 COPT Gene Family

High-affinity Cu carriers (CTRs, COPTs in other organisms other than plants) are essential components of the major pathway for cellular high-affinity Cu uptake. Six members of the putative Cu family (COPT1-6) have been identified in Arabidopsis, and three of them COPT1, COPT2 and COPT6 are localized in plasma membrane (Peñarrubia et al. 2010; Sancenón et al. 2003; Yuan et al. 2011). COPT1 is required in Cu acquisition when Cu availability decreases in the soil (Sancenón et al. 2004; Yamasaki et al. 2009), whereas COPY2 is a cell surface transporter and expressed mostly in all plant parts, especially in roots, young leaves, apical meristems, tri-chomes, and anthers (Perea-García et al. 2013). COPT5 is reported to be localized in both the tonoplast membrane (Klaumann et al. 2011) and the prevacuolar com-partment (Perea-García et al. 2013) and is believed to be involved in intracellular homeostasis. The plasma membrane-localized COPT6 functions in Cu distribution in plants and is expressed in seeds and leaves during Cu deficiency (Perea-García et al. 2013). In Cu deficiency, the increase in transcript levels of COPT1 and COPT2 is activated by the SPL7 transcription factor (Bernal et al. 2012; Perea-García et al. 2013). CTR/COPT family Cu transporters have also been identified in rice (Yuan et al. 2011).

8.4 Nramp Gene Family

Nramp gene family defines a new family of proteins involved in the transport of divalent metal ions. This gene family is highly conserved during evolution and is involved in the transport of metal ions in a wide range of living organisms, including bacteria, yeasts, insects, mammals, and also higher plants (Cellier 2012; Gruenheid et al. 1995). It encodes integral membrane polypeptide containing 10–12 transmembrane domains, 1–2 extracellular loops, and an intracellular domain with structural characteristics of a transporter protein (Bairoch 1993).

NRAMP genes play an important role in the transport several metal ions including Fe, Mn, Zn, and Cd (Nevo and Nelson 2006). In plants, this family was first described in rice, and three Nramp genes (OsNramp 1–3) were reported (Belouchi et al. 1997; Cellier et al. 1995; Hall and Williams 2003). OsNramp3 is expressed in both roots and shoots, whereas OsNramp2 is expressed only in the leaves (Belouchi et al. 1997). In *Arabidopsis thaliana*, six Nramp genes have been identified and classified into two subfamilies. AtNRAMP1 and AtNRAMP6 comprise the first group, and AtNRAMP2 to AtNRAMP5 belong to the second (Mäser et al. 2001). AtNRAMP1 is localized to both the plasma membrane and intracellular membranes including the Golgi apparatus (Agorio et al. 2017; Cailliatte et al. 2010) and can also transport iron (Fe), manganese (Mn), and cadmium (Cd). AtNRAMP2 is located in trans-Golgi network and is activated in plant roots when subjected to low Mn concentration (Gao et al. 2018). AtNRAMP3 and AtNRAMP4 are required for Mn transport from vacuolar compartment to chloroplast and are localized to the tonoplast membrane. Double mutant of these transporters reduced the function of photosystem II and impaired plant growth (Lanquar et al. 2010). OsNramp5 is involved in Mn uptake and is localized in the plasma membrane of rice roots (Ishimaru et al. 2012). In soybean genome, 13 NRAMP genes have been identified (Qin et al. 2017). Gene expression analysis showed that GmNRAMP is differentially regulated by deficiencies of major elements in the cell like N, P, K, Fe, S and also regulated by the toxicities of Fe, Cu, Cd, and Mn (Qin et al. 2017). *Brassica napus* genome contains 22 NRAMP transporter genes, and based on sequence identity, these transporters are classified into six subfamilies. Nineteen NRAMP transporters were confirmed by RNA-seq analysis, and 10 NRAMP genes were differentially expressed under Cd exposure (Meng et al. 2017). BnNRAMP1b was constitutively expressed throughout all developmental stages of *B. napus* plant and can be strongly induced in seedlings exposed to high Cd concentration. BnNRAMP1b was found to be cleaved by miR167, suggesting that it is posttranscriptionally regulated by miR167 in *B. napus* under Cd stress (Meng et al. 2017). A new rice line with extremely low Cd content has been developed by knocking out OsNramp5 transporter using the CRISPR/cas9 technique (Tang et al. 2017). This transgenic line can grow in Cd-contaminated paddy soils and accumulate very low Cd in their grain (0.05 mg/kg); in contrast, wild-type India rice had high Cd concentration in their grain ranging from 0.33 to 2.90 mg/kg when grown in the same soil (Tang et al. 2017). Excessive Cd in rice grain is a serious problem to health (e.g., itai-itai disease) of those who consume rice as a staple food. These mutant rice lines with very low Cd content in their grains could be useful material to develop new hybrid lines, which could be commercially used in the Cd-contaminated areas, and their grains can be consumed for dietary purpose.

8.5 ABC-Type Family

ATP-binding cassette (or ABC) proteins constitute one of the largest proteins families and are present in all organisms ranging from bacteria to humans (Henikoff et al. 1997). ABC transporters are ATP-driven pumps comprising two nucleotide-binding domains (NBDs) and two transmembrane domains (TMDs). Three structural type ABC transporters have been reported. One is full-type transporter containing two membrane domains (TMD) and two nucleotide-binding domains (NBD) and largely expressed as TMD–NBD–TMD–NBD or NBD–TMD–NBD–TMD. The second one is called as half transporters containing one TMD and one NBD domain and dimerizes in pairs to form virtual homodimers or heterodimers transporters. The third-type transporter contains two NBDs but no TMDs (Verrier et al. 2008). The NBD domain in all structural type of ABC transporters contains some conserved motifs, which are as follows: Walker A, Q-loop, Walker B, D-loop, switch H-loop, and a signature motif (LSGGQ). Each motif has its specific function, for example, the D-loop functions as holding dimers together, the switch H-loop interacts with transmembrane domain, the P-loop (Walker A and B motifs together) binds ATP, the other two loops Q- and H- are important for interaction with the γ-phosphate of the ATP (Davidson et al. 2008; Lane et al. 2016). The signature motif (LSGGQ) is specific to ABC transporter proteins and distinguishes them from ATPases (Davies and Coleman 2000).

In humans, seven to eight subfamilies of ABC transporters have been identified based on their primary sequence and domain structure (Dean and Allikmets 2001). In Arabidopsis and in rice, 130 and 133 ABC transporters, respectively, have been identified (Hwang et al. 2016; Lefèvre et al. 2015). Among them, only a few numbers of transporters have been functionally characterized, and a limited number of transporters have also been identified in other plant species, for example, wheat, maize, and *Vitis vinifera* (Bhati et al. 2015; Cakir et al. 2013; Kang et al. 2011a; Pang et al. 2013).

Recently, 314 ABC transporters in *Brassica napus* genome have been identified and are classified into eight subfamilies from A-G and I. The ABCG transporters constitute the largest subfamily with 116 members followed by ABCB (69 members) and ABCC (47 members) (Kang et al. 2011b; Lane et al. 2016; Yan et al. 2017; Yang et al. 2016), and the ABCB and ABCC subfamilies rank second and third with 69 and 47 members, respectively. Most of the ABC genes were validated by RNA sequencing. Among 132 genes which were differentially expressed, 84 genes were significantly expressed by Cd stress (Yan et al. 2017). Recently, AtABCC1 and AtABCC2 have been implicated in phytochelatin-mediated Cd and Hg detoxification by vacuolar sequestration (Park et al. 2012). Furthermore, OsABCB14 is shown to be responsible for Fe homeostasis in rice (Xu et al. 2014), and OsABCG43/PDR5 is induced in rice roots when subjected to Cd stress and may be involved in detoxification of Cd by compartmentalizing Cd into organelles (Oda et al. 2011; Xu et al. 2014). Above studies demonstrate that ABC transporters play a central role in detoxification and accumulation of toxic metals in hyperaccumulator/tolerant plants.

9 P_{1B}-Type Metal ATPases (HMAs)

P_{1B}-type metal ATPases are a subfamily of P-type ATPases and are responsible for the transport of a number of important and potentially toxic metals (Cu^{2+}, Zn^{2+}, Cd^{2+}, Pb^{2+}) along the cell membranes (Argüello 2003; Axelsen and Palmgren 2001; Østerberg and Palmgren 2018; Rensing et al. 1999). These ion pumps use active ATP energy to pump various charged substrates along biological membranes and share a common enzymatic mechanism in which ATP hydrolysis is carried out for the transport of ions from the membrane (Axelsen and Palmgren 2001). These ATPases can be classified into several subfamilies according to both DNA sequence and functional protein similarities: for example, H^+-ATPases (type 3A) in plants and fungi, Na^+/K^+-ATPases in animals, Ca^{2+}-ATPases (type 2C/D) and heavy metal ATPases (type 1B) in bacteria, plants, and mammalian system (Axelsen and Palmgren 2001; Rosenzweig and Argüello 2012). P_{1B}-ATPases, in addition to the conserved regions in the P-type ATPases, such as DKTGT, GDGxNDxP, PxxK, and S/TGE, possess six to eight transmembrane domains (TMs), a HP locus, and a CPx/SPC motif (Williams and Mills 2005) necessary for metal transport, and putative metal-binding sites in the N and/ or C terminus (Williams and Mills 2005). These metal ATPases based on their substrate specificity are dived into two subgroups, Cu/Ag (Cu^+-ATPases) and Zn/Cd/Co/Pb transporters (Zn^{2+}-ATPases) (Axelsen and Palmgren 2001). *Arabidopsis thaliana* and rice genomes possess eight and nine P_{1B}-ATPase genes, respectively, which have been classified into six groups (Williams and Mills 2005). Among these ATPases, HMA2, HMA3, and HMA4 are closely related in sequence and are specific to Zn/Cd transporters. HMA2 and HMA4 are located in the plasma membrane of pericycle and control root to shoot long-distance transport of Zn and Cd, while HMA3 is involved in vacuolar transport of Zn and Cd (Hanikenne et al. 2008; Hussain et al. 2004; Liu et al. 2017; Morel et al. 2009; Wong and Cobbett 2009). Lekeux et al. (2018) have reported that C terminus of HMA4 plays an important role in Zn binding, and this region of HMA4 substantially diverges between *A. thaliana* (nonaccumulator) and *A. halleri* (accumulator). Their data further showed that di-Cys motifs in C-terminal region of HMA4 contribute to high-affinity zinc binding in plants. In *B. juncea*, BjHMMA4 was upregulated by Zn and Cd in the roots, stems, and leaves (Wang et al. 2019). A detailed study with BjHMA4 protein showed that a repeat region named BjHMA4R in the C-terminal region of HMA4 is not far from the last transmembrane domain and is localized in the cytosol. This C-terminal region substantially binds Cd^{2+} and improves Cd tolerance and accumulation in *B. juncea.* AtHMA1 is located in chloroplast membrane and is involved in the translocation of Cu and Zn into and out of chloroplast, respectively (Boutigny et al. 2014; Kim et al. 2009; Moreno et al. 2008; Seigneurin-Berny et al. 2006). Recently, it has been shown that HMA1 in the leaves of *Sedum plumbizincicola* plays an important role in the detoxification of Cd in chloroplast by exporting Cd out of the chloroplast (Zhao et al. 2018). SpHMA1RNA interference transgenic plants and CRISPR/Cas9-induced HMA1 mutant lines showed significant increase in Cd accumulation in the chloroplasts compared with wild-type plants. Their data showed that SpHMA1 in *S. plumbizincicola* leaves is a chloroplast Cd exporter and protects photosynthesis by inhibiting the Cd accumulation in the chloroplast. The AtHMA5 to AtHMA8 ATPases are involved in Cu transport in the cell. AtHMA5 contributes in the detoxification of excess Cu in roots by increasing Cu translocation from roots to shoots (Kobayashi et al. 2008). AtHMA6 (PAA1) and AtHMA8 (PAA2) are localized in chloroplast envelope and thylakoids and deliver Cu into chloroplast (Abdel-Ghany et al. 2005; Shikanai et al. 2003). Recently, 20 HMA genes have been found in soybean and are annotated as GmHMA1 to GmHMA20 (Fang et al. 2016). Phylogenetically, these 20 HMAs were divided into six clusters. Based on the Arabidopsis and rice HMA function, six GmHMAs (5, 19,13,16,14 and 18) were classified as Zn^{2+}-ATPases, while the remaining HMAs were classified as Cu^{+}-ATPases (Fang et al. 2016). The expression pattern of these 20 genes in both wild and cultivated soybean lines suggests that they may be functionally conserved and divergent and possibly involved in Cd detoxification (Fang et al. 2016). Li et al. (2015) performed genome-wide of the *Populus trichocarpa* HMA gene family and identified 17 HMA genes, which were showed to be differentially regulated by excessive metal stress. Tissue-specific expression of HMA genes showed that HMA1 and HMA4 were highly expressed in the leaves of populous, whereas HMA 5-HMA8 genes were upregulated in roots when plants were subjected to high metal concentration (Li et al. 2015). Given the main functions of HMA proteins in metal transport in *A. thaliana* and other plants in *Brassicaceae,* we performed phylogeny analysis, multiple sequence alignments, 3D structure prediction, and validation, as well as subcellular localization prediction of these proteins, with the goal of investigating the interacting proteins present in different plant species in *Brassicaceae*. The main approach of generating interactome protein analysis is to check for the specificity in structure and function between *B. rapa* (genome is recently published) and *A. halleri* (a metal accumulator plant) and then compare to genomic data of *A. thaliana* (a model plant). Study on protein–protein interactions is important in order to understand the complexity of the function of HMA proteins. Detailed bioinformatics and comparative structural studies

Table 1 Protein–protein interactions of *A. thaliana*, with description and function and localization of each protein

Gene name	Gene ID	Protein	Localization in the cell
AT2G18196	AT2G18196	Heavy metal transport/detoxification superfamily protein, metal ion binding and transport	Extracellular region
AT4G13010	AT4G13010	Oxidoreductase, zinc-binding dehydrogenase family protein, oxidoreductase activity, binding, catalytic activity, zinc ion binding	Chloroplast, chloroplast envelope, chloroplast inner membrane, chloroplast thylakoid membrane, cytoplasm, plasma membrane, vacuole
ZIP6	AT2G30080	Cation transmembrane transporter/metal ion transmembrane transporter	Chloroplast thylakoid membrane, integral component of membrane, plasma membrane
AT2G01320	AT2G01320	ABC transporter family protein, ATPase activity	Chloroplast, chloroplast envelope, integral component of membrane, membrane
ZIP10	AT1G31260	Cation transmembrane transporter/metal ion transmembrane transporter	Integral component of membrane, plasma membrane
NAP8	AT4G25450	ATNAP8, ATPase, coupled to transmembrane movement of substances/transporter	Chloroplast, chloroplast envelope, integral component of membrane, membrane, plasma membrane

Analysis was done by STRING software the confidence view (String v9.1 software was used for interaction analysis, Franceschini et al. 2013) (Maida and Memon, Unpublished data)

with HMA1–HMA4 in *Arabidopsis thaliana, A. halleri, B. rapa*, and *B. juncea* have shown the similarities and dissimilarities in the structural component of these transporters and their interaction with other proteins. The differences have been seen in the domain analysis and subcellular localization of these proteins among plant species (Jusovic and Memon 2015, Unpublished data, see Table 1 and Fig. 3).

9.1 Interactome Analysis Between HMAs and Other Metal Transporters

The interactome analysis revealed the strong interactions of HMA4 with HMA1, HMA2, and HMA3 and also with other metal transporters like different ZIP (Guerinot 2000) and ZAT (van der Zaal et al. 1999), whereas all HMA proteins showed strong interactions with ATCCS proteins (copper chaperone for superoxide dismutase) (Chu et al. 2005). The interaction of HMAs with different metal transporters and also with other several heavy metal transporter domain-containing proteins (Fig. 1) suggests that the metal tolerance mechanism in the plants is a function of the expression of many transporter genes and the subcellular localization of the

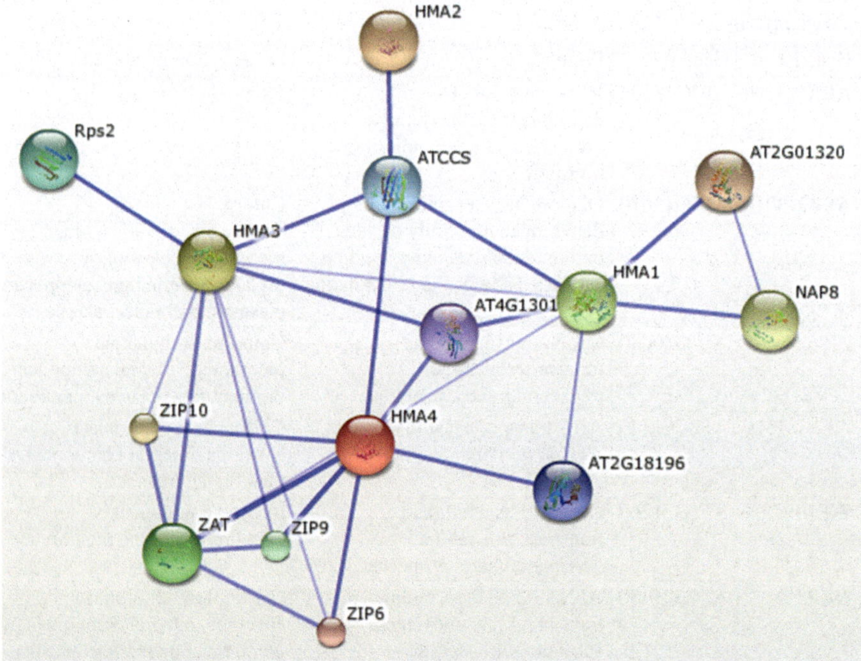

Fig. 3 Protein–protein interaction prediction, where the interaction of other proteins with HMA1, HMA2, HMA3, and HMA4 in *Arabidopsis thaliana* is shown. The interactome is generated by STRING software. Stronger protein associations are represented by thicker lines (String v9.1 software was used for interaction analysis; Franceschini et al. 2013) (Maida and Memon, Unpublished data)

different transporter proteins in accumulator plants. The detailed description of these transporters is shown in Table 1.

All previous bioinformatics analyses have given some information about their regulation, but not much work has been done on their structural analysis and their interactome. We have here clearly shown the HMA interaction with other proteins, and some structural differences have also been identified in our bioinformatics analysis (data not shown). These differences are seen in the domain analysis and subcellular localizations of these proteins. Experimental determination of 3D structures is important to better understand the function of these proteins, which is crucial for proper functioning of all cellular processes in plants. Especially, docking sites and domains need to be experimentally researched further in order to understand the function and role of these proteins in accumulator plants.

A significant advance in metal-induced gene expression and the role of different metal transporters in metal uptake transport, accumulation, and detoxification have been carried out in metal accumulator and nonaccumulator plant species. Genomic analysis of well-known metal accumulator species (e.g., *Arabidopsis halleri, Noccaea caerulescens, Brassica juncea, B. napus, B. nigra*) has shown the role of several metal transporters, including metal ATPases in metal accumulation and tolerances in plants (Cevher-Keskin et al. 2019; Dalyan et al. 2017; Memon 2016).

Above data show that P_{1B}-ATPase family plays an important role in heavy metal transport in plants. HMA genes have not only been identified in model plants like *Arabidopsis thaliana* and rice, but these HMAs have also been identified in *B. napus, B. rapa, B. juncea, Glycine max,* and *Populus trichocarpa.* The HMA1 to HMA4 in the model plants have been thoroughly characterized at both gene and protein levels. The role of HMA1 to HMA4 in Cu, Zn, and Cd transport in plat cell has been well documented, and especially, the importance of C terminus of HMA4 in metal binding and its role in metal tolerance and accumulation in plants have been suggested. Functional studies of these transporters have been carried out in yeast, and extremely useful information related to the transporter proteins and metal interaction has been obtained (Fang et al. 2016; Wang et al. 2019).

Furthermore, the plant genomic sequencing and bioinformatics analysis with Arabidopsis, rice, *Brassica napus,* and *B. rapa* have shown us that a large number of transporter gene families exist in the genomes of these plants. In the last decades, a range of plant transporter genes involved in metal uptake and translocation have been cloned and characterized, but there are many other transporter genes which are still not identified and await for identification and functional analysis. The future work should be focused on the expression and function of these transporter genes at cellular and subcellular levels in order to find out their specific roles in metal localization and compartmentalization in the cell. More studies with the structural analysis of the transporter proteins will reveal the basic mechanism of the substrates selection and transport activity.

From the information obtained from functional studies of these transporters together with their structural analyses, it is possible to assign their role in metal transport and accumulation at specific site and location of the cell and tissues. Two different kinds of strategies can be carried out with these transporters and their expression in plants. For accumulator plants, the transporters related to metal uptake and translocation can be overexpressed to optimize the translocation of toxic metals to aerial parts, which would be the target for phytoremediation. Another strategy could be used for edible crop plants where the low uptake transporters could be engineered to minimize the transport of toxic cation in edible crops.

10 Recent Advances and New Possibilities in Phytoremediation

microRNAs (miRNAs) are universal regulator for gene expression in both plants and mammalian system when encountered to biotic or abiotic stress. Recently, a high-throughput sequence analysis with metal accumulator and tolerant plants has shown the upregulation and downregulation of several miRNAs when plants were exposed to metals (e.g., Zn, Cd, Cu) (Memon 2016). Several studies have demonstrated that heavy metal-induced gene expression in the accumulator plants can be regulated by different miRNAs (Gielen et al. 2012; Memon 2016; Zhou et al. 2012). Genome wide identification of miRNAs and corresponding target genes that are induced in response to metal stress in *A. thaliana*, rice, tobacco, radish, *B. napus*, and *B. rapa* have been documented to some extent (Liu et al. 2015; Yuan et al. 2011). However, these studies are limited and need further attention in terms of miRNA-regulated heavy metal-related gene expression in hyperaccumulator plants and their impact on metal accumulation and tolerance in the cell.

Recently, several gene editing technologies have been developed such as customized homing nuclease (meganuclease), zinc-finger nuclease (ZNFs), and transcription activator-like effector nuclease (TALENs) (Zhu et al. 2017). There are several advantages and disadvantages associated with these techniques. Recently, CRISPR-Cas 9 technology (clustered regularly interspaced short palindromic repeats(CRISPR)-associated protein 9) has been successfully used for genome editing in bacteria, plants, mammalian, and human system (Barrangou and Doudna 2016; Doudna and Charpentier 2014; Park et al. 2017). This new technology has great potential to help molecular biologists and crop breeders to edit or modify the genome or genes of crop plants that are able to produce high yields under conditions of biotic/abiotic stress (Kumlehn et al. 2018; Scheben et al. 2017; Schindele et al. 2018; Wolter and Puchta 2018). CRISPR/Cas9 technique could be used to introduce point mutations in the sequences of the metal transporter genes in order to enhance their activity or modify their selectivity toward metal of interest. The increase in transporter activities (overexpression) will enhance the ability of the plant to uptake and translocate the metals in shoots efficiently, and this could be paramount importance for phytoremediation. In other case, the transporter activities could be reduced or knocked out by deleting the sequences of specific transporter gene/genes. In this case, the crop plants could be generated, which could be used in contaminated soils for human consumption. For example, OsNramp5, which is major transporter responsible for Mn and Cd uptake in rice (Sasaki et al. 2014), was successfully knockout in the rice by using CRISPR/Cas9 system. These OsNramp5 mutants showed a remarkable reduction in Cd content in rice grain while maintaining all important agronomical traits such as growth, yield, and test (Tang et al. 2017). Overall, this CRISPR/Cas9-mediated genome editing has a great potential for exploiting plant genomes to enhance the metal accumulation and metal tolerance capacities in plants which could be effectively used for phytoremediation.

11 Conclusion

In this review, I have summarized the recent advances in the field of phytoremediation. The obligate and facultative hyperaccumulator species have been identified, and their role in metal accumulation has been described. The metal distribution and localization in the leaf cells of hyperaccumulator plants have been analyzed by X-ray microprobe analyzer, and their distribution pattern at subcellular level has been examined. The global hyperaccumulator database (www.hyperaccumulators.

org) has been recently established by the Center for Mined Land Rehabilitation of The University of Queensland, Australia. The database is freely available and contains information about taxonomy, worldwide distribution, ecology, analytical data, and other relevant information of all known metal and metalloid hyperaccumulator plant species, which could be useful for biologists, environmentalists, and molecular biologists for their research. Recent developments in the genome analysis of *B. rapa* and *B. napus* have shown the expression of many metal transporters in these plant species when subjected to high metal concentration in the soil. These data have been compared with the data of well-known dicot plant species *Arabidopsis thaliana* and monocot species *Oryza sativa.* Recent development in functional analysis of metal transporters including ZIPs, ZIPs, MTPs, NRAMPs, ABC transporters, and HMAs in accumulator plants has been summarized. Interactome analyses of the different HMAs and their interaction with other metal transporters indicate that the hyperaccumulation is fascinating phenomenon controlled by multigene-regulated process. Furthermore, the role of the transporters in metal accumulation and detoxification in hyperaccumulator plants and their implication in phytoremediation have been discussed.

Acknowledgments The author thanks Ms. Fatma Kusur, Department of Molecular Biology and Genetics, Usak University, and Mr. Muhammet Memon, International Biomedical and Genomic Research Center, Dokuz Eylul University, for their help in reference arrangement and their valuable comments on some part of this review. This work is supported by BAP project F010.

References

Abdel-Ghany SE, Müller-Moulé P, Niyogi KK, Pilon M, Shikanai T (2005) Two P-type ATPases are required for copper delivery in Arabidopsis thaliana chloroplasts. Plant Cell 17:1233–1251

Agorio A et al (2017) Phosphatidylinositol 3-phosphate–binding protein AtPH1 controls the localization of the metal transporter NRAMP1 in Arabidopsis. PNAS 114:E3354–E3363

Ali H, Khan E, Sajad MA (2013) Phytoremediation of heavy metals—concepts and applications. Chemosphere 91:869–881

Alkorta I, Garbisu C (2001) Phytoremediation of organic contaminants in soils. Bioresour Technol 79:273–276

Anjum NA, Ahmad I, Pereira ME, Duarte AC, Umar S, Khan NA (2012) The plant family Brassicaceae: contribution towards phytoremediation, vol 21. Springer, Dordrecht

Anjum NA, Gill SS, Duarte AC, Pereira E, Ahmad I (2013) Silver nanoparticles in soil–plant systems. J Nanopart Res 15:1896

Argüello JM (2003) Identification of ion-selectivity determinants in heavy-metal transport P 1B-type ATPases. J Membr Biol 195:93–108

Axelsen KB, Palmgren MG (2001) Inventory of the superfamily of P-type ion pumps in Arabidopsis. Plant Physiol 126:696–706

Babst-Kostecka AA, Waldmann P, Frérot H, Vollenweider P (2016) Plant adaptation to metal polluted environments—physiological, morphological, and evolutionary insights from Biscutella laevigata. Environ Exp Bot 127:1–13

Bairoch A (1993) The PROSITE dictionary of sites and patterns in proteins, its current status. Nucleic Acids Res 21:3097

Baker AJM, McGrawth SP, Reeves RD, Smith JAC (2000). Metal hyperaccumulator plants: a review of the ecology and physiology of a biological resources for phytoremediation of metal-polluted soils. In: Phytoremediation of contaminated soil and water (Terry, N, Banuelos, G, eds) CRC press, Lewis Publishers, USA. ISBN-13:978-1566704502, pp. 85–107.
Baker A, Brooks RR (1989) Terrestrial higher plants which hyperaccumulate metallic elements. A review of their distribution, ecology and phytochemistry. Biorecovery 1:81–126
Barrangou R, Doudna JA (2016) Applications of CRISPR technologies in research and beyond. Nat Biotechnol 34:933–941. https://doi.org/10.1038/nbt.3659
Basic N, Salamin N, Keller C, Galland N, Besnard G (2006) Cadmium hyperaccumulation and genetic differentiation of Thlaspi caerulescens populations. Biochem Syst Ecol 34:667–677
Belouchi A, Kwan T, Gros P (1997) Cloning and characterization of the OsNramp family from Oryza sativa, a new family of membrane proteins possibly implicated in the transport of metal ions. Plant Mol Biol 33:1085–1092
Bernal M et al (2012) Transcriptome sequencing identifies SPL7-regulated copper acquisition genes FRO4/FRO5 and the copper dependence of iron homeostasis in Arabidopsis. Plant Cell 24:738–761
Bhargava A, Carmona FF, Bhargava M, Srivastava S (2012) Approaches for enhanced phytoextraction of heavy metals. J Environ Manag 105:103–120
Bhati KK et al (2015) Genome-wide identification and expression characterization of ABCC-MRP transporters in hexaploid wheat. Front Plant Sci 6:488. https://doi.org/10.3389/fpls.2015.00488
Boutigny S et al (2014) HMA1 and PAA1, two chloroplast-envelope PIB-ATPases, play distinct roles in chloroplast copper homeostasis. J Exp Bot 65:1529–1540. https://doi.org/10.1093/jxb/eru020
Boyd J, Banzhaf S (2007) What are ecosystem services? The need for standardized environmental accounting units. Ecol Econ 63:616–626
Cailliatte R, Schikora A, Briat J-F, Mari S, Curie C (2010) High-affinity manganese uptake by the metal transporter NRAMP1 is essential for Arabidopsis growth in low manganese conditions. Plant Cell 22:904–917
Cakir B, Kiliçkaya O, Olcay AC (2013) Genome-wide analysis of Aux/IAA genes in Vitis vinifera: cloning and expression profiling of a grape Aux/IAA gene in response to phytohormone and abiotic stresses. Acta Physiol Plant 35:365–377
Campbell P, Stokes P, Galloway J (1983) Effects of atmospheric deposition on the geochemical cycling and biological availability of metals. Heavy Metals Environ 2:760–763
Cellier MF (2012) Nramp: from sequence to structure and mechanism of divalent metal import. In: Current topics in membranes, vol 69. Elsevier, pp 249–293
Cellier M, Prive G, Belouchi A, Kwan T, Rodrigues V, Chia W, Gros P (1995) Nramp defines a family of membrane proteins. Proc Nat Acad Sci U S A 92:10089–10093
Cevher-Keskin B, Yıldızhan Y, Yüksel B, Dalyan E, Memon AR (2019) Characterization of differentially expressed genes to Cu stress in Brassica nigra by Arabidopsis genome arrays. Environ Sci Pollut Res 26:299–311
Chaney RL, Baklanov IA (2017) Phytoremediation and phytomining: status and promise. In: Advances in botanical research, vol 83. Elsevier, pp 189–221
Chaney RL, Baker AJ, Morel JL (2018) The long road to developing agromining/phytomining. In: Agromining: farming for metals. Springer, Cham, pp 1–17
Chen J-Q, Meng X-P, Zhang Y, Xia M, Wang X-P (2008) Over-expression of OsDREB genes lead to enhanced drought tolerance in rice. Biotechnol Lett 30:2191–2198
Chu C-C, Lee W-C, Guo W-Y, Pan S-M, Chen L-J, Li H-m, Jinn T-L (2005) A copper chaperone for superoxide dismutase that confers three types of copper/zinc superoxide dismutase activity in Arabidopsis. Plant Physiol 139:425–436
Clarkson DT, Lüttge U (1989) Mineral nutrition: divalent cations, transport and compartmentation. In: Progress in botany. Springer, Berlin, pp 93–112
Clemens S, Palmgren MG, Krämer U (2002) A long way ahead: understanding and engineering plant metal accumulation. Trends Plant Sci 7:309–315

Cohen CK, Garvin DF, Kochian LV (2004) Kinetic properties of a micronutrient transporter from Pisum sativum indicate a primary function in Fe uptake from the soil. Planta 218:784–792
Colangelo EP, Guerinot ML (2006) Put the metal to the petal: metal uptake and transport throughout plants. Curr Opin Plant Biol 9:322–330
Dalyan E et al (2017) The identification of genes associated with Pb and Cd response mechanism in Brassica juncea L. by using Arabidopsis expression array. Environ Exp Bot 139:105–115
Davidson AL, Dassa E, Orelle C, Chen J (2008) Structure, function, and evolution of bacterial ATP-binding cassette systems. Microbiol Mol Biol Rev 72:317–364. , table of contents. https://doi.org/10.1128/MMBR.00031-07
Davies T, Coleman J (2000) The Arabidopsis thaliana ATP-binding cassette proteins: an emerging superfamily. Plant Cell Environ 23:431–443
De Vos C, Schat H (1991) Free radicals and heavy metal tolerance. In: Ecological responses to environmental stresses. Springer, Dordrecht, pp 22–31
Dean M, Allikmets R (2001) Complete characterization of the human ABC gene family. J Bioenerg Biomembr 33:475–479
Diwan H, Khan I, Ahmad A, Iqbal M (2010) Induction of phytochelatins and antioxidant defence system in Brassica juncea and Vigna radiata in response to chromium treatments. Plant Growth Regul 61:97–107
Doudna JA, Charpentier E (2014) The new frontier of genome engineering with CRISPR-Cas9. Science 346:1258096
Eide D, Broderius M, Fett J, Guerinot ML (1996) A novel iron-regulated metal transporter from plants identified by functional expression in yeast. Proc Nat Acad Sci U S A 93:5624–5628
Fang X et al (2016) Genome-wide characterization of soybean P 1B-ATPases gene family provides functional implications in cadmium responses. BMC Genomics 17:376
Franceschini A et al (2013) STRING v9.1: protein-protein interaction networks, with increased coverage and integration. Nucleic Acids Res 41:D808–D815. https://doi.org/10.1093/nar/gks1094
Galeas ML, Klamper EM, Bennett LE, Freeman JL, Kondratieff BC, Quinn CF, Pilon-Smits EA (2008) Selenium hyperaccumulation reduces plant arthropod loads in the field. New Phytol 177:715–724
Gao H et al (2018) NRAMP2, a trans-Golgi network-localized manganese transporter, is required for Arabidopsis root growth under manganese deficiency. New Phytol 217:179–193
Gielen H, Remans T, Vangronsveld J, Cuypers A (2012) MicroRNAs in metal stress: specific roles or secondary responses? Int J Mol Sci 13:15826–15847. https://doi.org/10.3390/ijms131215826
Gleba D et al (1999) Use of plant roots for phytoremediation and molecular farming. Proc Nat Acad Sci U S A 96:5973–5977
Grotz N, Fox T, Connolly E, Park W, Guerinot ML, Eide D (1998) Identification of a family of zinc transporter genes from Arabidopsis that respond to zinc deficiency. Proc Nat Acad Sci U S A 95:7220–7224
Gruenheid S, Cellier M, Vidal S, Gros P (1995) Identification and characterization of a second mouse Nramp gene. Genomics 25:514–525
Guerinot ML (2000) The ZIP family of metal transporters. Biochim Biophys Acta Biomembr 1465:190–198
Guerinot ML, Salt DE (2001) Fortified foods and phytoremediation. Two sides of the same coin. Plant Physiol 125:164–167
Hall J, Williams LE (2003) Transition metal transporters in plants. J Exp Bot 54:2601–2613
Hanikenne M, Nouet C (2011) Metal hyperaccumulation and hypertolerance: a model for plant evolutionary genomics. Curr Opin Plant Biol 14:252–259
Hanikenne M et al (2008) Evolution of metal hyperaccumulation required cis-regulatory changes and triplication of HMA4. Nature 453:391
Henikoff S, Greene EA, Pietrokovski S, Bork P, Attwood TK, Hood L (1997) Gene families: the taxonomy of protein paralogs and chimeras. Science 278:609–614

Huitson SB, Macnair MR (2003) Does zinc protect the zinc hyperaccumulator Arabidopsis halleri from herbivory by snails? New Phytol 159:453–459
Hussain D et al (2004) P-type ATPase heavy metal transporters with roles in essential zinc homeostasis in Arabidopsis. Plant Cell 16:1327–1339. https://doi.org/10.1105/tpc.020487
Hwang JU et al (2016) Plant ABC transporters enable many unique aspects of a terrestrial plant's lifestyle. Mol Plant 9:338–355. https://doi.org/10.1016/j.molp.2016.02.003
Isaure M-P et al (2015) Evidence of various mechanisms of Cd sequestration in the hyperaccumulator Arabidopsis halleri, the non-accumulator Arabidopsis lyrata, and their progenies by combined synchrotron-based techniques. J Exp Bot 66:3201–3214
Ishimaru Y et al (2012) Characterizing the role of rice NRAMP5 in manganese, iron and cadmium transport. Sci Rep 2:286. https://doi.org/10.1038/srep00286
Ivanov R, Bauer P (2017) Sequence and coexpression analysis of iron-regulated ZIP transporter genes reveals crossing points between iron acquisition strategies in green algae and land plants. Plant Soil 418:61–73
Kang J, Park J, Choi H, Burla B, Kretzschmar T, Lee Y, Martinoia E (2011a) Plant ABC transporters vol 9. Arabidopsis Book, 2012/02/04 edn. https://doi.org/10.1199/tab.0153
Kang J, Park J, Choi H, Burla B, Kretzschmar T, Lee Y, Martinoia E (2011b) Plant ABC transporters. The Arabidopsis Book/Am Soc Plant Biol 9:e0153
Kim YY, Choi H, Segami S, Cho HT, Martinoia E, Maeshima M, Lee Y (2009) AtHMA1 contributes to the detoxification of excess Zn(II) in Arabidopsis. Plant J 58:737–753. https://doi.org/10.1111/j.1365-313X.2009.03818.x
Klaumann S, Nickolaus SD, Fürst SH, Starck S, Schneider S, Ekkehard Neuhaus H, Trentmann O (2011) The tonoplast copper transporter COPT5 acts as an exporter and is required for interorgan allocation of copper in Arabidopsis thaliana. New Phytol 192:393–404
Kobayashi Y et al (2008) Amino acid polymorphisms in strictly conserved domains of a P-type ATPase HMA5 are involved in the mechanism of copper tolerance variation in Arabidopsis. Plant Physiol 148:969–980. https://doi.org/10.1104/pp.108.119933
Kolaj-Robin O, Russell D, Hayes KA, Pembroke JT, Soulimane T (2015) Cation diffusion facilitator family: structure and function. FEBS Lett 589:1283–1295
Kumar PN, Dushenkov V, Motto H, Raskin I (1995) Phytoextraction: the use of plants to remove heavy metals from soils. Environ Sci Technol 29:1232–1238
Kumar V, Mahajan M, Yadav SK (2012) Toxic metals accumulation, tolerance and homeostasis in brassicaoilseed species: overview of physiological, biochemical and molecular mechanisms. In: The plant family Brassicaceae. Springer, Dordrecht, pp 171–211
Kumlehn J, Pietralla J, Hensel G, Pacher M, Puchta H (2018) The CRISPR/Cas revolution continues: from efficient gene editing for crop breeding to plant synthetic biology. J Integr Plant Biol 60:1127–1153. https://doi.org/10.1111/jipb.12734
Küpper H, Zhao FJ, McGrath SP (1999) Cellular compartmentation of zinc in leaves of the hyperaccumulator Thlaspi caerulescens. Plant Physiol 119:305–312
Küpper H, Lombi E, Zhao F-J, McGrath SP (2000) Cellular compartmentation of cadmium and zinc in relation to other elements in the hyperaccumulator Arabidopsis halleri. Planta 212:75–84
Lane TS et al (2016) Diversity of ABC transporter genes across the plant kingdom and their potential utility in biotechnology. BMC Biotechnol 16:47
Lanquar V, Ramos MS, Lelièvre F, Barbier-Brygoo H, Krieger-Liszkay A, Krämer U, Thomine S (2010) Export of vacuolar manganese by AtNRAMP3 and AtNRAMP4 is required for optimal photosynthesis and growth under manganese deficiency. Plant Physiol 152:1986–1999
Lasat MM (2000) The use of plants for the removal of toxic metals from contaminated soils. US Environmental Protection Agency, Washington, DC
Lasat MM (2002) Phytoextraction of toxic metals: a review of biological mechanisms. J Environ Qual 31:109–120
Lefèvre F, Baijot A, Boutry M (2015) Plant ABC transporters: time for biochemistry? Biochem Soc Trans 43:931–936
Lekeux G et al (2018) di-Cysteine motifs in the C-terminus of plant HMA4 proteins confer nanomolar affinity for zinc and are essential for HMA4 function in vivo. J Exp Bot 69:5547–5560

Li H et al (2013) Simultaneous saccharification and fermentation of broken rice: an enzymatic extrusion liquefaction pretreatment for Chinese rice wine production. Bioprocess Biosyst Eng 36:1141–1148

Li D et al (2015) Genome-wide analysis and heavy metal-induced expression profiling of the HMA gene family in Populus trichocarpa. Front Plant Sci 6:1149

Lin YF et al (2009) Arabidopsis IRT3 is a zinc-regulated and plasma membrane localized zinc/iron transporter. New Phytol:182, 392–404

Lin Y-F, Severing EI, te Lintel HB, Schijlen E, Aarts MG (2014) A comprehensive set of transcript sequences of the heavy metal hyperaccumulator Noccaea caerulescens. Front Plant Sci 5:261

Lin Y, Gritsenko D, Feng S, Teh YC, Lu X, Xu J (2016) Detection of heavy metal by paper-based microfluidics. Biosens Bioelectron 83:256–266

Liu W et al (2015) Transcriptome-wide analysis of chromium-stress responsive microRNAs to explore miRNA-mediated regulatory networks in radish (Raphanus sativus L.). Sci Rep 5:14024. https://doi.org/10.1038/srep14024

Liu H, Zhao H, Wu L, Liu A, Zhao FJ, Xu W (2017) Heavy metal ATPase 3 (HMA3) confers cadmium hypertolerance on the cadmium/zinc hyperaccumulator Sedum plumbizincicola. New Phytol 215:687–698. https://doi.org/10.1111/nph.14622

Lone MI, He Z, Stoffella PJ, Yang X (2008) Phytoremediation of heavy metal polluted soils and water: progresses and perspectives. J Zhejiang Univ 9:210–220

Lysak MA, Koch MA, Pecinka A, Schubert I (2005) Chromosome triplication found across the tribe Brassicaceae. Genome Res 15:516–525

Ma L, Komar K, Tu C, Zhang W, Cai Y, Kennelley E (2001) Addendum: a fern that hyperaccumulates arsenic. Nature 411:438

Marschner H (2011) Marschner's mineral nutrition of higher plants. Academic, Boston

Martens SN, Boyd RS (1994) The ecological significance of nickel hyperaccumulation: a plant chemical defense. Oecologia 98:379–384

Mäser P et al (2001) Phylogenetic relationships within cation transporter families of Arabidopsis. Plant Physiol 126:1646–1667

McGrath SP, Zhao F-J (2003) Phytoextraction of metals and metalloids from contaminated soils. Curr Opin Biotechnol 14:277–282

McGrath S, Zhao J, Lombi E (2002) Phytoremediation of metals, metalloids, and radionuclides. Adv Agron 75:1–56

Memon AR (2016) Metal hyperaccumulators: mechanisms of hyperaccumulation and metal tolerance. In: Phytoremediation. Springer, Berlin, pp 239–268

Memon AR, Schröder P (2009) Implications of metal accumulation mechanisms to phytoremediation. J Environ Sci Pollut Res 16:162–175

Memon AR, Yatazawa M (1982) Chemical nature of manganese in the leaves of manganese accumulator plants. J Soil Sci Plant Nutr 28:401–412

Memon AR, Yatazawa M (1984) Nature of manganese complexes in manganese accumulator plant-Acanthopanax sciadophylloides. J Plant Nutr 7:961–974

Memon AR, Zahirovic E (2014) Genomics and transcriptomics analysis of Cu accumulator plant Brassica nigra L. J Appl Biol Sci 8:1–8

Memon AR, Itô S, Yatazawa M (1979) Absorption and accumulation of iron, manganese and copper in plants in the temperate forest of Central Japan J. Soil Sci Plant Nutr 25:611–620

Memon AR, Aktoprakligil D, Özdemir A, Verti A (2001) Heavy metal accumulation and detoxification mechanisms in plants. Turk J Bot 25:111–121

Memon AR, Yildizhan Y, Kaplan E (2008) Metal accumulation in crops–human health issues. In: Trace elements as contaminants nutrients—consequences in ecosystems human health. Wiley, Hoboken, pp 81–98

Meng JG, Zhang XD, Tan SK, Zhao KX, Yang ZM (2017) Genome-wide identification of Cd-responsive NRAMP transporter genes and analyzing expression of NRAMP 1 mediated by miR167 in Brassica napus. Biometals 30:917–931. https://doi.org/10.1007/s10534-017-0057-3

Merlot S, de la Torre VSG, Hanikenne M (2018) Physiology and molecular biology of trace element hyperaccumulation. In: Agromining: farming for metals. Springer, Cham, pp 93–116
Milner MJ, Seamon J, Craft E, Kochian LV (2013) Transport properties of members of the ZIP family in plants and their role in Zn and Mn homeostasis. J Exp Bot 64:369–381
Morel M, Crouzet J, Gravot A, Auroy P, Leonhardt N, Vavasseur A, Richaud P (2009) AtHMA3, a P1B-ATPase allowing Cd/Zn/Co/Pb vacuolar storage in Arabidopsis. Plant Physiol 149:894–904. https://doi.org/10.1104/pp.108.130294
Moreno I et al (2008) AtHMA1 is a thapsigargin-sensitive Ca2+/heavy metal pump. J Biol Chem 283:9633–9641
Nagaharu U (1935) Genome analysis in Brassica with special reference to the experimental formation of B. napus and peculiar mode of fertilization. Jpn J Bot 7:389–452
Nevo Y, Nelson N (2006) The NRAMP family of metal-ion transporters. BBA – Mol Cell Res 1763:609–620
Noret N, Meerts P, Vanhaelen M, Dos Santos A, Escarré J (2007) Do metal-rich plants deter herbivores? A field test of the defence hypothesis. Oecologia 152:92–100
Oda K, Otani M, Uraguchi S, Akihiro T, Fujiwara T (2011) Rice ABCG43 is Cd inducible and confers Cd tolerance on yeast. Biosci Biotechnol Biochem 75:1211–1213
Østerberg JT, Palmgren M (2018) Heavy metal pumps in plants: structure, function and origin. In: Advances in botanical research. Elsevier, pp 57–89
Ovečka M, Takáč T (2014) Managing heavy metal toxicity stress in plants: biological and biotechnological tools. Biotechnol Adv 32:73–86
Ozturk M, Memon AR, Gucel S, Sakcali MS (2012) Brassicas in Turkey and their potential role for degraded habitats' remediation. In: The plant family Brassicaceae. Springer, Dordrecht, pp 265–287
Padmavathiamma PK, Li L (2007) Phytoremediation technology: hyper-accumulation metals in plants. Water Air Soil Pollut 184:105–126
Pang K, Li Y, Liu M, Meng Z, Yu Y (2013) Inventory and general analysis of the ATP-binding cassette (ABC) gene superfamily in maize (Zea mays L). Gene 526:411–428. https://doi.org/10.1016/j.gene.2013.05.051
Park J et al (2012) The phytochelatin transporters AtABCC1 and AtABCC2 mediate tolerance to cadmium and mercury. Plant J 69:278–288. https://doi.org/10.1111/j.1365-313X.2011.04789.x
Park JY, Moon BY, Park JW, Thornton JA, Park YH, Seo KS (2017) Genetic engineering of a temperate phage-based delivery system for CRISPR/Cas9 antimicrobials against Staphylococcus aureus. Sci Rep 7:44929
Paunov M, Koleva L, Vassilev A, Vangronsveld J, Goltsev V (2018) Effects of different metals on photosynthesis: Cadmium and zinc affect chlorophyll fluorescence in Durum Wheat. Int J Mol Sci 19:787
Pedas P, Ytting CK, Fuglsang AT, Jahn TP, Schjoerring JK, Husted S (2008) Manganese efficiency in barley: identification and characterization of the metal ion transporter HvIRT1. Plant Physiology 148:455–466
Peñarrubia L, Andrés-Colás N, Moreno J, Puig S (2010) Regulation of copper transport in Arabidopsis thaliana: a biochemical oscillator? J Biol Inorg Chem 15:29
Pence NS et al (2000) The molecular physiology of heavy metal transport in the Zn/Cd hyperaccumulator Thlaspi caerulescens. Proc Nat Acad Sci 97:4956–4960
Perea-García A, Garcia-Molina A, Andrés-Colás N, Vera-Sirera F, Pérez-Amador MA, Puig S, Peñarrubia L (2013) Arabidopsis copper transport protein COPT2 participates in the cross talk between iron deficiency responses and low-phosphate signaling. Plant Physiol 162:180–194
Peuke AD, Rennenberg H (2005) Phytoremediation: molecular biology, requirements for application, environmental protection, public attention and feasibility. EMBO Rep 6:497–501
Pollard AJ, Reeves RD, Baker AJ (2014) Facultative hyperaccumulation of heavy metals and metalloids. Plant Sci 217:8–17
Qin L et al (2017) Genome-wide identification and expression analysis of NRAMP family genes in soybean (Glycine Max L). Front Plant Sci 8:1436. https://doi.org/10.3389/fpls.2017.01436

Rascio N, Navari-Izzo F (2011) Heavy metal hyperaccumulating plants: how and why do they do it? And what makes them so interesting? Plant Sci 180:169–181

Reeves RD, Baker AJM (2000) Metal accumulating plants. In: Phytoremediation of Toxic Metals: Using Plants to Clean up the Environment (Raskin I, Finsley BD eds), Wiley, New York, 193–229.

Reeves RD, Schwartz C, Morel JL, Edmondson J (2001) Distribution and metal-accumulating behavior of Thlaspi caerulescens and associated metallophytes in France. Int J Phytoremediation 3:145–172

Reeves R, Baker A, Jaffré T, Erskine P, Echevarria G, van der Ent A (2017) A global database for hyperaccumulator plants of metal and metalloid trace elements. New Phytol:14907

Reeves RD, Baker AJ, Jaffré T, Erskine PD, Echevarria G, van der Ent A (2018) A global database for plants that hyperaccumulate metal and metalloid trace elements. New Phytol 218:407–411

Rensing C, Ghosh M, Rosen BP (1999) Families of soft-metal-ion-transporting ATPases. J Bacteriol 181:5891–5897

Ricachenevsky FK, Sperotto RA (2016) Into the wild: Oryza species as sources for enhanced nutrient accumulation and metal tolerance in rice. Front Plant Sci 7:974

Ricachenevsky FK, Menguer PK, Sperotto RA, Williams LE, Fett JP (2013) Roles of plant metal tolerance proteins (MTP) in metal storage and potential use in biofortification strategies. Front Plant Sci 4:144

Rogers EE, Eide DJ, Guerinot ML (2000) Altered selectivity in an Arabidopsis metal transporter. Proc Nat Acad Sci 97:12356–12360

Rosenzweig AC, Argüello JM (2012) Toward a molecular understanding of metal transport by P1B-Type ATPases. In: Current topics in membranes, vol 69. Elsevier, pp 113–136

Salt DE, Blaylock M, Kumar NP, Dushenkov V, Ensley BD, Chet I, Raskin I (1995) Phytoremediation: a novel strategy for the removal of toxic metals from the environment using plants. Bio/Technol 13:468

Sancenón V, Puig S, Mira H, Thiele DJ, Peñarrubia L (2003) Identification of a copper transporter family in Arabidopsis thaliana. Plant Mol Biol 51:577–587

Sancenón V, Puig S, Mateu-Andrés I, Dorcey E, Thiele DJ, Peñarrubia L (2004) The Arabidopsis copper transporter COPT1 functions in root elongation and pollen development. J Biol Chem 279:15348–15355

Sarma RK, Gowtham I, Bharadwaj R, Hema J, Sathishkumar R (2018) Recent advances in metal induced stress tolerance in plants: possibilities and challenges. In: Plants under metal and metalloid stress. Springer, Singapore, pp 1–28

Sasaki A, Yamaji N, Ma JF (2014) Overexpression of OsHMA3 enhances Cd tolerance and expression of Zn transporter genes in rice. J Exp Bot 65:6013–6021. https://doi.org/10.1093/jxb/eru340

Scheben A, Wolter F, Batley J, Puchta H, Edwards D (2017) Towards CRISPR/Cas crops–bringing together genomics and genome editing. New Phytol 216:682–698

Schindele P, Wolter F, Puchta H (2018) Transforming plant biology and breeding with CRISPR/Cas9, Cas12 and Cas13. FEBS Lett 592:1954–1967. https://doi.org/10.1002/1873-3468.13073

Seigneurin-Berny D et al (2006) HMA1, a new Cu-ATPase of the chloroplast envelope, is essential for growth under adverse light conditions. J Biol Chem 281:2882–2892

Shikanai T, Müller-Moulé P, Munekage Y, Niyogi KK, Pilon M (2003) PAA1, a P-type ATPase of Arabidopsis, functions in copper transport in chloroplasts. Plant Cell 15:1333–1346

Stein RJ et al (2017) Relationships between soil and leaf mineral composition are element-specific, environment-dependent and geographically structured in the emerging model Arabidopsis halleri. New Phytol 213:1274–1286

Stephan UW, Schmidke I, Stephan VW, Scholz G (1996) The nicotianamine molecule is made-to-measure for complexation of metal micronutrients in plants. Biometals 9:84–90

Stephens BW, Cook DR, Grusak MA (2011) Characterization of zinc transport by divalent metal transporters of the ZIP family from the model legume Medicago truncatula. BioMetals 24:51–58

Szczygłowska M, Bodnar M, Namieśnik J, Konieczka P (2014) The use of vegetables in the biomonitoring of cadmium and lead pollution in the environment. Crit Rev Anal Chem 44:2–15
Tang X et al (2017) A CRISPR–Cpf1 system for efficient genome editing and transcriptional repression in plants. Nature Plants 3:17018
Tangahu BV, Abdullah S, Rozaimah S, Basri H, Idris M, Anuar N, Mukhlisin M (2011) A review on heavy metals (As, Pb, and Hg) uptake by plants through phytoremediation. Int J Chem Eng 2011:1–31
Tiong J, McDonald G, Genc Y, Shirley N, Langridge P, Huang CY (2015) Increased expression of six ZIP family genes by zinc (Zn) deficiency is associated with enhanced uptake and root-to-shoot translocation of Zn in barley (Hordeum vulgare). New Phytologist 207:1097–1109
Vamerali T, Bandiera M, Mosca G (2010) Field crops for phytoremediation of metal-contaminated land. A review. Environ Chem Lett 8:1–17
van der Ent A, Echevarria G, Baker AJ, Morel JL (2018) Preface. In: Agromining: farming for metals extracting unconventional resources using plants. Springer, Cham, pp V–VI
van der Zaal BJ, Neuteboom LW, Pinas JE, Chardonnens AN, Schat H, Verkleij JA, Hooykaas PJ (1999) Overexpression of a novel Arabidopsis gene related to putative zinc-transporter genes from animals can lead to enhanced zinc resistance and accumulation. Plant Physiol 119:1047–1056
Vara Prasad MN, de Oliveira Freitas HM (2003) Metal hyperaccumulation in plants: biodiversity prospecting for phytoremediation technology. Electron J Biotechnol 6:285–321
Verbruggen N, Hermans C, Schat H (2009) Molecular mechanisms of metal hyperaccumulation in plants. New Phytol 181:759–776
Verbruggen N, Hanikenne M, Clemens S (2013) A more complete picture of metal hyperaccumulation through next-generation sequencing technologies. Front Plant Sci 4:388
Verrier PJ et al (2008) Plant ABC proteins–a unified nomenclature and updated inventory. Trends Plant Sci 13:151–159
von Wirén N et al (1999) Nicotianamine chelates both FeIII and FeII. Implications for metal transport in plants. Plant Physiol 119:1107–1114
Wang Z et al (2016) MoTe 2: a type-II Weyl topological metal. Physical Rev Lett 117:056805
Wang J, Liang S, Xiang W, Dai H, Duan Y, Kang F, Chai T (2019) A repeat region from the Brassica juncea HMA4 gene BjHMA4R is specifically involved in Cd2+ binding in the cytosol under low heavy metal concentrations. BMC Plant Biol 19:89. https://doi.org/10.1186/s12870-019-1674-5
Warwick SI, Black LD (1991) Molecular systematics of Brassica and allied genera (subtribe Brassicinae, Brassiceae)—chloroplast genome and cytodeme congruence. Theor Appl Genet 82:81–92
Weast RC, Astle MJ, Beyer WH (1988) CRC handbook of chemistry and physics vol 69. CRC Press, Boca Raton
Williams LE, Mills RF (2005) P1B-ATPases–an ancient family of transition metal pumps with diverse functions in plants. Trends Plant Sci 10:491–502
Wintz H et al (2003) Expression profiles of Arabidopsis thaliana in mineral deficiencies reveal novel transporters involved in metal homeostasis. J Biol Chem 278:47644–47653
Wolter F, Puchta H (2018) The CRISPR/Cas revolution reaches the RNA world: Cas13, a new Swiss Army knife for plant biologists. Plant J 94:767–775. https://doi.org/10.1111/tpj.13899
Wong CKE, Cobbett CS (2009) HMA P-type ATPases are the major mechanism for root-to-shoot Cd translocation in Arabidopsis thaliana. New Phytologist 181:71–78. https://doi.org/10.1111/j.1469-8137.2008.02638.x
Xu X, Shi J, Chen X, Chen Y, Hu T (2009) Chemical forms of manganese in the leaves of manganese hyperaccumulator Phytolacca acinosa Roxb. (Phytolaccaceae). Plant Soil 318:197
Xu Y et al (2014) Os ABCB 14 functions in auxin transport and iron homeostasis in rice (Oryza sativa L.). Plant J 79:106–117
Yamasaki H, Hayashi M, Fukazawa M, Kobayashi Y, Shikanai T (2009) SQUAMOSA promoter binding protein–like7 is a central regulator for copper homeostasis in Arabidopsis. Plant Cell 21:347–361

Yan C, Duan W, Lyu S, Li Y, Hou X (2017) Genome-wide identification, evolution, and expression analysis of the ATP-binding cassette transporter gene family in Brassica rapa. Front Plant Sci 8:349

Yang J et al (2016) The genome sequence of allopolyploid Brassica juncea and analysis of differential homoeolog gene expression influencing selection. Nat Genet 48:1225

Yuan M, Li X, Xiao J, Wang S (2011) Molecular and functional analyses of COPT/Ctr-type copper transporter-like gene family in rice. BMC Plant Biol 11:69

Zhang W, Cai Y, Tu C, Ma LQ (2002) Arsenic speciation and distribution in an arsenic hyperaccumulating plant. Sci Total Environ 300:167–177

Zhao H, Wang L, Zhao FJ, Wu L, Liu A, Xu W (2018) SpHMA1 is a chloroplast cadmium exporter protecting photochemical reactions in the Cd hyperaccumulator Sedum plumbizincicola. Plant Cell Environ 42:1112–1124

Zheng X, Chen L, Li X (2018) Arabidopsis and rice showed a distinct pattern in ZIPs genes expression profile in response to Cd stress. Bot Stud 59:22

Zhou ZS, Song JB, Yang ZM (2012) Genome-wide identification of Brassica napus microRNAs and their targets in response to cadmium. J Exp Bot 63:4597–4613. https://doi.org/10.1093/jxb/ers136

Zhu C et al (2017) Characteristics of genome editing mutations in cereal crops. Trends Plant Sci 22:38–52. https://doi.org/10.1016/j.tplants.2016.08.009

Nano-contaminants: Sources and Impact on Agriculture

Mrinalini Prasad, Anamika Gupta, and Rajiv Ranjan

Contents

1 Introduction

Agriculture is the backbone of most developing countries, and approximately 60% of the world population depends on it. In a worldwide survey, it has been stated that around the world, 795 million people are suffering from insufficient amount of food. According to 2014–2015 census, India's population is 1.27 billion. To supply sufficient quantity of food for huge population is a very tedious work. To overcome this problem, we have to use some novel technology, which provides good quality and quantity of agro products in a short time period. In the earlier period of agriculture, different technologies and processes were involved in the development of hybrid varieties, synthetic chemicals, and biotechnology-based products, etc. But now, most of the agricultural researchers prefer "nanotechnology." With the help of these technologies, we can improve our environment, sustainability, varieties, productivity, and much more. It is useful to increase the amount and quality of agro food (Kah and Hofmann 2014; Aschberger et al. 2015; Chhipa and Joshi 2016) and make it more sustainable by reduced population for agro-chemicals, while supporting climate resilience

M. Prasad · A. Gupta · R. Ranjan (✉)
Department of Botany, Dayalbagh Educational Institute (Deemed University), Agra, Uttar Pradesh, India

M. Naeem et al. (eds.), *Contaminants in Agriculture*,
https://doi.org/10.1007/978-3-030-41552-5_8

Sustainable Development Goal (SDG) (Adhikari et al. 2016; Dasgupta et al. 2016). In many countries, such as in Europe, USA, China, and India, research and development (R & D) works are being carried out on agricultural nanotechnology (Cozzens et al. 2001). Only 9% of researches focus on agricultural products (Peters et al. 2016) and some are involved in the control and efficient release of pesticides, agrichemicals, nutrients, and also microorganisms (Amenta et al. 2015; Makama 2016). Some are associated with work related to increased bioavailability and bioactivity, high reactivity, adherence, and surface effects (Gutiérrez et al. 2011). The key areas are nanogenetic manipulation, diagnostics of agricultural, nanofertilizer, nanocomplexes, nanobiosensors, nanopesticides, nanoherbicides, nanobiofarming, etc. (Parisi et al. 2015).

2 Nanotechnology and Its Application in Agriculture

In 1974, Norio Taniguchi of the Tokyo Science University was the first to invent the term "nanotechnology," which means the study of manipulating matter on an atomic and molecular scale and its size range from 1 to 100 nanometer (nm) (Arivalagan et al. 2011). It utilizes to make new resources and equipment for good and bright future, for example in medical streams innovations such as antibodies customization, implanted devices, stem-cell engineering; and in production of food, examples such as nanoadditives, and many more (Lu and Bowles 2013). In earlier days, nanotechnology was targeted in four domains: (a) semi-dwarf and high-yield variety of wheat and rice, (b) wider us of irrigation, (c) agrochemicals and fertilizers, and (d) increase in agro production (Agrawal and Rathore 2014; Phogat et al. 2016). Especially, this technique is used to diminish the quantity of sprayed chemical (only active ingredients), reduce fertilizers (Gogos et al. 2012; Feregrino-Perez et al. 2018), and enhance the yield by smart delivery of water and nutrients. This technique has a better role in crop production with safe environment, ecological importance, and economic stability (Fig. 1) (Tarafdar et al. 2012). Nanotechnology depends upon nanoparticles and it contains two basic and important properties: one is increased relative surface area and the second is quantum effect (Somasundaran et al. 2010).

This technology has some adverse effects on agriculture due to nanoparticles like carbon nanotube and metal or metal oxides and so they are called nanocontaminants. In a study of rice, zinc oxide shows stunted root length and less number of roots at the starting seedling stage. To see the effect of ZnO, SiO_2, CeO_2, and TiO_2 on soil bacteria at specific concentration (1.0 mg g^{-1}), a plan was established (Chai et al. 2015). Some important soil bacteria such as azotobacter and phosphorus and potassium-solubilizing bacteria are reduced in number and also inhibited enzymatic activity by ZnO and CeO_2. TiO_2 also shows adverse effects on enzymatic activity and population of functional bacteria.

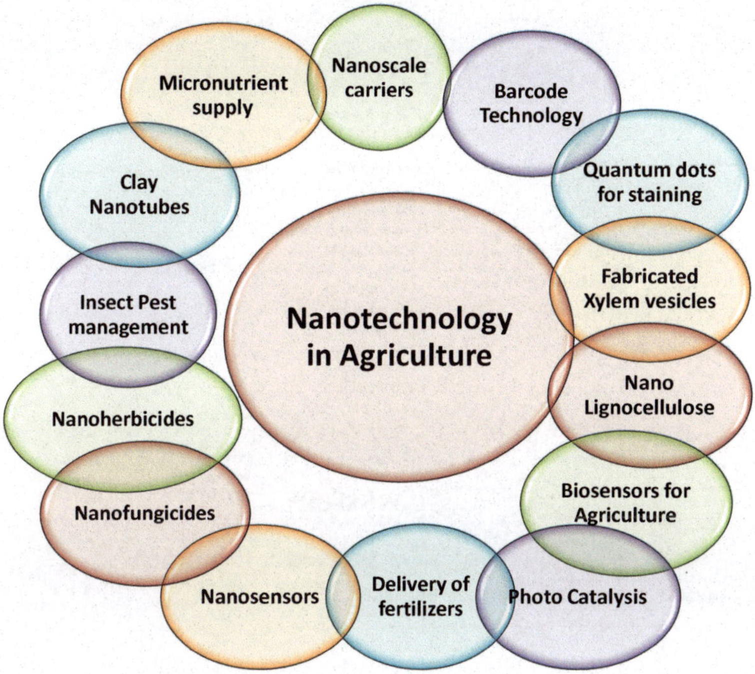

Fig. 1 Roles of nanotechnology in agriculture

3 Nanocontaminants and Their Effects, Phytotoxicity of Nanoparticles, or Nanocontaminants in Agriculture

The nanotechnology plays an important role in the plant productivity by controlling the nutrients, by monitoring water quality and pesticides for sustainable agriculture (Gruere 2012; Mukhopadhyay 2014, Prasad 2014). Although nano particles (NPs) are very useful, some nanocontaminants affect both humans and environment (Fig. 2) (Gruere et al. 2011). Plants are the important parts of ecosystems and help in the transportation of NPs to the environment (Xingmao et al. 2010). NPs are present in the soil, and it develops toxicity gradually (Shah and Belozerova 2009). For engineering of NPs, a large number of projects were established and more than 58,000 tons of NPs were produced for 2012–2020 (Maynard 2006). The waste generated from the industries directly affected the environment, soil, water, and air (Gottschalk et al. 2009; Klaine et al. 2008; Maurice and Hochella 2009; Tiede et al. 2009) and also produced synergistic or antagonistic toxicity (Newman and Unger 2003; Choi and Hu 2009). NPs have the ability to absorb contaminants, which are found in the environment (Ferguson et al. 2008). They act as a carrier or transporter of contaminants by forming a complex (Shen et al. 2012).

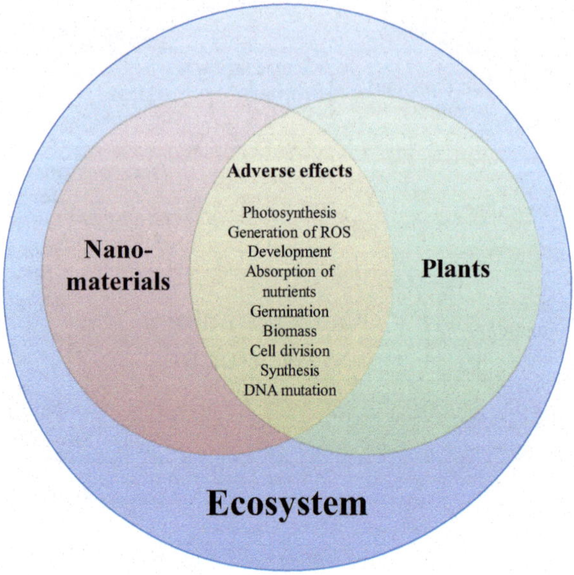

Fig. 2 Adverse effects of nanomaterials on plants

4 Classification of Nanoparticles as Nanocontaminates

Nano particles are classified into various categories (Fig. 3) and each of them is discussed below:

A. **Carbon Nanotubes (CNTs):** It is a new form of carbon that contains 2-D (two-dimensional) graphic sheet rolled into tube and they are two types: (a) single-walled nanotubes (SWNTs), (b) multi-walled nanotubes (MWNTs). It has a great attraction power of contaminants (Schwab et al. 2013) and their sites allow the contaminants for adsorption or desorption (Zhang et al. 2010). It is used in the labeling of plant proteins (Pyrzynska 2011; Chahine et al. 2014) and to reduce the agrochemicals found in the environment (Raliya et al. 2013; Hajirostamlo et al. 2015). CNT nano sponge contains sulfur and iron and it helps in the absorption of toxic dichlorobenzene, pesticides, oil, fertilizers, etc. But in some cases, CNTs may penetrate and accumulate in the cytoplasm (Porter et al. 2007). CNTs contaminants get accumulated and produce toxicity in an organism. In case of human health, lung epithelia are affected when they come into the contact of CNTs (Poland et al. 2009).
B. **Quantum Dot (QD):** This is fluorescence, quantum confinement of charge carrier particles and small-sized tunable band energy (Bulovic et al. 2004; Androvitsaneas et al. 2016). It converted water into hydrogen (Konstantatos and Sargent 2009) to detect the pathway of water in plant, especially in root system (Hu et al. 2010; Das et al. 2015). But at low concentration of QDs, no seed germination and seedling growth were detected.

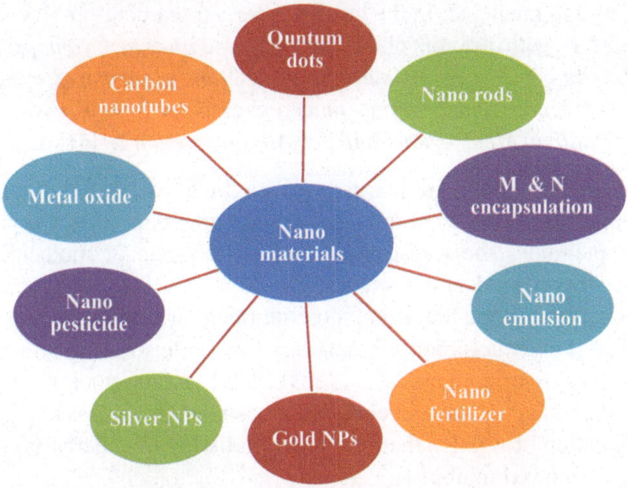

Fig. 3 Classification of nanomaterials

C. **Nanorod:** It performs as a multifunctional material such as sensing agent, regulation of size tunable energy, specific field response, and many more (Bulovic et al. 2004). It helps in the detection of phytotoxicity at a high concentration in watermelon plant and also capacity to transport auxin (2,4-D growth regulator) in tobacco cell culture (Nima et al. 2014).

D. **Micro and Nanoencapsulation:** Encapsulation means that a matter is enclosed by a layer of matrix, which is either homogeneous or heterogeneous or a polymeric membrane (Couvreur et al. 1995; Rodriguez et al. 2016). It useful in the drug delivery, protection of food components or nutraceuticals, for fortification, self-healing of materials, and many more (Ozdemir and Kemerli 2016). Capsules can be used as MRI-guided nanorobots or nanobots (Vartholomeos et al. 2011).

E. **Nanoemulsion:** Combination of nanoscaled droplets (oil/water system) is called nanoemulsion and has a size lower than 100 nm (Gutierrez et al. 2008; Anton and Vandamme 2011). By the mixing of two liquid phases one is lipophilic phase with drug and another hydrophilic aqueous phase with pure water are incorporated with each other (Tadros et al. 2004).

F. **Nanofertilizers (NF):** It is modified from traditional fertilizers to reduce the excess quantity of fertilizer. NF is a made from bulk material (Brunnert et al. 2006) and extracted from vegetative or reproductive part of plant with the help of nanotechnology to improve soil fertility and crop growth and also helps to improve the crop yield. Some NPs produce toxicity in agriculture in metal oxide form such as Al_2O_3, TiO_2, CeO_2, FeO, and ZnONPs (Dimkpa 2014; Zhang et al. 2016; Aziz et al. 2016; Patra and Baek 2017). Zinc is one of the main problems in agro products in the alkaline nature of soils (Sadeghzadeh 2013).

G. **Nanobiofarming:** With the help of NPs, farmers develop specific plants in a defined soil for industrial use and it is also mechanically separated from plant tissue (Misra et al. 2013). This kind of farming is useful in many areas like food, cosmetics, and medicine. In the last few years, researchers are focusing on gold and silver NPs with different plants such as *Vigna radiata, Cyamopsis tetragonolobus, Zea mays, Medicago sativa, Pennisetum glaucum, Sorghum vulgare, Arachis hypogaea,* and *Brassica junceaor* extracts from *Brassica juncea* and *Medicago sativa, Memecylon edule,* or *Allium sativum* L. It is of two types:

(i) **Gold NPs:** In the literature, it is reported that by inhibiting the function of aquaporins, AuNPs can induce the toxic effects in plants, a protein group that facilitates the transportation of wide range of molecules including water (Shah and Belozerova 2009).
(ii) **Silver NPs:** In the literature, more numbers of studies are reported on animal cells and microbial cells and very few studies were conducted on plants by using silver nanoparticles (AgNPs). Depending upon the plant species, AgNPs can affect the root elongation positively as well as negatively. By the application of AgNPs, the root length in barley was increased, while in lettuce, it showed inhibition.

H. **Nanopesticides:** Insect pests are dominant in the agricultural field and products, so NPs play an important role in the control of insect pests and host pathogens (Khota et al. 2012), as well as improve non-toxicity, increase food production, and reduce negative effects on environment (de Oliveira et al. 2014; Kah and Hofmann 2014; Bhattacharyya et al. 2016; Grillo et al. 2016). Nanoencapsulated pesticide has slow releasing property (Nair et al. 2010; Bhattacharyya et al. 2016) and is used for the construction of organic NPs with active agrochemicals (Gouin 2004). Few active ingredients are combined with the nano-encapsulated pesticides and they are used as weed controllers or herbicides at a specific condition of field (Gruere et al. 2011).
I. **Metal Oxide:** There are toxic or negative effects of NPs used in the industries (Yang and Watts 2005; Rana and Kalaichelvan 2013; Du et al. 2017; Tripathi et al. 2017a, b, c; Gottschalk and Nowack 2011). Few metal NPs of Au, Si, and Cu show root toxicity through soil system (Shah and Belozerova 2009). Metal NPs cause cytotoxicity on cell wall or membrane and accumulate in cytoplasm (Rana and Kalaichelvan 2011; Aziz et al. 2015, 2016). TiO_2, ZnO, SiO_2, and fullerenes NPs are photochemically active when exposed to light in the presence of oxygen; then excited electrons are produced and form superoxide radicals (Hoffmann et al. 2007; Rodriguez-Yanez et al. 2013; Li et al. 2016). In case of buckwheat, ZnO NPs show phototoxicity on seed germination and root growth (Lee et al. 2012a, b). TiO_2 and fullerenes NPs produce reactive oxygen species (ROS) (Sayes et al. 2004; Armstrong and Bharali 2013; Aziz et al. 2015; Wang et al. 2017), and collection of other NPs shows effects on respiration, transpiration, photosynthetic process and translocation of food material (Tripathi et al. 2016b; Du et al. 2017; Rana and Kalaichelvan 2013; Tripathi et al. 2016a; Bhatt and Tripathi 2011; Piccinno et al. 2012; Bumbudsanpharoke and Ko 2015; Prasad et al. 2017). By the help of soil microbes and plants, NPs are transported into food chain (Zhu et al. 2008; Priester et al. 2012) and also affect animal and human health (Nowack and Bucheli 2007; Kelly et al. 2007). Researchers have to develop some promising mechanisms to reduce the impact of NPs on the environment (Cox et al. 2017;

Singh et al. 2017). Iron (I, II) oxides (Fe_3O_4 NPs) are found as natural NPs and iron ore can deposit on plant like *Lepidium sativum* and *Pisum sativum*. There are many kinds of metal oxides found and few examples are as follows:

(i) **Silicon Dioxide NPs:** Seed germination, root elongation, and shoot emergence are the first stage of plant growth and development. It is reported in literature that seed germination is dependent on NPs concentration. When the tomato seeds were treated with low concentrations of nano-SiO_2, seed germination improved (Siddiqui and Al-Whaibi 2014). The plant growth and development are enhanced with nano-SiO_2 by increasing gas exchange and chlorophyll fluorescence parameters, such as net photosynthetic rate and stomatal conductance (Xie et al. 2011).

(ii) **Zinc Oxide NPs:** The concentrations of zinc oxide nanoparticles play a major role in plant growth and development. According to many reports, lower concentrations of ZnONPs showed beneficial effect on seed germination such as studies by Prasad et al. (2012) in peanut; Sedghi et al. (2013) in soybean; Ramesh et al. (2014) in wheat, and Raskar and Laware (2014) in onion, while higher dose of ZnONPs diminished the rate of seed germination. Plant germination depends on the rate of concentration of NPs, and it may vary from plant to plant. de la Rosa et al. (2013) performed the seed germination assay along with different concentrations of ZnONPs on cucumber, alfalfa, tomato, and they observed that seed germination enhanced only in cucumber.

(iii) **Titanium Dioxide NPs:** Very few articles have reported the impact of titanium dioxide nanoparticles (TiO_2NPs). In Canola seedlings, seed germination and radicle and plumule growth was enhanced by the application of TiO_2NPs (Mahmoodzadeh et al. 2013). TiO_2NPs showed both beneficial and toxic effects, depending upon the concentrations (Mukherjee et al. 2016). Recent reviews have shown the toxicity of TiO_2NP, such as effects on plant growth, genotoxicity, and effects of ion release (Chichiriccò and Poma 2015).

5 Nanobiosensors

To detect the activity of physical or chemical principle of plant/crop, a device was developed by the application of signal transduction technology called biosensor (Sagadevan and Periasamy 2014; Sertova 2015). It is cost-effective, sensitive, and a rapid technology and used in various human activities such as genome analysis, food and drink, health care, agriculture, industries, environmental monitoring, security, and defense. Nano biosensors are used to monitor moisture content, nutrient condition in soil and site-specific water and nutrient, nano fertilizer, nano herbicides supply in crop field. This technology is at early stage of development (Fogel and Limson 2016) and after improvement, it will be used to fabricate, measure, and image nanoscale objects. Thus, biosensor is an analytical device, which is used to convert biological activities due to some elements like antibody, enzyme, protein, and nucleic acid into an electrical signal. For the detection of low-level DNA concentration during hybridization, micro cantilever-based DNA biosensor was used with the help of AuNPs (Brolo 2012). Few biosensors are developed for the detection of plant breeding and genetic transformation (Torney et al. 2007). It is also helpful in the detection of contaminants, pollutants, microbes, and food freshness (Joyner and Kumar 2015).

6 Application of Nanosensor in Agriculture

(i) It is used to control soil conditions (e.g., moisture, soil pH), a wide variety of pesticides, herbicides, fertilizers, insecticides, pathogens, and crop growth as well.
(ii) Food-borne contaminants detection or to control the environmental conditions at the field.
(iii) Nanochips for identity preservation and tracking.
(iv) Nanocapsules for delivery of pesticides, herbicides, fertilizers, and vaccines.
(v) Nanosensors and nanobased smart delivery systems for efficient use of agricultural natural resources (e.g., water), nutrients, and chemicals through precision farming.
(vi) The role of nanoparticles in the delivery of the growth hormones or DNA to plants in a controlled manner.
(vii) The role of nanoparticles as smart nanosensors is used for early warning of changing conditions that can respond to different conditions.
(viii) Aptasensors for the determination of pesticides and insecticides (e.g., phorate, acetamiprid, isocarbophos).
(ix) Aptasensors for the determination of antibiotics, drugs, and their residues (e.g., cocaine, oxytetracycline, tetracycline, and kanamycin).
(x) Aptasensors for the determination of heavy metals (e.g., Hg2+, As3+, Cu2+).

7 Impact of Nanocontaminants in Agriculture

Although nanoparticles have greater impact on plants and environments, still many adverse effects have been reported and displayed in Table 1.

8 Some Existing Regulation of Agricultural Nanoproducts

Approximately 500 patents are available regarding nanotechnology, agriculture, nutrition, and biotechnology and about 36% contain metal oxide, fertilizer, pesticides, and drugs (Benckiser 2012). Few countries are more involved in this

Table 1 Impact of nanoparticles in plants

Sr. No.	Nanoparticles	Concentration	Effects Adverse effects	Plants/ Crops	References
1	TiO_2	–	Inhibition in cell growth and nitrogen fixation activity	*Anabaena variabilis*	Cherchi and Gu (2010)
2	TiO_2	–	Reduced germination	*Triticum aestivum L. var. Pishtaz*	Feizi et al. (2012)
3	TiO_2	–	Toxic to seed germination, decreased shoot length, decreased shoot biomass	*Mentha piperita*	Samadi et al. (2014)
4	TiO_2	–	Increases hill reaction, chloroplast activity, noncyclic photo phosphorylation and photosynthesis rate	*Spinacea oleracea*	Hong et al. (2005), (2013)
5	TiO_2	–	Increase in protein expression of Rubisco enzyme by 40%	*Spinacia oleracea*	Xuming et al. (2008)
6	TiO_2	–	Produce ROS (Reactive Oxygen Species), damage DNA	*Allium cepa, Nicotiana tabacum*	Ghosh et al. (2010)
7	TiO_2	–	Promotion of growth and photosynthesis	Spinach	Hong et al. (2005), Yang et al. (2006)
8	TiO_2	1000 mg/L	Reduce root hydraulic conductivity and leaf growth	Maize	Asli and Neumann (2009)
9	TiO_2	0–2 mM	Reduction in root growth and DNA damage	Onion & tobacco	Ghosh et al. (2010)
10	TiO_2	0–4.0‰	Germination delay and genotoxic effect	Maize, Narbon bean	Castiglione et al. (2011)
11	TiO_2, ZnO	Ti:4307 mg/kg, Zn: 214 mg/kg	Decrease biomass	Wheat	Du et al. (2011)
12	TiO_2	–	Significantly lowered carbon mineralization	Silty clay	Simonin et al. (2015)
13	TiO_2	–	Adverse impact on soil microbial community	Sandy loam	Simonin et al. (2015)
14	TiO_2	–	Significant reduction in soil microbial community	Alkaline soil	Simonin et al. (2015)
15	TiO_2	–	Toxic effect on microbial activity, carbon mineralization	High organic matter	Simonin et al. (2015)

(continued)

Table 1 (continued)

Sr. No.	Nanoparticles	Concentration	Effects Adverse effects	Plants/ Crops	References
16	TiO_2	–	Reduced bacterial diversity	Soil types	Ge et al. (2013)
17	TiO_2 and ZnO	–	Reduced microbial biomass and substrate-induced respiration	Soil types	Ge et al. (2011)
18	TiO_2,ZnO	–	Altered soil bacterial community with reduced taxa	Soil types	Ge et al. (2012)
19	Al	–	Decreased root length	*Lolium perenne*	Lin and Xing (2007)
20	Al	–	Reduced germination	*Lolium perenne*	Lin and Xing (2007)
21	Al	–	Reduced root length	*Zea mays, Lactuca sativa*	Lin and Xing (2007)
22	Ag	–	Reduced shoot and root length	*Triticum aestivum* L.	Dimkpa et al. (2013a, b)
23	Ag	–	Reduced germination	*(Hordeum vulgare L., cv. Annabell*	El-Temsah and Joner (2012)
24	Ag	–	Decreased mitosis, disturbed metaphase, sticky chromosome, cell wall disintegration and breaks	*Allium cepa*	Kumari et al. (2009)
25	Ag	–	Reduced shoot length	*Linum usitatissimum L., cv. Electra, Lolium perenne L., cv. Tove*	El-Temsah and Joner (2012)
26	Ag	–	Reduced transpiration	*Cucurbita pepo*	Stampoulis et al. (2009)
27	Ag	–	Decreased rate of mitotic index, chromosomal aberrations, irreversible DNA damage	*Vicia faba*	Patlolla et al. (2012)
28	Ag	–	Generation of ROS, cell death, mitotoic index, micronucleus and mitotic aberrations, DNA damage	*Allium cepa*	Panda et al. (2011)
29	Ag	800 μg kg^{-1}	Declined germination	Faba bean	Abd-Alla et al. (2016)
30	Ag	0, 5, 10, 20, 40 mg L^{-1}	Growth inhibition	Mung vean, Sorghum	Lee et al. (2012a, b)

31	Ag	0, 125, 250, 500 mg L^{-1}	Germination not affected	Radish	Zuverza-Mena et al. (2016)
32	Ag	20–100 mg/L	Decrease mitotic index and promotes/ cell wall disintegration	Onion	Kumari et al. (2009)
33	Ag & nZVI (Zero-valent iron)	Ag 0–100 mg/L nZVI 0–5000 mg/L	Inhibition of shoot growth and germination	Flax, barley, ryegrass	El-Temsah and Joner (2012)
34	Ag, Cu	100 mg/L & 500 mg/L	Ag NPs reduce biomass by 74–83%; 61% & 67% reduction in transpiration by ag and cu NPs	Squash	Musante and White (2012)
35	Ag, Cu, ZnO, Si, MWCNTs	1000 mg/L	Si NPs completely inhabit germination; cu NPs decrease root length after 24 h; ZnO NPs reduce biomass by 78–90%	Zucchini	Stampoulis et al. (2009)
36	AgNPs	–	Reduced microbial biomass	–	Hansch and Emmerling (2010)
37	AgNPs	–	Reduced soil enzymatic activities and substrate induced respiration	–	Colman et al. (2013)
38	AgNPs, ZnO	–	Enhanced toxicity toward *Eisenia fetida* Adverse effect on ammonification, respiration and dehydrogenase activity of soil microbes	Acidic soil	Shoults-Wilson et al. (2011); Shen et al. (2015)
39	AgNPs	–	Declined toxicity toward soil microbial activity	Alkaline soil	Schlich and Hund-Rinke (2015)
40	AgNPs	–	Reduced toxicity toward bio film forming communities	High organic matter	Sheng and Liu (2011), Wirth et al. (2012)
41	AgNPs	–	Reduced toxic impact on soil bacterium *Pseudomonas chlororaphis* O6	High cation exchange capacity	Calder et al. (2012)
42	AgNPs	–	Enhanced toxicity toward soil microbes	Low cation exchange capacity	Calder et al. (2012)
43	SiO_2 and TiO_2	–	Soybean	Promotor effect on germination	Prasad et al. (2012)

(continued)

Table 1 (continued)

Sr. No.	Nanoparticles	Concentration	Effects Adverse effects	Plants/ Crops	References
44	Carbon nanotubes (MWCNT)	10–40 μgml^{-1}	Tomato	Enhance germination and growth rates	Khodakovskaya et al. (2011), Canas et al. (2008)
				Inhibition of root elongation in tomato	
45	Carbon nanotubes (MWCNT)	–	Onion and cucumber	Enhancenment root elongation	Canas et al. (2008)
46	Zn	–	Reduced root growth and elongation	*Zea mays, Cucumis Sativus, Lactuca sativa, Raphanus sativus, Brassica napus, Lolium perenne*	Lin and Xing (2007)
47	Zn	–	Inhibits seed germination and root growth	*Lolium multiforum* (Rye grass)	Lin and Xing (2007)
48	Cu	–	Reduced seedling growth	–	Lee et al. (2008)
49	Cu	–	Reduced biomass and root growth	*Cucurbita pepo*	Stampoulis et al. (2009)
50	Al_2O_3	–	Reduced root growth	*Zea mays, Cucumis sativus, Brassica oleracea, Daucus carota*	Yang and Watts (2005)
51	Al_2O_3	–	Reduced root length	*Zea mays*	Lin and Xing (2007)
52	CeO_2	–	Reduced shoot growth	*Lycopersicon esculentum*	Lopez-Moreno et al. (2010a, b)
53	CeO_2	–	Reduced shoot growth	*Zea mays*	Lopez-Moreno et al. (2010a, b)
54	CeO_2	–	Enhancement in SOD (Superoxide Dismutase) activity and chloroplast ROS – scavenging activity	*Spinacia oleracia*	Giraldo et al. (2014)
55	CeO_2	125–4000 mgL^{-1}	Negative impact at biochemical and molecular level in plant	Cucumber, soybean, plants	Lopez-Moreno et al. (2016), Zhang et al. (2017)
56	CeO_2	1000–2000 mgKg^{-1}	Contamination of agricultural crops		Majumdar et al. (2016)

57	CeO_2	2000 mg/L	Reduction in corn, tomato and cucumber germination. Decrease in alfalfa and tomato root growth	Alfalfa, tomato, cucumber, corn	Lopez-Moreno et al. (2010a, b)
58	CeO_2	0.64 mg/mL	Reduce seed germination	Lettuce, cucumber, tomato, spinach	Garcia et al. 2011
59	CeO_2, La_2O_3, Gd_2O_3, Yb_2O_3	2000 mg/L	La_2O_3, Gd_2O_3, Yb_2O_3 inhibit root elongation at different growth stages	Radish, rape, tomato, lettuce, wheat, cucumber and cabbage	Ma et al. (2010)
60	CeO_2, ZnO	2000 mg/L	Appearance of new DNA bands in genomic DNA	Soybean	Lopez-Moreno et al. (2010a, b)
61	CeO_2,Fe_3O_4,SnO_2	–	No effect on microbial biomass C and N	–	Antisari et al. (2013)
62	CNTs	20 mg/L	Increase reactive oxygen species (ROS) in cells	Rice	Tan et al. (2009)
63	CNTs	40 µg/L	Sever reduction in the number of root hairs and their length	Mustard, bean	Ghodake et al. (2010)
64	Co, Zn	5–20 µg/mL	Inhibition of root elongation.	Onion	Ghodake et al. (2011)
65	Pd	0.4 mg/L	Sever inhibition of pollen tube emergence and elongation	Kiwi fruit	Speranza et al. (2010)
66	Si, Pd, Cu, Au.	0.013% and 0.066% (w/w)	Increase shoot/root ration after 15 days of incubation	Lettuce	Shah and Belozerova (2009)
67	ZnO	500–4000 mg/L	Increased the specific activity of stress enzymes CAT in root, stem, and leaves	*Velve tmesquite*	Hernandez-Viezcas et al. (2011)
68	ZnO	50 mg/L	Completely block root growth	Garlic	Shaymurat et al. (2012)
69	ZnO	0–2000 mg/L	Retardation of root seedlings growth (above 20 mg/L for mung and above1mg/L for gram)	Mung, gram	Mahajan et al. (2011)
70	ZnO	–	Toxic effect on *Triticum aestivum* in soil pH 7.36 and at concentration of 45.45 mg/kg	Loamy clay	Du et al. (2011)
71	ZnO	–	Nontoxic effect on *Lepidium sativum*	High cation exchange capacity	Josko and Oleszczuk (2013)
72	ZnO	–	Reduced germination	*Zea mays*	Lin and Xing (2007)

(continued)

Table 1 (continued)

Sr. No.	Nanoparticles	Concentration	Effects Adverse effects	Plants/ Crops	References
73	ZnO	–	Inhibits seed germination and root growth	*Zea mays*	Lin and Xing (2007)
74	ZnO	–	Shrunken root tips, collapsed cells in root epidermis and cortex	*Lolium perenne*	Lin and Xing (2008)
75	ZnO	500 mg kg^{-1}	Wheat	Reduction of root growth and increased production of reactive oxygen species	Dimkpa et al. (2012)
76	ZnO	1000 mg kg^{-1}	Peanut	Increased seed germination and vigor	Prasad et al. (2012)
77	ZnO	100–1000 mgL^{-1}	Garden pea	No impact on germination. Root length and nodulation was affected	Raliya et al. (2015)
78	ZnO	–	No toxicity on *Cucumis sativus* with soil pH 5.5 and at concentration of 2000 mg/kg	Soil types	Kim et al. (2011)
79	ZnO	–	Positive impact on *Zea mays* when alginate added at concentration of 400–800 mg/kg	High organic matter	Zhao et al. (2013)
80	Au	10 μgml^{-1}	Barley	Decreased root growth	Feichtmeier et al. (2015)
81	Au	0, 10, 25, 50 and 100 mg kg^{-1}	Mustard	Enhanced growth, increased number of leaves per plant	Arora et al. (2012)
82	Au	3.5 nM	Leaf necrosis after 14 days	Tobacco	Sabo-Attwood et al. (2012)
83	CdSe/ZnS QD	100 nM	Increase of ROS by 50% after 72 hours	Alfalfa	Santos et al. (2010)
84	Fe	–	Impact on structure of photosynthetic enzymes, small concentration increases growth of plantlets, high concentration decreases growth of plantlets	*Zea mays*	Racuciu and Creanga (2007)
85	CuO	–	Damages DNA	*Raphanus sativus, Lolium perenne, Lolium rigidum*	Atha et al. (2012)

86	CuO	2, 5, 10, 20, 30, 40, 50, 100 mgL^{-1}	Corn	Cu reduction from (II) to (I) no effects on germination, larger seedlings observed at 2 ppm	Wang et al. (2016)
87	CuO	0–500 mgL^{-1}	Mustard	Reduce shoot growth, root shortening	Nair and Chung (2015)
88	CuO, Fe_3O_4	–	Negative effect on soil microbial community	Soil types	Frenk et al. (2013)
89	CuO, Fe_3O_4	–	Enhanced toxicity toward microbial community	Low organic matter	Frenk et al. (2013)
90	CuO, ZnO	–	Toxic effect on *Triticum aestivum*	Soil types	Dimkpa et al. (2013a, b)

MWCNTs Multi walled carbon nanotubes, *CNTs* Carbon nano tubes, *QD* Quantum dots

Table 2 Regulation of agricultural nanoproducts in different countries

Country	Acts/Rules
USA	No specific rule but Federal Food. Drug, and Cosmetic Act (FFDCA) (US-FDA) are helpful.
	FIFRA (Federal Insecticide, Fungicide, and Rodenticide Act)
Canada	Food Inspection Agency (CFIA) and Public Health Agency of Canada (PHAC)
European Union	Regulation:
	(a) Provision of Food information of Consumers (1169/20119)
	(b) Plastic Food Contact Materials and articles (10/2011)
	(c) Active and Intelligent Materials and articles (450/2009)
	(d) Biocidal Products Regulation (528/2012)
	(e) The Cosmetic Products Regulation (1223/2009)
	(f) Registration, Evaluation, Authorization and Restriction of Chemicals Regulation (1907/2006)
	(g) Plant Protection Products (PPP) (1107/2009)
Switzerland, Turkey, and Russia	Federal Office of Public Health (FOPH)
Australia and New Zealand	Food Standards Australia New Zealand (FSANZ)
India, China, Japan, South Korea, Iran, and Taiwan	India: no guidelines
	China: National Centre for Nanoscience and Technology (NCNST)
	Japan: No legislation
	South Korea: National Nano safety Strategic Plan (2012/2016)
	Iran: Iran Nanotechnology Initiative Council (INIC) and Food and drug Organization (FDO)
	Taiwan: Taiwan Nanotechnology Industry Development Association (TANIDA)
	Brazil: no guidelines
	Africa: no guidelines

technology – China, France, Germany, Japan, South Korea, Switzerland, USA, and India (NAAS 2013). But there are no specific, uniform, or consistent governmental regulations in countries for nano materials (Adholeya et al. 2017). Some of the rules or acts that are found in different countries are mentioned in Table 2.

9 Conclusion and Future Aspects

Nanotechnology is an emerging field of science and technology, which has been introduced for various novel applications in agriculture and medicines. The studies among the interaction of nanomaterials and planting material have gaining interest because of their wide application, not only for better germination processes, or smart use of pesticides or nutrients delivery, but also for their potential negative relation that may affect the crop final product and climate too. Many research studies have shown positive effects of NPs on plant growth and development. Because of nanoparticles exposure, toxicity of plant has been affected by a host of factors, which are related to the nanoparticle characteristics. These characteristics are not limited to NP size, shape, coating, carrier, application, and experimental methods. It is difficult to observe toxicity in soil and hydroponic systems due to agglomeration of NPs in each medium. Nevertheless, most of these materials are accumulated in roots and leaves, leaving a serious problem of NMs accumulation and not a viable option for crop harvest due to the human health implications. Although this particular chapter explained various negative effects caused by nanoparticle to the plants, still more research is required to understand physiochemical, biochemical, and molecular mechanisms of plants in relation to nanoparticles. Also, large-scale modeling studies will be required at the level of different ecosystems and how these nanomaterials are transferred at the critical trophic levels affecting human health and biota. Finally, intensive work must be carried out for attaining advance research in the field of molecular biology, such as proteomic and transcriptomic studies, to provide information regarding the stress by NMs at the gene level.

Acknowledgment The authors are highly grateful to the Director, DEI and Head, Department of Botany, for their encouragement and continuous support.

References

Abd-Alla MH, Nafady NA, Khalaf DM (2016) Assessment of silver nanoparticles contamination on faba bean-Rhizobium leguminosarum bv. viciae-Glomus aggregatum symbiosis: implications for induction of autophagy process in root nodule. Agric Ecosyst Environ 218:163–177

Adhikari T, Kundu S, Rao AS (2016) Zinc delivery to plants through seed coating with nano-zinc oxide particles. J Plant Nutr 39(1):136–146

Adholeya A, Das RK, Dubey MK, Kochar M, Singh R (2017) Regulation of nanoproducts in agriculture. Zero Draft Policy Regul Nanoprod Agric:1–7

Agrawal S, Rathore P (2014) Nanotechnology pros and cons to agriculture: a review. Int J Curr Microbiol App Sci 3(3):43–55

Amenta V, Aschberger K, Arena M, Bouwmeester H, Moniz FB, Brandhoff P, Gottardo S, Marvin HJ, Mech A, Pesudo LQ, Rauscher H (2015) Regulatory aspects of nanotechnology in the agri/feed/food sector in EU and non-EU countries. Regul Toxicol Pharm 73(1):463–476

Androvitsaneas P, Young AB, Schneider C, Maier S, Kamp M, Hofling S, Knauer S, Harbord E, Hu CY, Rarity JG, Oulton R (2016) Charged quantum dot micropillar system for deterministic light-matter interactions. Phys Rev B 93(24):241409

Antisari LV, Carbone S, Gatti A, Vianello G, Nannipieri P (2013) Toxicity of metal oxide (CeO_2, Fe3O4, SnO_2) engineered nanoparticles on soil microbial biomass and their distribution in soil. Soil Biol Biochem 60:87–94

Anton N, Vandamme TF (2011) Nano-emulsions and micro-emulsions: clarifications of the critical differences. Pharm Res 28:978–985

Arivalagan K, Ravichandran S, Rangasamy K (2011) Nanomaterials and its potential applications. Int J Chem Tech Res 3:534–538

Armstrong D, Bharali DJ (2013) Oxidative stress and nanotechnology. Methods Protocol 1028:1–294

Arora S, Sharma P, Kumar S, Nayan R, Khanna PK, Zaidi MGH (2012) Gold nanoparticle induced enhancement in growth and seed yield of *Brassica juncea*. Plant Growth Regul 66(3):303–310

Aschberger K, Gottardo S, Amenta V, Arena M, Moniz FB, Bouwmeester H, Brandhoff P, Mech A, Pesudo LQ, Rauscher H, Schoonjans R (2015) Nanomaterials in food-current and future applications and regulatory aspects. J Phys Conf Ser 617(1):012032

Asli S, Neumann PM (2009) Colloidal suspensions of clay or titanium dioxide nanoparticles can inhibit leaf growth and transpiration via physical effects on root water transport. Plant Cell Environ 32(5):577–584

Atha DH, Wang H, Petersen EJ, Cleveland D, Holbrook RD, Jaruga P, Nelson BC (2012) Copper oxide nanoparticle mediated DNA damage in terrestrial plant models. Environ Sci Technol 46(3):1819–1827

Aziz N, Faraz M, Pandey R, Shakir M, Fatma T, Varma A, Barman I, Prasad R (2015) Facile algae-derived route to biogenic silver nanoparticles: synthesis, antibacterial, and photocatalytic properties. Langmuir 31(42):11605–11612

Aziz N, Pandey R, Barman I, Prasad R (2016) Leveraging the attributes of Mucor hiemalis-derived silver nanoparticles for a synergistic broad-spectrum antimicrobial platform. Front Microbiol 7:1984

Benckiser G (2012) Hot topic: nanotechnology and patents in agriculture, food technology, nutrition and medicine-advantages and risks. Recent Pat Food Nutr Agric 4(3):171–175

Bhatt I, Tripathi BN (2011) Interaction of engineered nanoparticles with various components of the environment and possible strategies for their risk assessment. Chemosphere 82:308–317

Bhattacharyya A, Duraisamy P, Govindarajan M, Buhroo AA, Prasad R (2016) Nano-biofungicides: emerging trend in insect pest control. In: Advances and applications through fungal nanobiotechnology. Springer, Cham, pp 307–319

Brolo AG (2012) Plasmonics for future biosensors. Nat Photonics 6:709–713

Brunnert I, Wick P, Manserp S, Grass RN, Limbach LK, Bruinink A, Stark WJ (2006) In vitro cytotoxicity of oxide nanoparticles: comparison to asbestos, silica, and the effect of particle solubility. Environ Sci Technol 40(14):4374–4381

Bulovic V, Mandell A, Perlman A (2004) Advanced Micro Devices Inc. Molecular memory device. U.S. Patent 6,781,868

Bumbudsanpharoke N, Ko S (2015) Nano-food packaging: an overview of market, migration research, and safety regulations. J Food Sci 80:R910–R923

Calder AJ, Dimkpa CO, McLean JE, Britt DW, Johnsonc W, Anderson AJ (2012) Soil components mitigate the antimicrobial effects of silver nanoparticles towards a beneficial soil bacterium, Pseudomonas chlororaphis O6. Sci Total Environ 429:215–222

Canas JE, Long M, Nations S, Vadan R, Dai L, Luo M, Ambikapathi R, Lee EH, Olszy D (2008) Effects of functionalized and nonfunctionalized single-walled carbon nanotubes on root elongation of select crop species. Environ Toxicol Chem 27(9):1922–1931

Castiglione MR, Giorgetti L, Geri C, Cremonini R (2011) The effects of nano-TiO_2 on seed germination, development and mitosis of root tip cells of *Vicia narbonensis* L. and *Zea mays* L. J Nanopart Res 13(6):2443–2449

Chahine NO, Collette NM, Thomas BC, Genetos DC, Loots GG (2014) Nanocomposite scaffold for chondrocyte growth and cartilage tissue engineering: effects of carbon nanotube surface functionalization. Tissue Eng Part A 20:2305–2315

Chai H, Yao J, Sun J, Zhang C, Liu W, Zhu M, Ceccanti B (2015) The effect of metal oxide nanoparticles on functional bacteria and metabolic profiles in agricultural soil. Bull Environ Contam Toxicol 94(4):490–495

Cherchi C, Gu AZ (2010) Impact of titanium dioxide nanomaterials on nitrogen fixation rate and intracellular nitrogen storage in Anabaena variabilis. Environ Sci Technol 44:8302–8307

Chhipa H, Joshi P (2016) Nanofertilisers, nanopesticides and nanosensors in agriculture. In: Nanoscience in food and agriculture, vol 1. Springer, Cham, pp 247–282

Chichiriccò G, Poma A (2015) Penetration and toxicity of nanomaterials in higher plants. Nano 5(2):851–873

Choi O, Hu Z (2009) Role of reactive oxygen species in determining nitrification inhibition by metallic/oxide nanoparticles. J Environ Eng 135(12):1365–1370

Colman BP, Arnaout CL, Anciaux S, Gunsch CK, Hochella MF Jr, Kim B, Lowry GV, McGill BM, Reinsch BC, Richardson CJ, Unrine JM (2013) Low concentrations of silver nanoparticles in biosolids cause adverse ecosystem responses under realistic field scenario. PLoS One 8(2):e57189

Couvreur P, Dubernet C, Puisieux F (1995) Controlled drug delivery with nanoparticles: current possibilities and future trends. Eur J Pharm Biopharm 41:2–13

Cox A, Venkatachalam P, Sahi S, Sharma N (2017) Reprint of silver and titanium dioxide nanoparticle toxicity in plants: a review of current research. Plant Physiol Biochem 110:33–49

Cozzens S, Cortes R, Soumonni O, Woodson T (2001) Nanotechnology and the millennium development goals: water, energy and agri-food. J Nanopart Res 15:2001

Das S, Wolfson BP, Tetard L, Tharkur J, Bazata J, Santra S (2015) Effect of N-acetyl cysteine coated CdS: Mn/ZnS quantum dots on seed germination and seedling growth of snow pea (*Pisum sativum* L.): imaging and spectroscopic studies. Environ Sci 2:203–212

Dasgupta N, Ranjan S, Rajendran B, Manickam V, Ramalingam C, Avadhani GS, Kumar A (2016) Thermal co-reduction approach to vary size of silver nanoparticle: its microbial and cellular toxicology. Environ Sci Pollut Res 23(5):4149–4163

de la Rosa G, López-Moreno ML, de Haro D, Botez CE, Peralta-Videa JR, Gardea-Torresdey JL (2013) Effects of ZnO nanoparticles in alfalfa, tomato, and cucumber at the germination stage: root development and X-ray absorption spectroscopy studies. Pure Appl Chem 85(12):2161–2174

de Oliveira JL, Campos EV, Bakshi M, Abhilash PC, Fraceto LF (2014) Application of nanotechnology for the encapsulation of botanical insecticides for sustainable agriculture: prospects and promises. Biotechnol Adv 32:1550–1561

Dimkpa CO (2014) Can nanotechnology deliver the promised benefits without negatively impacting soil microbial life? J Basic Microbiol 54:889–904

Dimkpa CO, McLean JE, Latta DE, Manangón E, Britt DW, Johnson WP, Boyanov MI, Anderson AJ (2012) CuO and ZnO nanoparticles: phytotoxicity, metal speciation, and induction of oxidative stress in sand-grown wheat. J Nanopart Res 14(9):1125

Dimkpa CO, Latta DE, McLean JE, Britt DW, Boyanov MI, Anderson AJ (2013a) Fate of CuO and ZnO nano-and microparticles in the plant environment. Environ Sci Technol 47:4734–4742

Dimkpa CO, McLean JE, Martineau N, Britt DW, Haverkamp R, Anderson AJ (2013b) Silver nanoparticles disrupt wheat (*Triticum aestivum* L.) growth in a sand matrix. Environ Sci Technol 47:1082–1090

Du W, Sun Y, Ji R, Zhu J, Wu J, Guo H (2011) TiO_2 and ZnO nanoparticles negatively affect wheat growth and soil enzyme activities in agricultural soil. J Environ Monit 13:822–828

Du W, Tan W, Peralta-Videa JR, Gardea-Torresdey JL, Ji R, Yin Y, Guo H (2017) Interaction of metal oxide nanoparticles with higher terrestrial plants: physiological and biochemical aspects. Plant Physiol Biochem 110:210–225

El-Temsah YS, Joner EJ (2012) Impact of Fe and Ag nanoparticles on seed germination and differences in bioavailability during exposure in aqueous suspension and soil. Environ Toxicol 27(1):42–49

Feichtmeier NS, Walther P, Leopold K (2015) Uptake, effects, and regeneration of barley plants exposed to gold nanoparticles. Environ Sci Pollut Res 22(11):8549–8558

Feizi H, Moghaddam PR, Shahtahmassebi N, Fotovat A (2012) Impact of bulk and nanosized titanium dioxide (TiO_2) on wheat seed germination and seedling growth. Biol Trace Elem Res 146:101–106

Feregrino-Perez AA, Magaña-López E, Guzmán C, Esquivel K (2018) A general overview of the benefits and possible negative effects of the nanotechnology in horticulture. Sci Hortic 238:126–137

Ferguson PL, Chandler GT, Templeton RC, Demarco A, Scrivens WA, Englehart BA (2008) Influence of sediment-amendment with single-walled carbon nanotubes and diesel soot on bioaccumulation of hydrophobic organic contaminants by benthic invertebrates. Environ Sci Technol 42(10):3879–3885

Fogel R, Limson J (2016) Developing biosensors in developing countries: South Africa as a case study. Biosensors 6(1):5

Frenk S, Ben-Moshe T, Dror I, Berkowitz B, Minz D (2013) Effect of metal oxide nanoparticles on microbial community structure and function in two different soil types. PLoS One 13 8(12):e84441

Garcia A, Espinosa R, Delgodo L, Casals E, Gonzalez E, Puntes V, Barata C, Font X, Sanchez A (2011) Acute toxicity of cerium oxide, titanium oxide and iron oxide nanoparticles using standardized tests. Desalination 269(1–3):136–141

Ge Y, Schimel JP, Holden PA (2011) Evidence for negative effects of $TiO_2$and ZnO nanoparticles on soil bacterial communities. Environ Sci Technol 45:1659–1664

Ge Y, Schimel JP, Holden PA (2012) Identification of soil bacteria susceptible to TiO_2 and ZnO nanoparticles. Appl Environ Microbiol 78:6749–6758

Ge Y, Priester JH, Van De Werfhorst LC, Schimel JP, Holden PA (2013) Potential mechanisms and environmental controls of TiO_2 nanoparticle effects on soil bacterial communities. Environ Sci Technol 47:14411–14417

Ghodake G, Seo YD, Park D, Lee DS (2010) Phytotoxicity of carbon nanotubes assessed by Brassica juncea and Phaseolus mungo. J Nanoelectron Optoelectron 5:157–160

Ghodake G, Seo YD, Lee DS (2011) Hazardous phytotoxic nature of cobalt and zinc oxide nanoparticles assessed using Allium cepa. J Hazard Mater 186(1):952–955

Ghosh M, Bandyopadhyay M, Mukherjee A (2010) Genotoxicity of titanium dioxide (TiO2) nanoparticles at two trophic levels: plant and human lymphocytes. Chemosphere 81(10):1253–1262

Giraldo JP, Landry MP, Faltermeier SM, McNicholas TP, Iverson NM, Boghossian AA, Strano MS (2014) Plant nanobionics approach to augment photosynthesis and biochemical sensing. Nat Mater 13(4):400–408

Gogos A, Knauer K, Bucheli TD (2012) Nanomaterials in plant protection and fertilization: current state, foreseen applications, and research priorities. J Agric Food Chem 60(39):9781–9792

Gottschalk F, Nowack B (2011) The release of engineered nanomaterials to the environment. J Environ Monit 13:1145–1155

Gottschalk F, Sonderer T, Scholz RW, Nowack B (2009) Modeled environmental concentrations of engineered nanomaterials (TiO_2, ZnO, Ag, CNT, fullerenes) for different regions. Environ Sci Technol 43(24):9216–9222

Gouin S (2004) Microencapsulation: industrial appraisal of existing technologies and trends. Trends Food Sci Technol 15:330–347

Grillo R, Abhilash PC, Fraceto LF (2016) Nanotechnology applied to bio-encapsulation of pesticides. J Nanosci Nanotechnol 16:1231–1234

Gruere GP (2012) Implications of nanotechnology growth in food and agriculture in OECD countries. Food Policy 37:191–198

Gruere G, Clare N, Linda A (2011) Agricultural, food and water nanotechnologies for the poor opportunities, constraints and role of the consultative group on International Agricultural Research. J Int Food Policy Res Inst 1:35

Gutierrez JM, Gonzalez C, Maestro A, Sole I, Pey CM, Nolla J (2008) Nano-emulsions: new applications and optimization of their preparation. Curr Opin Colloid Interface Sci 13:245–251

Gutiérrez FJ, Mussons ML, Gaton P, Rojo R (2011) Nanotechnology and food industry. Scientific, health and social aspects of the food industry. InTech 1:95–121

Hajirostamlo B, Mirsaeedghazi N, Arefnia M, Shariati MA, Fard EA (2015) The role of research and development in agriculture and its dependent concepts in agriculture [Short review]. Asian J Appl Sci Eng 4(1):78–80

Hansch M, Emmerling C (2010) Effects of silver nanoparticles on themicrobiota and enzyme activity in soil. J Plant Nutr Soil Sci 173:554–558

Hernandez-Viezcas JA, Castillo-Michel H, Servin AD, Peralta-Videa JR, Gardea-Torresdey JL (2011) ZnO NPs undergo differential biological transformations in Mesquite plants. Chem Eng J 170:346–352

Hoffmann M, Holtze EM, Wiesner MR (2007) Reactive oxygen species generation on nanoparticulate material. In: Environmental nanotechnology. Applications and impacts of nanomaterials. McGraw-Hill Education, New York, pp 155–203

Hong F, Yang F, Liu C, Gao Q, Wan Z, Gu F, Wu C, Ma Z, Zhou J, Yang P (2005) Influences of nano-TiO2 on the chloroplast aging of spinach under light. Biol Trace Elem Res 104(3):249–260

Hong J, Peralta-Videa JR, Gardea-Torresdey JL (2013) Nanomaterials in agricultural production: benefits and possible threats. In: Sustainable nanotechnology and the environment: advances and achievements, vol 1124. American Chemical Society, Washington, DC, pp 73–90

Hu Y, Li J, Ma L, Peng Q, Feng W, Zhang L, He S, Yang F, Huang J, Li L (2010) High efficiency transport of quantum dots into plant roots with the aid of silwet L-77. Plant Physiol Biochem 48(8):703–709

Josko I, Oleszczuk P (2013) Influence of soil type and environmental conditions on ZnO, TiO_2 and Ni nanoparticles phytotoxicity. Chemosphere 92:91–99

Joyner JR, Kumar DV (2015) Nanosensors and their applications in food analysis: a review. Int J Sci Technol 1(4):80–90

Kah M, Hofmann T (2014) Nanopesticide research: current trends and future priorities. Environ Int 63:224–235

Kelly BC, Ikonomou MG, Blair JD, Morin AE, Gobas FA (2007) Food web-specificbio-magnification of persistent organic pollutants. Science 317(5835):236–239

Khodakovskaya MV, de Silva K, Nedosekin DA, Dervishi E, Biris AS, Shashkov EV, Zharov VP (2011) Complex genetic, photothermal, and photoacoustic analysis of nanoparticle-plant interactions. Proc Natl Acad Sci 108(3):1028–1033

Khota LR, Sankarana S, Majaa JM, Ehsania R, Schuster EW (2012) Applications of nanomaterials in agricultural production and crop protection: a review. Crop Prot 35:64–70

Kim S, Kim J, Lee I (2011) Effects of Zn and ZnO nanoparticles and Zn^{2+} on soil enzyme activity and bioaccumulation of Zn in *Cucumis sativus*. Chem Ecol 27(1):49–55

Klaine SJ, Alvarez PJ, Batley GE, Lead JR (2008) Nanomaterials in the environment: behavior, fate, bioavailability, and effects. Environ Toxicol Chem 27(9):1825–1851

Konstantatos G, Sargent EH (2009) Solution-processed quantum dot photo detectors. Proc IEEE 97(10):1666–1683

Kumari M, Mukherjee A, Chadrasekaran N (2009) Genotoxicity of silver nanoparticle in *Allium cepa*. Sci Total Environ 407(19):5243–5246

Lee WM, An YJ, Yoon H, Kwbon HS (2008) Toxicity and bioavailability of copper nanoparticles to the terrestrial plants mung bean (*Phaseolus radiatus*) and wheat (*Triticum aestrivum*): plant agar test for water-insoluble nanoparticles. Environ Toxicol Chem 27(9):1915–1921

Lee S, Kim SY, Kim S, Lee I (2012a) Effects of soil-plant inter active system on response to exposure to ZnO nanoparticles. J Microbiol Biotechnol 22:1264–1270

Lee WM, Kwak JI, An YJ (2012b) Effect of silver nanoparticles in crop plants *Phaseolus radiatus* and Sorghum bicolor: media effect on phytotoxicity. Chemom Intell Lab Syst 86(5):491–499

Li H, Shan C, Zhang Y, Cai J, Zhang W, Pan B (2016) Arsenate adsorption by hydrous ferric oxide nanoparticles embedded in cross-linked anion exchanger: effect of the host pore structure. ACS Appl Mater Interfaces 8:3012–3020

Lin D, Xing B (2007) Phytotoxicity of nanoparticles: inhibition of seed germination and root growth. Environ Pollut 150(2):243–250

Lin D, Xing B (2008) Root uptake and phytotoxicity of ZnO nanoparticles. Environ Sci Technol 42(15):5580–5585

Lopez-Moreno ML, de la Rosa G, Hernandez-Viezcas JA, Castillo-Michel H, Botez C, Peralta-Videa JR, Gardea-Torresdey JL (2010a) Evidence of the differential biotransformation and genotoxicity of ZnO and CeO2 nanoparticles on soybean (Glycine max) plants. Environ Sci Technol 44(19):7315–7320

Lopez-Moreno ML, de la Rosa G, Hernandez-Viezcas JA, Peralta-Videa JR, Gardea-Torresdey JL (2010b) X-ray absorption spectroscopy (XAS) corroboration of the uptake and storage of CeO_2 nanoparticles and assessment of their differential toxicity in four edible plant species. J Agric Food Chem 58(6):3689–3693

Lopez-Moreno ML, Aviles LL, Perez NG, Irizarry BA, Perales O, Cedeno-Mattei Y, Roman F (2016) Effect of cobalt ferrite ($CoFe_2O_4$) nanoparticles on the growth and development of Lycopersicon lycopersicum (tomato plants). Sci Total Environ 550:45–52

Lu J, Bowles M (2013) How will nanotechnology affect agricultural supply chains? Int Food Agribus Manag Assoc 16(2):21–42

Ma Y, Kuang L, He X, Bai W, Ding Y, Zhang Z, Zhao Y, Chai Z (2010) Effects of rare earth oxide nanoparticles on root elongation of plants. Chemosphere 78(3):273–279

Mahajan P, Dhoke SK, Khanna AS (2011) Effect of nano-ZnO particle suspension on growth of mung (Vigna radiata) and gram (Cicer arietinum) seedlings using plant agar method. J Nanotechnol 2011:1

Mahmoodzadeh H, Nabavi M, Kashefi H (2013) Effect of nanoscale titanium dioxide particles on the germination and growth of canola (*Brassica napus*). J Ornamental Hortic Plants 3:25–32

Majumdar S, Trujillo-Reyes J, Hernandez-Viezcas JA, White JC, Peralta-Videa JR, Gardea-Torresdey JL (2016) Cerium biomagnification in a terrestrial food chain: influence of particle size and growth stage. Environ Sci Technol 50(13):6782–6792

Makama S. Private interview, 09-Dec-2016

Maurice PA, Hochella MF (2009) Nanoscale particles and processes: a new dimension in soil science. Adv Agron 100:123–153

Maynard AD (2006) Nanotechnology: a research strategy for addressing risk, vol 444. Woodrow Wilson International Center for Scholars, Washington, DC, pp 267–269

Misra AN, Misra M, Singh R (2013) Nanotechnology in agriculture and food industry. Int J Pure Appl Sci Technol 16(2):1–9

Mukherjee A, Majumdar S, Servin AD, Pagano L, Dhankher OP, White JC (2016) Carbon nanomaterials in agriculture: a critical review. Front Plant Sci 7:172

Mukhopadhyay SS (2014) Nanotechnology in agriculture: prospects and constraints. Nanotechnol Sci Appl 7:63–71

Musante C, White JC (2012) Toxicity of silver and copper to *Cucurbita pepo*: differential effects of nano and bulk-size particles. Environ Toxicol 27(9):510–517

NAAS (2013) Nanotechnology in agriculture: scope and current relevance. Policy paper no. 63. Nat Acad Agric Sci 20

Nair PMG, Chung IM (2015) Study on the correlation between copper oxide nanoparticles induced growth suppression and enhanced lignification in Indian mustard (*Brassica juncea* L.). Ecotoxicol Environ Saf 113:302–313

Nair R, Varghese SH, Nair BG, MaekawaT YY, Kumar DS (2010) Nanoparticulate material delivery toplants. Plant Sci 179:154–163

Newman MC, Unger MA (2003) Fundamentals of ecotoxicology, 2nd edn. Lewis Publishing, Albany, pp 320–358

Nima AZ, Lahiani MH, Watanabe F, Xu Y, Khodakovskaya MV, Biris AS (2014) Plasmonically active nanorods for delivery of bio-active agents and high-sensitivity SERS detection in planta. RSC Adv 4(110):64985–64993

Nowack B, Bucheli TD (2007) Occurrence, behavior and effects of nanoparticles in the environment. Environ Pollut 150:5–22

Ozdemir M, Kemerli T (2016) Innovative applications of micro and nanoencapsulation in food packaging. In: Encapsulation and controlled release technologies in food systems. Wiley Blackwell, Chichester, pp 333–378

Panda KK, Achary VMM, Krishnaveni R, Padhi BK, Sarangi SN, Sahu SN, Panda BB (2011) In vitro biosynthesis and genotoxicity bioassay of silver nanoparticles using plants. Toxicol In Vitro 25(5):1097–1105

Parisi C, Vigani M, Rodriguez-Cerezo E (2015) Agricultural nanotechnologies: what are the current possibilities? Nanotoday, Science Direct 10(2):124–127

Patlolla AK, Berry A, May L, Tchounwou PB (2012) Genotoxicity of silver nanoparticles in *Vicia faba*: a pilot study on the environmental monitoring of nanoparticles. Int J Environ Res Public Health 9(5):1649–1662

Patra JK, Baek KH (2017) Antibacterial activity and synergistic antibacterial potential of biosynthesized silver nanoparticles against food borne pathogenic bacteria along with its anticandidal and antioxidant effects. Front Microbiol 8:167

Peters RJ, Bouwmeester H, Gottardo S, Amenta V, Arena M, Brandhoff P, Marvin HJ, Mech A, Moniz FB, Pesudo LQ, Rauscher H (2016) Nanomaterials for products and application in agriculture, feed and food. Trends Food Sci Tech 54:155–164

Phogat N, Khan SA, Shankar S, Ansary AA, Uddin I (2016) Fate of inorganic nanoparticles in agriculture. Adv Mater Lett 7(1):03–12

Piccinno F, Gottschalk F, Seeger S, Nowack B (2012) Industrial production quantities and uses of ten engineered nanomaterials in Europe and the world. J Nanopart Res 14:1109

Poland CA, Duffin R, Kinloch I, Maynard A, Wallace WAH, Seaton A, Stone V, Brown S, MacNee W, Donaldson K (2009) Carbon nanotubes introduced into the abdominal cavity of mice show asbestos like pathogenicity in a pilot study. Nat Nanotechnol 3(7):423–428

Porter AE, Gass M, Muller K, Skepper JN, Midgley PA, Welland M (2007) Direct imaging of single-walled carbon nanotubes in cells. Nat Nanotechnol 2:713–717

Prasad R (2014) Synthesis of silver nanoparticles in photosynthetic plants. J Nanopart 2014:963961

Prasad TNVKV, Sudhakar P, Sreenivasulu Y, Latha P, Munaswamy V, RajaReddy K, Pradeep T (2012) Effect of nanoscale zinc oxide particles on the germination, growth and yield of peanut. J Plant Nutr 35(6):905–927

Prasad R, Bhattacharyya A, Nguyen QD (2017) Nanotechnology in sustainable agriculture: recent developments, challenges, and perspectives. Front Microbiol 8(1014):13

Priester JH, Ge Y, Mielke RE, Horst AM, Moritz SC, Espinosa K, Gelb J, Walker SL, Nisbet RM, An YJ, Schimel JP, Palmer RG, Hernandez-Viezcas JA, Zhao L, Gardea-Torresdey JL, Holden PA (2012) Soybean susceptibility to manufactured nanomaterials with evidence for food quality and soil fertility interruption. PNAS 109(37):E2451–E2456

Pyrzynska K (2011) Carbon nanotubes as sorbents in the analysis of pesticides. Chemosphere 83:1407–1413

Racuciu M, Creanga DE (2007) TMA-OH coated magnetic nanoparticles internalized in vegetal tissue. Romanian Journal of Physics 52(3/4):395

Raliya R, Tarafdar JC, Gulecha K, Choudhary K, Ram R, Mal P, Saan RP (2013) Review article; scope of nanoscience and nanotechnology in agriculture. J Appl Biol Biotechnol 1(03):041–044

Raliya R, Nair R, Chavalmane S, Wang WN, Biswas P (2015) Mechanistic evaluation of translocation and physiological impact of titanium dioxide and zinc oxide nanoparticles on the tomato (*Solanum lycopersicum* L.) plant. Metallomics 7(12):1584–1594

Ramesh M, Palanisamy K, Babu K, Sharma NK (2014) Effects of bulk & nano-titanium dioxide and zinc oxide on physio-morphological changes in *Triticum aestivum* Linn. J Glob Biosci 3:415–422

Rana S, Kalaichelvan PT (2011) Antibacterial effects of metal nanoparticles. Adv Biotech 2(2):21–23

Rana S, Kalaichelvan PT (2013) Ecotoxicity of nanoparticles. ISRN Toxicol 2013:574648

Raskar SV, Laware SL (2014) Effect of zinc oxide nanoparticles on cytology and seed germination in onion. Int J Curr Microbiol App Sci 3:467–473

Rodriguez J, Martin MJ, Ruiz AM, Clares B (2016) Current encapsulation strategies for bioactive oils: from alimentary to pharmaceutical perspectives. Food Res Int 83:41–59

Rodriguez-Yanez Y, Munoz B, Albores A (2013) Mechanisms of toxicity by carbon nanotubes. Toxicol Mech Method 23:178–195

Sabo-Attwood T, Unrine JM, Stone JW, Murphy CJ, Ghoshroy S, Blom D, Bertsch PM, Newman LA (2012) Uptake, distribution and toxicity of gold nanoparticles in tobacco (*Nicotiana xanthi*) seedlings. Nanotoxicology 6(4):353–360

Sadeghzadeh B (2013) A review of zinc nutrition and plant breeding. J Soil Sci Plant Nutr 13:905–927

Sagadevan S, Periasamy M (2014) Recent trends in nanobiosensors and their applications – a review. Rev Adv Mater Sci 36:62–69

Samadi N, Yahyaabadi S, Rezayatmand Z (2014) Effect of TiO2 and TiO2 nanoparticle on germination, root and shoot length and photosynthetic pigments of *Mentha piperita*. International Journal of Plant & Soil Science 3(4):408–418

Santos AR, Miguel A, Tomaz L, Malho R, Maycock C, Patto MV, Fevereiro P, Oliva A (2010) The impact of CdSe/ZnS quantum dots in cells of *Medicago sativa* in suspension culture. J Nanobiotechnol 8(1):24–37

Sayes CM, Fortner JD, Guo W, Lyon D, Boyd AM, Ausman KD, Tao YJ, Sitharaman B, Wilson LJ, Hughes JB, West JL (2004) The differential cytotoxicity of water-soluble fullerenes. Nano Lett 4(10):1881–1887

Schlich K, Hund-Rinke K (2015) Influence of soil properties on the effect of silver nanomaterials on microbial activity in five soils. Environ Pollut 196:321–330

Schwab F, Bucheli TD, Camenzuli L, Magrez A, Knauer K, Sigg L, Nowack B (2013) Diuron sorbed to carbon nanotubes exhibits enhanced toxicity to Chlorella vulgaris. Environ Sci Technol 47(13):7012–7019

Sedghi M, Hadi M, Toluie SG (2013) Effect of nano zinc oxide on the germination of soybean seeds under drought stress. Ann West Univ Timisoara Ser Biol 16(2):73–78

Sertova NM (2015) Application of nanotechnology in detection of mycotoxins and in agricultural sector. J Cent Eur Agric 16:117–130

Shah V, Belozerova I (2009) Influence of metal nanoparticles on the soil microbial community and germination of lettuce seeds. Water Air Soil Pollut 197(1–4):143–148

Shaymurat T, Gu J, Xu C, Yang Z, Zhao Q, Liu Y, Liu Y (2012) Nanotoxicology 6(3):241–248

Shen MH, Xia XH, Wang F, Zhang P, Zhao XL (2012) Influences of multi walled carbon nanotubes and plant residue chars on bioaccumulation of polycyclic aromatic hydrocarbons by *Chironomus plumosus* larvae in sediment. Environ Toxicol Chem 31(1):202–209

Shen Z, Chen Z, Hou Z, Li T, Lu X (2015) Ecotoxicological effect of zinc oxide nanoparticles on soil microorganisms. Front Environ Sci Eng 9:912–918

Sheng Z, Liu Y (2011) Effects of silver nanoparticles on wastewater biofilms. Water Res 45:6039–6050

Shoults-Wilson WA, Reinsch BC, Tsyusko OV, Bertsch PM, Lowry GV, Unrine JM (2011) Role of particle size and soil type in toxicity of silver nanoparticles to earthworms. Soils Sci Soc Am J 75:365–377

Siddiqui MH, Al-Whaibi MH (2014) Role of nano-SiO_2 in germination of tomato (*Lycopersicum esculentum* seeds Mill). Saudi J Biol Sci 21(1):13–17

Simonin M, Guyonnet JP, Martins JM, Ginot M, Richaume A (2015) Influence of soil properties on the toxicity of TiO_2 nanoparticles on carbon mineralization and bacterial abundance. J Hazard Mater 283:529–535

Singh S, Vishwakarma K, Singh S, Sharma S, Dubey NK, Singh VK, Liu S, Tripathi DK, Chauhan DK (2017) Understanding the plant and nanoparticle interface at transcriptomic and proteomic level: a concentric overview. Plant Gene 11:265–272

Somasundaran P, Fang X, Ponnurangam S, Li B (2010) Nanoparticles: characteristics, mechanisms and modulation of bio toxicity. Kona Powder Part J 28:38–49

Speranza A, Leopold K, Maier M, Taddei AR, Scoccianti V (2010) Pd-nanoparticles cause increased toxicity to kiwifruit pollen compared to soluble Pd (II). Environ Pollut 158(3):873–882

Stampoulis D, Sinha SK, White JC (2009) Assay-dependent phytotoxicity of nanoparticles to plants. Environ Sci Technol 43(24):9473–9479

Tadros TF, Izquierdo P, Esquena J, Solans C (2004) Formation and stability of nano emulsions. Adv Colloid Interf Sci 108:303–318

Tan XM, Lin C, Fugetsu B (2009) Studies on toxicity of multi-walled carbon nanotubes on suspension rice cells. Carbon 47(15):3479–3487

Tarafdar JC, Raliya R, Tathore I (2012) Microbial synthesis of phosphorus nano particles from Tricalcium phosphate using *Aspergillus tubingensis* TFR-5. J Bionanosci 6:84–89

Tiede K, Hassellov M, Breitbarth E, Chaudhry Q, Boxall ABA (2009) Considerations for environmental fate and ecotoxicity testing to support environmental risk assessments for engineered nanoparticles. J Chromatogr A1216(3):503–509

Torney F, Trewyn BG, Lin VSY, Wang K (2007) Mesoporous silica nanoparticles deliver DNA and chemicals into plants. Nat Nanotechnol 2(5):295–300

Tripathi DK, Shweta SS, Singh S, Pandey R, Singh VP, Sharma NC, Prasad SM, Dubey NK, Chauhan DK (2016a) An overview on manufactured nanoparticles in plants: uptake, translocation, accumulation and phytotoxicity. Plant Physiol Biochem 110:2–12

Tripathi DK, Singh S, Singh VP, Prasad SM, Chauhan DK, Dubey NK (2016b) Silicon nanoparticles more efficiently alleviate arsenate toxicity than silicon in maize cultiver and hybrid differing in arsenate tolerance. Front Environ Sci 4:46

Tripathi DK, Singh, Singh S, Srivastava PK, Singh VP, Singh S, Singh VP, Prasad SM, Singh PK, Dubey NK, Pandey AC, Chauhan DK (2017a) Nitric oxide alleviates silver nanoparticles (AgNps)-induced phytotoxicity in *Pisum sativum* seedlings. Plant Physiol Biochem 110:167–177

Tripathi DK, Mishra RK, Singh S, Singh S, Singh VP, Singh PK, Chauhan DK, Prasad SM, Dubey NK, Pandey AC (2017b) Nitric oxide ameliorates zinc oxide nanoparticles phytotoxicity in wheat seedlings: implication of the ascorbate-glutathione cycle. Front Plant Sci 8:1

Tripathi DK, Tripathi A, Singh S, Shweta SS, Singh Y, Vishwakarma K, Yadav G, Sharma S, Singh VK, Mishra RK, Upadhyay RG, Dubey NK (2017c) Uptake, accumulation and toxicity of silver nanoparticle in autotrophic plants, and heterotrophic microbes: a concentric review. Front Microbiol 8:07

Vartholomeos P, Fruchard M, Ferreira A, Mavroidis C (2011) MRI-guided nano robotic systems for therapeutic and diagnostic applications. Annu Rev Biomed Eng 13:157–184

Wang P, Lombi E, Zhao FJ, Kopittke PM (2016) Nanotechnology: a new opportunity in plant sciences. Trends Plant Sci 21(8):699–712

Wang L, Hu C, Shao L (2017) The antimicrobial activity of nanoparticles: present situation and prospects for the future. Int J Nanomedicine 12:1227–1249

Wirth SM, Lowry GV, Tilton RD (2012) Natural organic matter alters biofilm tolerance to silver nanoparticles and dissolved silver. Environ Sci Technol 46:12687–12696

Xie Y, He Y, Irwin PL, Jin T, Shi X (2011) Antibacterial activity and mechanism of action of zinc oxide nanoparticles against *Campylobacter jejuni*. Appl Environ Microbiol 77(7):2325–2331

Xingmao M, Geiser-lee J, Deng Y, Kolmakov A (2010) Interactions between engineered nanoparticles ENPs and plants: phytotoxcity, uptake and accumulation. Sci Total Environ 408(16):3053–3061

Xuming W, Fengqing G, Linglan M, Jie L, Sitao Y, Ping Y, Fashui H (2008) Effects of nano-anatase on ribulose-1, 5-bisphosphate carboxylase/oxygenase mRNA expression in spinach. Biol Trace Elem Res 126(1–3):280

Yang L, Watts DJ (2005) Particle surface characteristics may play an important role in phytotoxicity of alumina nanoparticles. Toxicol Lett 158:122–132

Yang F, Hong F, You W, Liu C, Gao F, Wu C, Yang P (2006) Influences of nano-anatase TiO2 on the nitrogen metabolism of growing spinach. Biol Trace Elem Res 110(2):179–190

Zhang SJ, Shao T, Kose HS, Karanfil T (2010) Adsorption of aromatic compounds by carbonaceous adsorbents: a comparative study on granular activated carbon, activated carbon fiber, and carbon nanotubes. Environ Sci Technol 44:6377–6383

Zhang Q, Han L, Jing H, Blom DA, Lin Y, Xin HL, Wang H (2016) Facet control of gold nanorods. ACS Nano 10(2):2960–2974

Zhang P, Ma Y, Liu S, Wang G, Zhang J, He X, Zhang Z (2017) Phytotoxicity, uptake and transformation of nano-CeO_2 in sand cultured romaine lettuce. Environ Pollut 220:1400–1408

Zhao L, Hernandez-Viezcas JA, Peralta-Videa JR, Bandyopadhyay S, Peng B, Munoz B, Keller AA, Gardea-Torresdey JL (2013) ZnO nanoparticle fate in soil and zinc bioaccumulation in corn plants (*Zea mays*) influenced by alginate. Environ Sci Processes Impacts 15(1):260–266

Zhu H, Han J, Xiao JQ, Jin Y (2008) Uptake, translocation, and accumulation of manufactured iron oxide nanoparticles by pumpkin plants. J Environ Monit 10:713–717

Zuverza-Mena N, Armendariz R, Peralta-Videa JR, Gardea-Torresdey JL (2016) Effects of silver nanoparticles on radish sprouts: root growth reduction and modifications in the nutritional value. Front Plant Sci 7:1–11

Heavy Metal Toxicity and Tolerance in Crop Plants

Mohammad Affan Baig, Sadia Qamar, Arlene Asthana Ali, Javed Ahmad, and M. Irfan Qureshi

Contents

1 Introduction

Production of crops is greatly influenced by abiotic stresses caused by environmental factors such as drought, salt, heavy metals and flooding (Ahmad et al. 2017). Industrialization and anthropogenic activities have severe and long-term consequences on our environment, which cause heavy metal accumulation in soil, water and air. Heavy metal-contaminated soil can be detrimental to plants and animals and pose a serious threat to the environment. Several physiological and molecular mechanisms were evolved by plants for tolerance, adaptation and survival in response to adverse environmental conditions (Baig et al. 2018). Plants evolve different mechanisms to withstand metal toxicity such as reduced uptake of heavy metals, vacuolar sequestration of heavy metals, activation of antioxidants and binding to metal chelators like phytochelatins and metallothioneins.

Arsenic (As), lead (Pb), cadmium (Cd) and mercury (Hg) are some of the highly toxic elements which can be taken up by plants inadvertently. The US agency for

M. A. Baig · S. Qamar · A. A. Ali · J. Ahmad · M. I. Qureshi (✉)
S. Ramanujan Block, Department of Biotechnology, Jamia Millia Islamia, New Delhi, India

M. Naeem et al. (eds.), *Contaminants in Agriculture*,
https://doi.org/10.1007/978-3-030-41552-5_9

toxic substance and disease registry (ATSDR) 2013 (Clemens and Ma 2016) ranked these elements in the top four priority list of hazardous substances. Elucidation of the mechanisms and pathways of metal accumulation can facilitate the development of crop plants with reduced noxious element content. Ionic forms of Cd, As, Hg and Pb are extremely toxic to both plants and animals. Methylated forms of Hg and As are also toxic. These elements can damage various cellular organizations, tissues and organs. Interaction with the sulphydryl group and interference with essential element homeostasis are some of the major causes of toxicity (Ahmad et al. 2018). Impairment of protein function due to thiol reactivity can cause oxidative stress.

2 Uptake of Heavy Metals and Their Translocation

Metal transporters play an important role in the uptake of heavy metals in plant cells. Several families of metal transporters are present in the plant cell plasma membrane. Heavy metal ions have a wide range of substrate specificity and are co-transported across the root cell plasma membrane along with other nutrients. There is little information regarding mechanisms of heavy metal acquisition developed in plants. However, various genes have been recently identified in plants, which encode potential heavy metal transporters. Previous studies have shown the significant role of P1B-type heavy metal ATPases (HMAs) in the translocation of specific heavy metal ions in *Arabidopsis* and certain other metal hyperaccumulator plants. The HMA transporters have been categorized on the basis of their metal-substrate specificity into two subgroups, such as the zinc/cobalt/cadmium/lead (Zn/Co/Cd/Pb) group and copper/silver (Cu/Ag) group. The genome of rice encodes for nine *HMA* genes. Among these, *OsHMA1–OsHMA3* are the members of the Zn/Co/Cd/Pb subgroup, while *OsHMA2* has also been known to play a vital role in translocation of Zn and Cd from the root to the shoot and transport of Cd and Zn to developing seeds (Takahashi et al. 2012). In root cells, after uptake of heavy metal ions through the transporters, they are consequently loaded onto the xylem elements and transported towards the shoots in the form of complexes formed along with several chelators of metal ions. Various transporter proteins have been shown to function for the heavy metal ion transportation from root to shoot. For exclusion of heavy metal ions from the cell, the heavy metal transporting P-type ATPase (HMAs) are required, which function as efflux pumps. However, HMAs are also known to act as internal transporters for loading Zn and Cd metals into the xylem elements from the adjacent regions. *AtHMA4* in *Arabidopsis thaliana* is a type of biomembrane-based P-type ATPase, which functions in transportation of divalent metal ions essential for sustaining homeostasis of Zn^{2+}ions and is also known to play an important role in detoxification of Cd, which otherwise leads to the disruption of cytosolic free Ca^{2+} (Mills et al. 2005).

3 Impact of Heavy Metals on Plant Metabolism

Anthropogenic activities such as the excessive use of pesticides and the fungicides, use of organic and inorganic fertilizers, and the residual waste during sewage treatment called as sewage sludge, contaminated irrigation water are the major sources which are responsible for the heavy metal contamination in the agricultural soil. Metal processing at a very high temperature, which includes casting of metal and also smelting for extraction of ore, results in exposure of atmosphere to the heavy metals in the form of vapour and particulate matter in areas near industrial areas.

In atmosphere, various heavy-metal contaminants such as Zn, As, Pb, Cu and Cd in their gaseous forms combine with the water present and lead to the formation of aerosols and are then consequently dispersed by the wind and this process is known as dry deposition or else it can be precipitated by rainfall and is known as wet deposition. These processes are responsible for polluting the water bodies and the agricultural soil. After entering into the plant cell, the heavy metals cause genotoxic and cytotoxic damage by attacking the thiol groups of the protein and therefore cause disruption in the structure of the protein, thereby triggering conformational changes in the structure of the protein. In addition, the photosynthetic apparatus and cellular macromolecules are damaged by the production of ROS by the heavy metal.

The effects of heavy metal toxicity could be observed at biochemical and physiological levels in plants with diminished photosynthetic yield and membrane stability, nutrient and hormonal imbalance, decreased photosynthetic pigment production and inhibition of division of cell division, replication of DNA and gene expression (Singh et al. 2009). The generation of different stress responses in plant cell depends on the heavy metal uptake, type of metal and its development stage and also on concentration of heavy metal. When present in excessive quantities, heavy metals result in inducing genotoxic as well as oxidative stress response and hence cause cytotoxicity and damage of different cellular constituents including disruption of protein, nucleic acid and membranes, thus creating abiotic stress in the plants.

The method of production of reactive oxygen species (ROS) through oxidative stress by heavy metals frequently disrupts the cellular redox system of plants as well as the preliminary protection strategies (Pandey et al. 2009; Mourato et al. 2012). Plant cells are known to develop two very essential antioxidant defence mechanisms. One such constituent includes the activation of antioxidant enzymes, such as catalase, (CAT), superoxide dismutase (SOD), glutathione reductase (GR) and ascorbate peroxidase (APX) for the direct inactivation of the reactive radicals.

Plants are equipped with two essential types of defence mechanisms in response to oxidative damage. One component includes the triggering and activation of antioxidant enzymes such as superoxide dismutase (SOD), glutathione reductase (GR), ascorbate peroxidase (APX) and glutathione reductase (GR), which lead to direct inactivation of free radicals.

Another important component involves non-enzymatic antioxidants, namely glutathione (GSH), phenylpropanoids (tannins, flavonoids and lignin), alkaloids and proline, carotenoids and ascorbate (AsA) to combat against the toxicity of heavy metals through detoxification of ROS produced by the heavy metal and exclusion of free radicals by scavenging (Hossain et al. 2012; Michalak 2006; Štolfa et al. 2015). It has been noted that exposure of seedlings of barley (*Hordeum vulgare*) to numerous heavy metals led to augmentation of lipid peroxidation and simultaneously results in reduction in the dry biomass of these seedlings (Juknys et al. 2012; Kacienė et al. 2015).

Anthropogenic activities result in the production of inorganic arsenic, which is highly toxic to plant cells. After plant uptake, the inorganic arsenic is changed into its reduced form As (III) with the help of glutathione (GSH). The As(III) results in the cellular cytotoxicity by binding to the thiol groups of the protein and distorting its structure. Different low molecular weight thiol compounds (LMWTs) in plants such as γ-glutamylcysteine, cysteine, phytochelatins and glutathione are known to accumulate whenever confronted with stressful conditions arising due to high concentration of arsenic, which in turn is responsible for detoxification of arsenic by binding to As (III) (Zenk 1996, Srivastava et al. 2007). It was found in in vitro studies conducted on *Vicia Faba* that the activities of CAT and SOD were amplified in response to arsenate-mediated oxidative stress (Lin et al. 2008). Cd has been considered as the foremost feature for the stimulation of oxidative stress through ROS generation and also diminishing of non-enzymatic antioxidant defence system (Cho and Seo 2005) Oxidative stress arising due to Cd led to interrupting the functioning of plasma membrane in plants via membrane lipid peroxidation (Gratão et al. 2005).

Various anomalies ranging from damage to the photosynthetic pigments and cell membrane distorting ultra-structure of chloroplast were observed due to chromium-induced oxidative stress, which ultimately results in the disrupting of the antioxidant metabolism and severe retardation of plant growth (Panda and Choudhury 2005). Smelting and mining activities led to the accumulation of lead (Pb) in the soil, which results in oxidative stress through the generation of reactive oxygen species. (Sharma et al. 2012). During endogenous metabolic processes, ROS are known to be formed as by-product and these ROS are responsible for the stimulation of oxidative damage in the plant cells. Recent research studies have shown that ROS have a dual function in plants. Apart from their damaging effect in plants, they are also known to play a significant role as a signalling molecule to control numerous essential processes such as responses towards abiotic and biotic stress, programmed cell death and induction of systemic signalling, cell growth and cell differentiation.

The oxidative stress induced due to the ROS production led to activation of signalling pathway which results in cell death. The major reason of heavy metal–induced oxidative stress in plants is the production of excessive ROS followed by disruption of the redox environment of cells (Nada et al. 2007). In spinach (*Spinacea oleracea*), differential antioxidant responses and oxidative damage have been detected due to exposure to several heavy metal ions (Pandey et al. 2009). Increased levels of Cu are responsible for damaging the macromolecules and also disrupting the metabolic pathways by ROS production. An elevated level of Hg^{2+} results in lipid peroxidation and hampers the functioning of mitochondria and finally disrupts the cellular metabolism in plants via oxidative-mediated stress and production of reactive oxygen species (Messer et al. 2005; Cargnelutti et al. 2006). In the shoots of rice (*Oryza sativa*), Ni^{2+} affects the H(+)-ATPase activity of the biomembrane and also affects the structural integrity of membrane lipids. (Ros et al. 1990). The augmented levels of Ni^{2+} result in increase of the malonaldehyde (MDA) content in wheat (Pandolfini et al. 1992). As a result, it affects the ion balance in the cytoplasm and also affects the function of plasma membrane.

4 Heavy Metal Toxicity and Impact on Plant Growth

According to the Agency for Toxic Substances and Disease Registry (ATSDR, 2003), As, Hg, Cd and Pb are considered the utmost toxic elements, which are based on their incidence of occurrence, toxicity and also on their ability to affect fauna and flora. The impact and the origin of these heavy metals on plant growth and on environment are briefly explained below.

4.1 Arsenic

Arsenic is naturally an environmentally toxic metal that is responsible for causing severe health threats to billions of people around the world (Kumar et al. 2015). The arsenic exposure arises as a result of rock erosion, volcanic eruptions and by various anthropogenic activities such as smelting and mining operations, excessive use of wood preservatives and pesticides (Wang and Mulligan 2006; Tripathi et al. 2007; Neumann et al. 2010).

Millions of people all over the world suffer from arsenic toxicity due to contamination of groundwater with arsenic, which is used for drinking purposes and irrigation. Arsenic not only affects the plant by affecting its crop productivity but also causes serious threat to humans, as arsenic can also enter into the food chain through its accumulation in various plant tissues, which include grains (Verma et al. 2016b). Numerous researches have been carried out to explore the molecular and physiological mechanism of action of its toxicity; level of tolerance in variety of plant species such as spinach, carrot, lettuce and rice; its accumulation in environment and the mechanism of its detoxification. Arsenic exists in nature in two forms, namely inorganic arsenite As(III) and arsenate As(V). According to the United States Environmental Protection Agency (USEPA) evaluation, both forms of arsenic As(III) and As(V) are considered major environmental contaminants (Tripathi et al. 2007; Verma et al. 2018). Arsenic in its arsenite form As(III) is much more toxic than its arsenate form As(V), and it is observed that it acts in different ways; it binds with the sulphydryl groups in the proteins and interferes with their activity and thus disturbs various biological functions in plants (Verma et al. 2016a). It also led to production of free radicals and ROS. It binds to the vicinal thiols present in enzymes 2-oxo-glutarate dehydrogenase and pyruvate dehydrogenase, and thus inhibits the respiration. It induces the process of homologous recombination and thus it indirectly acts as a mutagen (Helleday et al. 2000). Alternately, As(V) acts in plants by interfering with the electron transport chain by obstructing the oxidative phosphorylation as well as the ATP synthesis process during energy metabolism (Carbonell et al. 1998; Verma et al. 2016a).

4.2 Lead

Lead (Pb) is one of the most toxic, persistent and widespread heavy metal contaminants of soil (Fahr et al. 2013). The degree of Pb toxicity in plants depends on its concentration, salt type, soil pH and plant species. Lead toxicity can affect growth, development, mineral nutrition and enzymatic activities. Lead toxicity can cause accumulation of ROS, which can damage proteins, nucleic acids and membrane lipids (Qureshi et al. 2007). Inorganic lead salt can induce c-mitosis, strongly inhibit root growth and lower mitotic activity. Lead uptake and accumulation have been reported in roots, nodules, stems, leaves and seeds. Most of the Pb absorbed by plant remains in the roots and a small fraction is translocated upwards. Low soil pH can increase lead to uptake in roots (Patra et al. 2004). Lead can reduce the uptake of calcium and macronutrients from the soil and decrease their tissue content, which can affect mineral metabolism, leading to deficiency. Lead-induced inhibition of calcium channels might be caused due to competitive uptake of lead through a calcium channel or its blockage.

Lead has higher affinity for calcium-binding sites in cell structures due to which it uses calcium pathway to cross membranes (Antosiewicz 2005). In roots of *Lathyrus sativus*, lead ions cause no change in phytochelatin synthase transcript *LsPCS*, which suggests that lead sequestration by PCs was not regulated through transcriptional mechanism but overexpression of *LsGST* transcripts promotes sequestration of lead ions via formation of GSH-Pb complexes (Brunet et al. 2009; Gupta et al. 2010). Many cell transporters are known to be involved in detoxification of heavy metals. For the removal of heavy metals from the cytoplasm, they are extruded via transporters across plasma membranes or sequester them in organelles. *AtPDR12* is a member of *Arabidopsis* ABC transporters, which is involved in lead detoxification (Lee et al. 2005). Plant hormone ethylene is involved in different growth and developmental processes such as seed germination, formation of root hairs, root nodulation, senescence, abscission, ripening and provides protection against environmental stresses. *Arabidopsis* ethylene-insensitive 2 gene *EIN2* provides lead resistance via glutathione (GSH)-dependent and *AtPDR12*-mediated (GSH-independent) mechanism. Cao et al. (2009) and Jiang and Liu (2010) reported immobilization of lead ions and formation of cysteine-rich proteins in the cell wall of *Allium sativum.* Lead toxicity induces expression of genes involved in GSH metabolism, sulphur assimilation, jasmonic acid and indole-3-acetic acid (IAA) biosynthesis pathways and metal transporters (Liu et al. 2009). Pea plants treated with increasing concentration of lead showed effective accumulation in roots.

Pb accumulates in cell wall, plasma membrane, cytoplasm, vacuoles, peroxisomes and mitochondria (Małecka et al. 2008). *Arabidopsis* seedlings exposed to lead caused root growth inhibition and overproduction of ROS and lipid hydroperoxide (Phang et al. 2011). ROS production by lead-exposed *Vicia faba* excised roots were sensitive to low concentration of NADPH oxidase inhibitors, which suggests a prominent role of NADPH oxidase such as enzymes in ROS generation in lead-exposed *Vicia faba* roots (Pourrut et al. 2008). Overexpression of yeast protein YCF1 in transgenic *Arabidopsis thaliana* showed enhanced tolerance and accumulation of lead (Song et al. 2003). Such transgenic plants can be used for phytoremediation of heavy metals. Exogenous application of IAA significantly affected lead accumulation and both IAA and lead causes synergistic effect on oxidative stress response in maize seedlings (Wang et al. 2007).

4.3 Mercury

Mercury is a nonessential heavy metal for plants and is reported to be frequently released into the biosphere including air, water and soil (Zhou et al. 2009). The ionic form of mercury (Hg^{2+}) has high affinity for sulphydryl (-SH) group, leading to impaired protein function. High mercury concentration can cause protein precipitation. It can cause nonspecific inhibition of various intracellular enzymes. Prolonged exposure to mercury can cause decreased levels of photosynthetic pigments. It can inhibit membrane water channels in higher plants. Higher mercury concentrations (>1 mg/L) cause lipid peroxidation, disruption of membrane structural integrity and increase in membrane permeability. Mercuric chloride ($HgCl_2$) was found to decrease water conductivity and rapid depolarization of membrane potential in wheat root cells (Zhang and Tyerman 1999). Endogenous tolerance to mercury stress in plants can be achieved by increased activities of antioxidant enzymes (Baig et al. 2018).

High levels of mercury can induce oxidative stress. Elevated levels of ROS can activate signalling cascade for defence response, apoptosis and cell cycle. Glutathione (GSH) is a key metabolite in tolerance against heavy metals. *Arabidopsis* γ-glutamylcysteinesynthetase (γECS) mutants exposed to mercury showed lower levels of GSH with reduced levels of GR and NADPH-oxidase in roots and exhibited more pronounced phytotoxic effects. Mercury toxicity-induced biochemical mechanisms and modification in metabolic pathways of crop plants still need elucidation. Wheat plants exposed to mercury exhibit positive response to stress with increased antioxidant enzyme activities (Li et al. 2013). Rice seedlings exposed to mercury and arsenic showed reduced uptake, transformation and translocation of arsenic in roots, and low levels of arsenic inhibited mercury uptake. Both arsenic and mercury caused damage to roots, lipid peroxidation and formation of iron plaque (Ren et al. 2014). Mercury toxicity has been shown to reduce root and shoot growth as well as cell proliferation in different plants. Tolerance to toxic effects of heavy metals involves metabolic alterations. Mercury-affected cellular redox homeostasis caused cell necrosis and plasmolysis with loss of cell content in

Medicago sativa (Ortega-Villasante et al. 2005). Microarray analysis of mercury-induced genes in *Arabidopsis* showed induction of proteins encoding for antioxidant enzyme system, cell wall proteins, cytochrome P450 for possible protection against oxidative stress (Heidenreich et al. 2001).

Membrane transporters for mercury still need to be identified. Little information on root adaptation to mercury stress is available. Soils contaminated with mercury have permanent stress on plants; therefore, molecular response to long-term exposure is needed. Barley plants exposed to mercury showed accumulation mainly in roots with decreased root and shoot growth; microarray analysis showed overexpression of aquaporin, DNA synthesis, cell organization, secondary metabolism and abscisic acid-related transcripts. Changes in nitrogen metabolism-related transcripts were accompanied by decreased shoot nitrogen concentration (Lopes et al. 2013). Investigation of rice proteins involved in organic and inorganic mercury toxicity through metalloproteomic approach revealed proteins involved in antioxidant defence, carbohydrate and energy metabolism, sulphur and glutathione metabolism, pathogen defence and programmed cell death. Binding of mercury to proteins causes irreversible root growth inhibition (Chen et al. 2012; Li et al. 2016).

4.4 Cadmium

Cadmium is known to be a most phytotoxic heavy metal. It is easily taken up by plants, as it is extremely soluble in water, thus representing the main entry pathway into the food chain producing severe human health threats (Buchet et al. 1990). According to The International Agency for Research on Cancer (Gianazza et al. 2007; Gill and Tuteja 2011), Cd has been categorized as a potent human carcinogen. Interestingly, it has been reported that from industrial processes and farming practices, it is generally released into the arable soil and also that crops are the foremost source of Cd consumption by humans (Satarug et al. 2002; Gill and Tuteja 2011). Cd can rigorously alter several enzyme activities even at low concentrations including those that are involved in the Calvin cycle, CO2 fixation, metabolism of carbohydrate and phosphorus (Sandalio et al. 2001; Verma and Dubey 2001; Sharma and Dubey 2006; Gill and Tuteja 2011) eventually causing diminutive growth, leaf epinasty, chlorosis, inhibition of photosynthesis and tube growth, and thus obstructing pollen germination, alterations in chloroplast ultrastructure, stimulation of lipid peroxidation and alterations in the metabolism of nitrogen (N) and sulphur (S) and disruption of antioxidant apparatus (Gill and Tuteja 2011).

5 Heavy Metal Sequestration in Plants

Plants are equipped with several strategies to alleviate the detrimental effects caused by heavy metal ion-induced cytotoxicity. One of the important approaches includes transportation of the heavy metal ions out of the cell or by sequestering into the cell vacuoles and in intracellular compartments via several membrane transporters. Heavy metals are transported into the vacuole via ABC transporters, which represent one of the major families of transporters. AtMRP1 and AtMRP2 are two essential members of the ABC family found in *Arabidopsis*, which known to be involved in phytochelatin–Cd complex transportation into the vacuole. ABC transporters have also been associated in the efflux of metal ion from the plasma membrane (Mendoza-Cózatl et al. 2011; Song et al. 2014). In *Arabidopsis*, another member reported in ABC family transporter is AtPDR8, which has been shown to provide heavy metal tolerance by working as an effective efflux pump of Cd in the plasma membrane of epidermal cells and root hairs (Kim et al. 2007). In *Arabidopsis,* the plasma membrane–confined ZIP (ZRT, IRT-like protein) family transporters have also been associated in Cd detoxification and shown to be triggered and activated during Zn-limiting environment (Komal et al. 2015). The efflux of metal ions from the cytoplasm take place through HMAs, member of the P-type metal ATPase. AtHMA3 in *Arabidopsis* acts as transporter for Zn/Cd and has been reported to be involved in sustaining Zn and Cd homeostasis and also in the detoxification of several other heavy metals through sequestration into the vacuole (Mills et al. 2005; Krämer et al. 2007; Morel et al. 2009). Another important family of metal transporter includes the natural resistance-associated macrophage proteins (NRAMPs). In *Arabidopsis*, mobilization of vacuolar iron and Cd takes place via AtNRAMP3 and AtNRAMP4 (Lanquar et al. 2005; Molins et al. 2013).

Plant cells activate numerous interweaved biochemical pathways after sensing the presence of heavy metals in the intracellular environments and hence led to the accumulation and biosynthesis of different classes of metabolites for counteracting the heavy metal-triggered toxicity. These metabolites, which are activated during heavy metal stress, are low-molecular weight proteins that include the chelators or metallochaperones (spermine, spermidine, glutathione, putrescine, phytochelatins, nicotianamine, other organic acids, etc.), phenylpropanoid compounds (anthocyanins, flavonoids), metallothioneins, amino acids (histidine, proline, etc.), stress responsive phytohormones and heat-shock proteins (Hsps) (Sharma and Dietz 2006; Dalvi and Bhalerao 2013).

6 Role of Metal Chelators in Heavy Metal Sequestration

6.1 *Phytochelatins*

Phytochelatins (PCs) are characterized as an important family of thiol-rich peptides found in plants, which are mostly known for metal detoxification and homeostasis because of their strong binding with heavy metals. Their occurrence is not limited to plants only, but has been found in varied organisms like fungi and animals. The synthesis of PCs occurs in the cytosol by the enzymatic activity of phytochelatin synthase, the substrate being glutathione (GSH). From the cytosol, the PCs, in the form of high molecular weight PC–metal complexes, find their way into the vacuole with the help of ATP-binding cassette (ABC) transporter or Mg ATP-dependent carrier. Exposure of plant species like sunflower to higher concentrations of heavy metal ions like Cd^{2+}have been linked to the synthesis and accumulation of PCs in the roots formerly. Leaves were also the sites of higher PC accumulation when *Brassica juncea* was treated with Cd^{2+} for a considerable duration of time. It is suggested that the heavy metal detox mechanism by plants is the result of cumulative action of PCs and antioxidant enzymes, which ultimately improve the plant's resistance. Phytochelatin synthase genes are important candidate genes for improved metal tolerance. Such genes have been characterized in *Arabidopsis, Triticum aestivum, Oryza sativa* and *Brassica juncea.* Transgenic *Arabidopsis* plants expressing the garlic arsenic phytochelatin synthase 1 (AsPCS1) and yeast cadmium factor 1 (YCF1) and transgenic tobacco plants overexpressing such phytochelatin synthase genes have been found to have better resistance to heavy metals.

6.2 Metallothioneins

The metallothioneins (MTs) belong to another essential family of cysteine-rich, small polypeptides, which play a vital role in the detoxification of a variety of metal ions, namely Cu, Zn, Cd and As. Apart from plants, these low molecular weight polypeptides are efficient in prokaryotes, invertebrates, fungi and mammals. However, plant MTs are significantly different from the MTs found in the other life forms. In plants, MTs by means of transport and metal sequestration bring about homeostasis of intracellular metal ions, thereby playing a central role in alleviating heavy metal cytotoxicity. Also, MTs aid in the repair of stress-induced damage of plasma membrane and maintenance of redox balance in the cell by scavenging ROS (Wong et al. 2004; Kohler et al. 2004; Mishra and Dubey 2006). Depending on the array of cysteine residues, MTs found in plants have been categorized into four types, each type having distinct as well common functions for defence against heavy metal toxicity. There are prior reports of Cu^{2+} chelating ability of MT1, MT2 and MT3 and their sub-types in Arabidopsis. Zinc (Zn^{2+}) homeostasis in *Oryza sativa* roots has been found to be associated with type 1 MT (OsMT1a) (Yang et al. 2009). In *Glycine max*, MT1, MT2 and MT3 have been known to be concerned with Cd detoxification, while MT4 plays a significant role in Zn^{2+} chelation and homeostasis (Grennan 2011; Pagani et al. 2012).

6.3 Organic Acids and Amino Acids

Besides PCs and MTs, organic acids and amino acids are also known to confer substantial heavy metal tolerance to plants mainly by xylem transport of the ions and their sequestration into the vacuoles. The important organic acids found in plants involved in the detoxification process are malate, citrate and oxalate. Amino acids like histidine and their derivatives are efficient metal chelators that help plants overcome heavy metal toxicity and boost their metal metabolism (Sharma and Dubey 2007). The pattern of proline accumulation during stressed conditions is not unknown to us. Numerous studies specify that proline provides resistance against heavy metal cytotoxicity by scavenging ROS generated by heavy metal stress, which is achieved primarily by the removal of hydroxyl radicals and quenching of singlet oxygen species (Emamverdian et al. 2015)

7 Conclusion

The need of the hour is to enhance global food production in order to feed the ever-growing population, but heavy metal toxicity on plants has become a hindrance to this attempt. The risks posed by heavy metals on plants like genomic instability and deteriorated plant health are a matter of great concern and scientists have put their minds together since the last two decades to mitigate these risks by improving abiotic stress tolerance mechanisms in crops. This has been quite a challenge considering the complex nature of abiotic stresses. A few plant species have been known to act as hyperaccumulators of heavy metals compared to their relative species, as they exhibit normal growth in heavily metal polluted soils and have a high threshold to resist toxicity. This is possible because they are able to hold up high amounts of metals in their tissues without damage. Transcriptomic studies reveal the plants' ability to comprehend and defend themselves from heavy metal toxicity by changing the expression of some stress-responsive genes. These studies have paved the way for genetic regulation of such heavy metal stress-related genes, especially the involved transcription factors, proteins and metabolites to affect better stress tolerance in plants. However, a more holistic biotechnological approach should be adopted for successful transfer of such technology to the field for improved crop production to benefit the world population.

References

Ahmad J, Bashir H, Bagheri R, Baig A, Al-Huqail A, Ibrahim MM, Qureshi MI (2017) Drought and salinity induced changes in ecophysiology and proteomic profile of *Parthenium hysterophorus*. PLoS One 12(9):e0185118

Ahmad J, Baig MA, Ali AA, Al-Huqail AA, Ibrahim MM, Qureshi MI (2018) Differential antioxidative and biochemical responses to aluminium stress in *Brassica juncea* cultivars. Hortic Environ Biotechnol 59(5):615–627

Antosiewicz DM (2005) Study of calcium-dependent lead-tolerance on plants differing in their level of Ca-deficiency tolerance. Environ Pollut 134(1):23–34

Asli S, Neumann PM (2009) Colloidal suspensions of clay or titanium dioxide nanoparticles can inhibit leaf growth and transpiration via physical effects on root water transport. Plant, cell & environment 32(5): 577–584

Atha DH, Wang H, Petersen EJ, Cleveland D, Holbrook RD, Jaruga P, Nelson BC (2012) Copper oxide nanoparticle mediated DNA damage in terrestrial plant models. Environmental science & technology 46(3): 1819–1827

Baig MA, Ahmad J, Bagheri R, Ali AA, Al-Huqail AA, Ibrahim MM, Qureshi MI (2018) Proteomic and ecophysiological responses of soybean (*Glycine max* L.) root nodules to Pb and hg stress. BMC Plant Biol 18(1):283

Brunet J, Varrault G, Zuily-Fodil Y, Repellin A (2009) Accumulation of lead in the roots of grass pea (*Lathyrus sativus* L.) plants triggers systemic variation in gene expression in the shoots. Chemosphere 77(8):1113–1120

Buchet JP, Lauwerys R, Roels H, Bernard A, Bruaux P, Claeys F, Ducoffre G, de Plaen P, Staessen J, Amery A, Lijnen P, Thijs L (1990) Renal effects of cadmium body burden of the general population. Lancet 336:69–702

Cao S, Chen Z, Liu G, Jiang L, Yuan H, Ren G, Bian X, Jian H, Ma X (2009) The Arabidopsis Ethylene-Insensitive 2 gene is required for lead resistance. Plant Physiol Biochem 47(4):308–312

Carbonell AA, Aarabi MA, DeLaune RD, Gambrell RP, Patrick WH Jr (1998) Arsenic in wetland vegetation: availability, phytotoxicity, uptake and effects on plant growth and nutrition. Sci Total Environ 217(3):189–199

Cargnelutti D, Tabaldi LA, Spanevello RM, de Oliveira Jucoski G, Battisti V, Redin M et al (2006) Mercury toxicity induces oxidative stress in growing cucumber seedlings. Chemosphere 65(6):999–1006

Castiglione MR, Giorgetti L, Geri C, Cremonini R (2011) The effects of nano-TiO2 on seed germination, development and mitosis of root tip cells of Vicia narbonensis L. and Zea mays L. Journal of Nanoparticle Research 13(6): 2443–2449

Chen YA, Chi WC, Huang TL, Lin CY, Nguyeh TTQ, Hsiung YC et al (2012) Mercury-induced biochemical and proteomic changes in rice roots. Plant Physiol Biochem 55:23–32

Cho UH, Seo NH (2005) Oxidative stress in Arabidopsis thaliana exposed to cadmium is due to hydrogen peroxide accumulation. Plant Sci 168(1):113–120

Clemens S, Ma JF (2016) Toxic heavy metal and metalloid accumulation in crop plants and foods. Annu Rev Plant Biol 67:489–512

Dalvi AA, Bhalerao SA (2013) Response of plants towards heavy metal toxicity: an overview of avoidance, tolerance and uptake mechanism. Ann Plant Sci 2(09):362–368

de la Rosa G, López-Moreno ML, de Haro D, Botez CE, Peralta-Videa JR, Gardea-Torresdey JL (2013) Effects of ZnO nanoparticles in alfalfa, tomato, and cucumber at the germination stage: root development and X-ray absorption spectroscopy studies. Pure and Applied Chemistry 85(12): 2161–2174

Emamverdian A, Ding Y, Mokhberdoran F, Xie Y (2015) Heavy metal stress and some mechanisms of plant defense response. Sci World J 2015:1–18

Fahr M, Laplaze L, Bendaou N, Hocher V, El Mzibri M, Bogusz D, Smouni A (2013) Effect of lead on root growth. Front Plant Sci 4:175

Ghosh M, Bandyopadhyay M, Mukherjee A (2010) Genotoxicity of titanium dioxide (TiO2) nanoparticles at two trophic levels: plant and human lymphocytes. Chemosphere 81(10): 1253–1262.

Giraldo JP, Landry MP, Faltermeier SM, McNicholas TP, Iverson NM, Boghossian AA, Strano MS (2014) Plant nanobionics approach to augment photosynthesis and biochemical sensing. Nature materials 13(4): 400–408

Gianazza E, Wait R, Sozzi A, Regondi S, Saco D, Labra M, Agradi E (2007) Growth and protein profile changes in Lepidium sativum L. plantlets exposed to cadmium. Environ Exp Bot 59(2):179–187

Gill SS, Tuteja N (2011) Cadmium stress tolerance in crop plants: probing the role of sulfur. Plant Signal Behav 6(2):215–222
Gratão PL, Polle A, Lea PJ, Azevedo RA (2005) Making the life of heavy metal-stressed plants a little easier. Funct Plant Biol 32(6):481–494
Grennan AK (2011) Metallothioneins, a diverse protein family. Plant Physiol 155(4):1750–1751
Gupta DK, Huang HG, Yang XE, Razafindrabe BHN, Inouhe M (2010) The detoxification of lead in Sedum alfredii H. is not related to phytochelatins but the glutathione. J Hazard Mater 177(1–3):437–444
Heidenreich B, Mayer K, SandermannJr H, Ernst D (2001) Mercury-induced genes in Arabidopsis thaliana: identification of induced genes upon long-term mercuric ion exposure. Plant Cell Environ 24(11):1227–1234
Helleday T, Nilsson R, Jenssen D (2000) Arsenic [III] and heavy metal ions induce intrachromosomal homologous recombination in the hprt gene of V79 Chinese hamster cells. Environ Mol Mutagen 35(2):114–122
Hossain MA, Piyatida P, da Silva JAT, Fujita M (2012) Molecular mechanism of heavy metal toxicity and tolerance in plants: central role of glutathione in detoxification of reactive oxygen species and methylglyoxal and in heavy metal chelation. J Bot 2012:1
Jiang W, Liu D (2010) Pb-induced cellular defense system in the root meristematic cells of Allium sativum L. BMC Plant Biol 10(1):40
Juknys R, Vitkauskaitė G, Račaitė M, Venclovienė J (2012) The impacts of heavy metals on oxidative stress and growth of spring barley. Open Life Sci 7(2):299–306
Kacienė G, Milčė JŽE, Juknys R (2015) Role of oxidative stress on growth responses of spring barley exposed to different environmental stressors. J Plant Ecol 8(6):605–616
Kim DY, Bovet L, Maeshima M, Martinoia E, Lee Y (2007) The ABC transporter AtPDR8 is a cadmium extrusion pump conferring heavy metal resistance. Plant J 50(2):207–218
Kohler A, Blaudez D, Chalot M, Martin F (2004) Cloning and expression of multiple metallothioneins from hybrid poplar. New Phytol 164(1):83–93
Komal T, Mustafa M, Ali Z, Kazi AG (2015) Heavy metal uptake and transport in plants. In: Heavy metal contamination of soils. Springer, Cham, pp 181–194
Krämer U, Talke IN, Hanikenne M (2007) Transition metal transport. FEBS Lett 581(12):2263–2272
Kumar S, Dubey RS, Tripathi RD, Chakrabarty D, Trivedi PK (2015) Omics and biotechnology of arsenic stress and detoxification in plants: current updates and prospective. Environ Int 74:221–230
Lanquar V, Lelièvre F, Bolte S, Hamès C, Alcon C, Neumann D et al (2005) Mobilization of vacuolar iron by AtNRAMP3 and AtNRAMP4 is essential for seed germination on low iron. EMBO J 24(23):4041–4051
Lee M, Lee K, Lee J, Noh EW, Lee Y (2005) AtPDR12 contributes to lead resistance in Arabidopsis. Plant Physiol 138(2):827–836
Li Y, Sun H, Li H, Yang L, Ye B, Wang W (2013) Dynamic changes of rhizosphere properties and antioxidant enzyme responses of wheat plants (Triticumaestivum L.) grown in mercury-contaminated soils. Chemosphere 93(6):972–977
Li Y, Zhao J, Li YF, Xu X, Zhang B, Liu Y et al (2016) Comparative metalloproteomic approaches for the investigation proteins involved in the toxicity of inorganic and organic forms of mercury in rice (Oryza sativa L.) roots. Metallomics 8(7):663–671
Lin D, Xing B (2008) Root uptake and phytotoxicity of ZnO nanoparticles. Environmental science & technology 42(15): 5580–5585
Lin A, Zhang X, Zhu YG, Zhao FJ (2008) Arsenate-induced toxicity: effects on antioxidative enzymes and DNA damage in Viciafaba. Environ Toxicol Chem 27(2):413–419
Liu T, Liu S, Guan H, Ma L, Chen Z, Gu H, Qu LJ (2009) Transcriptional profiling of Arabidopsis seedlings in response to heavy metal lead (Pb). Environ Exp Bot 67(2):377–386
Lopes MS, Iglesia-Turino S, Cabrera-Bosquet L, Serret MD, Bort J, Febrero A, Araus JL (2013) Molecular and physiological mechanisms associated with root exposure to mercury in barley. Metallomics 5(9):1305–1315

Małecka A, Piechalak A, Morkunas I, Tomaszewska B (2008) Accumulation of lead in root cells of Pisum sativum. Acta Physiol Plant 30(5):629–637
Mendoza-Cózatl DG, Jobe TO, Hauser F, Schroeder JI (2011) Long-distance transport, vacuolar sequestration, tolerance, and transcriptional responses induced by cadmium and arsenic. Curr Opin Plant Biol 14(5):554–562
Messer RL, Lockwood PE, Tseng WY, Edwards K, Shaw M, Caughman GB et al (2005) Mercury (II) alters mitochondrial activity of monocytes at sublethal doses via oxidative stress mechanisms. J Biomed Mater Res B Appl Biomater 75(2):257–263
Michalak A (2006) Phenolic compounds and their antioxidant activity in plants growing under heavy metal stress. Pol J Environ Stud 15(4)
Mills RF, Francini A, da Rocha PSF, Baccarini PJ, Aylett M, Krijger GC, Williams LE (2005) The plant P1B-type ATPase AtHMA4 transports Zn and Cd and plays a role in detoxification of transition metals supplied at elevated levels. FEBS Lett 579(3):783–791
Mishra S, Dubey RS (2006) Heavy metal uptake and detoxification mechanisms in plants. Int J Agric Res 1(2):122–141
Molins H, Michelet L, Lanquar V, Agorio A, Giraudat J, Roach T et al (2013) Mutants impaired in vacuolar metal mobilization identify chloroplasts as a target for cadmium hypersensitivity in Arabidopsis thaliana. Plant Cell Environ 36(4):804–817
Morel M, Crouzet J, Gravot A, Auroy P, Leonhardt N, Vavasseur A, Richaud P (2009) AtHMA3, a P1B-ATPase allowing Cd/Zn/co/Pb vacuolar storage in Arabidopsis. Plant Physiol 149(2):894–904
Mourato M, Reis R, Martins LL (2012) Characterization of plant antioxidative system in response to abiotic stresses: a focus on heavy metal toxicity. In: Advances in selected plant physiology aspects. InTech
Nada E, Ferjani BA, Ali R, Bechir BR, Imed M, Makki B (2007) Cadmium-induced growth inhibition and alteration of biochemical parameters in almond seedlings grown in solution culture. Acta Physiol Plant 29(1):57–62
Neumann RB, Ashfaque KN, Badruzzaman ABM, Ali MA, Shoemaker JK, Harvey CF (2010) Anthropogenic influences on groundwater arsenic concentrations in Bangladesh. Nat Geosci 3(1):46
Ortega-Villasante C, Rellán-Alvarez R, Del Campo FF, Carpena-Ruiz RO, Hernández LE (2005) Cellular damage induced by cadmium and mercury in Medicago sativa. J Exp Bot 56(418):2239–2251
Pagani MA, Tomas M, Carrillo J, Bofill R, Capdevila M, Atrian S, Andreo CS (2012) The response of the different soybean metallothionein isoforms to cadmium intoxication. J Inorg Biochem 117:306–315
Panda KK, Achary VMM, Krishnaveni R, Padhi BK, Sarangi SN, Sahu SN, Panda BB (2011) In vitro biosynthesis and genotoxicity bioassay of silver nanoparticles using plants. Toxicology in vitro 25(5): 1097–1105
Panda SK, Choudhury S (2005) Changes in nitrate reductase activity and oxidative stress response in the moss Polytrichum commune subjected to chromium, copper and zinc phytotoxicity. Braz J Plant Physiol 17(2):191–197
Pandey N, Pathak GC, Pandey DK, Pandey R (2009) Heavy metals, Co, Ni, Cu, Zn and Cd, produce oxidative damage and evoke differential antioxidant responses in spinach. Braz J Plant Physiol 21(2):103–111
Pandolfini T, Gabbrielli R, Comparini C (1992) Nickel toxicity and peroxidase activity in seedlings of Triticumaestivum L. Plant Cell Environ 15(6):719–725
Patlolla AK, Berry A, May L, Tchounwou PB (2012) Genotoxicity of silver nanoparticles in Vicia faba: a pilot study on the environmental monitoring of nanoparticles. International journal of environmental research and public health 9(5): 1649–1662.
Patra M, Bhowmik N, Bandopadhyay B, Sharma A (2004) Comparison of mercury, lead and arsenic with respect to genotoxic effects on plant systems and the development of genetic tolerance. Environ Exp Bot 52(3):199–223

Phang C, Leung DW, Taylor HH, Burritt DJ (2011) Correlation of growth inhibition with accumulation of Pb in cell wall and changes in response to oxidative stress in Arabidopsis thaliana seedlings. Plant Growth Regul 64(1):17–25

Pourrut B, Perchet G, Silvestre J, Cecchi M, Guiresse M, Pinelli E (2008) Potential role of NADPH-oxidase in early steps of lead-induced oxidative burst in *Vicia faba* roots. J Plant Physiol 165(6):571–579

Qureshi MI, Abdin MZ, Qadir S, Iqbal M (2007) Lead-induced oxidative stress and metabolic alterations in *Cassia angustifolia* Vahl. Biol Plant 51(1):121–128

Racuciu M, Creanga DE (2007) TMA-OH coated magnetic nanoparticles internalized in vegetal tissue. Romanian Journal of Physics 52(3/4): 395

Ren JH, Sun HJ, Wang SF, Luo J, Ma LQ (2014) Interactive effects of mercury and arsenic on their uptake, speciation and toxicity in rice seedling. Chemosphere 117:737–744

Ros ROC, Cooke DT, Burden RS, James CS (1990) Effects of the herbicide MCPA, and the heavy metals, cadmium and nickel on the lipid composition, Mg2+-ATPase activity and fluidity of plasma membranes from rice, *Oryza sativa* (cv. Bahia) shoots. J Exp Bot 41(4):457–462

Samadi N, Yahyaabadi S, Rezayatmand Z (2014) Effect of TiO2 and TiO2 nanoparticle on germination, root and shoot Length and photosynthetic pigments of Mentha piperita. International Journal of Plant & Soil Science 3(4): 408–418

Sandalio LM, Dalurzo HC, Gomez M, Romero-Puertas MC, Del Rio LA (2001) Cadmium-induced changes in the growth and oxidative metabolism of pea plants. J Exp Bot 52(364):2115–2126

Satarug S, Baker JR, Reilly PE, Moore MR, Williams DJ (2002) Cadmium levels in the lung, liver, kidney cortex, and urine samples from Australians without occupational exposure to metals. Arch Environ Health: Int J 57(1):69–77

Sharma SS, Dietz KJ (2006) The significance of amino acids and amino acid-derived molecules in plant responses and adaptation to heavy metal stress. J Exp Bot 57(4):711–726

Sharma P, Dubey RS (2006) Cadmium uptake and its toxicity in higher plants. In: Cadmium toxicity and tolerance in plants. Narosa Publishing House, New Delhi, pp 63–86

Sharma P, Dubey RS (2007) Involvement of oxidative stress and role of antioxidative defense system in growing rice seedlings exposed to toxic concentrations of aluminum. Plant Cell Rep 26(11):2027–2038

Sharma P, Jha AB, Dubey RS, Pessarakli M (2012) Reactive oxygen species, oxidative damage, and antioxidative defense mechanism in plants under stressful conditions. J Bot 2012:1

Singh N, Ma LQ, Vu JC, Raj A (2009) Effects of arsenic on nitrate metabolism in arsenic hyperaccumulating and non-hyperaccumulating ferns. Environ Pollut 157(8–9):2300–2305

Song WY, Sohn EJ, Martinoia E, Lee YJ, Yang YY, Jasinski M et al (2003) Engineering tolerance and accumulation of lead and cadmium in transgenic plants. Nat Biotechnol 21(8):914

Song WY, Mendoza-Cózatl DG, Lee Y, Schroeder JI, Ahn SN, Lee HS et al (2014) Phytochelatin–metal (loid) transport into vacuoles shows different substrate preferences in barley and Arabidopsis. Plant Cell Environ 37(5):1192–1201

Srivastava S, Mishra S, Tripathi RD, Dwivedi S, Trivedi PK, Tandon PK (2007) Phytochelatins and antioxidant systems respond differentially during arsenite and arsenate stress in Hydrillaverticillata (lf) Royle. Environ Sci Technol 41(8):2930–2936

Štolfa I, Pfeiffer TŽ, Špoljarić D, Teklić T, Lončarić Z (2015) Heavy metal-induced oxidative stress in plants: response of the antioxidative system. In: Reactive oxygen species and oxidative damage in plants under stress. Springer, Cham, pp 127–163

Takahashi R, Bashir K, Ishimaru Y, Nishizawa NK, Nakanishi H (2012) The role of heavy-metal ATPases, HMAs, in zinc and cadmium transport in rice. Plant Signal Behav 7(12):1605–1607

Tripathi RD, Srivastava S, Mishra S, Singh N, Tuli R, Gupta DK, Maathuis FJ (2007) Arsenic hazards: strategies for tolerance and remediation by plants. Trends Biotechnol 25(4):158–165

Verma S, Dubey RS (2001) Effect of cadmium on soluble sugars and enzymes of their metabolism in rice. Biol Plant 44(1):117–123

Verma S, Verma PK, Meher AK, Dwivedi S, Bansiwal AK, Pande V et al (2016a) A novel arsenic methyltransferase gene of Westerdykella aurantiaca isolated from arsenic contaminated soil:

phylogenetic, physiological, and biochemical studies and its role in arsenic bioremediation. Metallomics 8(3):344–353
Verma PK, Verma S, Pande V, Mallick S, Deo Tripathi R, Dhankher OP, Chakrabarty D (2016b) Overexpression of rice glutaredoxin OsGrx_C7 and OsGrx_C2. 1 reduces intracellular arsenic accumulation and increases tolerance in Arabidopsis thaliana. Front Plant Sci 7:740
Verma S, Verma PK, Meher AK, Bansiwal AK, Tripathi RD, Chakrabarty D (2018) A novel fungal arsenic methyltransferase, WaarsM reduces grain arsenic accumulation in transgenic rice (Oryza sativa L.). J Hazard Mater 344:626–634
Wang S, Mulligan CN (2006) Occurrence of arsenic contamination in Canada: sources, behavior and distribution. Sci Total Environ 366(2–3):701–721
Wang HH, Shan XQ, Wen B, Owens G, Fang J, Zhang SZ (2007) Effect of indole-3-acetic acid on lead accumulation in maize (*Zea mays* L.) seedlings and the relevant antioxidant response. Environ Exp Bot 61(3):246–253
Wong HL, Sakamoto T, Kawasaki T, Umemura K, Shimamoto K (2004) Down-regulation of metallothionein, a reactive oxygen scavenger, by the small GTPase OsRac1 in rice. Plant Physiol 135(3):1447–1456
Xuming W, Fengqing G, Linglan M, Jie L, Sitao Y, Ping Y, Fashui H (2008) Effects of nano-anatase on ribulose-1, 5-bisphosphate carboxylase/oxygenase mRNA expression in spinach. Biological trace element research 126(1-3): 280.
Yang Z, Wu Y, Li Y, Ling HQ, Chu C (2009) OsMT1a, a type 1 metallothionein, plays the pivotal role in zinc homeostasis and drought tolerance in rice. Plant Mol Biol 70(1–2):219–229
Zenk MH (1996) Heavy metal detoxification in higher plants-a review. Gene 179(1):21–30
Zhang WH, Tyerman SD (1999) Inhibition of water channels by $HgCl_2$ in intact wheat root cells. Plant Physiol 120(3):849–858
Zhou ZS, Guo K, Elbaz AA, Yang ZM (2009) Salicylic acid alleviates mercury toxicity by preventing oxidative stress in roots of *Medicago sativa*. Environ Exp Bot 65(1):27–34

Part III
Management Strategies: Utilization, Engineered Applications and Bioremediation of Agricultural Contaminants

Planning and Engineering Strategies of Agricultural Wastes and Their Remediation Strategies

Hera Naheed Khan and Muhammad Faisal

Contents

1 Introduction

It has been estimated that by 2050 the world population shall be 50 times more than at present and this increase in the world population shall demand 50% more food than currently being produced (Alexandratos and Bruinsma 2012; Conforti 2011; Grafton et al. 2015). In order to feed these many people, agricultural practices need to be improved (Tilman et al. 2002). More extensive agriculture means higher yield and in turn higher waste generation.

H. N. Khan · M. Faisal (✉)
Department of Microbiology and Molecular Genetics, Quaid-e-Azam Campus, University of the Punjab, Lahore, Pakistan
e-mail: faisal.mmg@pu.edu.pk

M. Naeem et al. (eds.), *Contaminants in Agriculture*,
https://doi.org/10.1007/978-3-030-41552-5_10

A thorough understanding about the nature of agricultural wastes is the most important aspect in agricultural waste management. Planning strategies can only be fabricated once a complete and thorough understanding about the nature and composition of agricultural wastes is obtained (Obi et al. 2016). So, agro-waste can be interpreted as the remnants and leftovers that are produced as a result of growing, harvesting and processing of plants, food crops, cash crops, poultry, vegetables, fruits, dairy products and livestock (Nagendran 2011). These dregs, though not apparently beneficial, might be loaded with composites that can be beneficial and can be reutilized after some rounds of processing. The physical and chemical composition of agricultural waste tends to vary and is dependent on the type of agricultural waste that generates it. They can be classified broadly as liquid wastes, solid wastes or slurries (Obi et al. 2016). Food processing by-products usually generate liquid wastes, while waste generated from livestock comprises both liquid and solid wastes. The chemical nature of the waste is also very essential in the development of remediation strategies. Chemical composition covers, chemical oxygen demand, biological oxygen demand, amount of dissolved organic compound are all factors that determine the kind of management strategies employed to remediate it (Thomson 1991).

Major agricultural waste is generated in developing countries and if they continue to intensify their agricultural practices, the annual agricultural waste will rise significantly. An estimated 998 million tons of agro-waste is produced annually, 80% of which comprises organic wastes (Fauziah and Agamuthu 2009; Kadir et al. 2006). If workable management strategies are not available, these wastes can populate our environment, soils and water bodies. Before identifying planning and remediation of agro-wastes, we need to identify the sources, practices and different farming activities that result in production of agricultural wastes.

2 Waste Generation Activities

Major agricultural activities that contribute to generation of waste are intensive farming, cultivation of crops, food processing units, livestock and fertilizer applications. All these applications lead to generation of waste in the form of sludge; if it has 90% of water content, it is characterized as liquid waste, and if it has 30–50% of water content, it is categorized as solid waste (Andreoli et al. 2007). Furthermore, waste composition is extremely helpful in mediating planning and removal of agricultural wastes. We shall discuss the waste generation activities one by one.

2.1 Farming Practices

The most important aspect of agriculture is farming. It is the process of growing fruits, vegetables and crops or rearing animals for the sole purpose of food and raw materials (Gold 2016). Good farming practices confer healthy crops, increased yield, less crop damage due to outbreak of infections, improved soil fertility and promote biodiversity. Good farming leads to increased agricultural wastes, as the overall food production increases the crop, fruit and vegetable residues, and waste products also increase; the fertilizers and pesticides sprayed on the soil for protection from infections and pests increase, which leads to an increase in agricultural waste production (Huttunen and Soini 2018). These chemical agents are left or thrown in the nearby ponds, lakes or soils that act as a source of environmental pollution (Carvalho 2017). According to an estimate by the Plant Protection Department (PPD), 1.8% of the chemicals stay in the packaging. Extensive use of fertilizers for improvement of soil quality and enhancement of plant production than required tend to make the soils barren (Cardoso et al. 2013) and these chemicals run-off to the water bodies, again posing a serious threat to the ecosystem. These run-offs not only contaminate nearby rivers, ponds or lakes, but they also tend to leach into the soil and contaminate underground water channels (Seiler and Berendonk 2012). Poor agricultural practices pollute and pose a threat to not just agricultural land but also put at stake the ecosystems in close proximity. They can contaminate land and water at an equal rate and could be thoroughly involved in converting fertile land into barren acres.

Intensive farming is the need of the hour, as the increase in world population demands food security that can only be met if agricultural practices are efficient. However, this demands good sustainable agricultural practices that not only improve and increase food production but also protect and shelter the environment. It refers to practices that do not increase yield at the stake of environmental degradation (Swanson 2008).

2.2 Food Processing Units

Food after production is sent to processing units for refinement. This requires huge machines, processing plants and a lot of waste is generated during this process. The modern world requires fruits and vegetables throughout the year. With the evolution of supermarkets and advancement in the food industry, the consumer desires washed and packaged fruits, vegetables and meat products. Fruits and vegetables are processed into pickles and available as canned foods. Seafood and dairy products are also processed prior to selling them to consumers. All this processing generates waste in large volumes. Small food-processing units and industries like mushroom industry individually are responsible for the generation of billions of tons of waste annually (Loehr 2012). Depending on the type of processing unit, the waste generated also differs in quality and quantity. The effluents generated during processing enter the ecosystem and pose a threat to humans, animals and even plants (Girotto et al. 2015).

2.3 Livestock

Livestock rearing is an important activity and it usually includes rearing of farm animals, cows, pigs, chickens, etc. It is crucial to agriculture because livestock utilizes crop wastes and residues that are not used by humans, thus reducing waste quantity and converts these by-products to milk and meat. The animal excreta are used as compost and manure as fertilizers for agricultural soils to maintain their quality and restock it with lost nutrients (Twardowska et al. 2004). Moreover, this can also be used as biogas for energy consumption (Abdeshahian et al. 2016). So, the importance of livestock to agriculture is undeniable. Rearing of livestock is now becoming an industry where thousands of animals are fed and maintained in a single facility. But the waste generated from livestock rearing comprises manure, waste from slaughter houses, excrement waste (urine and excreta), animal feed waste and gases produced during the management of wastes like H_2S, CH_4 and NH_4 (Cantrell et al. 2008). To prevent infection and disease spread in animal farms regularly, animals are fed different antibiotics and insecticidal sprays are done that protect the animals but in the long run pose a serious threat to our environment by affecting the air, water and soil quality. The frequent use of antibiotics in animal feed results in antimicrobial resistance in bacteria and generates multi-drug resistance bacteria (Van Boeckel et al. 2015). This also seriously compromises the environment; moreover, these farms are usually established near residential areas, so spread of zoonotic infections and outbreaks is not very uncommon (Loehr 2012).

2.4 Use of Fertilizers

An estimated 6000 million people are fed by exploiting modern agricultural techniques (Tilman et al. 2002). According to reports, there has been a 50% increase in global cereal production in the past 40 years and all this is attributed to improved agriculture practices mainly due to applications of fertilizers and pesticides (http://apps.fao.org/).The increase in agricultural yield would not have been possible without the use of fertilizers and these on the one hand help to improve soil quality and replenish the lost nutrients (Stewart et al. 2005), while, on the other hand, their overuse has led to increased water pollution. Chemical fertilizers are extremely harmful and their incorporation in water can contaminate the soils that water irrigates and so in the long run, it also produces polluted plants and toxic crops. The heavy metal load in the waters and soil can increase due to extensive application of fertilizers and the toxicities associated with these metals can cause serious health threats (Savci 2012). Hence, for good sustainable agriculture, a limited use of fertilizers shall be done to minimize their harmful effects.

3 Removal Strategies of Agricultural Wastes

Agricultural waste residues are a rich source of organic compounds and their treatment usually involves recycling of those components to generate useful end products. The large amount of biomass can be strategically monitored and divided to convert them into useful products. The following treatment options can be employed for safe disposal and utilization of agricultural wastes.

3.1 *Effluent Flotation*

One method of removing solids or particulate matter from liquid is through air floatation as demonstrated in Fig. 1. Separation is mediated by the gas bubbles that attach to the suspended particles and decrease their specific gravity than water, which results in the flotation of the suspended particles on the surface. A pressurized tank supplies the liquid with air at 30–50 Psi from one end of the floatation tank, the gas passes through the liquid and air bubbles attach to the particulate matter and as the air comes out of the water, any gas can be used for this process, but the rate of travelling of CO_2 through solution is more than air at the same pressure. But CO_2 is more expensive, so is not usually preferred. The rate of bubble production is directly proportional to the product of absolute pressure and pressurized flow rate. Solubility of air in water increases as temperature is reduced; so wastewater treatment temperatures are significant in this process. Another factor that controls effluent flotation and removal is their size (Rubio et al. 2002).

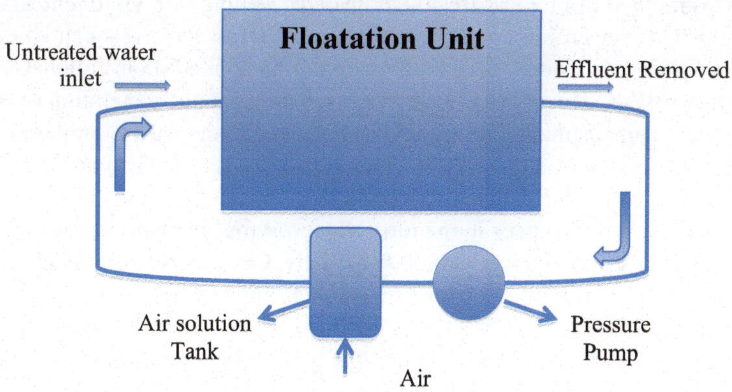

Fig. 1 Air floatation unit for the separation of effluents from wastewater

3.2 *Direct Combustion/Incineration*

This process involves directly incinerating the agricultural residues, crop residues, animal feed, fodder foliage, vegetable and fruit residues into incinerators and burning them at high temperatures. This process involves direct combustion and is carried out in an aerobic environment. In this process, waste is treated and exposed to high temperatures of above 1200–1800 °F. Operating incinerators at lower temperatures usually results in incomplete combustion and production of volatile compounds and odorous gases that contribute to air pollution (Obi et al. 2016). The energy-rich process leads to oxidation of the organic matter and converts them into CO_2 and H_2O as the end products. Energy is released in the form of heat and radiation. Depending on the waste, this could be dangerous also and usually contaminates the environment. The gases produced during this process lead to environmental pollution. However, this thermal conversion process can be utilized by using the generated heat for cooking, production of steam and electric and mechanical power applications. It is an old method of removing agro-waste, yet it is still applied and is the dominant method for treatment of agricultural wastes. However, the major disadvantage of this conventional approach is its cost. Incineration leads to the production of gases and aerosols that pollute the air; so to protect the environment, filters should be installed within the incinerators (Loehr 2012).

3.3 *Pyrolysis*

An ancient treatment strategy for conversion of organic matter-rich biomass is pyrolysis. This thermochemical process is applied to biomass for production of biofuel. This method of biomass treatment involves heating of it up to temperatures of 400–600 °C in anaerobic conditions (Obi et al. 2016). Pyrolysis can be classified into two basic types depending on the conditions applied: conventional pyrolysis and fast pyrolysis. The conventional process is useful in the generation of chemical end-products like methanol, acetic acid and charcoal, whereas fast or flash pyrolysis involves conversion of almost 75% of the organic waste to pyrolytic acid while the rest is converted to char. This process of waste degradation helps in the generation of renewable energy sources in the form of bio-oil that can help replace fossil fuels in the coming years (Yanik et al. 2007). Usually, this rich product is obtained from lingo-cellulosic rich biomass. However, though this technology is old, the bio-oil produced has certain undesirable characteristics that make it unsuitable as fuel, especially its thermally unstable nature that turns it into a gummy paste-like material, with high water quantity and components that are corrosive in nature. This process, however, helps in not just reducing and getting rid of agro-waste but also provides a renewable source of energy that can contribute to 19% of renewable energy sources that make up the total energy consumption. It not only provides a raw alternative for fossil fuels but also helps in the production of different chemicals like ammonia, glucose and alcohols (Zanzi et al. 2002).

4 Bioremediation of Agricultural Waste

First and foremost, we need to understand the remediation techniques that can be employed for safe disposal of large amounts of waste generated through various agricultural practices as already mentioned. Bioremediation of wastes has been in use for many years now. Exploiting microorganisms for treatment of waste is an intelligent strategy (Chandra and Singh 2015). Agricultural waste is a very rich source of nutrients. The by-products generated during cultivation, harvesting and processing of crops, fruits and vegetables and production of livestock can be utilized for production of many useful end products. However, the major threat posed by agro-waste is of chemical leachates that are key contaminators of surface and ground water. Bioremediation is categorized into the following three types: (i) fermentation to degrade organic solid wastes (Degueurce et al. 2016), (ii) aerobic and anaerobic degradation of organic wastes (Shou-liang et al. 2008) and (iii) bioelectrochemical removal of organic wastes (Iskander et al. 2016; Yang et al. 2016). While using bacteria for waste treatment, the waste is used as a source of electrons, carbon and energy for bacteria, and they in turn degrade the waste to less toxic forms.

The most common method of getting rid of excess organic waste is by composting. This process helps to decrease the volume of the waste and convert it into useable and marketable products. Composting is done with the help of microorganisms that decompose the organic compounds in the waste and not just reduce them in size but also convert them into less toxic forms.

4.1 Composting

Agriculture waste is rich in nutrients, especially organic compounds. It comprises manure generated from livestock, crop wastes and harvests, fruit peels vegetable peels, dead plants, etc. Decomposers are special class of bacteria and fungi that feed on organic waste. They utilize this rich waste to fulfil their carbon, nitrogen and energy requirements (Štursová et al. 2012). This process is carried out under aerobic conditions that convert the unwanted complex organic waste to more stable, short-chain molecular structures that are rich in humus and can be recycled for their beneficial nature to soil and crops. The compost generated following this process helps to return organic matter to the soil that can be used as natural fertilizers to improve soil fertility. Soil microbes carry out this process through the following chemical reaction:

$$\text{Organic matter} + O_2 \rightarrow \text{Compost} + CO_2 + H_2O + NO_{-3} + SO_4^{-2} + \text{Heat}$$

The heat generated during this process helps to kill harmful microorganisms and thus disinfects the resultant end product. Composting can be carried out in closed composting containers or units or in the open. Closed containers help to speed up the process and give the resultant end product within 5–7 days, while open windrows make sufficient product in 3–8 weeks. Compositing is divided into two major phases:

(i) the degradation phase and (ii) the maturation phase. The degradation phase is initiated by the degradation of the most easily degrading organic compounds largely composed of sugars and simple peptides, organic acids, etc. This action is performed by aerobic microbes that aerobically utilize organic-rich material and release CO_2 and energy in the form of heat. This phase takes weeks to months depending on the nature and characteristics of the substrates. The process requires continuous aeration to keep an adequate supply of oxygen and also to ensure temperature management. Mostly bacteria carry out this process because of the extreme physiological conditions (pH, temperature and humidity). During this phase, the temperature reaches 130–160 °F where most of the mesophilic pathogenic bacteria die. The end product of this process is fresh compost. Following degradation or stabilization phase, there is a decrease in temperature and conditions become mesophilic. The major microbial population is of actinomycetes that actively decompose complex carbohydrates (starch, cellulose and lignin) that produce humus, the process of humification is the final phase, that is, the maturation phase. No further increase in temperature occurs following this phase even upon aeration, indicating that composting is completed. Moisture content and decrease in percentage of volatile compounds mark the completion of this process (Loehr 2012).

This process of remediating agricultural waste not only reduces the volume of the waste, but it also provides an alternative for the treatment of unfertile soils. The microbial mediated removal of waste is a healthy and environment-friendly method employed for getting rid of agro-waste. Compost is an environment-friendly fertilizer that can be applied to degraded soils for improving their quality and producing good crops (Bernal et al. 2009).

4.2 Anaerobic Digestion

Another attractive strategy of treating agriculture waste and recycling it using biologic agents, especially bacteria, is anaerobic digestion. A schematic description of the process is represented in Fig. 2. This process not only helps to save energy compared to aerobic reduction methods but also produces almost 50% less sludge. On the bright side, this process helps in the generation of biogas, that is, methane gas, which can be used as a replacement of gas. The process has been classified into three steps and each step is conducted by bacterial and Archaeal species. Initially, bacteria decompose polymer-rich organic waste comprising sugars and amino acids into monomers. These monomers are than acted upon by bacteria and converted to fatty acids like formate. Finally, these organic acids (acetic acid, formic acid) are reduced to CO_2 and methane by methylotrophs, hydrogenotrophs and acetotrophic bacteria (Ziganshin et al. 2013). Depending on this, anaerobic digestion can be divided into four phases, namely hydrolysis (polymers to monomers), acidogenesis (formation of organic acids & acidic end products), acetogenesis (generation of acetate, CO_2 and hydrogen) and methanogenesis (production of methane, i.e. biogas).

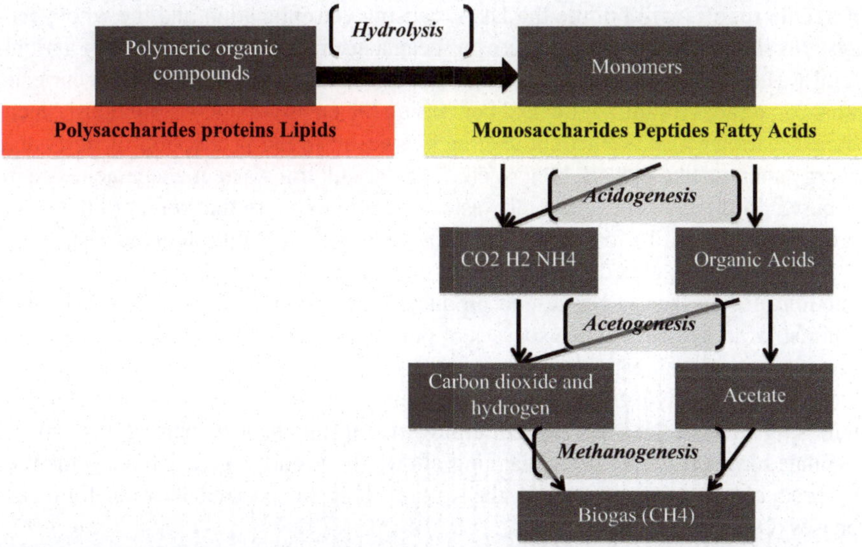

Fig. 2 A Schematic representation of the process of anaerobic digestion of agricultural waste that leads to biogas production

This process like any other process depends on physical factors like temperature, pH amount and quantity of organic biomass and most importantly, the bacterial flora. The process takes around 3–4 weeks during which microorganisms can adapt to the changing environmental conditions. Methanogens are most sensitive to thermal changes and the whole process largely depends on the optimal controlling of temperature and pH. This process results in controlling and disposing off large wastes generated via animal feeds poultry and dietary wastes. The biogas produced has 50–70% content of CH_4, 25–45% of CO_2, >5% of N_2 and H_2 and traces of hydrogen sulphide. The process, though costly, is favoured over other conventional process due to its various benefits, significantly biogas production (Merlin and Boileau 2013).

5 Agricultural Waste Management Strategies

For a sustainable agricultural environment, there is a need for proper development of management systems. This highlights that considering agricultural waste as undesirable and unwanted and discharging it into the environment without proper and prior treatment is unhealthy and damaging to our environment, but it is also the waste of organic-rich biomass that can be utilized for a number of applications. These systems consider agricultural waste also as potential for renewable energy resources. Release of waste from animal farms, diary farms, crop harvests directly to the environment will contaminate our water bodies, drop the fertility of the soil and will become a harvesting ground for pests and insects. Fertilizer, insecticide and pesticide runoffs will pollute the lakes, causing eutrophication and the whole ecosystems shall collapse. Hence, there has been a need for the development of agricultural waste management strategies and systems that can regulate the most enriched waste generated for utilization and benefitting the environment. As previously mentioned, various waste remediation strategies result in the production of renewable energy sources like bio-oil, biogas, etc. The agricultural waste management system is based on the concept of a total waste generation system that refers to the waste production to its utilization throughout the year in an agricultural enterprise.

Open and unregulated decomposition of organic-rich biomass results in the production of dangerous volatile end products like NH_4 that can result in acid rain. Animal farms are the biggest sources of outbreaks of zoonotic epidemics and rearing of drug-resistant pathogens due to overuse of antimicrobials in animal feeds. All these pollute our hydrosphere, atmosphere and lithosphere. Hence, sustainable system is the need of the hour. Agricultural waste management system refers to such a regulated system where prior planning allows the production and management of by-products for reuse to sustain the system. It is categorized into the following phases (Obi et al. 2016).

5.1 Functions of AWMS

All agricultural waste management systems operate on the above-mentioned (Fig. 3) regulation system. Waste generated either from farms, animal waste, crop waste diary waste is collected; then, it can be stored or treated; the end products are then utilized for sustaining the whole system. In this way, waste production can be easily controlled and reused.

5.1.1 Production

The first function of the management system refers to the production of waste. This refers to the quantity and quality of waste that is generated in any agriculture-related activities. The composition, kind, consistency, location and time of the waste are all categorized under the production section. This tells us about the waste type and what kind of treatment strategy could be employed to reduce and recycle it.

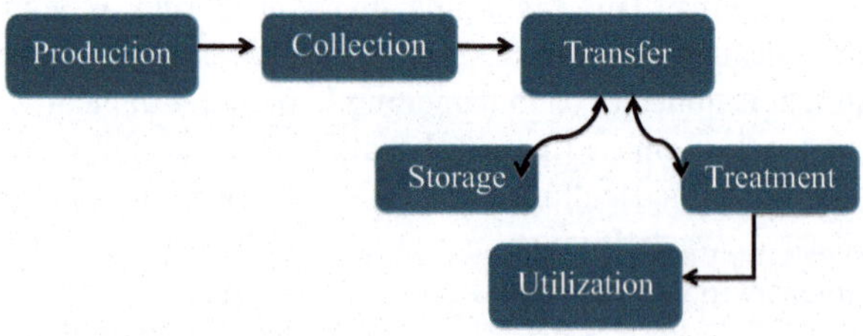

Fig. 3 Phases characterizing AWMS

5.1.2 Collection

The second section is involved in concentrating the waste at a single point from all the other production sites. Here, the waste generated via one kind of agricultural activity is collected and concentrated; for example, a crop harvest of different fields located at different locations shall be collected at a single location, making it ready for processing. The collected residues are then categorically divided according to their types, solids, liquids and slurries for making them available for treatment.

5.1.3 Storage

Following collection, the waste can be stored temporarily in storage sites until they are processed and treated. The storage facility ensures that the waste is contained without its degradation due to climatic changes and is just a momentary holding option until treatment. However, this storage shall not chemically or physically harm the waste and degrade it before treatment. After storage and containment, the waste is ready to be treated.

5.1.4 Treatment

Treatment options strongly depend on the type, composition and volume of the waste. These are designed to convert the pollutants to usable forms that can be utilized in the last step of the AWMS. Physical, chemical and biologic treatments are performed on the wastes and it is made ready to be converted to usable products.

5.1.5 Utilization

Finally, the treated waste is now in a new reusable form that can be utilized as bio-oil, biogas or production of manures, all again replenishing the agricultural sector. This ensures that the generated waste does not go without extracting the available nutrients and helps in energy conservation (https://directives.sc.egov.usda.gov/31493.wba).

5.2 *The "3R" Approach in AWM*

The 3R refers to reduce, reuse and recycle approach that is the slogan for waste management and environment protection agencies. This approach ensures that initially such agricultural practices shall be avoided, which increases the contaminant load and techniques that produce less contaminant are preferred. This approach is usually used for management of solid wastes. The produced waste is then reused or made reusable using the different waste treatment techniques mentioned in the previous section. Recycling the waste and applying it for energy production or other alternatives are the key to producing a sustainable agricultural system.

Governmental agencies and policy-makers play a significant role in setting up ground rules for sustainable agriculture and reducing waste production and implementation of the 3R approach. Policies shall be made to ensure that a regulated environment is operated. Threshold levels shall be set up for use of fertilizers, antimicrobials, pesticides, insecticides and all such chemicals that are dangerous for the environment in the long run. Regulated use of these compounds shall be ensured and most importantly, awareness shall be given to the farmers and farm owners. Environment quality shall never be compromised over anything. Each of the agricultural production sites (animal farms, dairy farms, fields and food processing units) shall be made such that the waste generated in each of these sites has its own agricultural waste management system and it is collected, stored, treated and utilized at the site where it is generated. AWMS ensures that minimum quantity of by-products is generated and benefits from the generated by-products are maximized (https://directives.sc.egov.usda.gov/31493.wba).

6 Conclusion

Though the benefits provided to us via agriculture cannot be denied, yet the modern world demands the developing of a sustainable agricultural environment to ensure less environment damage and more benefits. Various remediation techniques are utilized to treat agro-wastes and generate end products that can be reused. This recycling of waste in the form of manure, fuel, thermal energy and fertilizers accounts for the 3R approach that is now applied in other waste treatments also.

References

Abdeshahian P, Lim JS, Ho WS, Hashim H, Lee CT (2016) Potential of biogas production from farm animal waste in Malaysia. Renew Sust Energ Rev 60:714–723

Alexandratos N Bruinsma J (2012) World agriculture towards 2030/2050: the 2012 revision. ESA working paper number. 12–01. Rome

Andreoli CV, Von Sperling M, Fernandes F, Ronteltap M (2007) Sludge treatment and disposal. IWA Publishing Company, London

Bernal MP, Alburquerque JA, Moral R (2009) Composting of animal manures and chemical criteria for compost maturity assessment. Bioresour Technol 100(22):5444–5453

Cantrell KB, Ducey T, Ro KS, Hunt PG (2008) Livestock waste-to-bioenergy generation opportunities. Bioresour Technol 99(17):7941–7953

Cardoso EJBN, Vasconcellos RLF, Bini D, Miyauchi MYH, Santos CAD, Alves PRL, Nogueira MA (2013) Soil health: looking for suitable indicators. What should be considered to assess the effects of use and management on soil health? J Sci Agric 70(4):274–289

Carvalho FP (2017) Pesticides, environment, and food safety. Food Energy Sec 6(2):48–60

Chandra S, Singh N (2015) Enhanced bioremediation techniques for agricultural soils. Int J Curr Res Acad Rev 3(7):166–173

Conforti P (2011) Looking ahead in world food and agriculture: perspectives to 2050. Food and Agriculture Organization of the United Nations (FAO), Rome

Degueurce A, Trémier A, Peu P (2016) Dynamic effect of leachate recirculation on batch mode solid state anaerobic digestion: influence of recirculated volume, leachate to substrate ratio and recirculation periodicity. Bioresour Technol 216:553–561

Fauziah S, Agamuthu P (2009) Sustainable household organic waste management via vermicomposting. Malays J Sci 28(2):135–142

Girotto F, Alibardi L, Cossu R (2015) Food waste generation and industrial uses: a review. Waste Manag 45:32–41

Gold MV (2016) Sustainable agriculture: the basics. CRC Press, Boca Raton

Grafton RQ, Daugbjerg C, Qureshi ME (2015) Towards food security by 2050. Food Sec 7(2):179–183

Huttunen S, Soini K (2018) Cultivating cultural sustainability in farming practices. In: Cultural sustainability and the nature-culture interface. Routledge, Abingdon, pp 35–48

Iskander SM, Brazil B, Novak JT, He Z (2016) Resource recovery from landfill leachate using bioelectrochemical systems: opportunities, challenges, and perspectives Bioresour Technol 201:347–354

Kadi AA, Hinta H, Sarani NA (2006) The utilization of coffee waste into fired clay brick. J Eng Appl Sci 10(15):6289–6292

Loehr R (2012) Agricultural waste management: problems, processes, and approaches. Elsevier, Saint Louis

Merlin G, Boileau H (2013) Anaerobic digestion of agricultural waste: state of the art and future trends. Nova Science Publishers, New York, pp 6284–6292

Nagendran R (2011) Agricultural waste and pollution. In: Waste: a handbook for management. Academic/Elsevier, Amsterdam, pp 341–355

Obi F, Ugwuishiwu B, Nwakaire J (2016) Agricultural waste concept, generation, utilization and management. NIJOTECH 35(4):957–964

Rubio J, Souza ML, Smith RW (2002) Overview of flotation as a wastewater treatment technique. Miner Eng 15(3):139–155

Savci S (2012) An agricultural pollutant: chemical fertilizer. IJESD 3(1):73

Seiler C, Berendonk TU (2012) Heavy metal driven co-selection of antibiotic resistance in soil and water bodies impacted by agriculture and aquaculture. Front Microbiol 3:399

Shou-liang H, Bei-dou X, Hai-chan Y, Shi-lei F, Jing S, Hong-liang L (2008) In situ simultaneous organics and nitrogen removal from recycled landfill leachate using an anaerobic - aerobic process. Bioresour Technol 99 (14):6456–6463

Stewart WM, Dibb DW, Johnston AE, Smyth TJ (2005) The contribution of commercial fertilizer nutrients to food production. Agron J 97(1):1–6

Štursová M, Žifčáková L, Leigh MB, Burgess R, Baldrian P (2012) Cellulose utilization in forest litter and soil: identification of bacterial and fungal decomposers. FEMS Microbiol Ecol 80(3):735–746

Swanson BE (2008) Global review of good agricultural extension and advisory service practices: Citeseer. Food and Agricultural Organization of the UN, Roma

Thomson WT (1991) Agricultural chemicals book: III-miscellaneous agricultural chemicals: fumigants, growth regulators, seed safeners, repellents, fish toxicants, bird toxicants, pheromones, rodenticides and others, vol 206. Thomson Publications, Fresno. ISBN: 0913702374

Tilman D, Cassman KG, Matson PA, Naylor R, Polasky S (2002) Agricultural sustainability and intensive production practices. Nature 418(6898):671–677

Twardowska I, Allen HE, Kettrup A, Lacy WJ (2004) Solid waste: assessment, monitoring and remediation, vol 4. Gulf Professional Publishing, Elsevier

Van Boeckel TP, Brower C, Gilbert M, Grenfell BT, Levin SA, Robinson TP, Laxminarayan R (2015) Global trends in antimicrobial use in food animals. Proc Natl Acad Sci 112(18):5649–5654

Yang JL, Liu JZ, Wu CX, Kerr PG, Wong PK, Wu YH (2016) Bioremediation of agricultural solid waste leachates with diverse species of Cu (II) and Cd (II) by periphyton. Bioresour Technol 221:214–221

Yanik J, Kommayer C, Saglam M, Yuksel M (2007) Fast pyrolysis of agricultural wastes: characterization of pyrolysis products. Fuel Process Technol 88(10):942–947

Zanzi R, Sjostrom K, Bjornbom E (2002) Rapid pyrolysis of agricultural residues at high temperature. Biomass Bioenergy 23(5):357–366

Ziganshin AM, Liebetrau J, Proter J, Kleinsteuber S (2013) Microbial community structure and dynamics during anaerobic digestion of various agricultural waste materials. Appl Microbiol Biotechnol 97(11):5161–5174

The Use of Constructed Wetlands to Mitigate Pollution from Agricultural Runoff

Vassiliki Ioannidou and Alexandros I. Stefanakis

Contents

1 Introduction

Water pollution mitigation is a topic of high importance to address, having significant impact on the aquatic ecosystems and human health. Therefore, there is an increasing engagement to set and advance regulations and policies at a global level to prevent water quality deterioration and pollution, and to protect the environment (European Commission 2016).

A significant component of water pollution derives from diffuse or nonpoint source (NPS) pollution, closely associated with agricultural activities (Reichenberger et al. 2007). Modern agriculture utilises fertilisers and pesticides to secure high crop yields, which contribute to diffuse pollution. NPS pollution has been registered to have negative effects on the receiving water bodies, aquatic ecosystems, and human

V. Ioannidou
School of Engineering & the Built Environment, Faculty of Computing, Engineering & the Built Environment, Birmingham City University, Birmingham, UK

A. I. Stefanakis (✉)
Department of Engineering, German University of Technology, Athaibah, Oman
e-mail: Alexandros.Stefanakis@bauer.de

M. Naeem et al. (eds.), *Contaminants in Agriculture*,
https://doi.org/10.1007/978-3-030-41552-5_11

health (Bollmann et al. 2014; Dabrowski and Schulz 2003; Feng et al. 2011; Humenik et al. 1987; Line et al. 1998; Loague et al. 1998; Zhang et al. 2011). Problems associated with agricultural runoff and NPS pollution include contamination of surface water, eutrophication, harmful consequences to aquatic ecosystems and habitat, and human health risk due to exposure via drinking water (Borgvang and Tjomsland 2001; Brock et al. 2000; Kimbrough and Litke 1996; Kreuger 1998; Schulz 2004; Zhang and Zhang 2011). In particular, insecticide agricultural runoff ending up in the receiving water course causes severe impact on aquatic ecosystem life, including fish, invertebrates, birds, and other wildlife (Thompson 1996; Stefanakis and Becker 2016). Eutrophication, resulting from nutrient loads, deteriorates surface water quality intended for fisheries and water supply (Koskiaho et al. 2003).

Among the best management practices to moderate and control the impact of NPS agricultural runoff pollution is the application of constructed wetlands (CWs), as they constitute an environmentally friendly and promising means of water pollution purification (Stefanakis et al. 2014; Vymazal and Březinová 2015). This chapter summarises the current knowledge on the mitigation capacity of CWs in terms of various agricultural runoff pollutants, investigates the potential contribution of plants in contaminant removal and assesses their overall efficiency.

2 Agricultural Runoff and Associated Pollutants

Agricultural runoff is the runoff of agrochemicals, that is, fertilisers and pesticides discharged into surface waters, and is the major NPS pollution source (Fulton et al. 1999). Surface water contamination, spread via agricultural NPS runoff, is the leading source resulting in water quality challenges (EPA 2016). Modern agriculture invariably applies fertilisers (nutrients) and pesticides to secure high crop yields and to prevent crops from diseases and insects. Agricultural diffuse runoff results in large discharges of fertilisers, pesticides and suspended solids downstream agricultural catchments or into aquatic ecosystems adjacent to rural areas (Fulton et al. 1999). The impact of agricultural runoff has profound adverse effects on wildlife, aquatic ecosystems, and on human health via drinking water supply (EPA 2016), thus contributing to surface and ground water deterioration, with serious environmental and economic consequences (Wu et al. 2013a, b). As a result, there is a growing interest around agricultural NPS pollution research over the last two decades (Yanhua et al. 2012).

2.1 Fertilisers

Fertilisers contain various nitrogen (N) and phosphorous (P) forms, and constitute the principal pollutants in diffuse agricultural runoff (Mitsch et al. 2000; Poe et al. 2003). Problems associated with elevated loads of fertilisers are well documented that lead to eutrophication, excessive phytoplankton production and hypoxia (deficiency of dissolved oxygen (DO) of surface water courses), and hence in ecosystem disturbance (Beutel et al. 2009; Borgvang and Tjomsland 2001; Coffey 1997; Jordan et al. 2003; Koskiaho et al. 2003; Volkmar and Dahlgren 2006). Nutrients also account for groundwater contamination, especially in water intended for drinking purposes (Beutel et al. 2009; Coffey 1997; Stefanakis et al. 2017). The challenge with the treatment of drinking water due to algae growth involves costly procedures. As a consequence, eutrophication does not only impact the ecosystems function, but also the economic value of clean water (Stanton and Taylor 2012).

As regards N, free ammonia in particular might be toxic to aquatic life and fish, especially in water recipients with aquatic vegetation (Shilton 2005). Furthermore, nitrate (NO_3^-) is another form of N that is largely released through N fertiliser application and is liable for the aforementioned problems. Elevated NO_3^- levels are highly correlated with cause of methemoglobinaemia (or 'blue baby' syndrome) in infants (Horne 2002; Knobeloch et al. 2000). It is reported that 40–60% of the N fertiliser field application is used by crops, while the rest is lost as runoff or entrains groundwater (Coffey 1997). Therefore, the overapplication of N in crop fields is not a successful practice from a holistic perspective. Furthermore, it is reported that in Europe, 90% of N losses leached via agricultural runoff that originated from NO_3^- (Billy et al. 2013; Tournebize et al. 2015). Moreover, P is the principal cause of eutrophication, with all the previously mentioned associated ecological and financial consequences (Lu et al. 2009; Dunne et al. 2015). As a consequence of the nutrient implications on estuarine habitat and environment, the European Directive 2000/60/CE has designated regulation and goals related to good ecological standards in rivers and estuaries.

2.2 *Pesticides*

Pesticides are chemical compounds extensively used in modern agriculture to improve crop production and to secure high yields. Pesticides are classified into categories depending on the targeting enemy, including herbicides, insecticides, fungicides and several other categories. For the majority of pesticides applied, losses via runoff are deemed the most serious means of pollution transfer, followed by losses via erosion (Reichenberger et al. 2007).

The largest pesticide losses have been observed when intense storm events follow pesticides field application (Gregoire et al. 2009; Kladivko et al. 2001; Schulz 2004). There is increasing concern about the effects of pesticide agricultural runoff on human health, estuarine habitat, and groundwater (Bollmann et al. 2014; Feng et al. 2011; Stefanakis et al. 2017; Tao and Fletcher 2013; Runes et al. 2003; Tediosi et al. 2012; Tournebize et al. 2013; Zhang et al. 2011; Zhang and Zhang 2011). As a public health and groundwater quality prevention measure against pesticide runoff impact, the European Drinking Water Directive (98/83/EC) has established stringent maximum detection limit of 0.1 μg/L of any individual pesticide in potable water. The emerging problem so far is identified in some pesticides such as carbetamide, clopyralid, imazalil, metaldehyde, propyzamide, pendimethalin and tebuconazole whose detected concentration is greater than the allowed concentration (Tao and Fletcher 2013; Lv et al. 2016; Tediosi et al. 2012). The challenge arises to the drinking water supply companies, which either have to deal with high energy consumption and associated high costs for potable water treatment, or find it impossible to remove some of those pesticides with the state-of-the-art technological processes.

It is therefore inferred that the elevated costs and energy input involved in the agricultural runoff treatment are a great driver to search and investigate alternative options that are capable of mitigating the NPS pollution equally sufficiently, cheaper and on the source. One such technology is constructed wetlands systems.

3 Constructed Wetlands for Agricultural Runoff Mitigation

Constructed wetlands (CWs) are engineered systems using natural processes through plants, soil and related microbial consortia, to aid in treating polluted water (Stefanakis et al. 2014). With respect to agricultural NPS pollution, the major associated contaminants comprise nutrients, pesticides and particulate matter (Stefanakis et al. 2017). CWs provide a wide variety of regulatory functions, involving flood mitigation, storm water retention, control of pollutant transport, water quality enhancement and biodiversity productivity (Verhoeven and Setter 2010; Stefanakis et al. 2014). In particular, however, CW processes related to mitigation of eutrophication and toxicity from agricultural runoff involve adsorption, denitrification, sediment retention and plant uptake (Rodgers and Dunn 1992; Haygarth and Jarvis 2002). Wetlands provide a rich spectrum of values, linked to population, ecosystem and global perspectives. The ecosystem values of wetlands refer to storm abatement, flood moderation, groundwater recharge, aesthetics and water quality enhancement. In terms of global and local profits, wetlands are considered as potentially pivotal elements contributing to sustain the cycles of carbon dioxide, methane and nitrogen (Mitsch and Gosselink 2007; Shutes et al. 2010).

The majority of CWs applications refer to wastewater treatment (WWT) from urban and industrial activities (Locke et al. 2011; Stefanakis 2018). The scientific interest of CWs for purifying diffuse agricultural runoff commenced in 1980s (Schulz 2004; Locke et al. 2011; Bodin et al. 2012), with the first diffuse pollution-related publication in 1975 (McElroy et al. 1975). Complying with the stringent regulations of the framework directives about surface, ground, and drinking water quality and estuarine ecosystem standards, CWs have emerged as an increasingly popular measure against NPS agricultural pollution. The investigation and assessment of CWs' treatment efficiency to abate nutrients and/or pesticides have rocketed over the last years, with ample research being conducted on pilot-scale and full-scale units.

3.1 Types of CWs Applied for Agricultural Runoff

CWs are categorised into free-water surface (FWS) or subsurface-flow (SF) systems, where the SF systems are further subdivided into vertical (VSF) and horizontal (HSF), depending on the direction of the flow path (Stefanakis et al. 2014). Most of the current literatures about any wetland type pertain to wastewater treatment (Kadlec 2009; Kotti et al. 2010), with the most commonly employed types being FWS and HSF wetlands. FWS CWs afford a better habitat for particular flora and fauna species because of the water ponding result most of the year (Mitsch and Gosselink 2007; Kadlec 2009). On the other hand, HSF systems exhibit higher removal capacities for nitrate (NO_3^-) and pathogens, mainly due to the anoxic conditions that promote denitrification (Fennel et al. 2009; Stefanakis et al. 2016; Reed and Brown 1995).

3.2 CWs Efficiency for Mitigating Agricultural Runoff Pollutants

Studies carried out in CWs for agricultural runoff mitigation are summarised in Table 1. As it is obvious, the most popular CW type for agricultural runoff mitigation is FWS. Full-scale units are increasingly employed and assessed for their treatment performance. Small-scale units, such as mesocosm lab and pilot-scale units, undeniably serve successfully in the addition of current knowledge around various CW components, that is, plant species, substrate material, etc. Table 1 shows the variability in treatment rates, that is, between systems, seasons, soils, regions, indicating that there is a wide range of factors that affect the removal of agricultural runoff pollutants.

3.3 Factors Affecting CWs Treatment Efficiency

The physicochemical, environmental and biological processes occurring in CWs determine the removal efficiency of nutrients, suspended solids and pesticides (Stefanakis et al. 2014). Every pollutant is usually mitigated or removed by a combination of processes, which depends on the pollutant properties and climatic factors (i.e. temperature and season). For example, nutrients have demonstrated high dependence on the plant uptake and soil accumulation processes (Borin and Tocchetto 2007; Tanner et al. 2005; Wu et al. 2013a, b), and temperature (Beutel et al. 2009; Kadlec 2005; Tournebize et al. 2016). The presence of vegetation characteristics (i.e. plant morphology, species and biomass) is also a factor of the

Table 1 CWs studies on agricultural runoff mitigation

Type of CW	Size – scale, location	Dimensions	Pollutant removal%	References
VSF (EM)	Pilot-scale (Lab), China	0.5∗0.4 (D∗d)	TP: 3–21%	Wu et al. (2013a, b)
FWS (EM)	Full-scale, China	2800 m^2	TP: 59%	Lu et al. (2009)
FWS (EM)	Pilot-scale, Canada	9.29 m^2	TP: 41%	Yates and Prasher (2009)
FWS (EM)	Full-scale, Switzerland	2350 m^2, 0.6 (D)	TP: 23%	Reinhardt et al. (2005)
FWS (EM)[4 CWs]	Full-scale, Norway	350–900 m^2	TP: 21–44%	Braskerud (2002)
FWS (SUB)[9 CWs]	Mesocosm, USA	4.7∗0.8∗1 (L∗W∗D)	TP: 50–79%	Dierberg et al. (2002)
3 SF & 1 FWS (EM)	Full-scale, Norway	40∗3 (L∗W)	NO3: 0–90.5% (based on trench & season)	Søvik and Mørkved (2008)
FWS (EM)[10% coverage]	Full-scale	860 m^2, 50 (D)	NO3: 50 ± 18%	Tournebize et al. (2015)
FWS (EM)[6CWs]	Full-scale, USA	16,000 m^2	TN: > 60% NO3: 90%	Beutel et al. (2009)
FWS (EM)	Full-scale, France	12,000 m^2, 0.1–1 (D)	NO3: 90%	Mander et al. (2015)
FWS (EM)	Full-scale, USA	13,000 m^2	TN4+: 25% NO3: 52% TP: 27% TN: 14% TSS: 13%	Jordan et al. (2003)
FWS (EM)[7 CWs]	Full-scale, USA	23–150 ha, 0.5–1.5 m(D)	TSS: 31–96% NO3: 22–99%	Diaz et al. (2012)
FWS (EM)	Full-scale, South Africa	4400 m^2	TSS: 15–78% Orthophosphate: 54–75% NO3: 70–84% Toxicity: 89% OP pesticides: 100% (dry & wet weather conditions)	Schulz and Peall (2001)
FWS [in series: 1 unplanted& 1 planted systems]	Pilot-scale, Australia	Unplanted: 100 m^2; 1 m (D); Planted: 200 m^2; 0.5 m (D)	Herbicides: fluometuron: 0–34%; Diuron: 27–55%; aldicard: 15–39% Insecticides: endosulfan: 24% (unplanted), 27% (planted)	Rose et al. (2006)

(continued)

Table 1 (continued)

Type of CW	Size – scale, location	Dimensions	Pollutant removal%	References
FWS (EM)	Full-scale, Italy	3200 m^2	N: 90%	Borin and Tocchetto (2007).
FWS (EM) [Aulnoy: in stream, 10% coverage; Bray: 3*in-series CWs, off-stream, 70% coverage]	Full-scale, France	860m^2, 0.5 m(D); 1280m^2, 0.2–0.8 m (D)	Pesticides: Aulnoy: 54%; Bray: 45%	Tournebize et al. (2013)
FWS [4 CWs; 50% planted; 50% unplanted]	Full-scale, Korea	13,294 m^2	TSS: 38% TN: 37% TP: 60%	Lee et al. (2015)
FWS	Full-scale, Italy	3200 m^2	Herbicides: metolachlor, terbuthylazine: 98%	Pappalardo et al. (2016)
(SF) SFW & VSF (EM)	Pilot-scale, China	0.6*0.8*0.5 (L*W*D)	Insecticides: endosulfan, chlorpyrifos, fenvalerate>95%. Herbicides: diuron 45%	Tang et al. (2016)
FWS (EM)[2 CWs]	Full-scale, Norway	840 m^2, 100 m (L); 1200 m^2	7 pesticides: 3–67%	Blankenberg et al. (2006)

L Length, *W* Width, *D* Depth, *d* diameter, *EM* Emergent, *SUB* Submerged, *OP* Organophosphorus. All units refer to m

removal rate (Rose et al. 2006; Stefanakis and Tsihrintzis 2012; Stefanakis et al. 2014; Wu et al. 2013a, b). Wu et al. (2013a, b) suggest that selection of plants with grand biomass and of media with large adsorption ability for the particular pollutant to be treated is critical for the treatment.

Pesticides removal, on the other hand, depends on a combination of transport (solubility) and degradation processes, where, importantly, the transport and removal degree of a specific pesticide hinges mainly upon its physicochemical properties (Crossan 2002; Stangroom et al. 2000); as such, pesticides belonging to the same category (e.g. herbicides) are often found to be controlled by different processes. Sorption is the dominant reaction and transformation mechanism that affects pesticides' fate (Miller and Weber 1986). Environmental conditions such as organic carbon content, temperature and pH have considerable impact on sorption process (Nowell et al. 1999; Gao et al. 1998). Sorption is such a rigorous process that can also influence transport and degradation procedures (Gao et al. 1998). Overall, compounds with higher adsorption coefficient (K_{oc}) are considered as highly sorbing. In particular, based on the K_{oc}value, pesticides are classified as low sorbing when $K_{oc} < 400$ mL/g and highly sorbing when $K_{oc} > 1000$ mL/g (Tournebize et al. 2016). In addition to this, seasonality in terms of pesticide application period constitutes another factor of the CW removal efficiency, mainly because the pesticides' transfer and transformation processes are related to season. In other words, the first couple of heavy storm events occurring after pesticides application generally carry the most elevated pesticide concentrations, and hence, those events are related to highest pesticide transport risk (Gregoire et al. 2009; Tournebize et al. 2016). Important factors that affect degradation process include compound format, micro-climate (i.e. hydrology, precipitation and temperature) and biological activity (Blankenberg et al. 2006).

NO_3^- removal is principally achieved via denitrification, a process that requires anaerobic conditions to take place. As a biological process, denitrification is a function of temperature, DO levels, pH, and vegetation (Bachand and Horne 1999; Beutel et al. 2009; Stefanakis et al. 2014). Denitrification rates generally increase with temperature (Vymazal 2007). N removal is positively influenced in the presence of vegetation (Stefanakis and Tsihrintzis 2012; Stefanakis et al. 2014). In CWs, denitrification takes place in the vicinity of the substrate (Fennel et al. 2009; Reed and Brown 1995), which might explain why SF CWs display greater removal efficiency of NO_3^- (Reed and Brown 1995).

CWs efficiency is also dependent on hydrological and climatic factors (Persson and Wittgren 2003; Stefanakis et al. 2014), their actual shape (Persson 2000), the substrate material (Yates and Prasher 2009), and the agrochemical application and irrigation practices (Zhang et al. 2008; Bianchi and Harter 2002). The hydraulic residence time (HRT) dictates the removal efficiency and is principally linked to hydrological conditions, such as storm and drought weather conditions. Dierberg et al. (2002) remarked higher TP removal rates for longer HRTs. Lee et al. (2015) underlined that three climatic-related parameters – rainfall intensity and depth, and antecedent dry days – are pivotal in the removal mechanisms of NPS pollution. Johannesson et al. (2015) observed good dependence of P and TSS retention and CW aspect ratio, with higher aspect ratios to be recommended. Ioannidou and Pearson (2018) report that the width is a more important dimension for the overall efficiency compared to the depth. Therefore, it is apparent that the CW hydraulic design should not be overlooked. Concerning the substrate material, Yates and Prasher (2009) investigated how two types of substrates, namely sandy clay loam and sandy soil, affect P retention in pilot-scale units (Table 1). The authors obtained no notable difference in the holding P between the two soil types, but they drew the conclusion that sandy soil appears to be a more sustainable material in maintaining for longer its properties as a P sink.

3.4 *The Role of Plants*

Plants offer multiple and ancillary services in CWs. They promote pollutant mitigation and removal via uptake, sorption and degradation through the biofilm they support for microbial consortia activity, and via the oxygen supply through the root system (Stefanakis et al. 2014). In addition to this, vegetation offers bed surface stabilisation and channel bank erosion control, while it also reduces wind velocity and enhances sedimentation of suspended sediments (Nepf 2012; Vymazal 2013). There is still research required to investigate, elucidate and relate the removal efficiency of particular plant species on certain target pollutants removal, however. Gottschall et al. (2007) reported that wetlands planted with emergent macrophytes demonstrated higher N removal percentage than those planted with submersed macrophytes. On the contrary, Dierberg et al. (2002) examined and assessed mesocosm CW cells with submerged plant species, and found similar P removal efficiencies from the different tested plant species, but noticed that submerged aquatic plants aggregated twice higher P mass compared to soil accumulation process. However, to date, there are no definitive data regarding the removal capacity of different plant morphologies and species in relation to specific target NPS pollutants. Significant difference in pesticide concentration abatement between planted and unplanted ponds has been observed (Rose et al. 2006; Tang et al. 2016). It has been demonstrated that planted CWs show enhanced pesticide removal efficiencies (Schulz et al. 2003), and accelerated degradation rates (Sethunathan et al. 2004).

Wu et al. (2013a, b) scrutinised the seasonal P removal of four different plant species and of a non-vegetated cell in microcosm units. The presence of any plant species resulted overall in greater removal rates compared to the non-vegetated case, referring to the same seasons and testing conditions, which underpins the essential role of plants. Furthermore, it was observed that the dependence of plant contribution to the total removal was lower in spring compared to summer and autumn periods (Wu et al. 2013a, b). It is worth mentioning that all plant species demonstrated similar removal rates in spring, but *Trema orientalis* and *Phragmites australis* exhibited variability in summer and fall, while *Schoenoplectus validus* and *Iris pseudacorus* showed more stability and similarity in their removal trends in summer and autumn.

Borin and Tocchetto (2007) scrutinised the N retention of two plant species, namely *Phragmites australis* and *Typha latifolia*. The authors observed that *Phragmites* displayed higher partitioning degree than *Typha* on the N removal. Beutel et al. (2009) investigated temperature and DO variation in unplanned basins and in planted wetlands, and concluded that planted CWs showed abatement in temperature and DO levels, conditions which promote biological denitrification. Lu et al. (2009) noticed that the main P removal process was the plant harvest, attaining 58% reduction of the total P removal load. This result emphasises the importance of selecting plants with high biomass and P adsorption competence, and the fact that plant harvest is proved to prevent from release of the adsorbed P back in to the wetland.

In general, it is evidenced that the presence of vegetation is crucial in many aspects, as already registered by many researchers. Nevertheless, research ought to be oriented towards plant species and morphologies (e.g. emergent or submerged) with respect to specific agrochemical pollutants and testing conditions (e.g. field or lab, small- or full-scale units). This will allow optimisation of CW design for maximal treatment efficiencies.

4 Conclusion

The need for more sustainable strategies capable of removing nutrients and pesticides losses from croplands is indispensable due to the human and ecological impact, and the reason that constructed wetlands are at the forefront of environmental pollution research is their internationally demonstrated efficiency. To date, research results suggest that small CWs distributed over a wide farming area can offer fine treatment of agricultural NPS pollution. Although to date research has covered a wide range of pesticides, the unique nature/properties of each chemical compound rise the need for further research to better understand the related removal mechanisms, and possible advances to enhance CWs treatment efficiency for particular chemical compounds. The role of plants has emerged as pivotal in the treatment efficiency, as it enhances both nutrients plant uptake and offers ideal conditions for microbial biodegradation for agricultural runoff mitigation. However, plant species targeted for specific agrochemical pollutant removal is still an area with limited information, and further research is required to elucidate relationships between agrochemical compounds removal mechanisms and associated plant species. Additionally, there is promising and encouraging evidence related to the cost effectiveness and sustainability function of CWs, such as the application of Fe-Biochar as an enhancing factor of pesticide runoff abatement, deriving from the harvested plant waste of CWs. This chapter has identified that FWS is the most common and effective CW type implemented for agricultural runoff pollution, owing to the associated removal mechanisms. However, the increasing investigation and application of SF CWs could counterbalance the large space requirements of the FWS systems, and performance of SF CWs for NPS agricultural pollutant studies to date is encouraging. This indicates the high perspectives of further investigation of the treatment performance parameters of SF CW types.

References

Bachand PA, Horne AJ (1999) Denitrification in constructed free-water surface wetlands: II. Effects of vegetation and temperature. Ecol Eng 14(1):17–32

Beutel MW, Newton CD, Brouillard ES, Watts RJ (2009) Nitrate removal in surface-flow constructed wetlands treating dilute agricultural runoff in the lower Yakima Basin, Washington. Ecol Eng 35(10):1538–1546

Bianchi M, Harter T (2002) Nonpoint sources of pollution in irrigated agriculture: farm water quality planning series. Division of Agriculture and Natural Resources, University of California, USA

Billy C, Birgand F, Ansart P, Peschard J, Sebilo M, Tournebize J (2013) Factors controlling nitrate concentrations in surface waters of an artificially drained agricultural watershed. Landsc Ecol 28(4):665–684

Blankenberg AGB, Braskerud B, Haarstad K (2006) Pesticide retention in two small constructed wetlands: treating non-point source pollution from agriculture runoff. Int J Environ Anal Chem 86(3–4):225–231

Bodin H, Mietto A, Ehde PM, Persson J, Weisner SEB (2012) Tracer behaviour and analysis of hydraulics in experimental free water surface wetlands. Ecol Eng 49:201–211
Bollmann UE, Tang C, Eriksson E, Jönsson K, Vollertsen J, Bester K (2014) Biocides in urban wastewater treatment plant influent at dry and wet weather: concentrations, mass flows and possible sources. Water Res 60:64–74
Borgvang SA, Tjomsland T (2001) Nutrient supply to the Norwegian coastal areas (1999) calculated by the model TEOTIL. NIVA-report, pp 4343–2001
Borin M, Tocchetto D (2007) Five-year water and nitrogen balance for a constructed surface flow wetland treating agricultural drainage waters. Sci Total Environ 380(1):38–47
Braskerud BC (2002) Design considerations for increased sedimentation in small wetlands treating agricultural runoff. Water Science and Technology 45(9):77–85
Brock TCM, Van Wijngaarden RPA, Van Geest GJ (2000) Ecological risks of pesticides in freshwater ecosystems. Part II: insecticides. Alterra report-089, Wageningen, The Netherlands
Coffey S (1997) Selected agricultural best management practices to control nitrogen in the Neuse River basin. North Carolina State University, Raleigh, NC, USA. Technical Bulletin, pp 1–50
Crossan AN (2002) Remediation of pesticides on cotton farms: studies of the environmental distribution, transport and fate of five pesticides. PhD thesis. Department of Agricultural Chemistry and Soil Science, Faculty of Science, University of Sydney
Dabrowski JM, Schulz R (2003) Predicted and measured levels of azinphosmethyl in the Lourens River, South Africa: comparison of runoff and spray drift. Environ Toxicol Chem 22(3):494–500
Diaz FJ, O'Geen AT, Dahlgren RA (2012) Agricultural pollutant removal by constructed wetlands: implications for water management and design. Agric Water Manag 104:171–183
Dierberg FE, DeBusk TA, Jackson SD, Chimney MJ, Pietro K (2002) Submerged aquatic vegetation-based treatment wetlands for removing phosphorus from agricultural runoff: response to hydraulic and nutrient loading. Water Res 36(6):1409–1422
Dunne EJ, Coveney MF, Hoge VR, Conrow R, Naleway R, Lowe EF, Wang Y (2015) Phosphorus removal performance of a large-scale constructed treatment wetland receiving eutrophic lake water. Ecol Eng 79:132–142
EPA (2016) US Environmental Protection Agency. Polluted runoff: nonpoint source pollution. https://www.epa.gov/nps. Date accessed Nov 2016
European Commission (2016) Environment. http://ec.europa.eu/environment/index_en.htm. Date accessed 1 Nov 2016
Feng YZ, Xie XJ, Qin XW, Yang GH, Cao YC, Yang SQ (2011) Features and treatment of nonpoint source pollution in the Ningxia Yellow River area. Afr J Agric Res 6(24):5541–5550
Fennel K, Brady D, DiToro D, Fulweiler RW, Gardner WS, Giblin A, Tobias C (2009) Modeling denitrification in aquatic sediments. Biogeochemistry 93(1–2):159–178
Fulton MH, Moore DW, Wirth EF, Chandler GT, Key PB, Daugomah JW, Finley DB (1999) Assessment of risk reduction strategies for the management of agricultural nonpoint source pesticide runoff in estuarine ecosystems. Toxicol Ind Health 15(1–2):201–214
Gao JP, Maguhn J, Spitzauer P, Kettrup A (1998) Sorption of pesticides in the sediment of the Teufelweiher pond (Southern Germany). I: equilibrium assessments, effect of organic carbon content and pH. Water Resour 32:1662–1672
Gottschall N, Boutin C, Crolla A, Kinsley C, Champagne P (2007) The role of plants in the removal of nutrients at a constructed wetland treating agricultural (dairy) wastewater, Ontario, Canada. Ecol Eng 29:154–163
Gregoire C, Elsaesser D, Huguenot D, Lange J, Lebeau T, Merli A, Mose R, Passeport E, Payraudeau S, Schütz T, Schulz R, Tapia-Padilla G, Tournebize J, Trevisan M, Wanko A (2009) Mitigation of agricultural nonpoint-source pesticide pollution in artificial wetland ecosystems. Environmental Chemistry Letters 7(3):205–231
Haygarth PM, Jarvis SC (2002) Agriculture, hydrology, and water quality. CABI Publishing, Wallingford
Horne AJ (2002) Potential value of constructed wetlands for nitrate removal along some large and small rivers. Verh Int Ver Theor Angew Limnol 27(7):4057–4062

Humenik FJ, Smolen MD, Dressing SA (1987) ES&T feature: pollution from nonpoint sources. Environ Sci Technol 21(8):737–742

Ioannidou VG, Pearson JM (2018) Hydraulic and design parameters in full-scale constructed wetlands and treatment units: six case studies. Environ Process 5(1):5–22

Johannesson KM, Kynkäänniemi P, Ulén B, Weisner SEB, Tonderski KS (2015) Phosphorus and particle retention in constructed wetlands – a catchment comparison. Ecol Eng 80:20–31

Jordan TE, Whigham DF, Hofmockel KH, Pittek MA (2003) Nutrient and sediment removal by a restored wetland receiving agricultural runoff. J Environ Qual 32(4):1534–1547

Kadlec RH (2005) Nitrogen farming for pollution control. J Environ Sci Health A 40(6–7):1307–1330

Kadlec RH (2009) Comparison of free water and horizontal subsurface treatment wetlands. Ecol Eng 35:159–174

Kimbrough RA, Litke DW (1996) Pesticides in streams draining agricultural and urban areas in Colorado. Environ Sci Technol 30(3):908–916

Kladivko EJ, Brown LC, Baker JL (2001) Pesticide transport to subsurface tile drains in humid regions of North America. Crit Rev Environ Sci Technol 31(1):1–62

Knobeloch L, Salna B, Hogan A, Postle J, Anderson H (2000) Blue babies and nitrate-contaminated well water. Environ Health Perspect 108(7):675

Koskiaho J, Ekholm P, Räty M, Riihimäki J, Puustinen M (2003) Retaining agricultural nutrients in constructed wetlands – experiences under boreal conditions. Ecol Eng 20(1):89–103

Kotti IP, Gikas GD, Tsihrintzis VA (2010) Effect of operational and design parameters on removal efficiency of pilot-scale FWS constructed wetlands and comparison with HSF systems. Ecol Eng 36:862–875

Kreuger J (1998) Pesticides in stream water within an agricultural catchment in southern Sweden, 1990–1996. Sci Total Environ 216(3):227–251

Lee SH, Cha SM, Lee JC, Lee JY (2015) Performance evaluation of the free water surface constructed wetland treating nonpoint source pollutants in the agricultural area. J Environ Anal Toxicol 5(4):1

Line DE, McLaughlin RA, Osmond DL, Jennings GD, Harman WA, Lombardo LA, Spooner J (1998) Nonpoint sources. Water Environ Res 70(4):895–912

Loague K, Corwin DL, ELLSWORTH ATR (1998) Feature: the challenge of predicting nonpoint source pollution. Environ Sci Technol 32(5):130A–133A

Locke MA, Weaver MA, Zablotowicz RM, Steinriede RW, Bryson CT, Cullum RF (2011) Constructed wetlands as a component of the agricultural landscape: mitigation of herbicides in simulated runoff from upland drainage areas. Chemosphere 83:1532–1538

Lu SY, Wu FC, Lu YF, Xiang CS, Zhang PY, Jin CX (2009) Phosphorus removal from agricultural runoff by constructed wetland. Ecol Eng 35(3):402–409

Lv T, Zhang Y, Zhang L, Carvalho PN, Arias CA, Brix H (2016) Removal of the pesticides imazalil and tebuconazole in saturated constructed wetland mesocosms. Water Res 91:126–136

Mander Ü, Tournebize J, Soosaar K, Chaumont C, Hansen R, Muhel M, Teemusk A, Vincent B (2015) Nitrous oxide and methane emission in an artificial wetland treating polluted runoff from an agricultural catchment. In EGU general assembly conference abstracts 17, 13911

McElroy AD, Chiu SY, Nebgen JW, Aleti A, Vandegrift AE (1975) Water pollution from nonpoint sources. Water Res 9(7):675–681

Miller CT, Weber WJ (1986) Sorption of hydrophobic organic pollutants in saturated soil systems. J Contam Hydrol 1:243–261

Mitsch WJ, Gosselink JG (2007) Wetlands, 4th edn. Wiley, New York, p 582

Mitsch WJ, Horne AJ, Nairn RW (2000) Nitrogen and phosphorus retention in wetlands-ecological approaches to solving excess nutrient problems. Ecol Eng 14(1–2):1–7

Nepf HM (2012) Flow and transport in regions with aquatic vegetation. Annu Rev Fluid Mech 44:123–142

Nowell LH, Capel PD, Dileanis PD (1999) Pesticides in stream sediment and aquatic biota. CRC Press LLC, Boca Raton

Pappalardo SE, Otto S, Gasparini V, Zanin G, Borin M (2016) Mitigation of herbicide runoff as an ecosystem service from a constructed surface flow wetland. Hydrobiologia 774(1):193–202
Persson J (2000) The hydraulic performance of ponds of various layouts. Urban Water 2:243–250
Persson J, Wittgren HB (2003) How hydrological and hydraulic conditions affect performance of ponds. Ecol Eng 21(4):259–269
Poe AC, Piehler MF, Thompson SP, Paerl HW (2003) Denitrification in a constructed wetland receiving agricultural runoff. Wetlands 23(4):817–826
Reed SC, Brown D (1995) Subsurface flow wetlands – a performance evaluation. Water Environ Res 67(2):244–248
Reichenberger S, Bach M, Skitschak A, Frede HG (2007) Mitigation strategies to reduce pesticide inputs into ground-and surface water and their effectiveness; a review. Sci Total Environ 384(1):1–35
Reinhardt M, Gächter R, Wehrli B, Müller B (2005) Phosphorus retention in small constructed wetlands treating agricultural drainage water. J Environ Qual 34(4):1251–1259
Rodgers JH, Dunn A (1992) Developing design guidelines for constructed wetlands to remove pesticides from agricultural runoff. Ecol Eng 1:83–95
Rose MT, Sanchez-Bayo F, Crossan AN, Kennedy IR (2006) Pesticide removal from cotton farm tailwater by a pilot-scale ponded wetland. Chemosphere 63(11):1849–1858
Runes HB, Jenkins JJ, Moore JA, Bottomley PJ, Wilson BD (2003) Treatment of atrazine in nursery irrigation runoff by a constructed wetland. Water Res 37(3):539–550
Schulz R (2004) Field studies on exposure, effects, and risk mitigation of aquatic nonpoint-source insecticide pollution. J Environ Qual 33(2):419–448
Schulz R, Moore MT, Bennett ER, Farris JL, Smith S, Cooper CM (2003) Methyl parathion toxicity in vegetated and nonvegetated wetland mesocosms. Environ Toxicol Chem 22(6):1262–1268
Sethunathan N, Megharaj M, Chen ZL, Williams BD, Lewis G, Naidu R (2004) Algal degradation of a known endocrine disrupting insecticide, α-endosulfan, and its metabolite, endosulfan sulfate, in liquid medium and soil. J Agric Food Chem 52(10):3030–3035
Shilton AN (2005) Pond treatment technology. IWA Publishing, London
Shutes B, Revitt M, Scholes L (2010) Constructed wetlands for flood prevention and water reuse. In: Proceedings of the 12th international conference on wetland systems for water pollution control, 4–8 October 2010, Venice, Italy
Søvik AK, Mørkved PT (2008) Use of stable nitrogen isotope fractionation to estimate denitrification in small constructed wetlands treating agricultural runoff. Sci Total Environ 392(1):157–165
Stangroom SJ, Collins CD, Lester JN (2000) Abiotic behaviour of organic micropollutants in soils and the aquatic environment. A review: II. Transformations. Environ Technol 21(8):865–882
Stanton EA, Taylor M (2012) Valuing Florida's clean waters. Stockholm Environment Institute–US Center
Stefanakis AI (2018) Constructed wetlands for industrial wastewater treatment, 1st edn. Wiley, Chichester
Stefanakis AI, Becker JA (2016) A review of emerging contaminants in water: classification, sources and potential risks. In: McKeown AE, Bugyi G (eds) Impact of water pollution on human health and environmental sustainability. Information Science Reference (an imprint of IGI Global), Hershey, pp 55–80
Stefanakis AI, Tsihrintzis VA (2012) Effects of loading, resting period, temperature, porous media, vegetation and aeration on performance of pilot-scale vertical flow constructed wetlands. Chem Eng 181-182:416–430
Stefanakis AI, Akratos CS, Tsihrintzis VA (2014) Vertical flow constructed wetlands: eco-engineering systems for wastewater and sludge treatment, 1st edn. Elsevier, Amsterdam
Stefanakis AI, Seeger E, Dorer C, Sinke A, Thullner M (2016) Performance of pilot-scale horizontal subsurface flow constructed wetlands treating groundwater contaminated with phenols and petroleum derivatives. Ecol Eng 95:514–526

Stefanakis AI, Zouzias D, Marsellos A (2017) Groundwater pollution: human and natural sources and risks. In: Singhal JC, Sharma UC, Bhola RG, Govil JN (eds) Environmental science and engineering, Water pollution, vol 4. Studium Press LLC, Lanham, pp 82–102

Tang X, Yang Y, Tao R, Chen P, Dai Y, Jin C, Feng X (2016) Fate of mixed pesticides in an integrated recirculating constructed wetland (IRCW). Sci Total Environ 571:935–942

Tanner CC, Nguyen ML, Sukias JPS (2005) Nutrient removal by a constructed wetland treating subsurface drainage from grazed dairy pasture. Agric Ecosyst Environ 105(1):145–162

Tao B, Fletcher AJ (2013) Metaldehyde removal from aqueous solution by adsorption and ion exchange mechanisms onto activated carbon and polymeric sorbents. J Hazard Mater 244–245:240–250

Tediosi A, Whelan MJ, Rushton KR, Thompson TRE, Gandolfi C, Pullan SP (2012) Measurement and conceptual modelling of herbicide transport to field drains in a heavy clay soil with implications for catchment-scale water quality management. Sci Total Environ 438:103–112

Thompson HM (1996) Interactions between pesticides; a review of reported effects and their implications for wildlife risk assessment. Ecotoxicology 5(2):59–81

Tournebize J, Passeport E, Chaumont C, Fesneau C, Guenne A, Vincent B (2013) Pesticide decontamination of surface waters as a wetland ecosystem service in agricultural landscapes. Ecol Eng 56:51–59

Tournebize J, Chaumont C, Fesneau C, Guenne A, Vincent B, Garnier J, Mander Ü (2015) Long-term nitrate removal in a buffering pond-reservoir system receiving water from an agricultural drained catchment. Ecol Eng 80:32–45

Tournebize J, Chaumont C, Mander Ü (2016) Implications for constructed wetlands to mitigate nitrate and pesticide pollution in agricultural drained watersheds. Ecol Eng. https://doi.org/10.1016/j.ecoleng.2016.02.014

Verhoeven JTA, Setter TL (2010) Agricultural use of wetlands: opportunities and limitations. Ann Bot 105:155–163

Volkmar EC, Dahlgren RA (2006) Biological oxygen demand dynamics in the lower San Joaquin River, California. Environ Sci Technol 40(18):5653 5660

Vymazal J (2007) Removal of nutrients in various types of constructed wetlands. Sci Total Environ 380(1):48–65

Vymazal J (2013) Emergent plants used in free water surface constructed wetlands: a review. Ecol Eng 61:582–592

Vymazal J, Březinová T (2015) The use of constructed wetlands for removal of pesticides from agricultural runoff and drainage: a review. Environ Int 75:11–20

Wu M, Tang X, Li Q, Yang W, Jin F, Tang M, Scholz M (2013a) Review of ecological engineering solutions for rural non-point source water pollution control in Hubei Province, China. Water Air Soil Pollut 224(5):1–18

Wu H, Zhang J, Li C, Fan J, Zou Y (2013b) Mass balance study on phosphorus removal in constructed wetland microcosms treating polluted river water. Clean: Soil, Air, Water 41(9):844–850

Yanhua Z, Thuminh N, Beibei N, Song H (2012) Research trends in non-point source during 1975-2010. Phys Procedia 33:138–143

Yates CR, Prasher SO (2009) Phosphorus reduction from agricultural runoff in a pilot-scale surface-flow constructed wetland. Ecol Eng 35(12):1693–1701

Zhang X, Zhang M (2011) Modeling effectiveness of agricultural BMPs to reduce sediment load and organophosphate pesticides in surface runoff. Sci Total Environ 409(10):1949–1958

Zhang L, Scholz M, Mustafa A, Harrington R (2008) Assessment of the nutrient removal performance in integrated constructed wetlands with the self-organizing map. Water Res 42(13):3519–3527

Zhang J, Shen T, Liu M, Wan Y, Liu J, Li J (2011) Research on non-point source pollution spatial distribution of Qingdao based on L-THIA model. Math Comput Model 54(3):1151–1159

Role of Nitrogen and Its Agricultural Management in Changing Environment

Ram Kishor Fagodiya, Amit Kumar, Shilpi Kumari, Kristina Medhi, and Aftab A. Shabnam

Contents

R. K. Fagodiya
Division of Irrigation and Drainage Engineering, ICAR-Central Soil Salinity Research Institute, Karnal, Haryana, India

A. Kumar (✉) · A. A. Shabnam
Host Plant Division, Central Muga Eri Research & Training Institute, Jorhat, Assam, India

S. Kumari
Centre for Energy Studies, Indian Institute of Technology, Delhi, Delhi, India

K. Medhi
School of Environmental Sciences, Jawaharlal Nehru University, New Delhi, India

M. Naeem et al. (eds.), *Contaminants in Agriculture*,
https://doi.org/10.1007/978-3-030-41552-5_12

1 Nitrogen

Nitrogen (N) was first discovered by a Scottish physician Daniel Rutherford in 1772 and named as noxious gas Daniel 1772; Weeks 1932). The word nitrogen was coined by Jean-Antoine Chaptal in 1790 (Chaptal and Nicholson 1800; Smile 2001). The word nitrogen originated from the French word *nitrogene* (*Nitre* means '*saltpetre*' and *gene* means 'producing'), that is, the *saltpetre*-producing element. Nitrogen is the lightest element in the group 15 of the periodic table. This group is often called the pnictogens and nitrogen family. Two atoms of elemental N bind together and form dinitrogen at a standard temperature and pressure. It is a colourless and odourless diatomic gas having the molecular formula N_2. N_2 is the most abundant element in the universe and it forms about 78% of the Earth's atmosphere. Nitrogen occurs in all living organisms, and it is the fourth most abundant element in human body after oxygen, carbon, and hydrogen. It has second strongest triple bond ($N \equiv N$) after carbon monoxide (CO), which makes it very difficult to convert into useful compounds. The human body contains about 3% of nitrogen by its bodyweight. The plants, animals, and other organism cannot use N_2 directly from the atmosphere, as it is an inactive gas. N_2 is the most important constituent of the amino acids, proteins, nucleic acids (DNA and RNA), and adenosine triphosphate (ATP), the energy transfer molecule. Many industrially important compounds, namely, cyanides, ammonia, nitric acid, and organic nitrates, contain nitrogen. The synthetically produced ammonia and nitrates are the key components of several nitrogenous chemical fertilizers. It is also a constituent of the pharmacological drugs including antibiotics.

2 Background

N is one of the major controlling nutrients for ecosystem's structure and function (i.e. species diversity, species composition, dynamics, etc.) including agriculture. N is the key element for the life sustainability at an optimum concentration. Increasing anthropogenic activities such as agriculture, combustion of fossil fuels, and excessive use of fertilizer has substantially altered the nitrogen cycle (Adler et al. 2015). The reactive N species are more responsible for this alteration.

Nitrous oxide emissions (N_2O) from agriculture soil are the major contributors towards the global climate change due to the inefficient use of N fertilizers (Galloway et al. 2003). Application of N fertilizers (38%), manure management (38%), animal production systems (30%) majorly contributes to the N_2O emission (Crosson et al. 2011; Adler et al. 2015). N transformation in soil is mainly controlled by mineralization/ammonification and immobilization/assimilation. The balance between these two depends basically on the soil intrinsic C/N ratio and residue incorporation/addition into the soil (Kumar et al. 2016; Bhattacharyya et al. 2018). Increased soil organic carbon (SOC) content may increase N_2O emission (Brentrup et al.

2000), while soil moisture content and soluble C content provide significant correlation with N_2O in case of residue incorporation (Ciampitti et al. 2008) due to stimulated microbial activity (Cameron et al. 2013). Thus, the important change in the N cycle and production of Nr species, especially nitrous oxide, is the urgent need to control for minimizing the overall rate of global warming and changing environmental condition.

3 Nitrogen Cycle

Nitrogen (N) cycle starts with transformation of molecular nitrogen (N_2) to other forms of nitrogen through biological and industrial fixation and reverts back to its molecular state (N_2) through the denitrification process. The whole N cycle is carried out in five important steps: (i) nitrogen fixation, (ii) mineralization, (iii) assimilation, (iv) denitrification, and (v) immobilization (Fig. 1). It is a microbial mediated biogeochemical cycle in which several microbial species play an important role. N cycle is a complex and highly dynamic process and is affected by the different soil, plant, and climatic factors and their interactions (Ladha et al. 2003). Therefore, the proper information and understanding of the natural movement of N into different pools of soil, plant, and atmospheric systems are critical for the assessment of the N application, N-use efficiency and losses from agriculture.

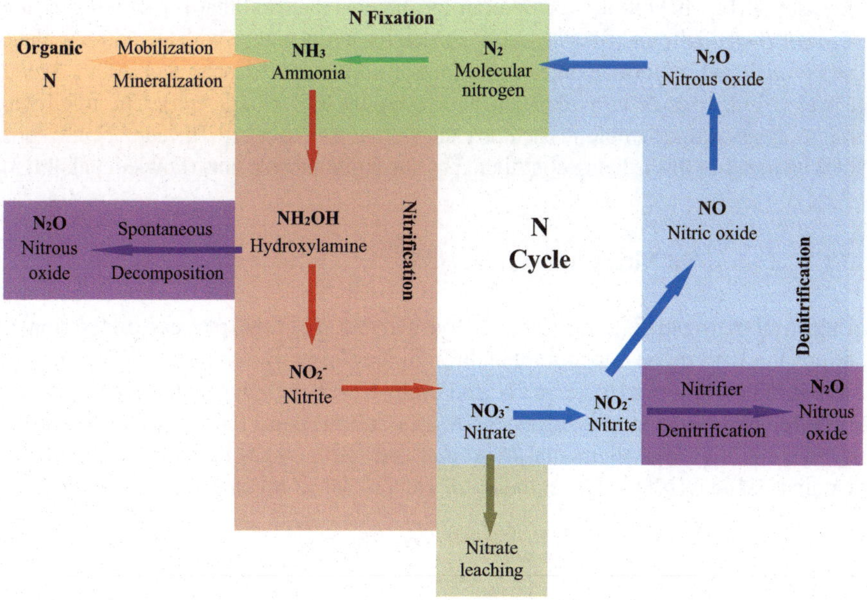

Fig. 1 The process of N cycle is shown in different steps

3.1 Nitrogen Fixation

The process of conversion of diatomic molecular nitrogen (N_2) into biologically available nitrogen form (NH_3) is called nitrogen fixation. Since N_2 has highly stable bond, it requires huge amount of energy to break it. Therefore, nitrogen fixation is an energy-intensive process that requires at least 16 ATP molecules to split the triple bond. Typically, the N fixation is carried out by microorganism naturally, which is known as biological nitrogen fixation (BNF). However, some N can be fixed by lightning or industrial processes known as industrial nitrogen fixation (INF).

3.1.1 Biological Nitrogen Fixation (BNF)

It is the process of transformation of atmospheric molecular nitrogen (N_2) into ammonia (NH_3) and others usable forms, which are easily available to plants and other living organisms into the soil (Postgate 1998). In other words, BNF is the process of reduction of nitrogen into ammonia by different microbial species. The overall reaction of BNF is given below:

$$N_2 + 3H_2 \rightarrow 2NH_3$$

$$N_2 + 8H^+ + 8e^- \rightarrow 2NH_3 + 2H_2$$

The BNF is predominated by cyanobacteria in aquatic and marine systems (Gaby and Buckley 2011). In soil, it is carried out by rhizobia bacteria, that is, *Azotobacter* and *Azospirillum* (Gaby and Buckley 2011), which form root nodule with leguminous plants and *Frankia,* which form nodules on some shrubs and trees (Franche et al. 2009). *Azolla,* a water fern, fixes N in association with cyanobacteria (*Anabaena*). The free-living microorganism and symbiotic bacteria can together fix about 118 Tg year^{-1} N into soils, of which half of this is being contributed by the leguminous plants (Fowler et al. 2013).

3.1.2 Industrial Nitrogen Fixation (INF)

Industrial nitrogen fixation (Haber–Bosch fixation) is the process of fixation of atmospheric nitrogen into ammonia through the Haber–Bosch process. This process was invented by Fritz Haber and scaled up by Carl Bosch during the early 1900s. It requires high temperature (around 500 °C) and pressure (about 150–200 atm) to synthesize NH_3 from elemental hydrogen and nitrogen gas in the presence of iron catalyst (Modak 2008). The equation of Haber–Bosch process is given below:

$$N_2 + H_2 \underset{\sim 500^\circ C\ \&\ \sim 200\,atm}{\overset{Fe}{\rightarrow}} 3NH_2$$

The resulting ammonia is used for the synthesis of nitrogenous fertilizers to fertigate the cropland. The Haber–Bosch reaction is exothermic and reversible in nature. Gu et al. (2013) estimated that the global N emission flux from the industrial N fixation has increased from 2.5 Tg N $year^{-1}$ during 1960 to 25.4 Tg N $year^{-1}$ during 2008.

3.2 *Nitrogen Mineralization*

The organic N accounts about 95% of total N in the agricultural soils. However, the mineral N accounts merely 3–5% of the total N. Plants uptake mineral nitrogen mostly in nitrate (NO_3^-) form, and some plants also prefer to uptake the ammonium (NH_4^+) form. Therefore, the organic N has to be converted into inorganic or mineral form for efficient plant uptake and utilization. Mineralization is the microbial mediated decomposition of organic N (amino acids, proteins, etc.) into mineral N or inorganic N, that is, NH_4^+, NO_2^-, and NO_3^- (Pathak et al. 2003).

N mineralization is essentially a two-step process: (i) ammonification and (ii) nitrification. Ammonification is enzymatically mediated through microbial hydrolysis of organic N compounds into inorganic N, that is, NH_4^+ (Regmi and Ladha 2006). Furthermore, nitrification is the two-step biological oxidative process in which the above-transformed NH_4^+ is first oxidized to NO_2^- and then into NO_3^-, with the help of soil microbes (Dobermann and Cassman 2004). The process of N mineralization is mentioned in Fig. 2.

The conversion of ammonia to nitrite is performed by autotrophic ammonia-oxidizing bacteria (AOB), that is, *Nitrosomonas* and *Nitrosococcus* (Prosser and Nicol 2012). However, the conversion of nitrite to nitrate is carried out by the nitrite oxidizing bacteria (NOB), that is, *Nitrobacter* and *Nitrospira* (Spieck et al. 2006). The AOB also carry the nitrite reductase genes and are able to produce nitrous oxide (N_2O) in anaerobic conditions (Jung et al. 2014; Poth and Focht 1985; Shaw et al. 2006). In most of the environments, both types of organisms are present that help to complete both steps of nitrification and yield nitrate as final product. Nitrification is a highly important process in agricultural systems, where generally fertilizer nitrogen is applied as ammonia and conversion of this ammonia to nitrate occurs through nitrification process only.

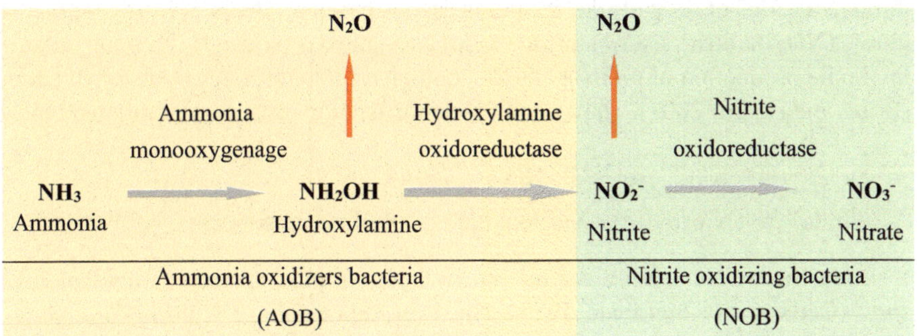

Fig. 2 Process of N mineralization

3.3 Nitrogen Assimilation

Plant assimilates mineralized (NH_4^+ and NO_3^-) N through their roots. This process is termed as nitrogen uptake. Mostly, plants uptake NO_3^-, however, rice plant can also uptake NH_4^+ from soils. Plants assimilate these nitrate or ammonia into amino acids for their needs. Usually, NO_3^- is the predominant form of plant-available nitrogen in aerobic soils (Xu et al. 2012; Nadelhoffer et al. 1984). However, in anaerobic and flooded rice soils, the NH_4^+ is the predominant form of the available nitrogen for the plants (Ishii et al. 2011). The nitrogen uptake in plant is processed through the roots and then it is further transported to the shoot via xylem (Scheurwater et al. 2002), where nitrogen assimilation is mostly carried out. Nitrogen assimilation is the process of reduction of nitrate into ammonia within higher plants and their incorporation for synthesis of protein. It is basically a two-step process as mentioned below.

$$\underset{\text{(Nitrate)}}{NO_3^-} \xrightarrow[\text{Nitrate}]{\text{Reductase}} \underset{\text{(Nitrite)}}{NO_2^-} \xrightarrow[\text{Nitrite}]{\text{Reductase}} \underset{\text{(Ammonia)}}{NH_3}$$

In the first step, the absorbed nitrate is reduced to nitrite with the help of enzyme nitrate reductase (Cassman et al. 2002). It is an energy-consuming process, and the energy is supplied through the oxidative respiration of sugars in plants. In the second phase, the nitrite is converted to ammonia with the help of nitrite reductase enzyme. Both the ammonia formed in the second phase and the ammonia absorbed by plant are incorporated into amino acids and further into proteins (Berntsen et al. 2003, Masclaux-Daubresse et al. 2006).

3.4 Denitrification

Denitrification is the reverse process of nitrification. It is a microbially mediated reduction process in which nitrate is converted to N_2 under anaerobic conditions, through a series of several intermediate gaseous oxide products of nitrogen, that is, nitrate (NO_3^-), nitrite (NO_2^-), nitric oxide (NO), nitrous oxide (N_2O), finally resulting in the production of dinitrogen (N_2), completing the nitrogen cycle nitrification (Bolan and Hedley 2003). The reaction of denitrification process is mentioned below.

$$\underset{\text{Nitrate}}{NO_3^-} \xrightarrow[\text{Nitrate}]{\text{reductase}} \underset{\text{Nitrite}}{NO_2^-} \xrightarrow[\text{Nitrite}]{\text{reductase}} \underset{\text{Nitric oxide}}{NO} \xrightarrow[\text{Nitric oxide}]{\text{reductase}} \underset{\text{Nitrous oxide}}{N_2O} \xrightarrow[\text{Nitrous oxide}]{\text{reductase}} \underset{\text{Nitrogen}}{N_2}$$

Denitrification is mainly carried out by heterotrophic bacteria (*Pseudomonads spp.*) (Carlson and Ingraham 1983). However, autotrophic bacteria (*Thiobacillus denitrificans*) also carry out this process (Baalsrud and Baalsrud 1954). Denitrification process typically occurs in anaerobic flooded soil conditions, where dissolved oxygen (O_2) concentration is depleted (Seitzinger et al. 2006) and it leads to conversion of nitrate to nitrogen (N_2) and nitrous oxide (N_2O), which is untimely lost to the atmosphere (Fageria 2002). The heavy textured soils having poor natural drainage are more susceptible to denitrification process (Mosier 2001). Denitrification returns most of the N applied to, or fixed in soil, to the atmosphere and is a major cause of fertilizer N losses from soil and greenhouse gas (GHG) emissions. The denitrification process is affected by several factors such as oxygen level, nitrate concentration, manures, and organic fertilizers, which tend to increase denitrification activity relative to mineral fertilizers (Hallin et al. 2009; Philippot et al. 2007).

3.5 Nitrogen Immobilization

The N immobilization is the reverse process of mineralization in which the mineral N compounds is converted into organic N compounds by microorganisms, which prevent the mineral N from being accessible to plants and reduce the availability of N for plant uptake (White 2005). The process of mineralization and immobilization depends on the C/N ratio of the organic N compounds (McLaren and Cameron 1996).

4 Loss of Nitrogen from Agricultural Soils

Nitrogen is the most widely used fertilizer nutrient. In agriculture, N can be applied through nitrogenous fertilizers, animal manure, green manuring, composting, atmospheric deposition, etc. Since green revolution, the use of fertilizer N has increased by many folds, and its use will be increased further to increase food production to feed the ever-growing human population. The improper use of fertilizer N alters the global nitrogen cycle, which results in decline in the total factor productivity, and reduces nitrogen-use efficiency (NUE). N is highly mobile nutrient and once fertilizer N is applied to soils, plant can uptake 30–35% of it and the remaining 65–70% of applied N can be lost from the soils through nitrate leaching, denitrification, ammonia volatilization and surface runoff mechanism (Pathak et al. 2016), causing climate change (N_2O), groundwater pollution (NO_3^- leaching), eutrophication (surface runoff), air pollution (NOx), and aerosols formation (NH_3) (Galloway et al. 2003; Kumar et al. 2020). Different mechanisms of loss of N from the agricultural soils are discussed below.

4.1 Leaching

Leaching of nitrogen is the process of loss of N with the deep percolating water. Nitrate leaching is one of the important pathways of N loss. The nitrate form of N is highly mobile and it cannot be strongly adsorbed on the soil surface, so it can be easily moved, beyond the soil profile with percolating water (Randall et al. 2003).

The process of N leaching is governed by many factors, namely soil texture, permeability, hydraulic conductivity, soil water content, rainfall amount of applied N, type of N fertilizers, and soil management (Aulakh and Bijay-Singh 1996; Brady and Weil 2002). Nitrogen leaching has special significance in light-textured soils coupled with high rainfall areas, where light-textured soils have sufficient water to cause movement of nitrate through the soil profile. In such cases, the loss of N could be as much as 25–50% of the applied N (Baligar et al. 2001 Bolan and Hedley 2003). The amount of nitrogen leached from rice fields may be around 15% of total applied N (Zhou et al. 2011). The N loss from light-textured soil of flooded rice field is about 30–50% of total N loss (Ghos and Bhat 1998). The IPCC default coefficient of nitrogen loss through nitrate leaching is about 30% of applied N (IPCC 2006). However, based on studies conducted in India, it has been estimated to be about 10% of applied N (Bhatia et al. 2013). Nitarte leaching causes deep percolation into groundwater aquifers and causes groundwater pollution (Galloway et al. 2003).

4.2 Denitrification

Denitrification is the mechanism of nitrate reduction into the gaseous form of nitrogen, that is, NO, N_2O, and N_2, under anaerobic conditions in the presence of denitrifying microorganisms (Bolan and Hedley 2003). This denitrification mechanism most commonly occurs in the water-logged soils, where oxygen deficiency leads to conversion of nitrate into nitrous oxide (N_2O), other oxide of nitrogen (NOx) and finally, molecular nitrogen (N_2) (Fageria 2002), which are ultimately lost to atmosphere, and cause climate change (N_2O) and air pollution (NOx) (Galloway et al. 2003). The mechanism of denitrification is affected by the soil water content, NH_4^+-N and NO_3^- -N concentration, soil temperature, and soil carbon content (Bolan and Hedley 2003). This mechanism is more common in heavy textured soil with a higher clay content and poor natural drainage (Mosier 2001). The denitrification mechanism of N loss is responsible for N losses up to 10–15% of applied nitrogen. Aulakh and Bijay-Singh (1996) estimated that in light-textured irrigated soils, 50% of the applied N may be lost through denitrification process. The IPCC default coefficient for N_2O-N emission from Indian soils is 1% of applied N (IPCC 2006). However, based on Indian studies, it has been estimated to be about 0.6% for N_2O-N emission (Fagodiya et al. 2019; Malyan et al. 2019; Jain et al. 2016; Bhatia et al. 2013b) and 0.5% for NOx emissions (Sharma et al. 2008).

4.3 Ammonia Volatilization

The ammonia volatilization is the process of conversion of ammonium ion nitrogen (NH_4^+-N) into gaseous ammonia (NH_{3g}) and further its emission into the atmosphere. Ammonia volatilization occurs from every source of nitrogen. However, this mechanism is found to be more severe when NH_4^+ containing synthetic nitrogen fertilizers and organic manures are generally applied through broadcasting on soil surface (Bolan and Hedley 2003). Application of urea and urea-based N fertilizers through broadcasting do not incorporated immediatedly into the soil, and prone to volatilization losses; 20% of applied N may be lost within a week into atmosphere through volatilization process, and this mechanism is even more severe under warm sunny weather and in alkaline soil reaction (Hutchinson et al. 2003). The NH_3 volatilization losses from the rice field ranged from 9.4% to 16.7% of applied N during the rice-growing periods (Lin et al. 2007). During 1990, the global ammonia N (NH_3-N) volatilization losses were about 54 Tg N $year^{-1}$, out of which 9 Tg N $year^{-1}$ NH_3-N volatilization were from the application of synthetic nitrogenous fertilizers only (Oliver et al. 1998). The IPCC default factor of nitrogen loss through ammonia volatilization for India is about 20% of applied N (IPCC 2006). However, based on Indian studies, it is about 15% of applied N (Aggarwal et al. 1987; Sarkar et al. 1991; Parashar et al. 1998). This loss of N through the volatilization mechanism caused air pollution through the aerosols formation, and acid rain (Galloway et al. 2003).

4.4 Soil Erosion and Surface Runoff

Soil erosion is a naturally occurring process, which refers to the loss of topsoil by the physical forces of water and wind. The soil erosion leads to loss of applied and native nitrogen, along with the potentially negative impacts on surface and groundwater, and on air quality also (Delgado et al. 2008; Mukherjee et al. 2018). The loss of nitrogen and phosphorous causes eutrophication of surface water bodies (Cardoso et al. 2012). The amount of nutrient loss through water erosion is influenced by several factors, namely rainfall amount, topography, soil type, soil management, and conservation practices (Cogo 1981). However, the rainfall and runoff are the most important climatic factors, with a harmful action (Beutler et al. 2006). After a heavy rain, surface-applied nitrate can be dissolved in water and lost through the process of runoff (Fageria 2002).

5 Nitrogen Use in Agriculture and Climate Change

Nitrogen in agricultural systems can be applied by various anthropogenic sources. These include (i) synthetic nitrogenous fertilizers, (ii) biological N fixation through legumes crops, (iii) animal manure, FYM, compost, (iv) crop residues incorporation and return to the field after harvest, and (v) atmospheric N deposition. Globally, agriculture mostly depends on organic manure. However, after industrial revolution, the development of low-cost N fertilizers, release of high-yielding crop varieties, accompanied by irrigation facilities, has led to steadily increased consumption of synthetic fertilizer N. Nitrogen is the most widely used fertilizer nutrient globally as well as in India. The global fertilizer N consumption has increased by 9.8 folds from 1961 to 2010. During 1961, the consumption of global fertilizer N was about 15% of total N

Table 1 Contribution of different sources to nitrogen used in global and Indian agriculture

	Nitrogen in Tg (%)			
	Global		Indian	
Sources of N	1961	2010	1961	2014
Fertilizer	11.59 (16)	113.40 (51)	0.25 (7)	16.93 (71)
Animal manure	24.20 (32)	34.02 (16)	1.4 (39)	2.42 (10)
Crop residue	14.05 (19)	31.79 (14)	1.77 (50)	4.15 (17)
Atmospheric deposition	3.12 (4)	14.33 (7)	0.12 (4)	0.52 (2)
Biological N fixation	21.98 (29)	27.16 (12)	–	–
Total	74.93 (100)	220.70 (100)	3.54 (100)	24 (100)

Where, *Tg* Teragram
Source: Fagodiya et al. (2017a) and Fagodiya et al. (2020)

consumption (0.11.59 Tg), which increased to about 51% of total N consumption (113.40 Tg) (FAOSTAT 2016). During 1961, the consumption of fertilizer N was only about 7% of total N consumption (0.25 Tg), which increased to 70% of total N consumption (16.93 Tg), and the per-hectare fertilizer N consumption increased by 68 folds (FAI 2017). The total N consumption in agriculture has increased 6.8 folds from 3.58 Tg during 1961 to 24.03 Mt. during 2014 (Table 1). The inefficient use of fertilizers up to 300 kg ha^{-1} in few intensive cropping systems like rice–wheat and maize–wheat in Indo-Gangetic Plains (IGP) of India has led to the loss of excessive Nr in the soil, water, and air (Pathak and Nedwell 2001), causing water pollution (NO_3^-), air pollution (NOx), and climate change (N_2O) (Galloway et al. 2003; Khan et al. 2019).

5.1 Global Temperature Change Potential of N_2O Emission from Agriculture

The N_2O is a potent GHG, having global temperature change potential (GTP) of 290 (Shine et al. 2005) and global warming potential of 310 (IPCC 2014) on a 100-year basis. The synthetic fertilizer N use in agriculture contributes to anthropogenic N_2O emissions to a greater extent (Davidson 2009). N_2O is mostly emitted from the agricultural soil during nitrification and denitrification process (Forster et al. 2007). The concentration of N_2O emission from agriculture has increased dramatically, particularly after industrial revolution (Davidson 2009). The N_2O emission from global and Indian agriculture shows an increasing trend. The N_2O emission from global agriculture has increased from 1.44 Tg (1961) to 4.25 Tg (2010), converting to CO_2e, the GTP_{20} (GTP on 20-year timescale) of N_2O emission increased from 396.67 to 1168.32 Tg CO_2e and the GTP_{100} (GTP on a 100-year timescale) increased from 439.94 to 1295.78 Tg CO_2e during 1961 and 2010, respectively, Fig. 3 (Fagodiya et al. 2017a). On the other hand, it is reported that the N_2O emission from Indian agriculture has increased from 0.04 Tg in 1961 to 0.24 Tg in 2014, resulting in GTP_{20} of N_2O emission, which increased from 9.64 to 65.41 Tg CO_2e and, GTP_{100} that increased from 10.70 to 72.54 Tg CO_2e during 1961–2014, respectively, Fig. 4 (Fagodiya et al. 2020).

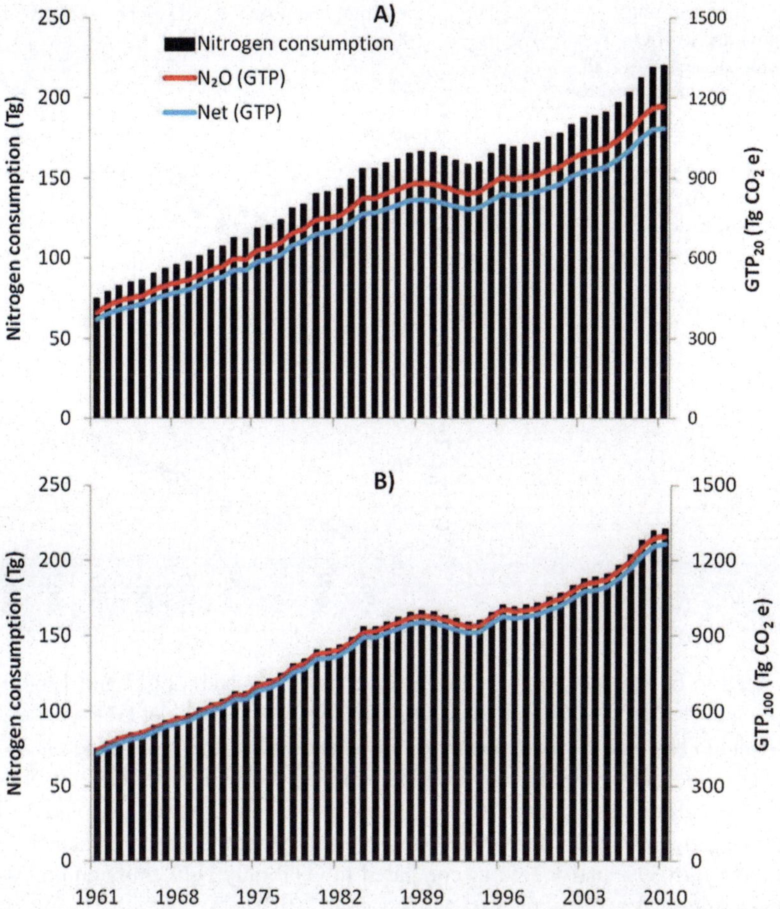

Fig. 3 The total nitrogen consumption, GTP due to N_2O emission and net GTP of N use in global agriculture (**a**) 20-year and (**b**) 100-year scales. (Source: Fagodiya et al. 2017a)

5.2 *Net Global Temperature Change Potential of Nitrogen Use in Agriculture*

Apart from the N_2O, the application of N in agricultural soils may also affect the fluxes of CH_4, CO_2, NO_x, and NH_3. The CH_4 and CO_2 are important GHGs causing direct global warming, whereas NH_3 formed light-scattering aerosols and contribute to global cooling (Bauer et al. 2007), whereas NO_x has both warming and cooling impacts. It contributes to global warming through the O_3 formation process (Berntsen et al. 2005) and to global cooling through the removal of atmospheric CH_4 (Derwent et al. 2001) and formation of light-scattering aerosols (Bauer et al. 2007). Therefore, to estimate the net warming impacts of N use in agriculture, all these impacts should be considered together (Fagodiya et al. 2017c). The net GTP of N use in global agriculture was

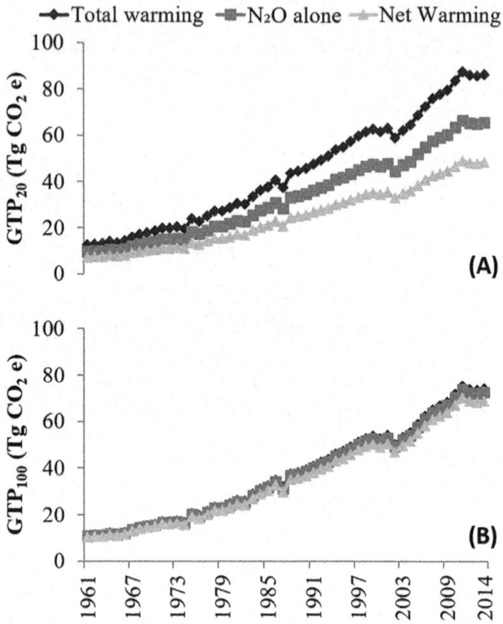

Fig. 4 Total warming, warming due to N_2O emission alone and overall net warming due to total N input used in Indian agriculture from 1961 to 2014 on (**a**) 20-year (GTP_{20}) and (**b**) 100-year (GTP_{100}) timesscales

estimated to be 369.44 and 1088.55 Tg CO_2e on GTP_{20} and 429.17 and 1264.06 Tg CO_2e on GTP100, in 1961 and 2010, respectively (Fig. 3). The net GTP_{20} lowered by 6.9% and GTP_{100} by 2.4% compared to the GTP of N_2O emission alone on 20- and 100-year timescales, respectively (Fagodiya et al. 2017a). The net GTP of N use in Indian agriculture was estimated to be 7.10 and 48.15 Tg CO_2e on GTP_{20}, and 0.19 and 68.62 Tg CO_2e on GTP_{100} during 1961 and 2014, respectively (Fig. 4). The net GTP is lowered by 26.39% and 4.77% as compared to GTP only N_2O emission on 20- and 100-year timescales, respectively (Fagodiya et al. 2020).

6 Nitrogen Loss from Agriculture and Environmental Consequences

In India, the use of nitrogen fertilizer is closely associated with food grain production. However, the nitrogen-use efficiency remained very low and the food grain production per unit nitrogen use has decreased. This anthropogenic application of fertilizer N has severely altered the N cycle, resulting in loss of reactive forms of N, that is, ammonia, nitrate, and nitrogen oxides into the environment, causing climate change (N_2O emission), aerosols formation and atmospheric N deposition (NH_3 and NOx emissions), groundwater pollution (nitrate leaching), eutrophication of surface water bodies (runoff loss of N), and loss of biodiversity (Ladha et al. 2016, Galloway et al. 2003). The climate change and environmental pollution associated with careless and high fertilizer N use have become a global concern. The principal environmental impacts associated with loss of N from agriculture and low nitrogen-use efficiency (NUE) of fertilizer N are summarized as follows:

6.1 Climate Change

The nitrous oxide (N_2O) is a potent greenhouse gas (GHG) with 310 times higher GWP than CO_2 (IPCC 2014; Gupta et al. 2016a, b). Since the pre-industrial era, the N_2O emission has increased by 20%. Globally, the N_2O emission contributed to 6.2% of total anthropogenic GHG emissions in 2010 (IPCC 2014; Bhatia et al. 2013a; Kumar et al. 2017). Agriculture is the major source of nitrous oxide (N_2O) emission with 70% of the total anthropogenic emission of N_2O (Bouwman 1996; Mosier et al. 1998). N_2O can be produced through the nitrification and denitrification process (Kool et al. 2010; Gupta et al. 2015). However, a major process of N_2O production following the application of N fertilizers occurs because of nitrification and denitrification process (Wrage et al. 2001; Snyder et al. 2009). The N_2O production through nitrification and denitrification process occurs simultaneously; however, the dominant process depends on the type of substrates, availability of substrates, soil moisture, and soil aeration (Mosier et al 1998; Khalil et al. 2004; Davidson 2009). Apart from this, N_2O can also lead to degradation of the ozone layer. Loss of soil-applied nitrogen through microbial reduction under higher moisture content is one of the major sources of N_2O .

6.2 Groundwater Pollution

Loss of nitrogen, beyond the crop root zone, with deep percolating water through leaching can cause groundwater pollution. The amount of N leaching depends on several soil properties, climatic factors, management practices, and methods and type fertilizer N (Brady and Weil 2002). A significant amount of N leaching is also reported in sandy soils (Baligar et al. 2001). According to the Environmental Protection Agency (EPA) of USA, the permissible limit of nitrate in drinking water is 45 mg L^{-1}. However, the Central Pollution Control Board (CPCB) and Indian Council of Medical Research have adopted 100 mg L^{-1} as the maximum permissible limit of nitrate in drinking. The higher quantity of nitrates than permissible limits in drinking water causes methemoglobinemia, popularly known as blue baby syndrome. The consumption of nitrate in drinking water can change haemoglobin to methemoglobin, decreasing the oxygen-carrying capacity of the blood (Majumdar Deepanjan 2003).

6.3 Eutrophication

Eutrophication is the big threat to surface water bodies, that is, ponds, lakes, and rivers in India and across the world. Eutrophication is the process of enrichment of nutrient, particularly phosphorus and nitrate into surface water bodies, causing excess algal growth, popularly known as algal blooms of blue–green algae (cyanobacteria) (Schindler 2006). The decomposition of algal growth leads to depletion of dissolved oxygen (hypoxia), contamination of the drinking water supply, and degradation of recreational opportunities and produces toxic substances, which are directly and indirectly toxic to aquatic life, livestock, and human beings (Baligar et al. 2001). The eutrophication causes water pollution in several freshwater and coastal marine ecosystems, and it is a fast-growing problem of water pollution in the developing countries (Smith and Schindler 2009). The annual cost of damage caused by eutrophication in the USA is estimated to be about 2.2 billion USD (Dodds et al. 2009). In modern-day intensive agriculture, the farmers are applying excessive nutrients, particularly nitrogen (Buol 1995) and the runoff loss of fertilizer N and detritus phosphorus from agricultural land with runoff water are major cause of eutrophication (Klimaszyk and Rzymski 2010).

6.4 Atmospheric N Deposition

The ammonia (NH_3) and oxides of nitrogen (NO_x) that are emitted through the volatilization process into the atmosphere from agricultural land and other anthropogenic sources can return back to the earth surface, causing atmospheric deposition (Buresh et al. 2004). Nitrogen can deposit on the earth surface through wet and dry deposition, depending on forms of N. In wet deposition, ammonium and nitrate are deposited by rain and snow. However, in dry deposition, the deposition of gases, that is, NH_3 and NO_2, and particulate matter of ammonium, nitrate, and nitrogen-containing aerosols, is carried out through the earth's gravity. Globally, the atmospheric deposition has contributed approximately 12% to the total reactive nitrogen addition to terrestrial and marine ecosystem (Kulshrestha et al. 2005). This deposited nitrogen can be a source of nutrient to the deposited land or it can have harmful or toxic effects, that is, acidification. The aquatic and forests ecosystems are more vulnerable to N deposition, and excess N deposition can cause disruption to the ecosystem functioning and it can also lead to global warming due to indirect N_2O emission, soil acidification due to acid rain, and can reduce the soil carbon stocks (Burns 2004). According to Singh and Kulshrestha (2014), the annual depositions of NH_4-N at rural, suburban, urban, and industrial sites of Indo-Gangetic Plains of India were of 2.38, 2.10, 2.38, and 5.04 kg ha^{-1}, respectively. However, the annual depositions of NO_3-N at rural, suburban, urban, and industrial sites were 4.06, 2.10, 4.48, and 3.92 kg ha^{-1}, respectively. This total atmospheric deposition of N can substitute the small amount of applied fertilizer N in agricultural land.

7 Nitrogen-Use Efficiency

The use efficiency is the ratio of output to input and it is measured as the output per unit use of an input. The nitrogen-use efficiency (NUE) is the ratio of N output, that is, biological or economic yield to the external N input used, that is, fertilizer N. The NUE is generally calculated as the agronomic efficiency (AE_N; kg kg^{-1} of fertilizer

N used), and it is the ratio of increase in the yield of crop due to amount of fertilizer N applied. During the last few decades, the research has been focused towards to improve NUE as well as minimizing nitrogen losses through a better synchronization of crop N demand and N supply. Nevertheless, the NUE remains quite low and it is in the range of 30–35% only, and presently, it is the one of the important priority areas for fertility management. The NUE can be improved by adopting advanced scientific technology of N management, which ensures effective use of N inputs, that is, fertilizers N, which will minimize its losses. Various scientific technologies of N management, that is, site-specific nutrient management, integrated nitrogen management (INM), inclusion of legumes and biofertilizers, and precision nitrogen management are discussed below:

7.1 Site-Specific Nitrogen Management (SSNM)

Systematic research on site-specific N management was initiated during 1980s–1990s to increase the NUE, fertilizer use by crop, to reduce N losses and to reduce the ground and surface water contamination by nitrate. SSNM was developed to increase NUE of irrigated rice (Dobermann et al. 2002). Presently in SSNM, to estimate total N rate based on indigenous N supply capacity and target yield, the following steps are being used: (1) setting an attainable yield target based on 85% of yield potential, (2) estimating indigenous N supply, (3) estimating N response (target yield–yield of N_0 treatment), and (4) estimating the N rate based on N response and agronomic NUE. In the case of rice crop, SSNM has been evaluated in farmers' fields in eight major irrigated rice domains in Asia (Dobermann et al. 2002; Wang et al. 2001, 2004). Across all sites in Asia, the average grain yield increased by 11% and the average recovery efficiency increased from 31% to 40%, with 20% of all farmers achieving more than 50% recovery efficiency (Dobermann et al. 2002). SSNM increases both grain yield and fertilizer N-use efficiency compared to conventional farming practices due to reducing total N rate and by reducing N rate during the early vegetative stage. The total N rate is reduced because of small N response and high target agronomic N-use efficiency. The reduction in the N rate during the early vegetative stage may be due to high indigenous N supply capacity of soil (Peng et al. 2010).

7.2 Integrated Nitrogen Management (INM)

Integrated nitrogen management (INM) includes the optimum use of different nitrogen sources such as chemical fertilizers, N with organic manures, crop residue, and biofertilizers (Olesen et al. 2004; Prasad 2008; Dwivedi et al. 2016). The experimental results of long-term fertilizer experiments (LTFEs) showed that the sustainable crop production cannot be achieved by the sole application of either fertilizer

N or organic manures. The combined applications of different sources of nitrogen have several advantages and have proved always superior to their sole application (Hegde and Dwivedi 1993). The integrated nitrogen management in India is not a new approach and the pre-green revolution agriculture largely depended on this with organic manures as leading source of nitrogen. The important sources of N in INM are nitrogenous fertilizer, organic manures, that is, farmyard manure (FYM), vermicompost, green manures, waste material and bio-product of industries, sewage sludge, and biofertilizers. The interest in INM has been increasing since past few years to revive soil health and to enhance nitrogen-use efficiency. The positive effects of INM are the improvement in soil physicochemical properties, enhancement in soil microbial activity, better root growth, and enhanced supply of other nutrients (Singh et al. 2012).

Green Manuring in INM The green manuring is the process of incorporation of tender undecomposed fresh plant parts, that is, leaves, stem, or whole plant into soils, either at the same place or brought from somewhere else (Pieters 1927; Kumar et al. 2013). The green manuring increases the sustainability of agriculture production by increasing the soil fertility (Fageria and Baligar 2005), improving nutrient status (Dinnes et al. 2002), reducing soil erosion and by reducing global warming (Robertson et al. 2000). Although the *Sesbania rostrata* has the highest N-fixing potential, Sunnhemp (*Crotolaria juncea*) and Dhaincha (*Sesbania aculeata*) are the two most commonly used green manuring crops (Chaudhury and Kennedy 2004). The legume green manure crops, that is, cluster bean, berseem, and clover, are also used for green manuring (Meena et al., 2018). The inclusion of legumes crops in the cropping system for grain or fodder purpose can meet the N demand by 50–60 kg N ha^{-1} of succeeding non-legume crop, thereby improving the productivity of succeeding crops (Singh and Dwivedi 2006).

Biofertilizers Biofertilizers are composed of living organism and latent cells which fix atmospheric N through colonization in the root zone of plants, popularly known as rhizosphere and provide it to the crop plants. N-fixing bio-fixers are basically of two types: (i) symbiotic and (ii) non-symbiotic. *Rhizobium* with legume (Sharma et al. 2011); *Cyanobacteria* with plant/fungi and *Frankia* with trees are some examples of symbiotic N-fixing microbes, which fix N with association of other plant and trees. On the other hand, the non-symbiotic N-fixing biofertilizers are aerobic and free-living in nature, which can fix N without association of plant and crops. *Azotobacter, Bejeirinkia, Azospirillum*; *Clostridium, Desulfovibrio,* etc., are the examples of non-symbiotic N-fixing microbes. Symbiosis association between legume and *Rhizobium* can fulfil >80% of legume N requirement. The microbial inoculation of *Azotobacter* and *Azospirillum* can contribute about 20 and 25 kg N ha^{-1}, respectively, in crop field conditions. The average N fixation by *cyanobacteria* and *Azolla* is estimated to be about 25–30 and 30–40 kg N ha^{-1}, respectively (Hegde and Dwivedi 1993). In India, the production of biofertilizers was merely 2000 tons during 1992–1993. Since then, it has increased by 332 times and reached to 65.5 × 1000 tons during 2013–2014 (FAI 2015).

7.3 Precision Nitrogen Management

Precision nitrogen management (PNM) is basically the N management strategy through which the N losses are minimized and N utilization efficiency of crop is increased. The precision nitrogen management simply states that the right type of fertilizer should be applied in right amount and at the right time using a right method of application. N fertilizers are predominantly available in ammonium, nitrate, or amide forms. The ammonium and amide forms are more vulnerable to volatilization loss than the nitrate form of fertilizers (Ladha et al. 2005). Therefore, the selection of right source of N fertilizers is greatly helpful in minimizing N losses and maximization of NUE for crops. Besides this, the NOCU, slow release N fertilizers, and use of nitrification inhibitors with urea are very helpful in efficient N management with reduced N losses (Prasad 2013; Fagodiya et al. 2017b; Fagodiya et al. 2020). Another approach for enhancing NUE is to match the timing and amount of N fertilizers application with the crop demand. To match the crop demand with N supply, the chlorophyll meter (SPAD) and leaf colour chart (LCC) can be used for monitoring of N in the crop field (Bijay-Singh et al. 2015). The LCC-based application (crop demand) of urea is superior to the conventional application of urea (in three splits) (Shukla et al. 2004). The right method of N fertilizers application could also be useful in minimizing N losses from the crop field. The broadcasting of N fertilizers causes high losses of N, particularly volatilization losses, which reduce the NUE. However, the deep placement of urea in the crop root zone improves the NUE (Majumdar et al. 2016). Humphreys et al. (1992) reported 37%, 46%, and 49% of N recovery efficiency of broadcasting, band placement, and deep placement of urea, respectively, in direct-seeded rice (DSR). Singh et al. (2014) achieved 80% of NUE through fertigation (fertilizers application with irrigation) in vegetables.

8 Conclusion and Recommendation

Globally, N is the key element for the agricultural productivity to feed the increasing human population. The inefficient uses of the N fertilizers are increasing the climate risk, and groundwater pollution through N_2O emission, and nitrate leaching, respectively. Intensive agriculture system is the need of the hour to feed the growing population. The efficient utilization of the N fertilizers for sustainable agricultural growth and production is highly required. In other words, the N-use efficiency of the crops could also be enhanced through conservation agriculture, decision support tools, precision N management, integrated nitrogen management (INM), site-specific nitrogen management (SSNM), etc. The application of the right form of N fertilizers at the right time in adequate quantity is important to use synthetic N fertilizers. Biofertilizers, organic manuring, and brown manuring can also substantially support sustainable agriculture growth and production.

Acknowledgements The authors are much thankful to the ICAR-Central Soil Salinity Research Institute, Karnal; Indian Institute of Technology, Delhi; Jawaharlal Nehru University, New Delhi; and Central Muga Eri Research and Training Institute, Jorhat, Assam for providing necessary support for this publication.

References

Adler AA, Doole GJ, Romera AJ, Beukes PC (2015) Managing greenhouse gas emissions in two major dairy regions of New Zealand: a system-level evaluation. Agric Syst 135:1–9

Aggarwal RK, Raina P, Kumar P (1987) Ammonia volatilization losses from urea and their possible management for increasing nitrogen use efficiency in an arid region. J Arid Environ 13:163–168

Aulakh MS, Bijay-Singh (1996) Nitrogen losses and fertilizer N use efficiency in irrigated porous soils. Nutr Cycl Agroecosyst 47(3):197–212. https://doi.org/10.1007/BF01986275

Baalsrud K, Baalsrud KS (1954) Studies on *Thiobacillus denitrificans*. Arch Microbiol 20:34–62

Baligar VC, Fageria NK, He H (2001) Nutrient use efficiency in plants. Commun Soil Sci Plant Anal 32:921–950

Bauer SE, Koch D, Unger N, Metzger SM, Shindell DT, Streets DG (2007) Nitrate aerosols today and in 2030: a global simulation including aerosols and tropospheric ozone. Atmos Chem Phys 7:5043–5059

Berntsen J, Petersen BM, Jacobsen BH, Olesen JE, Hutchings NJ (2003) Evaluating nitrogen taxation scenarios using the dynamic whole farm simulation model FASSET. Agric Syst 76:817–839

Berntsen TK, Fuglestvedt JS, Joshi MM, Shine KP, Stuber N, Ponater M, Sausen R, Hauglustaine DA, Li L (2005) Response of climate to regional emissions of ozone precursors: sensitivities and warming potentials. Tellus B Chem Phys Meteorol 57:283–304

Beutler AN, Centurion JF, Centurion MAPC, Silva AP (2006) Efeito da compactação na produtividade de cultivares de soja em Latossolo Vermelho. R Bras Ci Solo 30:787–794

Bhatia A, Jain N, Pathak H (2013) Methane and nitrous oxide emissions from Indian rice paddies, agricultural soils and crop residue burning. Greenhouse Gases Sci Technol 3(3):196–211. https://doi.org/10.1002/ghg.1339

Bhatia A, Kumar A, Kumar V, Jain N (2013a) Low carbon option for sustainable agriculture. Indian Farming 63(2):18–22

Bhatia A, Kumar AK, Das TK, Singh J, Jain N, Pathak H (2013b) Methane and nitrous oxide emissions from soils under direct seeded rice. Int J Agric Sci Stat 9(2):729–736

Bhattacharyya R, Bhatia A, Das TK, Lata S, Kumar A, Tomer R, Singh G, Kumar S, Biswas AK (2018) Aggregate-associated N and global warming potential of conservation agriculture-based cropping of maize-wheat system in the north-western Indo-Gangetic Plains. Soil Tillage Res 182:66–77

Bijay-Singh, Varinderpal-Singh, Purba J, Sharma RK, Jat ML, Yadvinder-Singh, Thind HS, Gupta RK, Choudhary OP, Chandna P, Khurana HS, Kumar A, Jagmohan-Singh, Uppal HS, Uppal RK, Vashistha M, Gupta RK (2015) Site-specific nitrogen management in irrigated transplanted rice (Oryza sativa) using an optical sensor. Precis Agric 16:455–475

Bolan NS, Hedley MJ (2003) Role of carbon, nitrogen, and sulfur cycles in soil acidification. In: Rengel Z (ed) Handbook of soil acidity. Marcek Dekker, New York, pp 29–56

Bouwman AF (1996) Direct emission of nitrous oxide from agricultural soils. Nutr Cycl Agroecosyst 46:53–70

Brady NC, Weil RR (2002) The nature and properties of soils, 13th edn. Prentice Hall, Upper Saddle River

Brentrup F, Kusters J, Lammel J, Kuhlmann H (2000) Methods to estimate on-field nitrogen emissions from crop production as an input to LCA studies in the agricultural sector. Int J Life Cycle Assess (Berlin) 5(6):349–357

Buol SW (1995) Sustainability of soil use. Annu Rev Ecol Syst 26:25–44

Buresh RJ, Peng SB, Huang JL, Yang JC, Wang GH, Zhong XH, Zou YB (2004) Rice systems in China with high nitrogen inputs. In: Mosier AR, Syers JK, Freney JR (eds) Agriculture and the nitrogen cycle: assessing the impacts of fertilizer use on food production and the environment. Island Press, Washington, DC, pp 143–153

Burns D (2004) The effects of atmospheric nitrogen deposition in the Rocky Mountains of Colorado and southwestern Wyoming, USA: a critical review. Environ Pollut 127:257–269

Cameron KC et al (2013) Nitrogen losses from the soil/plant system: a review. Ann Appl Biol (Warwick) 162(2):145–173

Cardoso D, Silva MLN, Freitas DAF, Avanzi JC (2012) Plantas de cobertura no controle das perdas de solo, água e nutrientes por erosão hídrica. R Bras Eng Agric Amb 16:632–638

Carlson CA, Ingraham JL (1983) Comparison of denitrification by *Pseudomonas stutzeri*, *Pseudomonas aeruginosa*, and *Paracoccus denitrificans*. Appl Environ Microbiol 45:1247–1253

Cassman KG, Dobermann A, Walters D (2002) Agro-ecosystems, nitrogen-use efficiency, and nitrogen management. Ambio 31:132–140

Chaptal JA, Nicholson WT (1800) Elements of chemistry, vol 1, 3rd edn. CC and J Robinson, London, pp xxxv–xxxvi

Chaudhury ATMA, Kennedy IR (2004) Prospects and potentials for systems of biological nitrogen fixation in sustainable rice production. Biol Fertil Soils 39:219–227

Ciampitti IA, Ciarlo EA, Conti ME (2008) Nitrous oxide emissions from soil during soybean (Glycine max (L.) Merrill) crop phenological stages and stubbles decomposition period. Biol Fertil Soils 44(4):581–588

Cogo NP (1981) Effect of residue cover, tillage induced roughness, and slope length on erosion and related parameters [tese]. Purdue University, West Lafayette

Crosson P, Shalloo L, O'brien D, Lanigan GJ, Foley PA, Boland TM, Kenny DA (2011) A review of whole farm systems models of greenhouse gas emissions from beef and dairy cattle production systems. Anim Feed Sci Technol 166(167):29–45

Daniel R (1772) Dissertatio Inauguralis de aere fixo, aut mephitico (Inaugural dissertation on the air [called] fixed or mephitic). MD dissertation, University of Edinburgh, Scotland. English translation: Dobbin L (1935) Daniel Rutherford's inaugural dissertation. J Chem Educ 12(8):370–375. https://doi.org/10.1021/ed012p370

Davidson EA (2009) The contribution of manure and fertilizer nitrogen to atmospheric nitrous oxide since 1860. Nat Geosci 2(9):659–662

Delgado JA, Shaffer MJ, Lal H, Mckinney S, Gross CM, Cover H (2008) Assessment of nitrogen losses to the environment with a Nitrogen Trading Tool (NTT). Comput Electron Agric 63:193–206

Derwent RG, Collins WJ, Johnson CE, Stevenson DS (2001) Transient behaviour of tropospheric ozone precursors in a global 3-D CTM and their indirect greenhouse effects. Clim Change 49:463–487

Dinnes DL, Karlen DL, Jaynes DB, Kaspar TC, Hatfield JL, Colvin TS, Cambardella CA (2002) Nitrogen management strategies to reduce nitrate leaching in tile-drained Midwestern soils. Agron J 94:153–171

Dobermann A, Cassman KG (2004) Environmental dimensions of fertilizer nitrogen: what can be done to increase nitrogen use efficiency and ensure global food security? In: Mosier AR, Syers JK, Freney JR (eds) Agriculture and the nitrogen cycle: assessing the impacts of fertilizer use on food production and the environment. Island Press, Washington, DC/Paris, pp 261–278

Dobermann A, Witt C, Dawe D, Gines HC, Nagarajan R, Satawathananont S, Son TT, Tan PS, Wang GH, Chien NV, Thoa VTK, Phung CV, Stalin P, Muthukrishnan P, Ravi V, Babu M, Chatuporn S, Kongchum M, Sun Q, Fu R, Simbahan GC, Adviento MAA (2002) Site-specific nutrient management for intensive rice cropping systems in Asia. Field Crop Res 74:37–66

Dodds WK, Bouska WW, Eitzmann JL, Pilger TJ, Pitts KL, Riley AJ, Schloesser JT, Thornbrugh DJ (2009) Eutrophication of U.S. freshwaters: analysis of potential economic damages. Environ Sci Technol 43:12–19

Dwivedi BS, Singh VK, Meena MC, Dey A, Datta SP (2016) Integrated nutrient management for enhancing nitrogen use efficiency. Indian J Fertil 12(4):62–71

Fageria NK (2002) Soil quality vs. environmentally based agriculture. Commun Soil Sci Plant Anal 33:2301–2329

Fageria NK, Baligar VC (2005) Role of cover crops in improving soil and row crop productivity. Commun Soil Sci Plant Anal 36:2733–2757

Fagodiya RK, Pathak H, Kumar A, Bhatia A, Jain N (2017a) Global temperature change potential of nitrogen use in agriculture: a 50-year assessment. Sci Rep 7:4928

Fagodiya RK, Pathak H, Bhatia A, Kumar A, Singh SD, Jain N, Harith R (2017b) Simulation of maize (*Zea mays L.*) yield under alternative nitrogen fertilization using infocrop-maize model. Biochem Cell Arch 17(1):65–71

Fagodiya RK, Pathak H, Meena BL, Meena RK, Nagdev R (2017c) Need to Estimate the Net Global Warming Potential of Nitrogenous Fertilizers. Adv Plants Agric Res 6(4):00220. https://doi.org/10.15406/apar.2017.06.00220

Fagodiya RK, Pathak H, Bhatia A, Jain N, Gupta DK, Kumar A, Malyan SK, Dubey R, Radhakrishanan S, Tomer R (2019) Nitrous oxide emission and mitigation from maize–wheat rotation in the upper Indo-Gangetic Plains. Carbon Manag 10(5):489–499

Fagodiya RK, Pathak H, Bhatia A, Jain N, Gupta DK (2020) Global warming potential and its cost of mitigation from maize (Zea mays) - wheat (Triticum aestivum) cropping system. Indian J Agr Sci 90(1):69–74

Fagodiya RK, Pathak H, Bhatia A, Jain N, Kumar A, Malyan SK (2020) Global warming impacts of nitrogen use in agriculture: An assessment of Indian since 1960. Carbon Management (In press)

FAI (2015) Fertiliser statistics, 4th edn. The Fertiliser Association of India, New Delhi

FAI (2017) The Fertilizers Association of India, Delhi, India. http://www.faidelhi.org

FAOSTAT (2016) Food and Agriculture Organization of the United Nations, Rome, Italy. Available at http://faostat.fao.org

Fowler D, Pyle JA, Raven JA, Sutton MA (2013) The global nitrogen cycle in the twenty-first century: introduction. Philos Trans R Soc B Biol Sci 36:20130165

Forster P, Ramaswamy V, Artaxo P, Berntsen T, Betts R, Fahey DW, Haywood J, Lean J, Lowe DC, Myhre G, Nganga J (2007) Changes in atmospheric constituents and in radiative forcing. Chapter 2. In Climate Change 2007. The Physical Science Basis 2007

Franche C, Lindstrom K, Elmerich C (2009) Nitrogen-fixing bacteria associated with leguminous and non-leguminous plants. Plant Soil 321:35–59

Gaby JC, Buckley DH (2011) A global census of nitrogenase diversity. Environ Microbiol 13:1790–1799

Galloway JN, Aber JD, Erisman JW, Seitzinger SP, Howarth RW, Cowling EB, Cosby BJ (2003) The nitrogen cascade. Bioscience 53:341–356

Ghosh BC, Bhat R (1998) Environmental hazards of nitrogen loading in wetland rice fields. Environ Pollut 102(S1):123–126

Gu B, Chang J, Min Y, Ge Y, Zhu Q, Galloway JN, Peng C (2013) The role of industrial nitrogen in the global nitrogen biogeochemical cycle. Sci Rep 3:2579. https://doi.org/10.1038/srep02579

Gupta DK, Bhatia A, Kumar A, Chakrabati B, Jain N, Pathak H (2015) Global warming potential of rice (*Oryza sativa*)-wheat (*Triticum aestivum*) cropping system of the Indo-Gangetic Plains. Indian J Agric Sci 85(6):807–816

Gupta DK, Bhatia A, Das TK, Singh P, Kumar A, Jain N, Pathak H (2016a) Economic analysis of different greenhouse gas mitigation technologies in rice–wheat cropping system of the Indo-Gangetic Plains. Curr Sci 110(5):867–873

Gupta DK, Bhatia A, Kumar A, Das TK, Jain N, Tomer R, Malyan SK, Fagodiya RK, Dubey R, Pathak H (2016b) Mitigation of greenhouse gas emission from rice–wheat system of the

Indo-Gangetic plains: through tillage, irrigation and fertilizer management. Agric Ecosyst Environ 230:1–9

Hallin S, Jones CM, Schloter M, Philippot L (2009) Relationship between N-cycling communities and ecosystem functioning in a 50-year-old fertilization experiment. ISME J 3:597–605

Hegde DM, Dwivedi BS (1993) Integrated nutrient supply and management as a strategy to meet nutrient demand. Fertil News 38(12):49–59

Humphreys E, Chalk PM, Muirhead WA, White RJG (1992) Nitrogen fertilization of dry-seeded rice in southeast Australia. Fertil Res 31:221–234

Hutchinson C, Simonne E, Solano P, Meldrum J, Livingston-Way P (2003) Testing of controlled release fertilizer programs for seep irrigated Irish potato production. J Plant Nutr 26:1709–1723

IPCC (2006) Guidelines for national greenhouse gas inventories (Eggleston HS, Buendia L, Miwa K, Ngara T and Tanabe K (eds)). Prepared by the National Greenhouse Gas Inventories Programme. IGES, Japan

IPCC (2014) Climate change 2014: impacts, adaptation, and vulnerability working group II contribution to the fifth assessment report. Cambridge University Press, Cambridge/New York

Ishii S, Ikeda S, Minamisawa K, Senoo K (2011) Nitrogen cycling in rice paddy environments: past achievements and future challenges. Microbes Environ/JSME 26(4):282–292. https://doi.org/10.1264/jsme2.me11293

Jain N, Arora P, Tomar R, Mishra SV, Bhatia A, Pathak H, Chakraborty D, Kumar V, Dubey DS, Harit RC, Singh JP (2016) Greenhouse gases emission from soils under major crops in Northwest India. Sci Total Environ 542:551–561

Jung MY, Well R, Min D, Giesemann A, Park SJ, Kim JG, Kim SJ, Rheet SK (2014) Isotopic signatures of N2O produced by ammonia-oxidizing archaea from soils. ISME J 8:115–1125

Khalil K, Mary B, Renault P (2004) Nitrous oxide production by nitrification and denitrification in soil aggregates as affected by O_2 concentration. Soil Biol Biochem 36(4):687–699

Khan SA, Sharma GK, Malla FA, Kumar A, Rashmi GN (2019) Microalgae based biofertilizers: a biorefinery approach to phycoremediate wastewater and harvest biodiesel and manure J Clean Prod 211:1412–1419

Klimaszyk P, Rzymski P (2010) Surface runoff as a factor determining trophic State of Midforest Lake. Pol J Environ Stud 20(5):1203–1210

Kool DM, Wrage N, Boltenstern Z, Pfeffer S, Brus M, Oenema D (2010) Nitrifier denitrification can be a source of N_2O from soil: a revised approach to the dual isotope labelling method. Eur J Soil Sci 61:759–772

Kulshrestha UC, Granat L, Engardt M, Rodhe H (2005) Review of precipitation monitoring studies in Indiada search for regional patterns. Atmos Environ 39:7403–7419. https://doi.org/10.1016/j.atmosenv.2005.08.035

Kumar A, Gupta DK, Kumar M (2013) Green manure crops: a boon for agricultural soil. Int J Agric Environ Biotechnol 6:193

Kumar A, Tomer R, Bhatia A, Jain N, Pathak H (2016) Greenhouse gas mitigation in Indian agriculture. In: Pathak H, Chakrabarti B (eds) Climate change and agriculture technologies for enhancing resilience. ICARIARI, New Delhi, pp 137–149

Kumar A, Bhatia A, Fagodiya RK, Malyan SK, Meena BL (2017) Eddy covariance flux tower: a promising technique for greenhouse gases measurement. Adv Plant Agric Res 7:00263

Kumar SS, Kumar A, Singh S, Malyan SK, Baram S, Sharma J, Singh R, Pugazhendhi A (2020) Industrial wastes: Fly ash, steel slag and phosphogypsum- potential candidates to mitigate greenhouse gas emissions from paddy fields. Chemosphere 241:124824

Ladha JK, Dawe D, Pathak H, Padre AT, Yadav RL, Bijay S, Singh YS, Singh Y, Singh P, Kundu AL, Sakal R, Ram N, Regmi AP, Gami SK, Bhandari AL, Amin R, Yadav CR, Bhattarai EM, Das S, Aggarwal HP, Gupta RK, Hobbs PR (2003) How extensive are yield declines in long term rice-wheat experiments in Asia? Field Crop Res 81:159–180

Ladha JK, Pathak HP, Krupnik TJ, Six J, van Kessel C (2005) Efficiency of fertiliser nitrogen in cereal production: retrospect and prospect. Adv Agron 87:85–156

Ladha JK, Tirol-Padre A, Reddy CK, Cassman KG, Verma S, Powlson DS, van Kessel C, Richter D de B, Chakraborty D, Pathak H (2016) Global nitrogen budgets in cereals a 50 year-assessment. Sci Rep 6:19355

Lin DX, Fan XH, Hu F, Zhao HT, Luo JF (2007) Ammonia volatilization and nitrogen utilization efficiency in response to urea application in rice fields of the Taihu Lake region, China. Pedosphere 17(5):639–645

Majumdar D (2003) The blue baby syndrome. Resonance 8:20–30. https://doi.org/10.1007/BF02840703

Majumdar K, Sanyal SK, Dutta SK, Satyanarayana T, Singh VK (2016) Nutrient mining: addressing the challenges to soil resources and food security. In: Singh U, Praharaj CS, Singh SS, Singh NP (eds) Biofortification of food crops. Springer, New Dehli, pp 177–198

Malyan SK, Bhatia A, Kumar SS, Fagodiya RK, Pugazhendhi A, Duc PA (2019) Mitigation of greenhouse gas intensity by supplementing with Azolla and moderating the dose of nitrogen fertilizer. Biocatal Agric Biotechnol 20:101266. https://doi.org/10.1016/j.bcab.2019.101266

Masclaux-Daubresse C, Reisdorf-Cren M, Pageau K, Lelandais M, Grandjean O, Kronenberger J, Valadier MH, Feraud M, Jouglet T, Suzuki A (2006) Glutamine synthetase-glutamate synthase pathway and glutamate dehydrogenase play distinct roles in the sink-source nitrogen cycle in tobacco. Plant Physiol 140(2):444–456

Meena BL, Fagodiya RK, Prajapat K, Dotaniya ML, Kaledhonkar MJ, Sharma PC, Meena RS, Mitran T, Kumar S (2018) Legume green manuring: an option for soil sustainability. In: Meena R, Das A, Yadav G, Lal R (eds) Legumes for Soil Health and Sustainable Management. Springer, Singapore, pp 387–408

McLaren RG, Cameron K (1996) In: 2nd (ed) Soil science: sustainable production and environmental protection. Oxford University Press, Auckland. ISBN 0-19-558345-0

Modak JM (2008) Haber process for ammonia synthesis. Resonance 7(9):69–77

Mosier AR (2001) Exchange of gaseous nitrogen compounds between agricultural systems and the atmosphere. Plant Soil 228(1):17–27

Mosier A, Kroeze C, Nevison C, Oenema O, Seitzinger S, Van Cleemput O (1998) Closing the global N_2O budget: nitrous oxide emissions through the agricultural nitrogen cycle. Nutr Cycl Agroecosyst 52:225–248

Mukherjee J, Mridha N, Mondal S, Chakraborty D, Kumar A (2018) Identifying suitable soil health indicators under variable climate scenarios: a ready reckoner for soil management. In: Bal S, Mukherjee J, Choudhury B, Dhawan A (eds) Advances in crop environment interaction. Springer, Singapore

Nadelhoffer KJ, John DA, Jerry MM (1984) Seasonal patterns of ammonium and nitrate uptake in ten temperate forest ecosystems. Plant Soil 80(3):321–335

Olesen JE, Sorensen P, Thomsen IK, Eriksen J, Thomsen AG, Berntsen J (2004) Integrated nitrogen input systems in Denmark. In: Mosier AR, Syers JK, Freney JR (eds) Agriculture and the nitrogen cycle: assessing the impacts of fertilizer use on food production and the environment. Island Press, Washington, DC/Paris, pp 129–140

Oliver JGJ, Bouwman AF, Van der Hoek KW, Berdowski JJM (1998) Global air emission inventories for anthropogenic sources of NOx, NH_3 and N_2O in 1990. Environ Pollut 102(S1):135–148

Parashar DC, Kulshreshtha UC, Sharma C (1998) Anthropogenic emissions of NOx, NH_3 and N_2O in India. Nutr Cycl Agroecosyst 52:255–259

Pathak H, Nedwell DB (2001) Strategies to reduce nitrous oxide emission from soil with fertiliser selection and nitrification inhibitor. Water Air Soil Pollut 129(4):217e228

Pathak H, Aggarwal PK, Roetter R, Kalra N, Bandyopadhaya SK, Prasad S, Van Keulen H (2003) Modeling the quantitative evaluation of soil nutrient supply, nutrient use efficiency, and fertilizer requirements of wheat in India. Nutr Cycl Agroecosyst 65:105–113

Pathak H, Jain N, Bhatia A, Kumar A, Chatterjee D (2016) Improved nitrogen management: a key to climate change adaptation and mitigation. Indian J Fertil 12(11):51–162

Peng S, Buresh RJ, Huang J, Zhong X, Zou Y, Yang J, Wang G, Liu Y, Hu R, Tang Q, Cui K, Zhang F, Dobermann A (2010) Improving nitrogen fertilization in rice by site-specific N management: a review. Agron Sustain Dev 30:649–656

Philippot L, Hallin S, Schloter M (2007) Ecology of denitrifying prokaryotes in agricultural soil. Adv Agron 96:249–305

Pieters JA (1927) Green manuring, principles and practice. Wiley, New York, p 365

Postgate J (1998) Nitrogen fixation, 3rd edn. Cambridge University Press, Cambridge. Proc Indian Natl Sci Acad 80(2):359–378

Poth M, Focht DD (1985) ^{15}N kinetic analysis of N^2O production by nitrosomonas europeae: an examination of nitrifier denitrification. Appl Environ Microbiol 49:1134–1141

Prasad R (2008) Integrated plant nutrient supply (IPNS) for sustainable agriculture. Indian J Fertil 4(12):71–90

Prasad R (2013) Fertiliser nitrogen, food security, health and the environment. Proc Indian Natl Sci Acad 79(4):997–1010. Spl. Issue, Part B

Prosser JI, Nicol GW (2012) Archaeal and bacterial ammonia-oxidisers in soil: the quest for niche specialisation and differentiation. Trends Microbiol 20:523–531

Randall GW, Vetsch JA, HuVman JR (2003) Corn production on a subsurface drained Mollisol as affected by time of nitrogen application and nitrapyrin. Agron J 95:1213–1219

Regmi AP, Ladha JK (2006) Enhancing productivity of rice-wheat system through integrated crop management in the Eastern-Gangetic plains of South Asia. J Crop Improv 15:147–170

Robertson GP, Paul EA, Harwood RR (2000) Greenhouse gases in intensive agriculture: contributions of individual gases to the radioactive forcing of the atmosphere. Science 289:1922–1925

Sarkar MC, Banerjee NK, Rana DS, Uppal KS (1991) Field measurements of ammonia volatilization losses of N from urea applied to wheat. Fertil News 33:25–29

Scheurwater I, Koren M, Lambers H, Atkin OK (2002) The contribution of roots and shoots to whole plant nitrate reduction in fast- and slow-growing grass species. J Exp Bot 53(374):1635–1642. https://doi.org/10.1093/jxb/erf008

Schindler DW (2006) Recent advances in the understanding and management of eutrophication. Limnol Oceanogr 51:356–363

Seitzinger S, Harrison JA, Bohlke JK, Bouwman AF, Lowrance R, Peterson B, Tobias C, Drecht GV (2006) Denitrification across landscapes and waterscapes: a synthesis. Ecol Appl 16(6):2064–2090

Sharma C, Tiwari MK, Pathak H (2008) Estimates of emission and deposition of reactive nitrogenous species for India. Curr Sci 94:1439–1446

Sharma A, Kumar A, Dhaka TS (2011) Impact on nodulation parameters of *Cicer arietinum Trigonella* and *foenum-graecum* of sugar factory effluent. Biochem Cell Arch 11(2):465–468

Shine KP, Fuglestvedt JS, Hailemariam K, Stuber N (2005) Alternatives to the Global Warming Potential for Comparing Climate Impacts of Emissions of Greenhouse Gases. Clim Change 68:281–302

Shaw LJ, Nicol GW, Smith Z, Fear J, Prosser JI, Baggs EM (2006) *Nitrosospira spp.* can produce nitrous oxide via a nitrifier denitrification pathway. Environ Microbiol 8:214–222

Shukla AK, Ladha JK, Singh VK, Dwivedi BS, Balasubramanian V, Gupta RK, Sharma SK, Singh Y, Pathak H, Pandey PS, Padre AT, Yadav RL (2004) Calibrating the leaf color chart for nitrogen management in different genotypes of rice and wheat in a systems perspective. Agron J 96:1606–1621

Singh VK, Dwivedi BS (2006) Yield and nitrogen use-efficiency in wheat, and soil fertility status as influenced by substitution of rice with pigeon pea in a rice-wheat cropping system. Aust J Exp Agric 46:1185–1194

Singh S, Kulshrestha UC (2014) Rural versus urban gaseous inorganic reactive nitrogen in the indo-Gangetic plains (IGP) of India. Environ Res Lett 9(12):125004

Singh B, Singh V, Singh Y, Thind HS, Kumar A, Gupta RK, Kaul A, Vashistha M (2012) Fixed-time adjustable dose site-specific nitrogen management in transplanted irrigated rice (*Oryza sativa L.*) in South Asia. Field Crop Res 126:63–69

Singh VK, Mishra RP, Rani M, Singh SK (2014) Water and nutrient interaction. I. Irrigated agriculture. In: Prasad R, Kumar D, Rana DS, Shivay YS, Tewatia RK (eds) Textbook of plant nutrient management. Indian Society of Agronomy, New Delhi, pp 249–267

Smil V (2001) Enriching the earth. The MIT Press, Cambridge, MA

Smith VH, Schindler DW (2009) Eutrophication science: where do we go from here? Trends Ecol Evol 24:201–207

Snyder CS, Bruulsema TW, Jensen TL, Fixen PE (2009) Review of greenhouse gas emissions from crop production systems and fertilizer management effects. Agric Ecosyst Environ 133:247–266

Spieck E, Hartwig C, McCormack I, Maixner F, Wagner M, Lipski A, Daims H (2006) Selective enrichment and molecular characterization of a previously uncultured Nitrospira-like bacterium from activated sludge. Environ Microbiol 8:405–415

Wang GH, Dobermann A, Witt C, Sun QZ, Fu RX (2001) Performance of site-specific nutrient management for irrigated rice in Southeast China. Agron J 93:869–878

Wang GH, Sun Q, Fu R, Huang XH, Ding XH, Wu J, He YF, Dobermann A, Witt C (2004) Site-specific nutrient management in intensive irrigated rice systems of Zhejiang province, China. In: Dobermann A et al (eds) Increasing productivity of intensive rice systems through site-specific nutrient management. Science Publishers, Inc/International Rice Research Institute, Enfield/Los Baños, pp 243–263

Weeks ME (1932) The discovery of the elements. IV. Three important gases. J Chem Educ 9(2):215

White RE (2005) Principles and practice of soil science: the soil as a natural resource, 4th edn. Blackwell Publishing, Malden, 384 pp. ISBN 0-632-06455-2

Wrage N, Velthof GL, Beusichem ML, van Oenema O (2001) Role of nitrifier denitrification in the production of nitrous oxide. Soil Biol Biochem 33:1723–1732

Xu G, Fan X, Miller AJ (2012) Plant nitrogen assimilation and use efficiency. Annu Rev Plant Biol 63:153–182

Zhou S, Sugawara S, Riya S, Sagehashi M, Toyota K, Terada A, Hosomi M (2011) Effect of infiltration rate on nitrogen dynamics in paddy soil after high load nitrogen application containing ^{15}N tracer. Ecol Eng 37:685–692

Efficient Transformation of Agricultural Waste in India

Mrinalini Prasad, Rajiv Ranjan, Akbar Ali, Deepika Goyal, Arti Yadav, Teg Bahadur Singh, Preksha Shrivastav, and Prem Kumar Dantu

Contents

1 Introduction

Global warming is an alarming issue, and to control this condition the best way is to reuse resources. After 1947, agro-waste contains waste which is produced from agricultural products, agro-industries, animal feed, horticulture, aquaculture, etc. (Ungureanu et al. 2017; Sindhu et al. 2015). Throughout the world, approximately 140 billion metric tons of biomass is produced from agriculture every year, and in India about 500 metric tons of agro-waste is generated every year (Ministry of New and Renewable Energy, MNRE 2009) (Shehrawat and Sindhu 2012; Sindhu et al. 2015; Mahawar et al. 2015; Singh and Prabha 2017). However, there is an urgent need to reduce and reuse the agro-waste in proper and systemic manner like bio-composting, mushroom production, energy production, animal fodder, and many more (Lim and Matu 2015). In India, we still use the basic waste treatment processes that include draining the waste in the septic pond, open ground, burning, etc.

Agro-wastes are the leftovers after harvesting of crops, and it includes leaves, stem, etc. These wastes are large in size and low in protein and fat contents. Examples of waste-generating crops are wheat, paddy, sugarcane, mustard, bagasse, vegetable waste, tea, jute fiber, food products, wooden mill waste,

M. Prasad · R. Ranjan · A. Ali · D. Goyal · A. Yadav · T. B. Singh
P. Shrivastav · P. K. Dantu (*)
Department of Botany, Dayalbagh Educational Institute (Deemed University), Agra, Uttar Pradesh, India

M. Naeem et al. (eds.), *Contaminants in Agriculture*,
https://doi.org/10.1007/978-3-030-41552-5_13

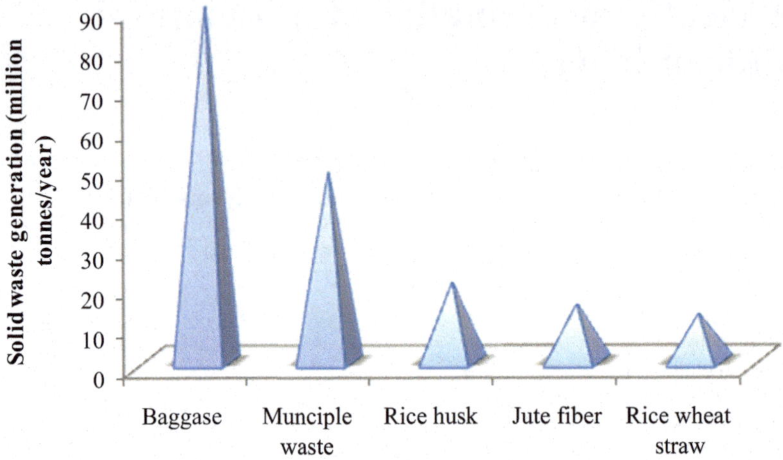

Fig. 1 Current situation of solid waste generation in India (million tons/year)

Table 1 Agricultural wastes and their utilization

Agro-waste	Reuses
Rice bran	Bran oil, de-oiled cake, cattle feed, wax, tar
Rice husk	Rice husk cement, husk board, furfural, silica, and black ash for bricks
Paddy straw	Straw board, straw paper, straw bags, handicraft products, shampoo, packing material
Whole banana plant	Feed, food, pharmaceutical, packaging, and many other industrial application
Fiber from pseudostem	Biodegradable ropes
Pith	Color absorber and even for food
Leaves	Feed, wrapping material, and thatching material

groundnut shell, coconut husk, cotton stalk, etc. (Maudgal 1995; Rai 1998; Sen 2002). Approximately 20% of agro-products are damaged due to poor postharvesting facility, and 10% are eaten by rodents (Lakshmi et al. 2017). The current position of solid wastes in India is shown in Fig. 1.

The choosing of any waste management process must be based on less environmental impact and maximum safety. Another main purpose of management is to reduce the quantity of agro-waste and recycle the organic matter (Scaglia and Adani 2008) (Table 1). Anaerobic and aerobic processes could be used to be convert waste into manure/compost or disposed of in landfills that will have less impact on the environment (Adani et al. 2004). Under waste management, there are many processes including collection, transport, processing, recycling or disposal, and checking of waste. For the recycling and processing of fruits and vegetables, waste microbial technology is used. In case of banana, a major crop in Maharashtra is cultivated in around 46,900 hectares area and produces a large quantity of waste after the harvest.

Another waste treatment technology is to prepare organic fertilizers by using a composting method. This technology helps in the improvement of soil fertility and crop production (Amoding 2007). Agriculture waste from mustard crop has been used for power generation by Kalpataru Power Transmission Ltd. in Rajasthan. The plan produces almost 8 MW of power (Hofman et al. 2004).

2 Agricultural Waste Characterization

Increasing population, progressing urbanization, and escalating living standards due to technology development day by day and its support to enhance the quantity and variety of solid wastes were produced by domestic, agricultural, mining, and industrial activities (Obi et al. 2016). The classification (Fig. 2) and categorization of agro-waste based on the agricultural activity is shown in Table 2.

There are different types of agriculture wastes:

I. *Cultivation activities*

   (a) *Rice hulls*: Generally, it is used as a fuel or as an abrasive character.
   (b) *Paddy waste*: Produced from paddy crop and their by-products are paddy husk, paddy straw, etc. They are used as animal feed, animal bed and shelter, mulching purpose, composting, and fuel purpose.
   (c) *Wheat waste*: The by-product of wheat crop is straw, and it is used as animal feed, particle board, dry flowers, briquettes, mats, hats, carpets, and many more handcrafts.
   (d) *Cotton waste*: Cotton sticks are the leftover of cotton crop. Utilization of sticks in power plant, plywood industries, particle board industries, and also in composting.

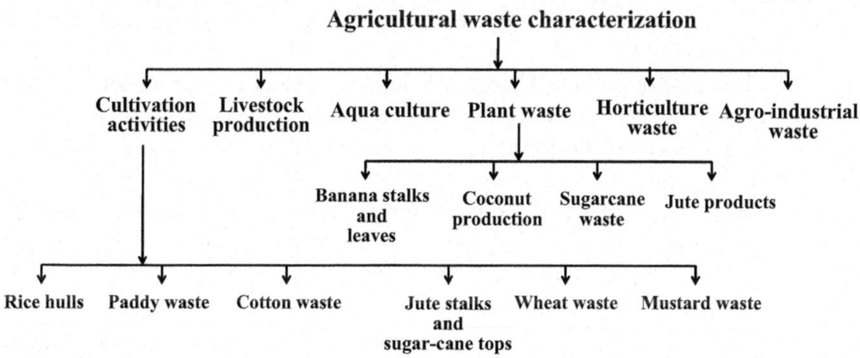

Fig. 2 Characterization of agricultural wastes

Table 2 Agricultural wastes based on the agro-activity

S. no.	Agricultural activity	Wastes
1.	Crop production and harvest	Straw, stover
2.	Fruit and vegetable processing	Biological sludges, trimmings, peels, leaves, stems, soil, seeds, and pits
3.	Sugar processing	Biological sludges, pulp, lime mud
4.	Animal production	Blood, bones, feather, litter, manures, liquid effluents
5.	Dairy product processing	Biological sludges
6.	Leather tanning	Fleshings, hair, raw and tanned trimmings, lime and chrome sludge, grease
7.	Rice production	Bran, straw, hull
8.	Coconut production	Stover, cobs, husk, leaves, coco meal

(e) *Jute stalks and sugarcane tops*: These by-products are utilized in chipboard, soft board, fabric, and matrices for acoustic purposes.
(f) *Mustard waste*: Two major by-products of mustard crop are mustard sticks and husk. Both products are sold to brick industries.

II. *Aquaculture*: Growth in aquaculture fully depends on feeds. But excess amount of feed becomes solid waste. Enhancement in temperature results in increased feeding which shows rising in waste generation. The very important aspect in aquaculture is water flow patterns. It reduces the fragmentation of fish feces and allows for rapid settling and concentration of the settleable solids.
III. *Livestock production*: Farmers used the livestock waste to make dung cakes, and it is better option for fuel generation. Some of them are used as biogas, composting, and bioinsecticide.
IV. *Plant waste* (fruit and vegetable processing)

(a) *Banana stalks and leaves*: These by-products are used as painting and waterproofing agent. Water after boiling with banana stalks and leaves mixed with lateritic clay has been used in Ghana as a waterproofing agent. This mixture is also used as painting, and it protects from heavy rains.
(b) *Coconut production*: There are many by-products of coconuts like husks, coir fiber, unretted coconut pith, retted pith, coconut shell, straw, etc. They are used as board making, fiber making, wood particles, roof, roads, mats, etc.
(c) *Sugarcane waste*: The leftover from the crop is sugarcane trash, growing green fodder, and bagasse. Sugarcane waste is also used in ethanol production, sugar manufacture, etc. (Figs. 3 and 4).
(d) *Jute products*: Jute stalks are the main by-product of the crop, and it is utilized in soft board making, matrices, and fabric for acoustic purpose. There are few factories which used jute stalks at a large scale: Bangalore Jute Factory and Narayanganj, Bangladesh.

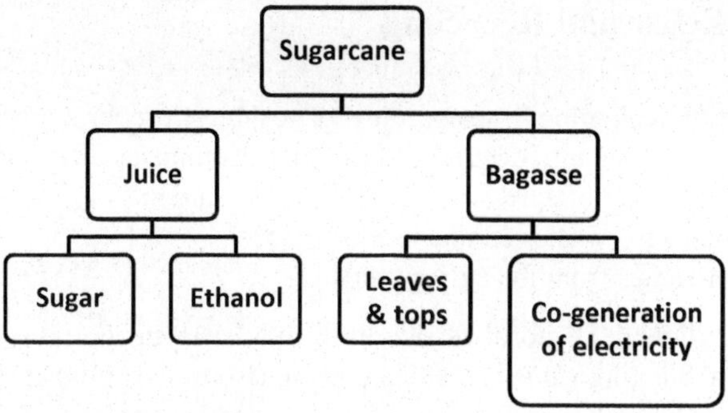

Fig. 3 Source of green energy (sugarcane)

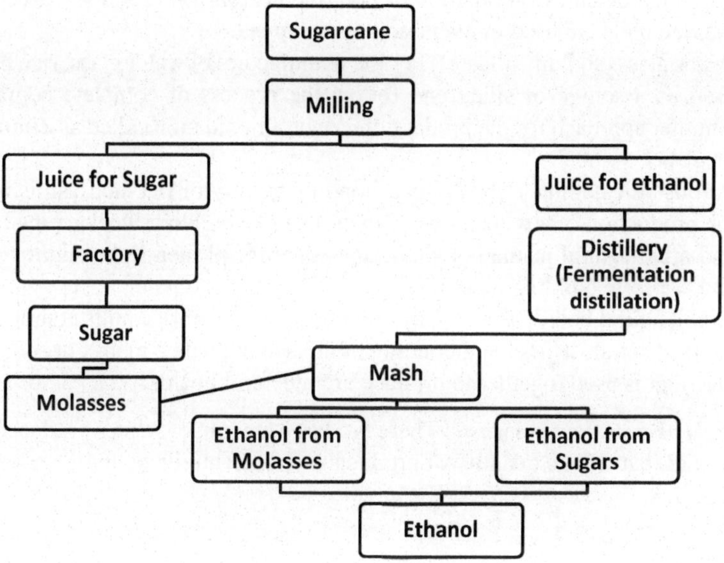

Fig. 4 Production of ethanol from molasses and juice

V. *Agro-industrial waste* (sugar processing): Sugar industry produces a large amount of bagasse per year that is being used in wall panels, insulation boards, printing paper, and also corrugating medium (Sen 2002; BMTPC 2005).
VI. *Horticulture waste*: Unused, spoiled, and damaged vegetables and fruits, branches, leaves, and dead plants are the example of horticulture wastes (Zhang et al. 2011). Generally, these wastes are converted into compost, animal feed. Unsold fruits and vegetables are also used as animal feed.

3 Waste Reuse and Recycling

Waste utilization technology means reuse of residues, proper storage system, and conversion of the waste into desired end product (Komnitsas 2012). However, there is a predicted life cycle of agro-waste (Fig. 7). There are lots of processes and applications to which agro-wastes can be used (Fig. 5).

Useful approaches from the agro-wastes

A. *Rice hulls*: For the disposal of rice hull, two basic categories were defined as follows: (a) Heating value and silica content (from rice hulls) are used in preparation of industrial products like Portland cement, porous silicate structural material (Pappua et al. 2007), and water glass. (b) Using chemical and physical properties, the hulls are converted into board and panel board formation.

Some other useful and beneficial uses of rice hulls:

(i) Portland cement: Combination of both the heating value and silica content of the rice hulls are used in manufacture of cement.
(ii) Water glass (sodium silicate): For the manufacture of water glass, rice hulls are used as a source of silica and follow the process of complete combustion. Another approach for the production is wet-air oxidation called as Zimmerman process.
(iii) Porous silicate materials: A good bonding quality of rice hull ash is used for the production of a wide range of materials like building blocks, pipe lagging, and architectural insulating slabs, aggregate for plaster, lightweight concrete, and concrete building blocks.
(iv) Architectural board: The physical bonding process of rice hulls comprises vulcanized cellulose fiber and bonding of hulls along with a matrix of sodium silicate that is used together to produce architectural boards.

B. *Products from coconut waste*: There is a large amount of wastes produced by the coconut, but a huge quantity of products is manufactured by these wastes.

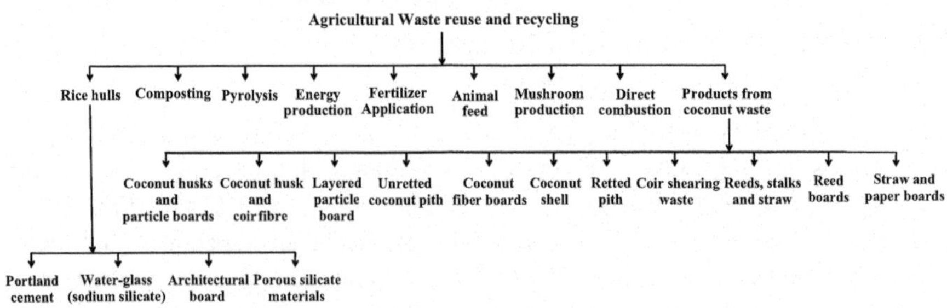

Fig. 5 Cataloging of agricultural waste reuse and recycle

(i) Coconut husks and particle boards: Particle boards made by coconut husks are more resistant to termites and wood-rotting fungi and marginally resistant to burning. The manufacture of board by husks is less costly as compared to wood.
(ii) Coconut husk and coir fiber: The combination of coconut husks, coir fiber, and Portland cement is used as strong building panel, corrugated roofing sheet, etc. The sheets are used in partition, walling, roofing, and many more.
(iii) Coconut fiber boards: The boards bonded with cashew nut glue are the best option for waterproofing. They are widely used for ancillary and temporary kind of construction.
(iv) Layered particle board: Normal proportion of adhesive with wood particle and coconut husks made a cost-effective board.
(v) Unretted coconut pith: Less density insulation boards are prepared with the combination of pith and wood fiber or banana stem fiber.
(vi) Retted pith: Coconut pith combined with cashew nutshell and liquid resin is used as filler in between roads, runways, and concrete slabs. It is resistant to termites, fungi, and moisture.
(vii) Coir-shearing waste: Coir fiber is used in preparation of mats and with the combination of pith and dust converted into particle boards.
(viii) Coconut shell: Building boards has been manufactured by the combination of coconut shell and urea formaldehyde or phenol formaldehyde glue.
(ix) Reeds, stalks, and straw: A low-cost house can be prepared by the assembly of dry reeds, stalks, and straw. Wall and boards can also be made with straw and cement.
(x) Straw and paper boards: Using Stramit process, the straw, vegetable fibers, and cereal straws are compressed under heat and pressure and converted into boards and craft paper.
(xi) Reed boards: These boards are produced for commercially building boards and wall making. They are very strong, lightweight, and less in cost.

C. *Manure application*: The animal manures are utilized as manures which supply 19% nitrogen, 38% phosphorus, and 61% potassium as compared to chemical fertilizer at the farm level (Pratt 1975). However, formation of fertilizer through animal manures has some consequences like high costs for transport, distribution, storage, odor, and groundwater contamination. Manure combined with soil increases its fertility, nutrient maintenance capacity, soil texture stability, and water-holding capacity (CAST Report No. 41. 1975).
D. *Adsorbents in the removal of heavy metals*: A large amount of heavy metals is produced due to industrialization and urbanization which is toxic to many life forms. Agricultural wastes are cost-effective alternative for the treatment of effluents having heavy metals through the adsorption process. Some agricultural wastes that have been used for elimination of heavy metals are bagasse (Mohan and Singh 2002), rice husk (Ayub et al. 2002), sawdust (Ajmal et al. 1996), coconut husk (Tan et al. 1993), oil palm shell (Khan et al. 2003), neem bark (Ayub et al. 2001), etc.

E. *Pyrolysis*: In this process, an agricultural waste is heated up to a temperature of 400–600 °C in the absence of oxygen resulting in a char material. Pyrolysis of agricultural waste yields oil, char, and low-heating-value gas. The main purpose of the process is alcohol for fuel, ammonia for fertilizers, glucose for food and feed, and biochar for soil supplement.
F. *Animal feed*: Imperfect quantity of protein sources is found in animal feed. To improve the quality of feed, some forage, pasture with grains, and protein-concentrated grains are supplied.
G. *Direct combustion*: Mainly the agricultural wastes are utilized in thermal conversion process, burning for heating, charcoal production, cooking, steam generation, and mechanical and electrical power generation, which are used as fuels and produce CO_2 and water. Of the wastes, 95% are converted into energy creation unit in the presence of oxygen (oxidation).
H. *Composting*: In this process, decomposition of organic materials under aerobic condition with the help of microorganism is performed. The main factor affecting the decomposed organic matter is oxygen and moisture. The best size of the decomposed matter is 0.5–1.5 inches (Geisel 2001).
I. *Wastes from mushroom production*: Mushroom is produced on natural materials from agriculture, manufacturing industries, and woodlands. The leftover of mushroom is used as manure, gardening, nursery making, vegetable growing, vermicomposting, animal feed, and biogas. Another use of wastes in *Agaricus bisporus* production is potting soil mixture, enriching soil, vermiculture, remediation of contaminated water, bioremediation, fuel, bedding for animal, animal feed, and many more. A mushroom processing unit is being run by a farmer in village Aterna of Sonipat district; his processing unit works using biomass energy.
J. *Biogas plant waste*: Manure is a useful and important element in biogas waste (Fig. 6). It includes dead stock, waste forage, and milk house waste and silage effluent (Lin et al. 2011; Liu 1995). This technology has remarkable way to improve energy release from agro residues, save plant nutrients, and develop health circumstances and quality of life in the villages (Fig. 7).

There are many reuses of agricultural waste but less awareness regarding the recycling of wastes in farmers. More farmers are using agro-wastes as a fuel but not

Fig. 6 Biogas production

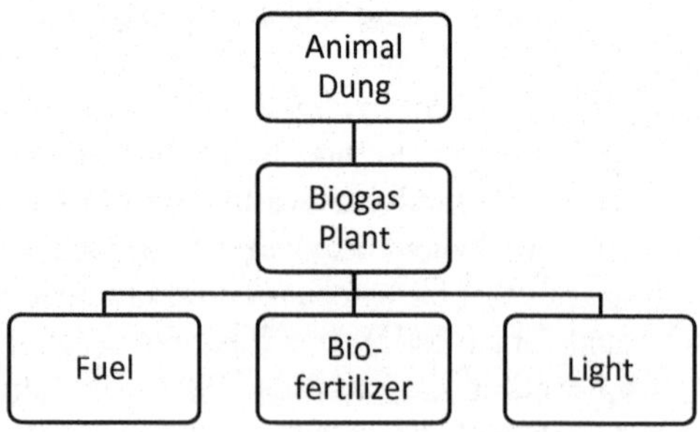

in other purposes. In Fig. 8, there is a survey with farmers' responsiveness regarding the management of agro-waste (Shehrawat et al. 2015).

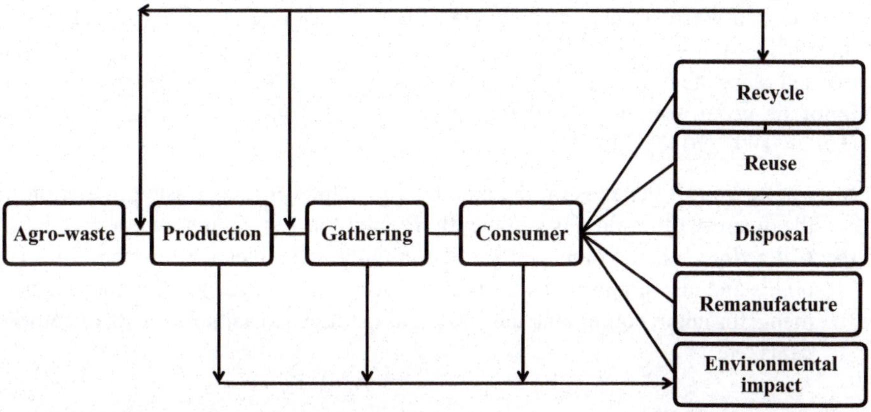

Fig. 7 Agricultural waste life cycle and reuse

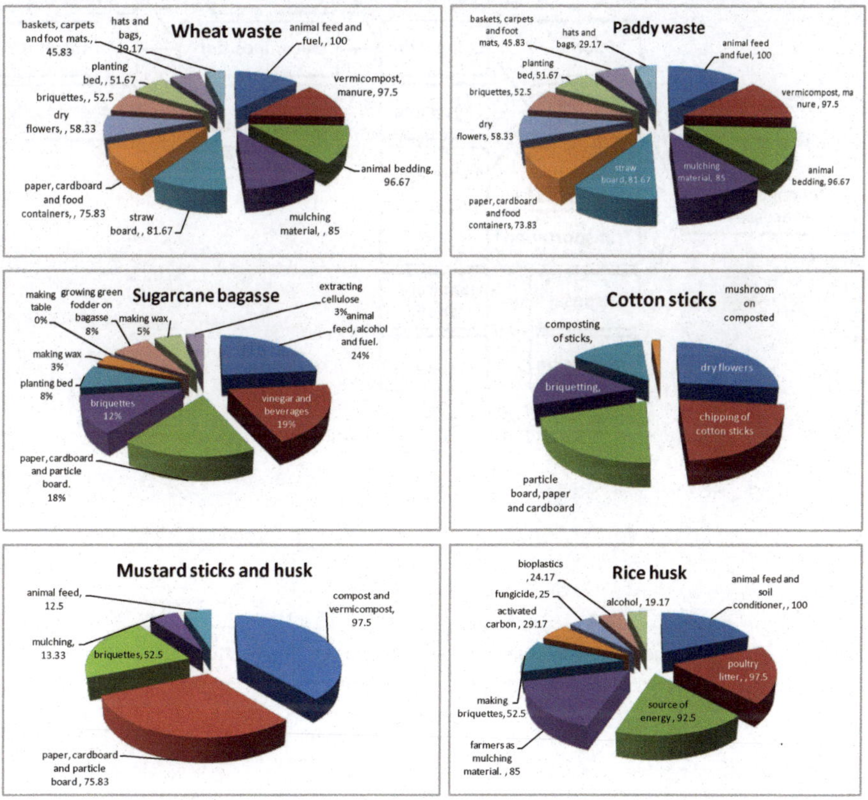

Fig. 8 Awareness regarding agricultural wastes

4 Waste Management and Optimization

A complete set of process of agro-waste management is shown in Fig. 9. An Agricultural Waste Management Organization (AWMO) (Hai and Tuyet 2010) is an intended system in which all the required systems are present and arranged to control and reuse the by-products of agro-wastes in a better manner to develop the quality of air, water, soil, plant, and animal resources. It consists of six basic functions (USDA 1992, 2012) (Fig. 10).

(a) *Production*: It depends on the amount and nature of agro-waste production. If the quantity of product is sufficiently formed for user, then no wastes are left.
(b) *Collection*: It is the management that initially assembles wastes from different points and also prepares a system for collection, method of collection, requirements for labor, equipment and structural facilities, installation costs of components, etc.

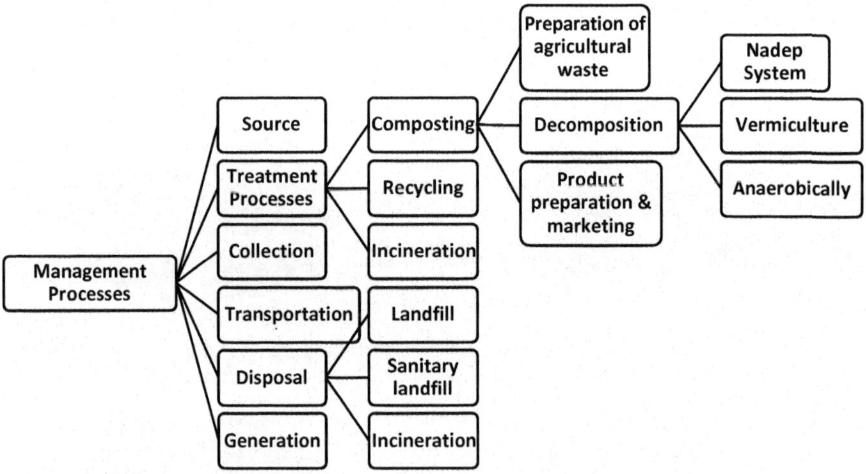

Fig. 9 A complete process of agricultural waste management

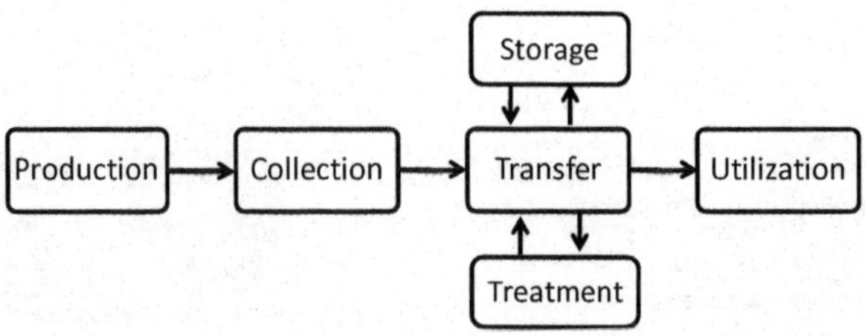

Fig. 10 Basic function of Agricultural Waste Management Organization (AWMO)

(c) *Storage*: It is a temporary solution to holding of the waste. This management organization contains storage volume, type, estimated quantity, site, period, and cost of the facility.
(d) *Treatment*: With the help of this management, pollution or toxic materials are reduced using physical, chemical, and biological treatment. It includes treatment type, specific size, site, cost, and process.
(e) *Transfer*: It refers to the transportation of wastes from collection point to the utilization site in a form of solid, liquid, slurry, etc.
(f) *Utilization*: It involves the conversion of wastes into valuable and reusable products. And it is also helpful in the renovation of nonreusable waste into environment (USDA 2012).

The management of agricultural waste is also an important aspect of the study. There are many factors which affect the management process (Fig. 10). For controlling the management, there is some support to control the management process in a proper manner (Fig. 11).

A. *Forest biomass*: It includes tree, biofuel biomass, farm animals, fast-growing woods, forest, etc. The renewable carbon sources are forest and agriculture. Agriculture is a part of bio-based economy, and it reduces greenhouse gas emission (Chandra et al. 2012). Wastes from forest are bark, sawdust, wood chips, timber slash, and mill scrap (Brar et al. 2013). Both have organic material like slurry, manure from livestock, silage, effluent, and crop waste (Bilgen and Sarıkaya 2016) (Table 3).

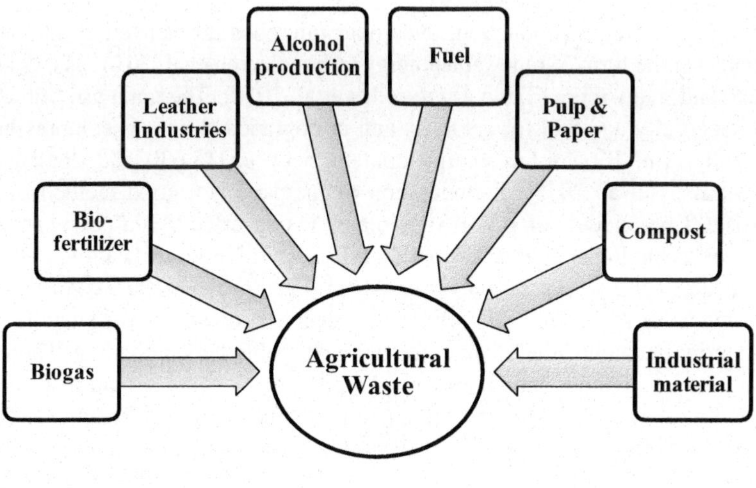

Fig. 11 Shows management of agro-wastes

Table 3 Carbon sources of agriculture and forest wastes

Wastes	C (%)	References
Coconut shell	63.45	Tsai et al. (2006)
Coffee grounds	54.61	Bok et al. (2012)
Corn stalks	50.02	Uzun and Sarioğlu (2009)
Cotton stalk	49.35	Fu et al. (2012)
Ordinary plywood	45.40	Jung et al. (2012)
Palm bark	50.10	Ngo et al. (2014)
Peanut shell	46.15	Wang et al. (2012)
Rice straw	50.93	Tsaia et al. (2006)
Sawdust	52.71	Lin and Kuo (2012)
Square timber	47.20	Jung et al. (2012)
Sugarcane bagasse	58.14	Tsaia et al. (2006)
Wood chip	51.07	Cao et al. (2011)

Table 4 A composition of biogas

Constituents	Percentage (%)
Methane, CH_4	55–75
Carbon dioxide, CO_2	25–45
Carbon monoxide, CO	0–0.3
Nitrogen, N_2	1–5
Hydrogen, H_2	0–3
Hydrogen sulfide, H_2S	0.1–0.5
Oxygen, O_2	Traces

B. *Biogas*: In biogas production, anaerobic microbes are converted into organic matter in the form of renewable energy (Abouelenien et al. 2014) as CO_2, methane, and many more (Table 4) (Karellas et al. 2010). The main use of biogas is in electricity and fuel for vehicles, and in upgraded form it is compressed into liquid form. Biogas has energy content between 18,630 and 26,081 kJ/m^3 (Matsakas et al. 2015). Fermentation of methane is a good method of using waste (Chmielewski et al. 2013). Biomass is converted to biofuel by using different technologies like anaerobic digestion, pyrolysis, gasification, carbonization, etc.
C. *Composting and briquetting*: Biological degradation and conversion of organic material in a controlled manner to support aerobic decomposition. There are a number of methods which explain the process of compost and also improve quantity. Examples are microorganism inoculation, bio-surfactants, and nutrition controllers (Zhang et al. 2011). Bacteria compost has a high thermal tolerance capacity. Briquette is a dense solid fuel made from loosely bound biomass. A uniform size of solid pieces of highly dense biomass is used as a fuel. It produces from bagasse, ground nutshell, dust, mustard stalk, cotton stalk, and press mud (Gangil 2014). Briquetting is a most sought-after fuel in industrial furnace systems using wood fuels. Uses of this biomass increase calorific value, charging of furnace, decrease of entrained particulate emis-

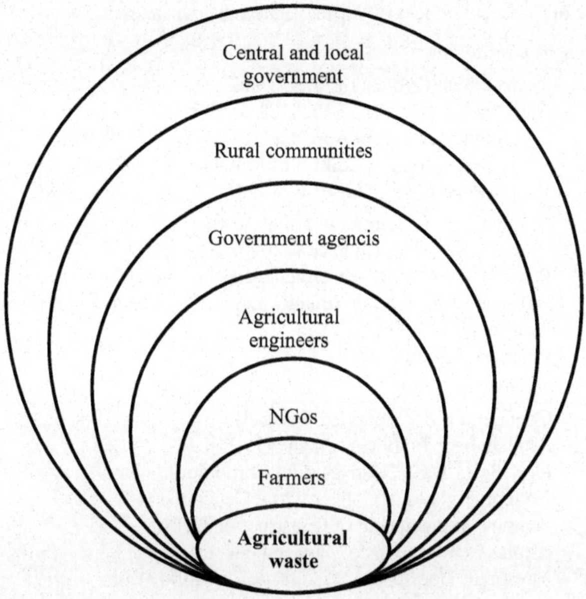

Fig. 12 Helping hands to improve agricultural waste management

sion, increase in combustion, equal size and shape, and replacement of natural wood fuel (Ndiema et al. 2002).

D. *Methane production*: Natural source of methane is wetlands, and it produces approximately 24.8% of the global atmosphere (Yu et al. 2013). It is produced by anaerobic digestion of forest and agro-wastes. Methane gas is generated from renewable resources, and it does not contain any greenhouse gases when released in the atmosphere. It is the most effective and useful technology for energy production from biomass. Methane gas is one of the greenhouse gases, and it is produced due to anaerobic digestion of inner portion of highly wet biomass (Fig. 12).

For waste management process, there are many rules and acts in India. These activities are very much important and useful for the protection of the environment. Some of other policies regarding the agro-waste management in India are mentioned in Policy Paper 49, 2010 (Agarwal et al. 2015). Sources of these regulations are given in Table 5.

5 Conclusion

Agricultural wastes are thrown in dumping lands and not reused properly leading to cause in various environmental problems. A prominent agro-waste utilization will help in developing agricultural and biofuel supply for many people.

Table 5 Rules, project, and polices of wastes related to environment

Law, policy, and regulation	
Legal framework	Environment (Protection) Act, 1986 National Environment Tribunal Act, 1995 National Environment Appellate Authority Act, 1997 Water (Prevention & Control of Pollution) Act, 1974 Water (Prevention & Control of Pollution) Cess Act, 1977
Policy initiatives	National Urban Sanitation Policy, 2008 National Environment Policy, 2006 Policy Statement for Abatement of Pollution, 1992 National Conservation Strategy and Policy Statement on Environment and Development, 1992 Ecomark Scheme, 1991
Technology and practices	
Key initiatives	Chennai: GPRS Equipped Waste Bin Ahmedabad: Tapping Methane Gas Goa: Solid Waste Management Corporation Nagpur: Bylaws to Collect Waste Generated in Hotels Nagpur: Management of Construction Debris Akola: CBO for Waste Management Yavatmal: Door-to-Door Collection of Solid Waste
Rural waste management	
Key projects	Tamil Nadu: Zero Waste Mgt. at Vellore District Maharashtra: SLWM at Dhamner Village Gujarat: Greywater Mgt. at Fathepura Village Maharashtra: Greywater Mgt. at Wadgaon Village Nashik: Wastepaper to Pepwood Kerala: Post-NGP Initiatives at Kattappana Village

For recycling the agro-waste, there is a huge scope for industries to come forward to recycle and manufacture the useful products from wastes. With the help of some lab processes and science and technology, we can create new era based on proper waste management. Many farmers even do not know how to utilize the agro-based wastes, and some of the farmers are aware regarding the waste reuses and management; they properly utilize the wastes. For the control of these wastes, some strategies are developed to reduce the waste, generate money, and develop employment. For soil fertility, forest and agricultural wastes are important and also used in bioenergy and industrial biotechnology. These waste materials are used as a raw material for building material, thermal production, engineering process, and many more. Some new technologies are used to recycle the agricultural and forest wastes for useful purposes. To develop a smart and proper management of agricultural wastes, farmers and public at large should be made aware of various practices for waste management. With the help of NGOs, private companies are creating attentive campaigns for the awareness and wakefulness of agro-waste uses and recycling as a useful product. Few examples are:

- Keep the people informed about the surrounding environment such as polluted river, air, and land.
- Attentiveness regarding the new policies that affect environment.
- Especially stop burning wastes.
- About the informal agro-waste management activities and its linkages.
- Information about the recycling, marketing, and distribution of the product.
- Separate system for waste and also proper collection, transportation, and recycling of waste.
- Say "no" to plastic bags.
- Encourage the people to start a vermiculture bin for the organic wastes.

A healthy and fresh environment can be developed by reducing the agro-wastes. By using prominent strategies for agro-wastes, well-planned management process and strict execution of the plan can help in the development of clean and bright India.

References

Abouelenien F, Namba Y, Kosseva MR, Nishio N, Nakashimada Y (2014) Enhancement of methane production from co-digestion of chicken manure with agricultural wastes. Bioresour Technol 159:80–87

Adani F, Tambone F, Gotti A (2004) Biostabilization of municipal solid waste. Waste Manag (New York, NY) 24(8):775–783

Agarwal R, Chaudhary M, Singh J (2015) Waste management initiatives in India for human well being. Eur Sci J 11(10):105–127

Ajmal M, Rao RAK, Siddiqui BA (1996) Studies on removal and recovery of Cr (VI) from electroplating wastes. Water Res 30(6):1478–1482

Amoding A (2007) Supply potential and agronomic value of urban market crop wastes. PhD thesis, Makerere University

Ayub S, Ali SI, Khan NA (2001) Efficiency evaluation of neem (*Azadirachta indica*) bark in treatment of industrial wastewater. Environ Pollut Control J 4(4):34–38

Ayub S, Ali SI, Khan NA (2002) Adsorption studies on the low cost adsorbent for the removal of Cr (VI) from electroplating wastewater. Environ Pollut Control J 5(6):10–20

Bilgen S, Sarıkaya I (2016) Utilization of forestry and agricultural wastes. Energy Sources Part A Recover Util Environ Eff 38(23):3484–3490

BMTPC (2005) Home page. http://www.bmtpc.org/fibre.pdf

Bok JP, Choi HS, Choi YS, Park HC, Kim SJ (2012) Fast pyrolysis of coffee grounds: characteristics of product yields and biocrude oil quality. Energy 47:17–24

Brar SK, Dhillon GS, Soccol CR (eds) (2013) Biotransformation of waste biomass into high value biochemicals. Springer, New York

Cao JP, Xia XB, Zhang SY, Zhao XY, Sato K, Ogawa Y, Wei XY, Takarada T (2011) Preparation and characterization of bio-oils from internally circulating fluidized-bed pyrolyses of municipal, livestock, and wood waste. Bioresour Technol 102:2009–2015

Chandra R, Takeuchi H, Hasegawa T (2012) Methane production from lignocellulosic agricultural crop wastes: a review in context to second generation of biofuel production. Renew Sust Energ Rev 16:1462–1476

Chmielewski AG, Urbaniak A, Wawryniuk K (2013) Membrane enrichment of biogas from two-stage pilot plant using agricultural waste as a substrate. Biomass Bioenergy 58:219–228

Fu P, Hu S, Xiang J, Sun L, Su S, An S (2012) Study on the gas evolution and char structural change during pyrolysis of cotton stalk. J Anal Appl Pyrolysis 97:130–136

Gangil S (2014) Beneficial transitions in thermo gravimetric signals and activation energy levels due to briquetting of raw pigeon pea stalk. Fuel 128:7–13

Geisel PM (2001) Compost in a Hurry. Compost in Hurry University of California, Agriculture and Neutral Recourses. Publication No 8037, pp 1–4

Handling Of Agricultural Wastes In Apmcs. Research report 2011–12. Niam Research Study

Hofman Y, Phylipsen GJM, Janzic R, Ellenbroek R (2004) Small scale project design document biomass Rajasthan ICC 30076

Jung SH, Kim SJ, Kim JS (2012) Characteristics of products from fast pyrolysis of fractions of waste square timber and ordinary plywood using a fluidized bed reactor. Bioresour Technol 114:670–676

Karellas S, Boukis I, Kontopoulos G (2010) Development of an investment decision tool for biogas production from agricultural waste. Renew Sust Energ Rev 14:1273–1282

Khan NA, Shaaban MG, Hassan MH (2003) Removal of heavy metal using an inexpensive adsorbent. In: Proceedings of UM research seminar 2003 organized by Institute of Research Management and Consultancy (IPPP), University of Malaya, Kuala Lumpur

Komnitsas K (2012) Best practices for agricultural wastes (AW) treatment and reuse in the Mediterranean countries, Project Number: LIFE10 ENV/GR/594

Lakshmi MV, Goutami N, Kumari AH (2017) Agricultural waste concept, generation, utilization and management. Int J Multidiscip Adv Res Trends IV(1(3)):1–4

Lim SF, Matu SU (2015) Utilization of agro-wastes to produce biofertilizer. Int J Energy Environ Eng 6(1):31–35

Lin TY, Kuo CP (2012) Study products yield of bagasse and sawdust via slow pyrolysis and iron-catalyze. J Anal Appl Pyrolysis 96:203–209

Lin J, Zuo J, Gan L, Li P, Liu F, Wang K, Chen L, Gan H (2011) Effects of mixture ratio on anaerobic co-digestion with fruit and vegetable waste and food waste of China. J Environ Sci 23(8):1403–1408

Liu JX (1995) Supplementing rice straw-based diets for ruminants. In: Ivan M (ed) Animal research and development: moving toward a new century. Proceeding of the 75th anniversary meeting of the Canadian Society of Animal Science, Ministry of Supply & Services Canada, Ottawa, Canada, pp 281–293

Mahawar N, Goyal P, Lakhiwal S, Jain S (2015) Agro waste: a new eco- friendly energy resource. Int Res J Environ Sci 4(3):47–49

Matsakas L, Rova U, Christakopoulos P (2015) Sequential parametric optimization of methane production from different sources of forest raw material. Front Microbiol 6:1163

Maudgal SC (1995) Waste management in India. J IAEM 22:203–208

Mohan D, Singh KP (2002) Single and multi-component adsorption of cadmium and zinc using activated carbon derived from bagasse – an agricultural waste. Water Res 36:2304–2318

Ndiema CKW, Manga PN, Ruttoh CR (2002) Influence of die pressure on relaxation characteristics of briquetted biomass. Energy Convers Manag 43:2157–2161

Ngo TA, Kim J, Kim SS (2014) Characteristics of palm bark pyrolysis experiment oriented by central composite rotatable design. Energy 66:7–12

Obil FO, Ugwuishiwu BO, Nwakaire JN (2016) Agricultural waste concept, generation, utilization and management. Niger J Technol 35(4):957–964

Pappua A, Saxenaa M, Asolekar SR (2007) Solid wastes generation in India and their recycling potential in building materials. Build Environ 42:2311–2320

Pratt PF CAST Report (1975) Utilization of animal manures and sewage sludges in food and fiber production. Report No. 41 of the Council for Agricultural Science and Technology. Ames, Iowa: Headquarters Office: Department of Agronomy, Iowa State University. 50010

Rai M (1998) Building materials in India, 50 years: a commemorative volume. Building Materials & Technology Promotion Council, New Delhi

Scaglia B, Adani F (2008) An index for quantifying the aerobic reactivity of municipal solid wastes and derived waste products. Sci Total Environ 394(1):183–191

Sengupta J (2002) Recycling of agro-industrial wastes for manufacturing of building materials and components in India. An over view. Civil Eng Constr Rev 15(2):23–33

Shehrawat PS, Sindhu N (2012) Agricultural waste utilization for healthy environment and sustainable lifestyle. Third international scientific symposium "Agrosym Jahorina 2012", pp 393–399

Shehrawat PS, Sindhu N, Devi P (2015) Agricultural waste awareness and utilization for healthy environment and sustainable livelihood. Sci Pap Ser Manag Econ Eng Agric Rural Dev 15(2):371–376

Sindhu NP, Seharawat SP, Malik JS (2015) Strategies of agricultural waste management for better employment and environment. Int J Curr Res 7(12):24604–24608

Singh DP, Prabha R (2017) Bioconversion of agricultural wastes into high value biocompost: a route to livelihood generation for farmers. Adv Recycl Waste Manage 2(3):1–5

Tan WT, Ooi ST, Lee CK (1993) Removal of chromium (VI) from solution by coconut husk and palm pressed fibre. Environ Technol 14:277–282

Tsaia WT, Lee MK, Chang YM (2006) Fast pyrolysis of rice straw, sugarcane bagasse and coconut shell in an induction-heating reactor. J Anal Appl Pyrolysis 76:230–237

Ungureanu G, Ignat G, Vintu CR, Diaconu CD, Sandu IG (2017) Study of utilization of agricultural waste as environmental issue in Romania. Rev Chim 1(3):570–575

USDA (1992) Agricultural waste management field handbook. Part 651 of National engineering handbook, USA. Department of Agriculture (Washington).Soil Conservation Service, Washington, DC, pp 1–431

USDA (2012) Agricultural waste management field handbook. United States Department of Agriculture, Soil Conservation Service, Washington, DC

Uzun BB, Sarioglu N (2009) Rapid and catalytic pyrolysis of corn stalks. Fuel Process Technol 90(5):705–716

Wang X, Chen H, Yang H, Wang J, Xin F (2012) The influence of alkali and alkaline earth metal compounds on pyrolysis of peanut shell. Asia Pac J Chem Eng 7:463–468

Yu B, Stott P, Li HYX (2013) Methane emissions and production potentials of forest swamp wetlands in the eastern great Xing'an Mountains, Northeast China. Environ Manag 52(5):1149–1160

Zhang F, Gu W, Xu P, Tang S, Xie K, Huang X, Huang Q (2011) Effects of alkyl polyglycoside (APG) on composting of agricultural wastes. Waste Manag 31:1333–1338

Impact of Biofertilisers on Crop Production Under Contaminated Soils

Indu Rialch, B. S. Bhople, and Anil Kumar

Contents

1 Introduction

1.1 *Biofertiliser: Significance in Sustaining Crop Productivity and Soil Health*

Soil provides a vital habitat for various organisms including microbes such as bacteria, actinomycetes, fungi, etc. Soil microbes play a crucial role in regulating various soil reactions, organic matter decomposition, nutrient recycling and soil health improvement, thereby influencing crop quality and productivity. The soil microorganisms convert organic residues into biomass or mineralise them to CO_2, H_2O, inorganic nitrogen,

I. Rialch
Department of Plant Breeding & Genetics, Punjab Agricultural University, Ludhiana, India

B. S. Bhople (✉)
Regional Research Station, Punjab Agricultural University, Ballowal Saunkhri, Punjab, India
e-mail: bsbhople@pau.edu

A. Kumar
Farm Science Centre, Guru Angad Dev Veterinary & Animal Sciences University, Tarn Taran, Punjab, India

M. Naeem et al. (eds.), *Contaminants in Agriculture*,
https://doi.org/10.1007/978-3-030-41552-5_14

phosphorus and many other nutrients including trace elements. These beneficial microorganisms also benefit soil–plant system in several ways through production of various biomolecules like enzymes, vitamins, antibiotics, hormones, organic acids, etc., that have the ability to bind soil particles leading to aggregate formation and improved soil structure (Harrier and Watson 2003; Kumar et al. 2015a, b, 2017; Suri et al. 2013).

In the present context, there is a growing concern about environmental hazards and threats to sustainable agriculture. The studies involving biofertilisers revealed that the long-term use of biofertilisers is economical, eco-friendly, more efficient, productive and accessible especially to marginal and small farmers in comparison with chemical fertilisers. Several researchers have evaluated that utilisation of soil microbes as biofertiliser in crop production not only improves crop quality and production (Kumar et al. 2017) but also exhibits significant influences on soil physical, chemical and biological properties (Bhardwaj et al. 2014; Bai et al. 2016a, b). Soil microbes such as *Rhizobium*, *Azotobacter*, etc., associated with legume roots or free living in soil supplements N supply to plants through biological N_2-fixation, while phosphate-solubilising bacteria, mycorrhizal fungi, etc., enhance the P availability and other nutrients especially the immobile ones from the soil (Suri et al. 2011; Bhat et al. 2015; Bai et al. 2017; Yadav et al. 2015a, b; Kumar et al. 2016a, b, c). The AM fungi also play an important role in P transformation (Kumar et al. 2014). Similar to the aforementioned microorganisms, there are several groups of microbes in soil that benefit from soil–plant system in different ways either directly or indirectly.

Primarily, soil microbes are responsible for organic matter decomposition. Several groups of soil microbes work on organic matter decomposition to humus formation, which is very fine material having very high surface area, possesses the ability to hold positively charged nutrients and retains soil moisture. Soil microbes such as phosphate-solubilising microorganisms play a crucial role in conversion of organic forms of nutrients in inorganic ones (mineralisation) by secreting various types of organic acids and enzymes (Rodriguez and Fraga 1999; Puente et al. 2004; Sharma et al. 2013). These inorganic or mineral forms of nutrients are then easily absorbed by growing plants. Certain groups of soil microorganisms such as AM fungi secrete polysaccharides and glycoproteins that have the ability to bind soil particles and form aggregates, thereby improving soil structure and overall physical properties of the soil (Wright and Upadhyaya 1998; Wright et al. 1998; Harrier and Watson 2003). Moreover, hyphae of AM fungi develop an extensive extra-radical hyphal network that grows into the soil matrix and holds primary soil particles together via physical entanglement. This hyphal network plays a crucial role in soil texture improvement and, in turn, water relations (Hamblin 1985; Tisdall 1991; Staddon et al. 2003; Rillig 2004). Nitrogen fixation is an important process carried out by soil microbes, that is, bacteria, especially by *Rhizobium* (symbiont) and *Azotobacter* (free living). The above bacteria has the ability to convert atmospheric nitrogen to ammoniacal form, thereby enriching soil with plant available nitrogen (Kass et al. 1971; Mila and Shamsuddin 2010). Certain soil fungi (*Trichoderma*) serve as biocontrol agents against fungal root diseases of plants (Harman 2006).

Moreover, some genera of fungi are utilised to control insect pests (Sahoo et al. 2013). Thus, the use of soil beneficial microorganisms as biofertiliser not only enhances nutrient and water use efficiencies of the crops but also improves overall soil health, crop quality and productivity in the long term.

2 Mechanism of Action of Various Biofertilisers

2.1 Arbuscular Mycorrhizal Fungi

2.1.1 Mechanism of Nutrient Absorption by AM Fungi

The arbuscular mycorrhizal fungi (AM fungi) expand the surface area of plant root system through ramification of hyphae and thus extend the exploratory area of plant roots for harnessing nutrients and water (Marschner and Dell 1994). The researchers have explored that in mycorrhizal plants, numbers of extension hyphae are usually far more in number as compared to root hairs of plant; the area of surface where AM fungi, plant and soil interacted increased greatly (Fig. 1), resulting into more nutrient and water absorption (Suri and Choudhary 2013a; Bai et al. 2016b; Kumar et al. 2016b). The AM fungi release low molecular weight organic acids such as oxalic, malic acids, etc., that have the ability to solubilise inorganic forms of phosphates; thus, P is released into soil solution and absorbed by the plants (Zou et al. 1995; Choudhary et al. 2013). Moreover, AM fungi attack complex organic compounds through secretion of various enzymes (*chitinase, peroxidase, cellulase, protease, phosphatase,* etc.) and converting them into simple ones, which can be taken up and utilised by fungi/host plants to fulfil their energy requirements for growth and reproduction (Chen et al. 2007).

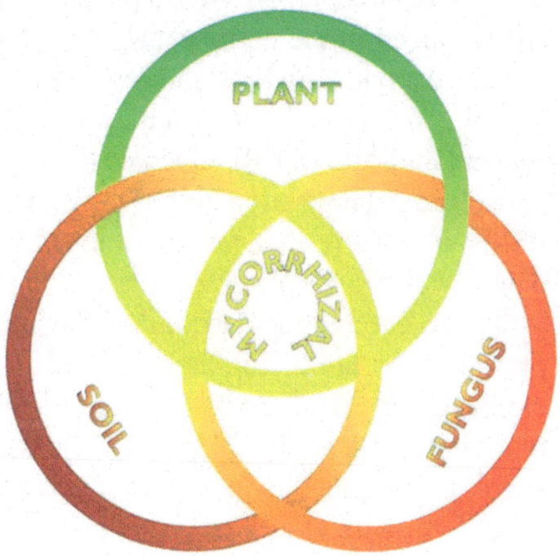

Fig. 1 Association showing interactions between AM fungi, plant and soil (Brundrett et al. 1996)

2.1.2 Mechanism of Water Absorption by AM Fungi

The AM fungi-inoculated plants explore larger volume of soil profile through extension of root system by developing higher order laterals by ramification of fungal hyphae associated with it (Song 2005; Suri and Choudhary 2013b), thus absorbing water from larger area of soil profile as well as from deeper soil layers. The main absorption apparatus of mycorrhizal fungi is extension hyphae having a diameter of 2–5 μm, which penetrate soil pores inaccessible to root hairs (10–20 μm) and hence absorb water from these pores which otherwise is not available to non-mycorrhizal plants (Gong et al. 2000). In addition, colonisation of plant roots with AM fungi might change the root architecture and enhance the interaction area of root and soil (Atkinson et al. 1994). Studies undertaken by Hamblin (1985), Tisdall (1991), Staddon et al. (2003) and Rillig (2004) revealed that AM fungi inoculation improves soil structure by binding of soil aggregates with their hyphal network and enhances moisture retention capacity of the soil. The fungal hyphae have a unique capacity of producing glomalin (a glycoprotein) that has ability to bind soil particles and form aggregate. The aggregation improves soil structure and moisture retention capacity (Wright and Upadhyaya 1998; Wright et al. 1998). As per reports in the literature, AM symbiosis also enhances resistance of plants towards various biotic and abiotic stresses (Harrier and Watson 2003).

2.2 *Phosphate-Solubilising Microorganisms (PSMs)*

2.2.1 Mechanisms of Inorganic Phosphate Solubilisation by PSMs

Several theories elaborated the mechanism of inorganic phosphate solubilisation by PSMs; however, most of theories primarily put emphasis on mechanism involving production of siderophores, organic acids, hydroxyl ions, protons, etc., that dissolve mineral compounds and make them available for plant use (Rodriguez and Fraga 1999; Sharma et al. 2013). As per the concept of Zhao et al. (2014), organic acids are produced in the periplasmic space by direct oxidation pathway. Organic acids produced along with their carboxyl and hydroxyl ions reduce the pH or cause chelation of cations to release P in the soil solution (Seshachala and Tallapragada 2012). As per Goldstein (2000), gluconic acid is one of the most frequent agents of mineral phosphate solubilisation amongst different organic acids produced and released by PSMs; it actually chelates the cations bound to phosphate and in turn makes phosphate available for plant use.

The researchers also explained the mechanisms, where PSMs solubilise mineral phosphate by producing inorganic acids, namely, carbonic, sulphuric, nitric acids, etc., and certain chelating substances. With time, however, the researchers found that the organic acids released by PSMs are more effective in releasing phosphorus in soil as compared to inorganic acids and chelating substances produced by PSMs. Therefore, Kim et al. (1997) suggested that organic acid production by PSMs for

P solubilisation is not only the cause for increased phosphorus concentration in culture medium; rather, liberation of enzymes or enzymolysis by PSMs also plays a critical role in phosphate solubilisation (Zhu et al. 2011).

2.2.2 Mechanisms of Organic Phosphorus Mineralisation by PSMs

Halvorson et al. (1990) proposed sink theory of solubilisation of organic P, where they highlighted that the continuous removal of P results in dissolution of Ca-P compounds. As per the concept proposed by Dighton and Boddy (1989), phosphorus decomposition in organic substrates is directly related with P content in the biomass of phosphate-solubilising microorganisms. The studies carried out by researchers elaborated that several groups of enzymes are associated with biological process of organic phosphorus mineralisation by PSMs. One group of enzymes has dephosphorylate, phosphor-ester or phosphoanhydride bond of organic compounds, which are nonspecific acid phosphatases (NSAPs). Amongst various NSAP enzymes released by phosphate-solubilising microorganisms, phosphomonoesterases are mostly studied, which are also called as phosphatases (Nannipieri et al. 2011). The aforementioned NSAP enzymes can either be acid or alkaline phosphomonoesterases (Jorquera et al. 2011). Another enzyme produced by PSMs in organic-P mineralisation process is phytase, which is responsible for the release of P from organic material stored in the form of phytate and makes it available for plant use (Richardson and Simpson 2011).

2.2.3 Mechanism of N-Fixation by Rhizobium

As we know, legume crops such as pea, lentil, berseem, pulses, clovers, etc., form a symbiotic relationship with soil-dwelling bacteria that takes gaseous nitrogen from the air present in soil pores and feeds it to the legume crop plants, and in turn the plant provides carbohydrates to the bacteria for its growth and reproduction; due to this reason, legume crops are said to 'fix' atmospheric nitrogen (*N-fixation*). Likewise other beneficial soil microbes, *Rhizobacteria*, are also present naturally in the soil, but due to their low populations, they did not maximise nitrogen fixation. Hence, inoculation of seed with *Rhizobium* biofertiliser culture is usually recommended to attain a maximum potential of N-fixation by legumes.

The actual process of N-fixation starts with nodule formation in the root of legume plant. *Rhizobia* (bacteria) invade legume root and multiply within cortex cells. The plant supplies all the necessary nutrients and energy for growth and multiplication of the bacteria. Within 6–7 days after infection, small nodules appear and are visible with naked eyes. Depending on legume species and germination conditions, small nodules can be seen within 2–3 weeks after sowing. Initially at younger stage, nodules are usually white or grey inside (*yet not started fixing N*), but as nodules grow in size, they gradually turn pink or reddish in colour (*N-fixation started*). Leghemoglobin (*controls oxygen flow to the bacteria*) imparts pink or red colour to the nodules. The root nodules that are no longer able to fix nitrogen turn green and may usually be discarded by the plant. In general, pink nodules must predominate on the roots of legume plant during mid of growing season, as pink nodules are considered to be most efficient and active in N-fixation. However, the predominance of white, grey or green nodules in the roots of legume indicated inefficient *Rhizobia* strain that led to poor N-fixation and in turn resulted in poor plant nutrition, pod filling and susceptibility of plant to various stresses.

The legume–*Rhizobium* symbiosis is a mutual association, and several researchers have widely exploited symbiotic N-fixation as a medium for increasing crop yields (Boholool 1990; Sharma et al. 1993). There are several genera of *Rhizobia* that belong to the *Rhizobiales*. They are characterised by their unique ability to infect root hairs of legume plant and bring out effective N_2-fixing nodules (Mila and Shamsuddin 2010). In soil, leguminous plants usually secrete dicarboxylic acid exudates that attract *Rhizobium* bacteria. As evident from studies, flavonoids play a crucial role in attracting the bacteria as they are easily absorbed through the membrane of organisms (Maj et al. 2010). Once the bacteria detect these chemicals, they actively move towards legume root and attach to it. Besides attracting bacteria, exudates and flavonoids also play a crucial role in activating genes involved in producing 'Nod factors' (Maj et al. 2010).

For the preparation of symbiotic relationship, *Rhizobium* attraction towards legume roots is usually followed by transcription of 'Nod genes'. Nod factors in turn stimulate the branching of root hair, hydrolysis and deformation of cell wall. In addition to attraction of *Rhizobium*, exudates and flavonoids also change the plant roots making it easier for the *Rhizobium* to enter the cells of the root hair for symbiosis. When the *Rhizobium* bacteria come in contact with root hair, they invade plasma membrane of the cells. As the bacterium penetrates the cell, the plant produces new cell wall material at the site that covers the bacteria as well as allows them to enter deeper into the root hairs (Gage 2017). Similar explanations have also been provided by Matiru and Dakora (2004), Dakora (1995) and Lhuissier et al. (2001), where they highlighted that different species such as *Rhizobium*, *Mesorhizobium*, *Bradyrhizobium*, *Azorhizobium*, *Allorhizobium*, *Sinorhizobium*, etc., respond chemotactically to flavonoid molecules released as signals by legume plant (host) and form intimate symbiotic relationships with them. Above plant compounds induce the expression of nodulation (nod) genes in *Rhizobia* and produce lipo-chitooligosaccharide signals that trigger mitotic cell division in roots and lead to nodule formation.

2.2.4 Mechanism of N-Fixation in Anabaena azollae

Azolla is a freshwater floating fern (a pteridophyte) and lives in symbiotic relationship with a diazotrophic cyanobacterium. All the species of this genus harbour a filamentous nitrogen-fixing cyanobacterium in their fronds that is usually referred to as *Anabaena azollae* (*Nostocaceae*) (Papaefthimiou et al. 2008). The *Azolla* occurs naturally on the surface of the lakes, slow-moving rivers, canals, ponds, etc., and in warm temperate to tropical climates. The *Azolla* has the ability to fix atmospheric nitrogen; hence *Anabaena–Azolla* association holds the potential to substitute application of nitrogenous fertilisers, if used as biofertiliser. Besides the above benefit, *Azolla* is also used as 'green manure' in several countries to fertilise paddy field and play a significant role in enhancing yield (Van Hove and Lejeune 2002) as it has the potential to fix more nitrogen as compared to plants. As per reports in literature, *Anabaena–Azolla* has the capacity to fix nearly about 1 kg N ha^{-1} day^{-1} in paddy field, thereby providing sufficient nitrogen for sustainable rice cultivation. Moreover, owing to its faster multiplication rate, *Azolla* covers the surface of water bodies very rapidly, thus helping to reduce the volatilisation of water and ammonia in rice fields.

The *Azolla* sporophyte bears a multibranched rhizome originating, on ventral surface, adventitious roots hanging down into the water in order to absorb nutrients directly. Further, rhizome has small leaves (about 1 mm in length) consisting of an aerial chlorophyllous dorsal lobe and a partially submerged colourless ventral lobe, which is cup-shaped to provide buoyancy. Dorsal lobe contains a specialised cavity, where cyanobiont is permanently housed. The interior surfaces of the mature cavities which are ellipsoid in shape are covered with mucilaginous layer, where usually 2000–5000 cyanobacterial cells are embedded and immobilised. There are several trichomes (hairs) that extend from the cavity surface into the mucilage layer and establish an intimate contact between the symbiotic partners, thus helping in the exchange of metabolites. Hence, leaf cavity is one type of natural microcosm having a self-organisation and an ecological well-defined structure. This behaves as both physiological and dynamic interface units of symbiotic relationship, where main metabolic and energetic flows occur (Peters and Perkins 1993; Rai 2000).

The *Anabaena* filaments lack heterocyst in the younger leaves of the water fern, whereas in mature leaf cavity, these gradually increase in frequency to 30–40%, relative to photosynthetic cells, reaching the cyanobacterial cells in *Anabaena* population of mature leaf cavities. The 50–90% of fixed nitrogen in the form of ammonia is delivered to the fern by *Anabaena*. Carbohydrate is synthesised in vegetative cells probably in the form of glucose and moves into heterocysts. In this way, nitrogen fixed in heterocysts moves to the vegetative cells in the form of amino acids (Herrero and Flores 2008).

2.3 Soil Contamination and Agriculture

Soil contamination is becoming a major confront that we need to overcome for establishing a healthy environment (Okrent 1999). A large part of bacterial biodiversity, other microscopic and macroscopic living organisms occur in the soil. In general, soil contamination is a major problem at several stages. The groundwater which interacts with and goes underneath the soil could also become contaminated due to soil contamination. Further, the contaminant (*heavy metals and pesticides*) passes to animals feeding on vegetation grown in contaminated soil, and similar is the case with humans (Kirpichtchikova et al. 2006).

2.4 Sources of Soil Contamination

Several sources are responsible for the contamination of soils. Past land use that has used substances may have probably entered the soil as contaminant (Raymond and Okieimen 2011). A gas station or mechanics garage is a perfect example for this, where different fuels and lubricants may have entered the soil inadvertently through poor storage practices or spillage onto the ground leading to contamination of the soil. There are several other good examples highlighting different sources of contamination that affect soils directly or indirectly such as microplastics, oil spills, intensive farming systems, agrochemicals (pesticides, herbicides, fertilisers, etc.), petrochemicals, industrial accidents, waste disposal, etc. The different techniques to overcome soil pollution are as follows:

2.5 Physical Soil Remediation Techniques

Physical soil remediation techniques in general involve soil washing, vitrification and encapsulation of contaminated soils/areas by impermeable vertical and horizontal layers, electrokinesis and permeable barrier systems (Audrone and Vasarevicius 2005). Encapsulation of contaminated areas is commonly used for remediation by pollution prevention or by containment. Most of above techniques have been adapted for the use in the field of environmental engineering from the watertight encapsulation of construction pits. There is extensive literature available on most of these techniques, available for further reading.

2.6 Biological Soil Remediation Techniques

The biological remediation techniques are performed in situ and include microbial remediation, phytoremediation, fungal remediation and composting techniques.

2.6.1 Microbial Remediation

In microbial remediation, microbes degrade the contaminants into a less toxic form. The microbial remediation technique proved to be very effective in the treatment of hydrocarbons and pesticides. The cost of this technique is relatively low and less time consuming as compared to other techniques of soil remediation; however, there is possibility of increased toxicity of certain metals.

2.6.2 Phytoremediation

The process of using plants to extract contaminants or to degrade them in the soil is known as phytoremediation. Effectiveness in bringing soil up to agricultural standard varies because for one type of contaminant only one plant species is generally used, potentially leaving a range of contaminants behind. Moreover, contaminated plants used for extraction must be disposed of.

2.6.3 Fungal Remediation

The use of certain species of fungus to degrade contaminants is known as fungal remediation. Remediation of contaminants following different species of fungus is still in the development phase and is not commercially available till now.

2.6.4 Compost Remediation

This remediation technique involves the addition of compost to the soil. This is a cheaper and quick method of remediation of contaminated soils. However, this technique is not considered a true remediation technique because the contaminants usually remained intact in the soil. The addition of compost in soil, however, could be used to create a raised bed, where plant roots cannot reach the contaminated soil. The bioremediation techniques in general are conditionally effective in bringing soil up to agricultural standard. Phytoremediation may take longer time to show effects, and the plants used must be disposed of after the completion of the project. However, these techniques are inexpensive and easy to implement and are environmentally friendly (Azubuike et al. 2016).

The following mechanisms are involved in soil contamination:

- Deposition of solid waste
- Accumulation of non-biodegradable materials
- Toxification of chemicals into poisons
- Alteration in soil chemical composition, that is imbalance of chemical equilibrium

Some of Agriculture Measures to Control Soil Contamination

(i) Reduction in the usage of pesticides
(ii) Judicious use of chemical fertilisers along with organic ones
(iii) Improved crop production techniques to ensure less weed growth
(iv) Dumping of wastes in garbage pit to prevent soil pollution
(v) Controlled grazing of animals and ensuring best forest management

(vi) Reduction in wind erosion through plantation of wind breaks and wind shield
(vii) Afforestation and reforestation

3 Role of Biofertiliser in Bioremediation

Pesticides are regarded as one of the indispensible means of agricultural production. Soil-applied as well as foliar-applied pesticides contaminate soil directly and after wash off crop stands. There are several microorganisms which have been used to improve the supply of nutrients to crop plants for their vigorous growth as well as to restrict the activity of plant pathogens. They also play an important role to improve the physical health of the soils in numerous ways. Other more recent objectives for the introduction of microorganisms into soil are the mineralisation of organic pollutants (bioremediation of polluted soils, Van Veen et al. 1997).

As a detrimental consequence of environmental stresses, productivity of crops is declining at an unprecedented rate. Our too much dependence on chemical fertilisers and pesticides has encouraged the industries to produce life-threatening chemicals as a form of pesticides or fertilisers. To tackle this adverse condition, biofertiliser can put aside the agriculture from the severity of various environmental stresses (Mahanty et al. 2016).

Although PGPR are mainly considered for promoting the plant growth and disease control, much attention has recently been focused on xenobiotic bioremediation using PGPR (Bishnoi 2015). As bio-inoculants, PGPR are widely used to support survival of plants under stressed conditions, such as pesticide contamination of soil.

4 Case Studies on Bioremediation Using Biofertilisers

The isolated new bacterium (*P. rhizophila* S211), from an agricultural contaminated soil, displayed both pesticide solubilising and plant-growth-promoting activities and genes involved in xenobiotic biodegradation (Hassen et al. 2018).

The three *Pseudomonas* strains (K03, Y04 and N05) isolated from tobacco seeds that could produce siderophores, indole-3-acetic acid and 1-aminocyclopropane-1-carboxylate deaminase fix nitrogen, dissolve phosphorus and potassium and tolerate heavy metals. The Pb stabilisation in soil and reduction of Pb in tobacco content might be due to the rational application of the above species (Li et al. 2019).

The Indian mustard in conjunction with rhizospheric bacteria can be used for enhancing plant Se accumulation, and volatilisation can be used for the removal of heavy metals such as Se from contaminated soils in the San Joaquin Valley and other places where Se contamination is a problem (Mark et al. 1999).

The overall 11 cadmium-tolerant bacterial strains were isolated from the root zone of Indian mustard (*Brassica juncea L. Czern.*) seedlings grown in

Cd-supplemented soils as well as sewage sludge and mining waste highly contaminated with Cd. The ability of these bacteria to protect plants against the inhibitory effects of high concentrations of heavy metals is related to the bacteria providing the plants with adequate iron (Belimov et al. 2005).

A pot experiment was conducted with bioremediation strategies: natural attenuation, phytoremediation with alfalfa (*Medicago sativa* L.), bioaugmentation with *Pseudomonas aeruginosa* and bioaugmentation-assisted phytoremediation, for the treatment of a co-contaminated soil presenting moderate levels of heavy metals and petroleum hydrocarbons. The conclusion focused on the combined use of plant and bacteria was the most advantageous option for the treatment of the co-contaminated soil, as compared to natural attenuation, bioaugmentation or phytoremediation applied alone (Agnello et al. 2016).

B. alba can be considered as a Cr hyperaccumulator plant, based on Cr concentration recorded in its shoots which exceeds the standard values of hyperaccumulator plants (1000 mg kg^{-1}). In particular, both compost and *B. licheniformis* MBBL1 strains are able to induce a significant metal accumulation in shoots and/or roots of tested Brassicaceae. Due to the low bioconcentration factors of tested species (less than 1), these cannot be considered the appropriate choice for metal phytoextraction from the polluted soils examined (Brunetti et al. 2012).

The General Organization of Agriculture Fund, Ministry of Agriculture, Egypt, suggest that bioremediate may act as potential candidates for soil inoculation (phosphoren, microbien, cerealin and azospirillum) to bioremediate pesticide (organophosphate, carbamate and chlorinated organic compounds)-contaminated soil (El-Kabbany 1999).

The plant-growth-promoting bacteria supports in improving agricultural yields, maintaining the soil health by improving physical and chemical properties of soil. Several microorganisms such as bacteria and fungi play a key role in providing conducive environment to the plants to flourish in a healthy way as well as diminishing the pollution possibilities (Fig. 2).

5 Future Perspectives and Way Forward

The biofertilisers have shown the impact and need throughout the world keeping in view the economic and environmental factors. In developing countries such as India, we always put forward to save the economy of the nation (Al-Masri 2001; Santra et al. 2015). Plant-growth-promoting rhizobacteria, having multiple activities directed towards plant growth promotion vis-à-vis exhibiting bioremediating potentials by detoxifying pollutants like heavy metals and pesticides and controlling a range of phytopathogens as biopesticides, have shown spectacular results in different crop studies. The productive efficiency of a specific PGPR may be further enhanced with the optimisation and acclimatisation according to the prevailing soil conditions. Further research and understanding of mechanisms of PGPR-mediated phytostimulation would pave the way to find out more competent rhizobacterial

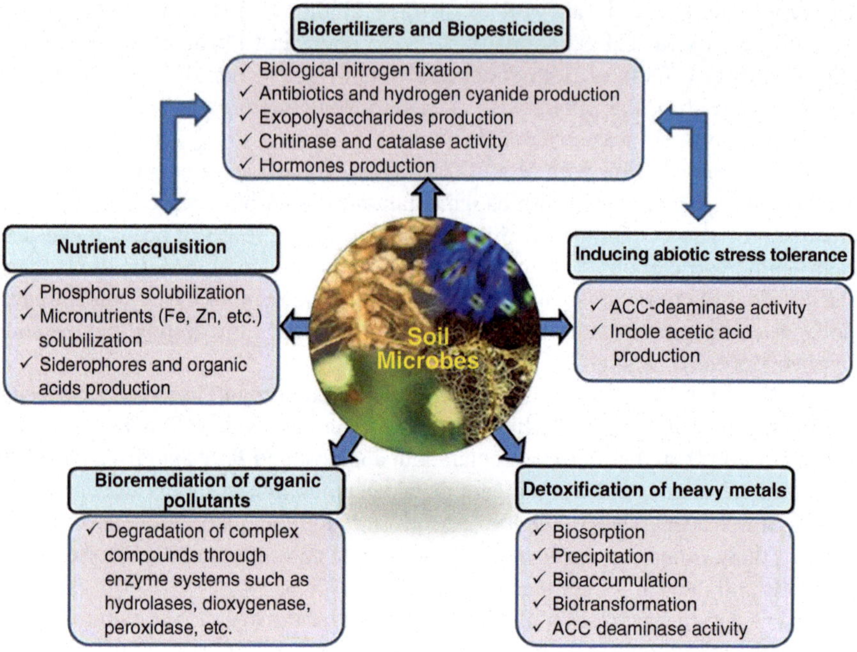

Fig. 2 Biofertilizers and Biopesticides. (Source adapted from: Ahmad et al. 2018)

strains which may work under diverse agro-ecological conditions (Ahemad and Kibret 2014). There is a need to tackle the contaminated soils with the biofertilisers as key member and bioremediation as key process.

References

Agnello A, Bagard M, van Hullebusch E, Esposito G, Huguenot D (2016) Comparative bioremediation of heavy metals and petroleum hydrocarbons co-contaminated soil by natural attenuation, phytoremediation, bioaugmentation and bioaugmentation-assisted phytoremediation. Sci Total Environ 2:693–703

Ahemad M, Kibret M (2014) Mechanisms and applications of plant growth promoting rhizobacteria: current perspective. J King Saud University Sci 26:1–20

Ahmad M, Pataczek L, Hilger TH, Zahir A, Hussain A, Rasche F, Schafleitner R, Solberg S (2018) Perspectives of microbial inoculation for sustainable development and environmental management. Front Microbiol 9:2992–2995. https://doi.org/10.3389/fmicb.2018.02992

Al-Masri MR (2001) Changes in biogas production due to different ratios of some animal and agricultural wastes. Bioresour Technol 77(1):97–100

Atkinson D, Berta G, Hooker JE (1994) Impact of mycorrhizal colonisation on root architecture, root longevity and the formation of growth regulators. In: Gianinazzi S, Schüepp H (eds) Impact of arbuscular mycorrhizas on sustainable agriculture and natural ecosystems. ALS Advances in Life Sciences. Birkhäuser, Basel

Audrone J, Vasarevicius S (2005) Remediation technologies for soils contaminated with heavy metals. J Environ Eng Landsc Manag 13(2):109–113

Azubuike CC, Chikere CB, Okpokwasili GC (2016) Bioremediation techniques-classification based on site of application: principles, advantages, limitations and prospects. World J Microbiol Biotechnol 32(11):180–192

Bai B, Suri VK, Kumar A, Choudhary AK (2016a) Influence of dual–inoculation of AM fungi and *Rhizobium* on growth indices, production economics and nutrient use efficiencies in garden pea (*Pisum sativum* L.). Commun Soil Sci Plant Anal 47(8):941–954

Bai B, Suri VK, Kumar A, Choudhary AK (2016b) Influence of *Glomus–Rhizobium* symbiosis on productivity, root morphology and soil fertility in garden pea in Himalayan acid Alfisol. Commun Soil Sci Plant Anal 47(6):787–798

Bai B, Suri VK, Kumar A, Choudhary AK (2017) Tripartite symbiosis of *Pisum–Glomus–Rhizobium* lead to enhanced productivity, nitrogen and phosphorus economy, quality and biofortification in garden pea in a Himalayan acid Alfisol. J Plant 40(4):600–613

Belimov AA, Hontzeas N, Safronova VI, Demchinskaya SV, Piluzza G, Bullitta S, Glick BR (2005) Cadmium-tolerant plant growth-promoting bacteria associated with the roots of Indian mustard (Brassica juncea L. Czern.). Soil Biol Biochem 37(2):241–250. https://doi.org/10.1016/j.soilbio.2004.07.033

Bhardwaj D, Ansari MW, Sahoo RK, Tuteja N (2014) Biofertilizers function as key player in sustainable agriculture by improving soil fertility, plant tolerance and crop productivity. Microb Cell Factories 13:1–10

Bhat TA, Ahmad GMA, Haq S, Khan OA (2015) Nitrogen fixing biofertilizers; mechanism and growth promotion: a review. J Pure Appl Microbiol 9(2):1675–1690

Bishnoi U (2015) PGPR interaction: an ecofriendly approach promoting the sustainable agriculture system. Adv Bot Res 75(33):81–113

Boholool B (1990) Introduction to nitrogen fixation in agriculture and industry: contribution of BNF to sustainability of agriculture. In: Chalk PM, Gresshoff MM, Roth LE, Stacey G, Newton WE (eds) Nitrogen fixation: achievements and objectives. Chapman and Hall, New York, pp 613–616

Brundrett M, Beegher N, Dell B, Groove T, Malajczuk N (1996) Working with mycorrhizas in forestry and agriculture. ACIAR Monogr 32:374

Brunetti G, Farrag K, Soler-Rovira P, Ferrara M, Nigro F, Senesi N (2012) The effect of compost and Bacillus licheniformis on the phytoextraction of Cr, Cu, Pb and Zn by three brassicaceae species from contaminated soils in the Apulia region, Southern Italy. Geoderma 170:322–330

Chen CR, Condron LM, Xu ZH (2007) Impacts of grassland afforestation with coniferous trees on soil phosphorus dynamics and associated microbial processes: a review. Forest Ecol Manag 255:396–409

Choudhary AK, Thakur SK, Suri VK (2013) Technology transfer model on integrated nutrient management technology for sustainable crop production in high value cash crops and vegetables in north-western Himalayas. Commun Soil Sci Plant Anal 44(11):1684–1699

Dakora FD (1995) Plant flavonoids: biological molecules for useful exploitation. Aust J Plant Physiol 22:7–99

Dighton J, Boddy L (1989) Role of fungi in nitrogen, phosphorus and sulphur cycling in temperate forest ecosystems. In: Boddy L, Marchant R, Read D (eds) Nitrogen, phosphorus and sulfur utilization by fungi. Cambridge University Press, Cambridge, pp 269–298

El-Kabbany S (1999) Evaluation of Four Bio fertilizers for Bioremediation of Pesticide contaminated Soil. In: Proceedings of the international conference on hazardous waste sources, effects and management, Egypt, p 1555

Gage DJ (2017) Infection and invasion of roots by symbiotic, nitrogen-fixing rhizobia during nodulation of temperate legumes. Microbiol Mol Biol Rev 68(2):280–300

Goldstein AH (2000) Bioprocessing of rock phosphate ore: essential technical considerations for the development of a successful commercial technology. In: Proceedings of the 4th international fertilizer association technical conference, IFA, Paris, p 220

Gong Q, Xu D, Zhong C (2000) Study on biodiversity of mycorrhizae and its application. Chinese Forest Press, Beijing, pp 51–61

Halvorson HO, Keynan A, Kornberg HL (1990) Utilization of calcium phosphates for microbial growth at alkaline pH. Soil Biol Biochem 22:887–890

Hamblin AP (1985) The influence of soil structure on water movement, crop root growth, and water uptake. Adv Agron 38:95–158

Harman GE (2006) Overview of mechanisms and uses of *Trichoderma* spp. Phytopathology 96(2):190–194

Harrier LA, Watson CA (2003) The role of arbuscular mycorrhizal fungi in sustainable cropping systems. Adv Agron 42:185–225

Hassen W, Neifar M, Cherif H, Najjari A, Chouchane H, Driouich RC, Salah A, Naili F, Mosbah A, Souissi Y, Raddadi N, Ouzari HI, Fava F, Cherif A (2018) Pseudomonas rhizophila S211, a new plant growth-promoting rhizobacterium with potential in pesticide-bioremediation. Front Microbiol 9:34. https://doi.org/10.3389/fmicb.2018.00034

Herrero A, Flores E (eds) (2008) The cyanobacteria: molecular biology, genomics and evolution, 1st edn. Caister Academic Press, Norfolk. ISBN 978-1-904455-15-8

Jorquera MA, Crowley DE, Marschner P, Greiner R, Fernandez MT, Romero D (2011) Identification of b-propeller phytase-encoding genes in culturable Paenibacillus and Bacillus sp. from the rhizosphere of pasture plants on volcanic soils. FEMS Microbiol Ecol 75:163–172

Kass DL, Drosdoff M, Alexander M (1971) Nitrogen fixation by Azotobacter paspali in association with Bahiagrass (*Paspalum notatum*). Soil Sci Soc Am J 35(2):286–289

Kim KY, McDonald GA, Jordan D (1997) Solubilization of hydroxyapatite by Enterobacter agglomerans and cloned Escherichia coli in culture medium. Biol Fertil Soils 24:347–352

Kirpichtchikova TA, Manceau A, Spadini L, Panfili F, Marcus MA, Jacquet T (2006) Speciation and solubility of heavy metals in contaminated soil using X-ray microfluorescence, EXAFS spectroscopy, chemical extraction, and thermodynamic modelling. Geochim Cosmochim Acta 70(9):2163–2190

Kumar A, Suri VK, Choudhary AK (2014) Influence of inorganic phosphorus, VAM fungi, and irrigation regimes on crop productivity and phosphorus transformations in okra (*Abelmoschus esculentus* L.)–pea (*Pisum sativum* L.) cropping system in an Acid Alfisol. Commun Soil Sci Plant Anal 45(7):953–967

Kumar A, Choudhary AK, Suri VK (2015a) Influence of AM–fungi and applied phosphorus on growth indices, production efficiency, phosphorus–use efficiency and fruit–succulence in okra (*Abelmoschus esculentus*)–pea (*Pisum sativum*) cropping system in an acid Alfisol. Indian J Agric Sci 85(8):1030–1037

Kumar A, Suri VK, Choudhary AK, Yadav A, Kapoor R, Sandal S, Dass A (2015b) Growth behavior, nutrient harvest index and soil fertility in okra–pea cropping system as influenced by AM fungi, applied phosphorus and irrigation regimes in Himalayan acid Alfisol. Commun Soil Sci Plant Anal 46(17):2212–2233

Kumar A, Choudhary AK, Suri VK (2016a) Influence of AM fungi, inorganic phosphorus and irrigation regimes on plant water relations and soil physical properties in okra (*Abelmoschus esculentus* l.) – pea (*Pisum sativum* l.) cropping system in Himalayan acid Alfisol. J Plant Nutr 39(5):666–682

Kumar A, Choudhary AK, Suri VK (2016b) Influence of AM fungi and inorganic phosphorus on fruit characteristics, root morphology, mycorrhizal colonization and soil phosphorus in okra–pea production system in Himalayan acid Alfisol. Indian J Horticult 73(2):213–218

Kumar A, Choudhary AK, Suri VK, Rana KS (2016c) AM fungi lead to fertilizer phosphorus economy and enhanced system productivity and profitability in okra (*Abelmoschus esculentus*) – pea (*Pisum sativum*) cropping system in Himalayan acid Alfisol. J Plant Nutr 39(10):1380–1390

Kumar A, Choudhary AK, Suri VK (2017) Agronomic bio–fortification and quality enhancement in okra–pea cropping system through arbuscular mycorrhizal fungi at varying phosphorus and irrigation regimes in Himalayan acid Alfisol. J Plant Nutr 40(8):1213–1229

Lhuissier FGP, de Rujiter NCA, Sieberer BJ, Esseling JJ, Emons AMC (2001) Time of cell biological events evoked in root hairs by *Rhizobium* nod factors: state of the art. Ann Bot 87:289–302

Li J, Zheng B, Hu R, Liu Y, Jing Y, Xiao Y, Sun M, Chen W, Zhou Q (2019) Pseudomonas species isolated from tobacco seed promote root growth and reduce lead contents in Nicotiana tabacum K326. Can J Microbiol 65(3):214–223. https://doi.org/10.1139/cjm-2018-0434

Mahanty T, Bhattacharjee S, Goswami M, Bhattacharyya P, Das B, Ghosh A, Tribedi P (2016) Biofertilizers: a potential approach for sustainable agriculture development. Environ Sci Pollut Res:1–21. https://doi.org/10.1007/s11356-016-8104-0

Maj D, Wielbo J, Marek-Kozaczuk M, Skorupska A (2010) Response to flavonoids as a factor influencing competitiveness and symbiotic activity of Rhizobium leguminosarum. Microbiol Res 165(1):50–60. https://doi.org/10.1016/j.micres.2008.06.002

Mark P, de Souza CD, Zhao M, Zayed AM, Steven ER, Schichnes D, Terry N (1999) Rhizosphere bacteria enhance selenium accumulation and volatilization by Indian mustard. Plant Physiol 119:565–573

Marschner H, Dell B (1994) Nutrient uptake in mycorrhizal symbiosis. Plant Soil 159:89–102

Matiru VN, Dakora FD (2004) Potential use of rhizobial bacteria as promoters of plant growth for increased yield in landraces of African cereal crops. Afr J Biotechnol 3(1):1–7

Mila MA, Shamsuddin ZH (2010) *Rhizobium* as a crop enhancer and biofertilizer for increased cereal production. Afr J Biotechnol 9(37):6001–6009

Nannipieri P, Giagnoni L, Landi L, Renella G (2011) Role of phosphatise enzymes in soil. In: Bunemann E, Oberson A, Frossard E (eds) Phosphorus in action: biological processes in soil phosphorus cycling. Soil biology. Springer, Berlin, pp 215–243

Okrent D (1999) On intergenerational equity and its clash with intragenerational equity and on the need for policies to guide the regulation of disposal of wastes and other activities posing very long time risks. Risk Anal 19:877–901

Papaefthimiou D, van Hove C, Lejeune A, Rasmussen U, Wilmotte A (2008) Diversity and host specificity of Azolla cyanobionts. J Phycol 44:60–70

Peters GA, Perkins SK (1993) The Azolla-Anabaena symbiosis: endophyte continuity in the Azolla life-cycle is facilitated by epidermal trichomes. II. Re-establishment of the symbiosis following gametogenesis and embryogenesis. New Phytol 123:65–75

Puente ME, Bashan Y, Li CY, Lebsk VK (2004) Microbial populations and activities in the rhizoplane of rock weathering desert plants root colonization and weathering of igneous rocks. Plant Biol 6:629–642

Rai AK (2000) Response of NaCI-adapted and Unadapted Azolla pinnata-Anabaena azollae complex to salt-stress: partial photosynthetic processes and respiration. Symbiosis 29(3):249–261

Raymond AW, Okieimen FE (2011) Heavy metals in contaminated soils: a review of sources, chemistry, risks and best available strategies for remediation. ISRN Ecol. https://doi.org/10.5402/2011/402647

Richardson AE, Simpson RJ (2011) Soil microorganisms mediating phosphorus availability. Plant Physiol 156:989–996

Rillig MC (2004) Arbuscular mycorrhizae, glomalin and soil aggregation. Can J Soil Sci 84:355–363

Rodriguez H, Fraga R (1999) Phosphate solubilizing bacteria and their role in plant growth promotion. Biotechnol Adv 17:319–339

Sahoo RK, Bhardwaj D, Tuteja N (2013) Biofertilizers: a sustainable eco-friendly agricultural approach to crop improvement. In: Tuteja N, Gill SS (eds) Plant acclimation to environmental stress. Springer Science plus Business Media, New York, pp 403–432

Santra SC, Mallick A, Samal AC (2015) Biofertilizer for bioremediation. In: Recent trends in biofertilizers, I K International Publishing House Pvt. Ltd, pp 205–234

Seshachala U, Tallapragada P (2012) Phosphate solubilizers from the rhizosphere of Piper nigrum L. in Karnataka. India Chil J Agric Res 72:397–403

Sharma PK, Kundu BS, Dogra RC (1993) Molecular mechanism of host specificity in legume-*Rhizobium* symbiosis. Biotechnol Adv 11:714–779

Sharma SB, Sayyed RZ, Trivedi MH, Gobi TA (2013) Phosphate solubilizing microbes: sustainable approach for managing phosphorus deficiency in agricultural soils. Springerplus 2:587–600

Song H (2005) Effects of VAM on host plant in the condition of drought stress and its mechanisms. Electron J Biol 1:44–48

Staddon PL, Ramsey CB, Ostle N, Ineson P, Fitter AH (2003) Rapid turnover of hyphae of mycorrhizal fungi determined by AMS microanalysis of C-14. Science 300:1138–1140

Suri VK, Choudhary AK (2013a) Effect of vesicular arbuscular mycorrhizae and applied phosphorus through targeted yield precision model on root morphology, productivity and nutrient dynamics in soybean in an Acid Alfisol. Commun Soil Sci Plant Anal 44(17):2587–2604

Suri VK, Choudhary AK (2013b) *Glycine-Glomus*-Phosphate Solubilizing Bacteria interactions lead to fertilizer phosphorus economy in soybean in a Himalayan Acid Alfisol. Commun Soil Sci Plant Anal 44(20):3020–3029

Suri VK, Choudhary AK, Chander G, Gupta MK, Dutt N (2011) Improving phosphorus use through co-inoculation of vesicular arbuscular mycorrhizal (VAM) fungi and phosphate solubilizing bacteria (PSB) in maize in an acid Alfisol. Commun Soil Sci Plant Anal 42(18):2265–2273

Suri VK, Choudhary AK, Kumar A (2013) VAM fungi spore populations in different farming situations and their effect on productivity and nutrient dynamics in maize and soybean in Himalayan acid Alfisol. Commun Soil Sci Plant Anal 44(22):3327–3339

Tisdall JM (1991) Fungal hyphae and structural stability of soil. Aust J Soil Res 29:729–743

Van Hove C, Lejeune A (2002) The Azolla-Anabaena symbiosis. Biol Environ Proc R Irish Acad 102(1):23–26. https://doi.org/10.3318/BIOE.2002.102.1.23

Van Veen JA, Leonard O, van Elsas J, Van JA, Overbeek LS, Van Elsas JD (1997) Fate and activity of microorganisms introduced into soil. Microbiol Mol Biol Rev 61:121–135. Microbiology and molecular biology reviews: MMBR. 61:121–35.

Wright SF, Upadhyaya A (1998) A survey of soils for aggregate stability and glomalin, a glycoprotein produced by hyphae of arbuscular mycorrhizal fungi. Plant Soil 198:97–107

Wright SF, Upadhyaya A, Buyer JS (1998) Comparison of n-linked oligosaccharides of glomin from arbuscular mycorrhizal fungi and soils by capillary electrophoresis. Soil Biol Biochem 30:1853–1857

Yadav A, Suri VK, Kumar A, Choudhary AK, Meena AL (2015a) Enhancing plant water relations, quality and productivity of pea (*Pisum sativum* L.) through AM fungi, inorganic phosphorus and irrigation regimes in a Himalayan acid Alfisol. Commun Soil Sci Plant Anal 46(1):80–93

Yadav A, Suri VK, Kumar A, Choudhary AK (2015b) Influence of AM fungi and inorganic phosphorus on growth, green pod yield and profitability of pea (*Pisum sativum* L.) in Himalayan acid Alfisol. Indian J Agron 60(1):163–167

Zhao K, Penttinen P, Zhang X, Ao X, Liu M, Yu X (2014) Maize rhizosphere in Sichuan, China, hosts plant growth promoting Burkholderia cepacia with phosphate solubilizing and antifungal abilities. Microbiol Res 169:76–82

Zhu F, Qu L, Hong X, Sun X (2011) Isolation and characterization of a phosphate solubilizing halophilic bacterium Kushneria sp. YCWA18 from Daqiao Saltern on the coast of yellow sea of China. Evid Based Complement Alternat Med 2011:615032

Zou X, Binkley D, Caldwell BA (1995) Effects of dinitrogen-fixing trees on phosphorus biogeochemical cycling in contrasting forests. J Am Soc Soil Sci 59:1452–1458

Intimidating Effects of Heavy Metals on *Mentha* Species and Their Mitigation Using Scientific Approaches

Aarifa Nabi, M. Naeem, Tariq Aftab, and M. Masroor A. Khan

Contents

1 Introduction

As plants are sessile, they cannot escape the unfavorable environmental alterations to which they remain exposed during their life cycle that include biotic, climatic, and mechanical stresses (Tuteja 2007), but the plants have developed complex mechanisms for perceiving the stress-related signals. These mechanisms ultimately permit them to decelerate their growth and metabolism, thus escalating their ability to survive under stress conditions. They have developed the antioxidant system to tackle with the toxic effects of reactive oxygen species (ROS), which are generated in response to the stresses (Kanazawa et al. 2000; Parihar et al. 2015). However, the response of a plant to stress is a highly dynamic process which is dependent on the

A. Nabi · M. Naeem (✉) · T. Aftab · M. M. A. Khan
Plant Physiology Section, Department of Botany, Aligarh Muslim University, Aligarh, Uttar Pradesh, India

M. Naeem et al. (eds.), *Contaminants in Agriculture*,
https://doi.org/10.1007/978-3-030-41552-5_15

duration, severity of the stress, and the developmental stage and preparedness of the plant (Claeys and Inzé 2013). The various active oxygen species (AOS) like oxygen-free radicals cause membrane peroxidation and ultimately lead to tissue damage. The ramification includes inhibition of enzymes, peroxidation of lipids, protein oxidation, activation of the pathway leading to programmed cell death, and finally cell death (Mittler 2002).

The pollution of soils caused by heavy metals (HMs) is a major concern nowadays, and the HM pollution has affected a large part of agricultural land worldwide, making it barren and unproductive for the crop plants. The term HMs refers to a group of metals and metalloids having an atomic density greater than 5 g/cm^3 (Singh and Kalamdhad 2011; Alloway 2011; Tchounwou et al. 2012; Edelstein and Ben-Hur 2018). Some HMs like nickel (Ni), chromium (Cr), zinc (Zn), copper (Cu), molybdenum (Mo), and iron (Fe) are required by plants in minute quantities as they are essential trace elements and the plant cannot complete their life cycle if the deficiency persists. These trace elements are required by plants during many structural and biochemical functions including electron transport reactions, oxidation–reduction reactions, growth, and metabolic processes and are also the components of various enzymes (Hänsch and Mendel 2009, López and Magnitski 2011;Tripathi et al. 2015). Nonessential HMs such as cadmium (Cd), arsenic (As), silver (Ag), mercury (Hg), and lead (Pb) have unknown biological functions, and they are toxic to plants even at lower concentrations (Gaur and Adholeya 2004). Heavy metals originate from natural sources like volcanism and weathering of rocks that are always present in the soil at a background concentration (Lasat et al. 2000; Ghiyasi et al. 2010). However, the concentration of HMs has increased dramatically from past few decades due to the exasperation of anthropogenic intrusion, and these HMs are not biodegraded though they get accumulated in living systems and hamper the basic fundamental processes of living organisms (Pehlivan et al. 2009). These toxic HMs enter into the environment through anthropogenic disturbances like mining, smelting, use of pesticides, herbicides, and other industrial activities. The impact of abiotic stress on the environment ultimately affects plant life by altering their morphological, physiological, and developmental processes finally diminishing their productivity.

Contamination of soil and water over the years by toxic HMs has become a great concern (Ikenaka et al. 2010; Sayyed and Sayadi 2011; Raju et al. 2011; Prajapati 2014; Zojaji 2014). Heavy metals cause severe damage to plants due to oxidative stress. The excessive concentration of these HMs induces the formation of ROS such as superoxide radical (O_2^-), hydroxyl radical ($OH^·$), singlet oxygen ($^1O_2$), and hydrogen peroxide (H_2O_2) by creating the oxidative stress. These are produced during various metabolic pathways as well as membrane-linked activities as by-products (Lajayer et al. 2017; Berni et al. 2018; Soares et al. 2019), and these ROS are responsible for peroxidase damages to DNA, RNA, proteins, nucleic acids, and fatty acids, disrupt DNA synthesis, and alter mitotic activity and transcriptional processes and chlorophyll content in plants (Burzyński 1985; Gallego et al. 2002). However, plant cells use different antioxidants like superoxide dismutase (SOD), catalase (CAT), ascorbate peroxidase (APX), and peroxidase (POX) by employing their antioxidant machinery which helps in scavenging the ROS by redox homeostasis (Noctor and Foyer 1998).

The pollution due to HMs across the globe has affected not only the crop plants but also the medicinal and aromatic plants (MAPs). The MAPs constitute a large segment of flora providing raw materials for use in pharmaceutical industries for therapeutic and aromatic purposes and culinary purposes. The WHO has estimated that 80% of people worldwide rely on herbal medicines, and 1500 species are known for their aroma and flavor. The genus *Mentha* is one among them and is an important member of the Lamiaceae family, and it has high medicinal and aromatic value (Šarić-Kundalić et al. 2009). The genus includes 18 species and 11 hybrids (some reports show the occurrence of more than 27 species). It is mostly grown around the temperate areas of the world, but nowadays it is cultivated throughout the world (Singh et al. 2015; Bhattacharya 2016). Due to its antioxidant properties, production of essential oil, and various other biological activities, it is used for the treatment of sinusitis and bronchitis and has antibacterial, antifungal, and antiviral activities besides their use in confectionery, perfumery, and pharmaceuticals (Liu and Lawrence 2007; Mahboubi and Haghi 2008; Nickavar et al. 2010; Chibane 2012; Benabdallah et al. 2016). The *Mentha* species are also known for their free radical scavenger properties, and most of these activities of *Mentha* species are due to the presence of essential oil obtained from different plant parts (leaves, aerial parts) (Ahmad et al. 2012; Singh et al. 2015). The EO consists of terpenoids and phenylpropanoids, and other components may also be present. The concentration of each component in EO varies and depends on the plant part used for the oil extraction (Figueredo et al. 2015).

2 The Genus *Mentha*

The genus *Mentha* grows well in the tropical and subtropical climate of Australia, Asia, Europe, China, Africa, Brazil, and North America. It consists of about 25 species and fewer hybrids. Various taxonomic names ascribed by the taxonomists to mint plants during the past 200 years reflect their great morphological variation (Kokkini 1992, Gupta et al. 2017). Different uses of genus *Mentha* are reported in the literature and found that its herbal products can cure bronchitis, nausea, anorexia, liver problems, and flatulence, owing to its antispasmodic, inflammatory, analgesic, antiemetic, sudorific, and stimulating effects (Cowan 1999; İşcan et al. 2002; Moreno et al. 2002). *Mentha arvensis* is an aromatic herb which belongs to the family Lamiaceae, and out of stupendous essential oil-bearing plants, its oil constitutes an important source for pharmaceutical, flavoring, and agrochemical industries worldwide (Misra et al. 2000; Tassou et al. 2004). There are about eight species of *Mentha* grown in India, namely, *M. arvensis*, *M. piperita*, *M. spicata*, *M. aquatica*, *M. sylvestris*, and *M. citrata*. Among the different species, *M. arvensis*, *M. piperita*, and *M. spicata* are widely grown in Uttar Pradesh, India.

3 Effects of HMs on *Mentha* Species

3.1 Effect of Cadmium (Cd)

Heavy metals are hazardous to the agricultural lands worldwide that impede the plants to outreach their full genetic potential and cause a greater loss by reducing their productivity (Yadav 2010). Among the various HMs, Cd is a nonessential and most pernicious heavy metal pollutant commonly released into the cultivable soil from various anthropogenic activities like industrial, mining, and farming operations (Wagner 1993). It is an overwhelming metal of extensive natural and world-related concern. It is broadly disseminated in the world outside at a normal grouping of about 0.1 mg kg^{-1}. It poses unpropitious effects on plant physiological and developmental processes. If it gets accumulated above the threshold levels, it causes the induction of various toxic responses in plants. The profound effect of Cd treatment was observed in different *Mentha* species. Amirmoradi et al. (2012) experimented on peppermint (*M. piperita*) by applying different concentrations of Cd (10, 20, 40, 60, 80, 100 ppm) on the plant. A decrease in fresh and dry weights, number of leaf and area per plant, main stem height, and number of nodes per main stem and EO of the plant was found as compared to control. Cadmium at 100 ppm proved very toxic for the plants as a maximum reduction was noticed in the above-studied parameters. Peyvandi et al. (2016) also reported a decrease in the growth parameters of *M. piperita* owing to the application of different concentrations of Cd, and there was a slight difference in the EO concentration of control and highest treated plants.

Furthermore, some findings regarding the effect of Cd were reported by Ahmad et al. (2018) and Zaid and Mohammad (2018) in the case of *M. piperita* and *M. arvensis*, respectively. A significant reduction occurred in various physiological and morphological parameters except for enzymatic and nonenzymatic antioxidants with the increasing concentration of Cd. The highest applied concentration of Cd decreased the yield of EO also. However, the content of EO in *M. piperita* was increased at the lowest concentration and decreased at the highest concentration.

3.2 Effect of Lead (Pb)

Lead is a naturally occurring (pale blue, dark) metal occurring in little sums in the world's outside layer. As Pb is already present in nature but due to the various anthropogenic activities like the burning of fossil fuels, mining and manufacturing of phosphate based fertilizers leads to the accumulation of these heavy metals in the environment (Casas and Sordo 2011). Contamination of soils with this HM affects the EO composition and production (Zheljazkov et al. 2006) and crop productivity also (Sharma and Dubey 2005). Various experiments were carried out to evaluate the toxic effects of Pb on crop productivity and EO composition of both MAPs and cash crops. Toxic effects of Pb and Zn on *M. spicata* in nutrient solution were also reported (Bekiaroglou and Karataglis 2002). They reported that chlorophyll content and root growth of plants decreased with the increasing concentrations of Zn and Pb. Another experiment was carried out by Prasad et al. (2010) on mint species to evaluate the toxic effects of Pb and Cr (30 and 60 mg kg^{-1} soil) on the chemical composition and yield of EO as well as to find out the phytoaccumulation of these HMs by using three mint species (*M. arvensis*, *M. piperita*, and *M. citrata*). The growth and yield of *M. citrata* were significantly reduced by the application of Pb and Cr. However, the above parameters were significantly enhanced in the case of *M. piperita*, and there was no significant effect of these HMs on *M. arvensis*. Furthermore, the effect of these HMs caused a decline in the EO content of all the tested mint species. The effects of absorption and localization of Pb and Cd on *M. arvensis* was reported by Jezler et al. (2015). *Mentha* plants were treated with 0, 8, 16, 32, 64, and 128 mg Pb kg^{-1} of soil. However, little changes were brought about by applied Pb and Cd and were insufficient to influence the oil yield and composition of *M. arvensis.*

3.3 *Effect of Mercury (Hg)*

Mercury is a widespread environmental pollutant and strong phytotoxic HM ion that causes plant growth inhibition and has long-term effects on the fertility of the soil. The toxic HM enters into the food chain and causes body tissue alterations besides having a wide range of adverse health effects (Suszcynsky and Shann 1995; Bhan and Sarkar 2005). Sources of mercury pollution are numerous industrial processes including coal industry, dentistry (dental amalgams), untreated batteries, nuclear reactors, industrial and waste disposal, solvent for metal, mining of silver and gold, and the electrical industry (switches, thermostats, batteries) (Pilon-Smits and Pilon 2000; Tchounwou et al. 2003). As Hg is widely present in the environment, plants, animals, and humans are unable to avoid exposure to its different forms (Holmes et al. 2009). Manikandan et al. (2015) tested the effect of different concentrations of Hg (5, 10, 15, 20, and 25 mgL^{-1}) on *M. arvensis* plants. They noted that at the highest applied dose after 12 h, Hg declined the seedling growth and biomass. A similar decline in growth parameters was also observed by Mitchell and Barre (1995), Suszcynsky and Shann (1995), and Zhou et al. (2007). Another similar experiment was also carried out by Manikandan and Venkatachalam (2011) on *M. arvensis* by applying different concentrations of Hg (10, 20, and 40 mg L^{-1}) to the plant. Treatments of Hg reduced the root and shoot growth and activities of ascorbate peroxidase (APX) and catalase (CAT) beyond 40 mg L^{-1}. They also observed that different polymorphic bands were formed owing to the highest applied Hg (40 mg L^{-1}).

3.4 Effect of Chromium (Cr)

Among the various HMs, Cr is abundantly present on the Earth's surface and is the seventh abundant element of the Earth's crust, and it exists in different valence states from −2 to +6. The stable and common forms in the environment are its trivalent [Cr(III)] and hexavalent forms [Cr(IV)] (Katz and Salem 1994; Kimbrough et al. 1999; Jacobs and Testa 2005). The concentration of Cr in soils is 10–150 mg kg^{-1} (McGrath 1995). Chromium enters into the environment from several anthropogenic and natural sources with the colossal release coming from industries like stainless steel welding and metallurgical industries (Tchounwou et al. 2012). Elemental Cr enters into the environment at a rate of 2000–3000 tons in India alone through tanning industries annually, and the effluent concentration ranges between 200 and 500 mg L^{-1} (Chandra et al. 1997). Chromium is also an essential component of diet (Anderson 1997) and has ecological significance in soils, but the toxic form of Cr is hexavalent, a toxic carcinogen, and if ingested in large doses may cause death of animals, and this form of Cr has been estimated 10–1000 times more toxic than trivalent form (Ajmal et al. 1984; Bishnoi et al. 1993; Syracuse Research Corporation 1993). The Cr is accumulated by plants, and at different trophic levels, it gets biomagnified through the food chain (Rai et al. 2002). Chromium interferes with several metabolic processes like photosynthesis, water relations, nutrient status and enzymatic activities, induces chlorosis, generates toxicity in plants, and ultimately causes a reduction in growth, biomass, and finally plant death (Sharma 1995). Furthermore, Barouchas et al. (2014) studied the effect of different concentrations of trivalent and hexavalent Cr (0, 1, 5, 10, 15 mg kg^{-1}) on *M. piperita*, *M. Spicata*, and *Lippia citriodora*. They concluded that Cr affected the uptake of mineral elements in the studied plants and the total Cr concentration in vegetative parts was more in *M. piperita* than the other two plants.

3.5 Effect of Copper (Cu)

Copper (Cu) is an essential micronutrient required by the plants in minute quantities (Ghorbanpour et al. 2016; Lafmejani et al. 2018), but the excess amount of Cu persistent nowadays in the environment mainly in the soil has created loss of productivity in both agricultural and MAPs (Panou-Filotheou et al. 2001; Mostofa et al. 2015; Ibrahim et al. 2017). A study was conducted to find out the effect of copper sulfate ($CuSO_4$) and Cu nanoparticles with the concentrations of 0, 0.5, 1.0, and 1.5 g L^{-1} (Lafmejani et al. 2018) on *M. piperita*. The foliar application of 0.5 g L^{-1} of $CuSO_4$ and 1.0 g $L^{-1}$of Cu nanoparticles enhanced the dry matter and EO content of the crop as observed by them.

A field experiment was conducted to study the response of *M. arvensis* plant toward six micronutrients, namely, copper, boron, molybdenum, zinc, iron, and manganese (Rajput et al. 2002). These micronutrients are required by plants in minute quantities and are important for the normal growth and maintenance of plants and enhance the essential oil production in *M. arvensis* L. f. *piperascens* Malinv. ex Holmes. The application of the micronutrients increased the plant height, leaf/stem ratio values, biomass, and essential oil yield (Rajput et al. 2002). Similarly, a correlation was studied between the activities of antioxidant enzymes and the level of lipid peroxidation in case of in *M. pulegium* under Cu^{2+}, Zn^{2+}, Mg^{2+}, Mn^{2+}, and Ca^{2+}(Candan and Tarhan 2003). According to their findings, the lipid peroxidation levels in *M. pulegium* organs, except roots, were higher (Ca^{2+} stress), and roots showed maximum increase under Cu^{2+}, Zn^{2+}, and Mn^{2+} stress. In the absence of Ca^{2+} and Mg^{2+}, the maximum lipid peroxidation levels were observed in leaves.

3.6 *Effect of Vanadium (V)*

Vanadium is a transition metal whose concentration in soil depends on the industrial pollution as well as on parent material, and its fate in soil depends on hydroxides, iron, and aluminum oxides that determine the mobility of this metal in soil and waters (Peterson and Girling 1981; WHO 1987; Naeem et al. 2007). Barouchas et al. (2019) studied the effect of V and Ni on *M. villosa* and *Lavandula angustifolia* and found that there were no visible symptoms of toxicity on the plants. However, the shoot and root dry matter of both plants decrease with increasing V concentration.

3.7 *Effect of Arsenic (As)*

Arsenic is ubiquitously present in the natural environment and is a nonessential toxic metalloid, and its presence in the soil above permissible limit adversely affects the plant growth, development, and productivity of crops (Sharma 2012; Srivastava et al. 2014; Chandrakar et al. 2018). Among various *Mentha* species, some are tolerant to HMs like *M. aquatica* and considered as an As-tolerant species. It was observed that *M. aquatica* had grown without any phytotoxicity symptoms in the contaminated soils (Száková et al. 2011) and a comparison in the mobility of arsenic species with *Phaseolus vulgaris*, *M. aquatica*, and *Pteris cretica* was also studied (Száková et al. 2009). *Mentha aquatica* was able to accumulate less As in comparison to *Pteris cretica*.

3.8 *Effect of Nickel (Ni)*

Nickel nowadays has become a toxic pollutant because its concentration is alarmingly increasing in the environment, notably in the soil and water across the world. It is a trace element which is required by the plants in minute quantities for their normal growth and development as it is an important constituent of the enzyme urease. Hence, with the level of increasing Ni pollution in the environment, it is essential to understand both the functional roles and toxic effects of Ni in plants. Two *Mentha* species (*M. aquatica* and *M. sylvestris*) were evaluated for their phytoaccumulation potential under Ni exposure (1, 2, 4, and 8 mg Ni L^{-1}). Both the species accumulated higher levels of Ni in their roots and therefore can be used in case of phytoremediation (Zurayk et al. 2002). Similarly, *M. spicata* grown on soil is treated with sewage sludge (SS) and municipal solid waste (MSW), and it was observed that with the increase in concentration of the said treatments and if applied long term, there was found an increase in metal content in various parts of spearmint (Sorboni et al. 2013).

3.9 Effect of Cobalt (Co)

Cobalt is a nonessential transition metal, and in nature it occurs in various transition states. It is beneficial for plants at lower concentrations; however, if applied at higher concentrations, it proved deleterious for the growth of plants and its toxicity is very rare (Zaborowska et al. 2016; Lange et al. 2017; WaLwalaba et al. 2017), although the literature is very scanty in this regard. A study on the effect of Co and As was conducted on different herbs in Poland (*Achillea millefolium* L., *Comarum palustre* L., *Lysimachia vulgaris* L., *Lycopus europaeus* L., *Potentilla anserina* L., and *M. arvensis*). Workers observed that As was highly accumulated in the herbs as compared to Co and the maximum Co content was present in *M. arvensis* (Malinowska et al. 2018). Another study was carried out by Aziz et al. (2011) on *M. piperita* to find out the effect of Co on plant's growth and chemical composition. The lower doses of Co increased both the fresh and dry herbage yields besides increasing the essential oil yield and micro- and macronutrients. However, the higher doses of Co decreased the menthol content, while it increased the menthone and isomenthone content.

4 Mitigation of HMs Using Scientific Approaches

Several MAPs belonging to the family Lamiaceae that bear chemical compounds are used as medicinal herbs and spices and also play a significant role in plant defense (Gautam et al. 2012). Plants synthesize the primary metabolites (amino acids, carbohydrates, and lipids) which play an important role in the fundamental processes of plants, besides synthesizing the secondary metabolites like terpenoids, alkaloids, flavonoids, and steroids that are vital in the mechanism of plant defense under various environmental stresses (Yazaki 2006; Mazid et al. 2011; Lajayer et al. 2017). Therefore, in such stressful conditions, a process of elicitation is induced in plants owing to the application of elicitors. The elicitors are those substances which, when applied in small amounts, induces the biosynthesis of secondary metabolites

(Naik and Al-Khayri 2016; Thakur et al. 2018). Elicitors can be biotic and abiotic, depending on their nature, and can get utilized during the stressful conditions in plants to elevate the toxic response. The menthol or menthone and limonenes are the terpenoids that are involved in the secondary metabolism of plants besides their role in respiration and photosynthesis. As there is an increasing demand for menthol worldwide, therefore, new scientific approaches are employed by researchers all over the world to enhance its production. However, synthetic biology is an approach which leads to the increased production of menthol by developing and optimizing the complex metabolic pathways (Ribeiro and Shapira 2019). But due to its limitations, there is an urgent need to find out the constructive approach which can be used to obtain the menthol in large production directly from the plants. The consumption of menthol is more than 7000 tons per year (Heydari et al. 2018). Hence, there is a need to find out the possible link between the HMs and their effects on proteins and genes involved in the biosynthetic pathway of secondary metabolites.

As various abiotic stresses cause widespread loss of crop productivity worldwide, there is a necessity to investigate how abiotic stress affects plant growth and developmental processes at the biochemical, physiological, and molecular level so that the productivity of crops may be enhanced (Kazan 2015). The major environmental factors imposing stress on plants are HMs, salinity, drought, chilling, pathogens, and heat stress. The stress response on plants is dependent on the stages of plants, period and development of stress, and biotic and abiotic factors. Therefore, plants need to respond to various internal and external stimuli by regulating their growth and development (Wolters and Jürgens 2009, Feller and Vaseva 2014). It has already been reported that the exogenous application of plant hormones enhances stress tolerance in HM-affected plants (Rubio-Wilhelmi et al. 2011; Peleg and Blumwald 2011; Elobeid et al. 2011; Srivastava et al. 2012; Zhu et al. 2012; Krishnamurthy and Rathinasabapathi 2013). Additionally, the application of different kinds of elicitors improves the secondary metabolite production in MAPs under stress conditions. Some species of *Mentha* acts as hyperaccumulators and can be used for phytoremediation in the HM-contaminated soils.

Plants have developed the capability to perceive the stress signals among them and recruit various signaling molecules for the transduction of these signals systemically or locally. According to the definition of the Environmental Protection Agency (EPA), a plant growth regulator is "any substance or mixtures of substances intended, through physiological action, to accelerate or retard the rate of growth or maturation or otherwise alter the behavior of plants". Plants naturally produce these hormones and are indispensable for regulating their growth, and they perform by modifying or controlling plant growth processes like the formation of flowers and leaves, elongation of stems, development, and ripening of fruits (Wani et al. 2016; Jagodzik et al. 2018; Jamwal et al. 2018).

The "classical" phytohormones that were identified during the first half of the twentieth century are abscisic acid (ABA), auxin, cytokinin, ethylene, and gibberellin (GA) (Bahyrycz and Konopińska 2007; Wheeler and Irving 2010). Recently, additional compounds are being added to the category of plant hormones and include brassinosteroids, salicylic acid, jasmonic acid, nitric oxide, and strigolactones (Grün et al. 2006; Browse 2005; Gomez-Roldan et al. 2008). Furthermore, phytosulfokines (PSKs), S-locus cysteine-rich proteins (SCPs), ENOD40, CLAVATA3 (CLV3), polaris, plant natriuretic peptides (PNPs), and systemin are biologically active peptides which have been found to be key signaling players in different aspects of plant life (Bahyrycz and Konopińska 2007; Wheeler and Irving 2010).

The PGRs have a positive effect on the overall growth and development of a plant as they escalate the various physiological and biochemical responses. Various phytohormones, biomolecules, and chemicals are assigned diverse roles to cope with different kinds of stresses. They are the most important endogenous molecules for modifying various molecular and physiological reactions and are critically required for the survival of plants under HM stress (Fahad et al. 2015). Besides having a direct role, it is now well accepted that these molecules work through crosstalk among them and culminate the stress (Gaosheng and Jingming 2012; Dar et al. 2015; Rahimi et al. 2017). They can control and signal the response, growth, and development like regulation of secondary metabolites, enzyme activity, and cell membrane permeability (Wani et al. 2016) via circulating through the whole plant or part of it. The pivotal role of plant hormones in promoting plant acclimatization to ever-changing environments has been well established by mediating fundamental processes like growth, development, source–sink transitions, and nutrient allocation (Fahad et al. 2015; Jamwal et al. 2018; Wasternack and Strnad 2019).

Here, we have summarized the effects of some phytohormones, plant growth regulators, irradiated polysaccharides, and micro- and macronutrients on *Mentha* species toward mitigation of the HM stress. It has been observed that these PGRs have also been proven beneficial for the MAPs in various studies. Kavina et al. (2011) studied the effect of traditional PGRs, namely, GA_3 and ABA, and nontraditional PGRs (Difenoconazole (DIZ) on *M. Piperita*. Most of the parameters (fresh and dry weights, growth of roots, and photosynthetic pigments) are enhanced by the application of DIZ and ABA; however, the length of the stem was decreased. Different concentrations (1, 10, and 100 mM) of GA_3 and calliterpenone (CA, a phyllocladane diterpenoid) were applied on *M. arvensis* (Bose et al. 2013). They found that the exogenously applied CA was found better over GA_3 in improving plant biomass, leaf area, branching, leaf/stem ratio, and stolon yield. Higher number and density of glandular trichome were observed in CA-treated plants. Additionally, they noticed that the transcript level of menthol dehydrogenase/menthone reductase was found highly upregulated in CA-treated plants as compared to GA_3 treated, with increased content of both menthol and menthone in oil. Finally, they postulated that both PGRs positively regulated the yield by enhancing the density of trichomes and branching, resulting in a higher accumulation of essential oil.

Parić et al. (2017) tested different PGRs (auxin, indole-3-butyric acid (IBA), and the cytokinin, N6-benzyladenine (BAP), both individually and in combination with *M. piperita* plants. Authors reported that an increase in several shoots and roots occurred by applying the treatment with BAP and IBA, but the highest concentration of BAP affected the production of phenolic compounds. However, concentrations of BAP and IBA affected the antioxidants and antimicrobial activities of the plant.

Few reports are available about the use of rhizospheric microorganisms on *Mentha* species. In this regard, Kumar et al. (2015) experimented on *M. piperita* and *M. arvensis* under two HMs (Cd and Ni), to find out their interaction with arbuscular mycorrhizae (AM). The AM fungal inoculation had no significant effect on the oil yield of *M. piperita* under control or soils amended with Cd as they reported. However, the oil yield significantly enhanced under the soils amended with Ni. More toxicity of these HMs was observed in case of *M. arvensis* as the yield of this species decreased in addition to application of the AM fungi inoculation. Another experiment was carried out by Kunwar et al. (2015) on *Ocimum basilicum* and *M. spicata* under Cu, Cd, and Pb. A significant variation in the EO of *O. basilicum* was observed, while there was no change in the EO composition of *M. spicata* recorded in the HM-amended soils.

Moreover, Cd-induced reduction on various growth, physiological and biochemical parameters, and EO production was studied by Ahmad et al. (2018) and Zaid and Mohammad (2018) in case of *M. piperita* and *M. arvensis*, respectively. *M. piperita* plants were treated with salicylic acid, and the *M. arvensis* plants were treated with methyl jasmonate (with and without nitrogen). Application of both PGRs significantly alleviated the Cd stress in case of *Mentha* species.

5 Radiation-Processed Polysaccharides Act as Plant Growth Promoters

Natural bioactive polymers have transmogrified the agricultural field because of the meagreness of toxicity and propensity to act as effective plant growth promoters. Ionizing radiation-mediated depolymerization of polymers has emerged as a recent and promising technology for boosting the productivity of crops (Hien et al. 2000; Nagasawa et al. 2000; Kume et al. 2002; Naeem et al. 2012a, b). Application of these radiation-processed polysaccharides (oligomers of sodium alginate, carrageenan, and chitosan) through foliar sprays on plants promotes various biological activities like plant growth (shoot and root growth, seed germination), production of flowers, enhanced content and yield of oil in aromatic plants, induction of phytoalexins, and amelioration of HM stress (Aftab et al. 2011; Dela Rosa et al. 2002; Hegazy et al. 2009; Khan et al. 2011; Sarfaraz et al. 2011; Idrees et al. 2013, Naeem et al. 2012a, b; Sadiq et al. 2017; Ahmad et al. 2019). Furthermore, application of these oligomers can abbreviate the harvesting period, and the use of chemical fertilizers and insecticides can be minimized by the application of oligomers (Hafiz et al. 2003; Luan et al. 2003). A considerable work has been done till now regarding the beneficial effects of the irradiated polysaccharides on a number of MAPs including *M. arvensis*, *Catharanthus roseus*, *Trigonella foenum-graecum*, *Cymbopogon flexuosus* Steud, *M. piperita*, *Foeniculum vulgare*, and *M. spicata* by various workers (Sarfaraz et al. 2011, Naeem et al. 2011, 2012a, b, 2015a, b, 2019, Dar et al. 2015, Singh et al. 2017, Sadiq et al. 2017; Ahmad et al. 2019).

6 Crosstalk Mechanism of the Biosynthetic Pathway in Plants

The volatile organic compounds are mainly derivatives of fatty acids, benzenoids, amino acids, and terpenoids and synthesized from different pathways. These organic compounds can be monoterpenes, diterpenes, tetraterpenes, sesquiterpenes, triterpenes, and sterols. The different pathways involved in their synthesis are 2-C-methyl-D-erythritol 4-phosphate (MEP) pathway, mevalonate (MVA) pathway, lipoxygenase (LOX) pathway, and shikimate–phenylalanine pathway. Through crosstalk mechanism, these pathways result in the formation of precursors of various secondary metabolites. Among these pathways, the two pathways involved in the formation of secondary metabolites in case of MAPs are MEP and MVA pathways. These two pathways through crosstalk and in combination are responsible for the synthesis of precursors of different kinds of terpenes. MEP and MVA are the two compartmentally separated pathways which are responsible for the formation of five-carbon isoprenoid precursors, isopentenyl diphosphate (IPP), and dimethylallyl diphosphate (DMAPP), and these provide the precursors for geranylgeranyl diphosphate (GGPP) and geranyl diphosphate (GPP). IPP, the precursor for farnesyl diphosphate (FPP), is obtained from the MVA pathway (Sharma et al. 2003; Orlova et al. 2009; Mishra et al. 2017; Heydari et al. 2018). These precursors ultimately result in the formation of various kinds of terpenes. By overexpressing the genes responsible for the activity of the enzymes which are accountable for the production of secondary metabolites in plants, the overall production of essential oil and their active constituents could be enhanced. The hypothetical model shows the synthesis of various kinds of terpenes given in Fig. 1.

7 Conclusion

The HMs strongly affect the primary and secondary metabolite production in the plants either by increasing or decreasing their production depending on the period, concentration, and the plant type. Heavy metals alter the biosynthetic pathways in case of essential oil-bearing plants by generating the harmful reactive oxygen species (ROS), and by the application of biotic or abiotic elicitors using different scientific strategies, the deleterious effects on plants might be declined. The application of various plant growth hormones that combat the deleterious effects of HMs is comprehensively discussed in this chapter. A few studies carried out on *Mentha* species depict that the growth, herbage, and productivity of essential oils including the content or yield are negatively affected by the HMs. Only a few species of *Mentha* could be used as hyperaccumulators but up to a certain limit of HMs, and by achieving the threshold limit and beyond, these species also show decline in growth and photosynthetic, enzymatic, and quality of essential oils. The studies on nanotechnology, radiation biology, proteomics, metabolomics, and genomics could promote

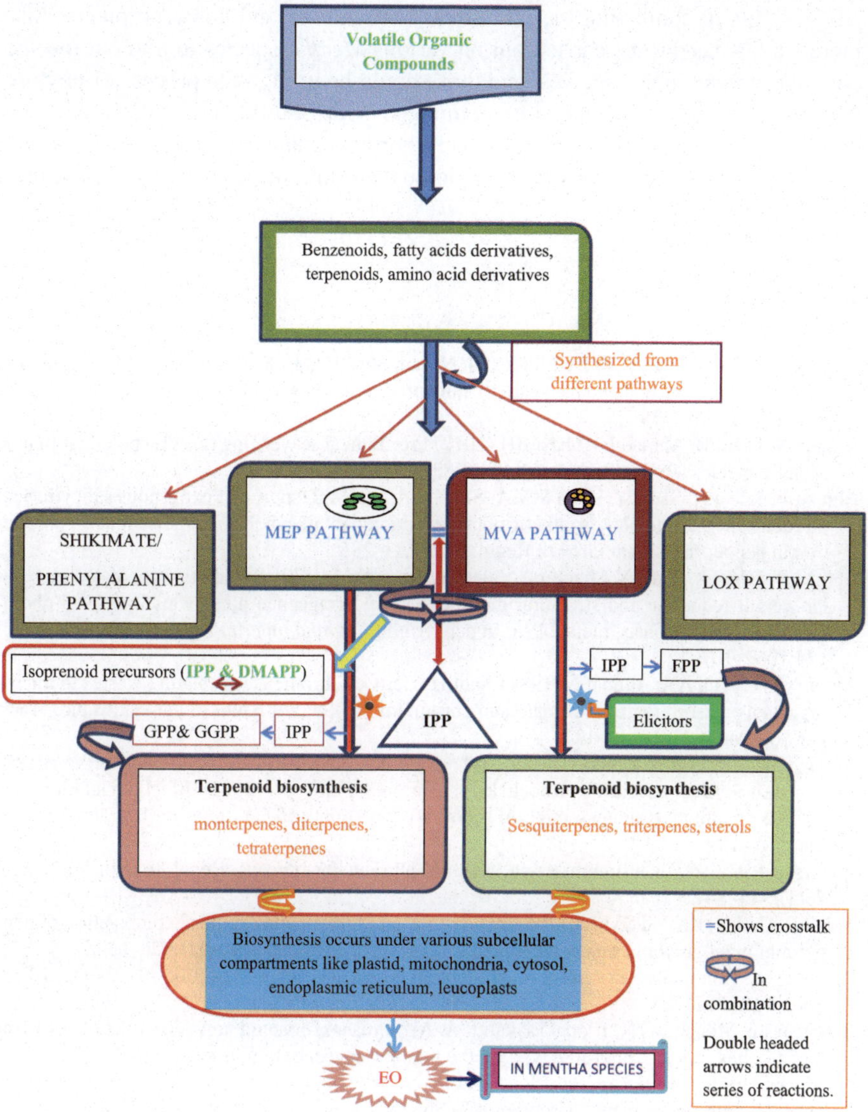

Fig. 1 Showing the synthesis of terpenoids (EO) from the two pathways methylerythritol (MEP) and mevalonic acid (MVA) occurring in different cellular compartments. In *Mentha* species, the synthesis occurs from precursors isopentenyl diphosphate (IPP) and dimethylallyl diphosphate (DMAPP). IPP derived from the MVA acts as precursor for farnesyl diphosphate (FPP), and the one from MEP pathway is a precursor for geranyl diphosphate (GPP) and geranylgeranyl diphosphate (GGPP) and finally results in the formation of terpenes. Lipoxygenase (LOX) and shikimate pathway leads to the formation of other organic compounds

new insights in combating the HM stress both in crop and non-crop plants. Also, more studies need to be carried out on various *Mentha* species to find out the specific hyperaccumulator species, and those could be used in the process of phytoremediation in the heavy metal-contaminated soils. Furthermore, the studies on crosstalk mechanism of newly discovered hormones interacting with each other in the biosynthetic pathway need to be explored in the modification of stress conditions.

References

Aftab T, Khan MMA, da Silva JAT, Idrees M, Naeem M (2011) Role of salicylic acid in promoting salt stress tolerance and enhanced artemisinin production in *Artemisia annua* L. J Plant Growth Regul 30(4):425–435

Ahmad N, Fazal H, Ahmad I, Abbasi BH (2012) Free radical scavenging (DPPH) potential in nine Mentha species. Toxicol Ind Health 28(1):83–89

Ahmad B, Jaleel H, Sadiq Y, Khan MMA, Shabbir A (2018) Response of exogenous salicylic acid on cadmium induced photosynthetic damage, antioxidant metabolism and essential oil production in peppermint. Plant Growth Regul 86(2):273–286

Ahmad B, Jahan A, Sadiq Y, Shabbir A, Jaleel H, Khan MMA (2019) Radiation-mediated molecular weight reduction and structural modification in carrageenan potentiates improved photosynthesis and secondary metabolism in peppermint (*Mentha piperita* L.). Int J Biol Macromol 124:1069–1079

Ajmal M, Nomani AA, Ahmad A (1984) Acute toxicity of chrome electroplating wastes to microorganisms: adsorption of chromate and chromium (VI) on a mixture of clay and sand. Water Air Soil Pollut 23(2):119–127

Alloway BJ (2011) Heavy metals in soil, 3rd edn. Wiley, New York

Amirmoradi S, Moghaddam PR, Koocheki A, Danesh S, Fotovat A (2012) Effect of cadmium and lead on quantitative and essential oil traits of peppermint (*Mentha piperita* L.). Not Sci Biol 4(4):101–109

Anderson RA (1997) Chromium as an essential nutrient for humans. Regul Toxicol Pharmacol 26(1):S35–S41

Aziz EE, Gad N, Khaled SM (2011) Effect of cobalt on growth and chemical composition of peppermint plant grown in newly reclaimed soil. Aust J Basic Appl Sci 5(11):628–633

Bahyrycz A, Konopińska D (2007) Plant signalling peptides: some recent developments. J Pept Sci 13(12):787–797

Barouchas P, Moustakas N, Liopa-Tsakalidi A, Akoumianaki-Ioannidou A (2014) Effect of trivalent and hexavalent Chromium (Cr) on the total Cr concentration in the vegetative plant parts of spearmint (*'Mentha spicata'* L.), lemon verbena (*'Lippia citriodora*' L.) and peppermint (*Mentha piperita* L.). Aust J Crop Sci 8(3):363

Barouchas PE, Akoumianaki-Ioannidou A, Liopa-Tsakalidi A, Moustakas NK (2019) Effects of vanadium and nickel on morphological characteristics and on vanadium and nickel uptake by shoots of mojito (*Mentha*× *villosa*) and lavender (*Lavandula angustifolia*). Not Bot Horti Agrobot Cluj-Napoca 47(2):487–492

Bekiaroglou P, Karataglis S (2002) The effect of lead and zinc on *Mentha spicata*. J Agron Crop Sci 188(3):201–205

Benabdallah A, Rahmoune C, Boumendjel M, Aissi O, Messaoud C (2016) Total phenolic content and antioxidant activity of six wild Mentha species (Lamiaceae) from northeast of Algeria. Asian Pac J Trop Biomed 6(9):760–766

Berni R, Luyckx M, Xu X, Legay S, Sergeant K, Hausman JF et al (2018) Reactive oxygen species and heavy metal stress in plants: impact on the cell wall and secondary metabolism. Environ Exp Bot

Bhan A, Sarkar NN (2005) Mercury in the environment: effect on health and reproduction. Rev Environ Health 20(1):39–56

Bhattacharya S (2016) Cultivation of essential oils. In: Essential oils in food preservation, flavor and safety. Academic, Amsterdam, pp 19–29

Bishnoi NR, Dua A, Gupta VK, Sawhney SK (1993) Effect of chromium on seed germination, seedling growth and yield of peas. Agric Ecosyst Environ 47(1):47–57

Bose SK, Yadav RK, Mishra S, Sangwan RS, Singh AK, Mishra B, Sangwan NS (2013) Effect of gibberellic acid and calliterpenone on plant growth attributes, trichomes, essential oil biosynthesis and pathway gene expression in differential manner in *Mentha arvensis* L. Plant Physiol Biochem 66:150–158

Browse J (2005) Jasmonate: an oxylipin signal with many roles in plants. Vitam Horm 72:431–456

Burzyński M (1985) Influence of lead on the chlorophyll content and on initial steps of its synthesis in greening cucumber seedlings. Acta Soc Bot Pol 54(1):95–105

Candan N, Tarhan L (2003) The correlation between antioxidant enzyme activities and lipid peroxidation levels in *Mentha pulegium* organs grown in Ca2+, Mg2+, Cu2+, Zn2+ and Mn2+ stress conditions. Plant Sci 165(4):769–776

Casas JS, Sordo J (eds) (2011) Lead: chemistry, analytical aspects, environmental impact and health effects. Elsevier

Chandra, P., Sinha, S., &Rai, U. N. (1997). Bioremediation of chromium from water and soil by vascular aquatic plants

Chandrakar V, Dubey A, Keshavkant S (2018) Modulation of arsenic-induced oxidative stress and protein metabolism by diphenyleneiodonium, 24-epibrassinolide and proline in *Glycine max* L. Acta Bot Croat 77(1):51–61

Chibane M (2012) Optimization of solvent extraction of antioxidants (phenolic compounds) from Algerian mint (*Mentha spicata* L). Pharmacogn Commun 2:72–86

Claeys H, Inzé D (2013) The agony of choice: how plants balance growth and survival under water-limiting conditions. Plant Physiol 162(4):1768–1779

Cowan M (1999) Plant products as antimicrobial agents. Clin Microb Rev 12(4):564–582

Dar TA, Uddin M, Khan MMA, Hakeem KR, Jaleel H (2015) Jasmonates counter plant stress: a review. Environ Exp Bot 115:49–57

Dela Rosa A, Abad L, Relleve L, Aranilla C (2002) Radiation-modified carrageenan for agricultural and health care applications. In: Proceedings of IAEA Project Coordination Mtg on Radiation Processing of Chitin/Chitosan, Bangkok

Edelstein M, Ben-Hur M (2018) Heavy metals and metalloids: sources, risks and strategies to reduce their accumulation in horticultural crops. Sci Hortic 234:431–444

Elobeid M, Göbel C, Feussner I, Polle A (2011) Cadmium interferes with auxin physiology and lignification in poplar. J Exp Bot 63(3):1413–1421

Fahad S, Hussain S, Bano A, Saud S, Hassan S, Shan D, Tabassum MA (2015) Potential role of phytohormones and plant growth-promoting rhizobacteria in abiotic stresses: consequences for changing environment. Environ Sci Pollut Res 22(7):4907–4921

Feller U, Vaseva II (2014) Extreme climatic events: impacts of drought and high temperature on physiological processes in agronomically important plants. Front Environ Sci 2:39

Figueredo G, Chalchat JC, Dogu S, Chalard P, Özcan MM, Bagci Y, Juhaimi FA (2015) Chemical composition of the essential oil of *Pimpinella isaurica* Matthews subsp. isaurica. J Essent Oil Bear Plants 18(3):739–743

Gallego S, Benavides M, Tomaro M (2002) Involvement of an antioxidant defence system in the adaptive response to heavy metal ions in *Helianthus annuus* L. cells. Plant Growth Regul 36(3):267–273

Gaosheng H, Jingming J (2012) Production of useful secondary metabolites through regulation of biosynthetic pathway in cell and tissue suspension culture of medicinal plants. In: Recent advances in plant in vitro culture. IntechOpen

Gaur A, Adholeya A (2004) Prospects of arbuscular mycorrhizal fungi in phytoremediation of heavy metal contaminated soils. Curr Sci:528–534

Gautam MK, Gangwar M, Singh A, Rao CV, Goel RK (2012) In vitro antioxidant properties of *Murraya paniculata* (l.) Leaves extract. Inventi Rapid: Ethnopharmacology

Ghiyasi S, Karbassi A, Moattar F, Modabberi S, Sadough MB (2010) Origin and concentrations of heavy metals in agricultural land around aluminium industrial complex. J Food Agric Environ 8(3&4):1237

Ghorbanpour M, Asgari LH, Hadian J (2016) Influence of copper and zinc on growth, metal accumulation and chemical composition of essential oils in sweet basil (*Ocimum basilicum* L.). J Med Plants 3:132–144

Gomez-Roldan V, Fermas S, Brewer PB, Puech-Pagès V, Dun EA, Pillot JP et al (2008) Strigolactone inhibition of shoot branching. Nature 455(7210):189

Grün S, Lindermayr C, Sell S, Durner J (2006) Nitric oxide and gene regulation in plants. J Exp Bot 57(3):507–516

Gupta AK, Mishra R, Singh AK, Srivastava A, Lal RK (2017) Genetic variability and correlations of essential oil yield with agro-economic traits in Mentha species and identification of promising cultivars. Ind Crop Prod 95:726–732

Hafiz M, Muhamad O, Muhammad SY, Ahmad MS (2003) Irradiated chitosan as plant promoter. In: ProsidingSimposium Biologic Gunaan Ke-7, Mines Beach Resort & Spa, Seri Kumbangan, Selangor, Malaysia, pp 3–4

Hänsch R, Mendel RR (2009) Physiological functions of mineral micronutrients (cu, Zn, Mn, Fe, Ni, Mo, B, cl). Curr Opin Plant Biol 12(3):259–266

Hegazy EA, Abdel-Rehim H, Diaa DA, El-Barbary A (2009) Controlling of degradation effects in radiation processing of polymers. International Atomic Energy Agency Vienna International Centre, Vienna, 234 p

Heydari M, Zanfardino A, Taleei A, ShahnejatBushehri A, Hadian J, Maresca V, Rigano D (2018) Effect of heat stress on yield, monoterpene content and antibacterial activity of essential oils of *Mentha x piperita* var. Mitcham and *Mentha arvensis* var. piperascens. Molecules 23(8):1903

Hien NQ, Nagasawa N, Tham LX, Yoshii F, Dang VH, Mitomo H, Kume T (2000) Growth-promotion of plants with depolymerized alginates by irradiation. Radiat Phys Chem 59(1):97–101

Holmes P, James KAF, Levy LS (2009) Is low-level environmental mercury exposure of concern to human health? Sci Total Environ 408(2):171–182

Ibrahim M, Chee Kong Y, Mohd Zain N (2017) Effect of cadmium and copper exposure on growth, secondary metabolites and antioxidant activity in the medicinal plant Sambung Nyawa (*Gynuraprocumbens* (Lour.) Merr). Molecules 22(10):1623

Idrees M, Naeem M, Aftab T, Khan MMA (2013) Salicylic acid restrains nickel toxicity, improves antioxidant defence system and enhances the production of anticancer alkaloids in *Catharanthus roseus* (L.). J Hazard Mater 252:367–374

Ikenaka Y, Nakayama SM, Muzandu K, Choongo K, Teraoka H, Mizuno N, Ishizuka M (2010) Heavy metal contamination of soil and sediment in Zambia. Afr J Environ Sci Technol 4(11):729–739

İşcan G, Kírímer N, Kürkcüoğlu M, Başer HC, Demirci F (2002) Antimicrobial screening of *Mentha piperita* essential oils. J Agric Food Chem 50(14):3943–3946

Jacobs JA, Testa SM (2005) Overview of chromium (VI) in the environment: background and history. In: Chromium (VI) handbook. CRC Press, Boca Raton, pp 1–21

Jagodzik P, Tajdel-Zielińska M, Cieśla A, Marczak M, Ludwikow A (2018) Mitogen-activated protein kinase cascades in plant hormone signaling. Front Plant Sci 9:1387

Jamwal K, Bhattacharya S, Puri S (2018) Plant growth regulator mediated consequences of secondary metabolites in medicinal plants. J Appl Res Med Aromat Plants 9:26–38

Jezler CN, Mangabeira PAO, Almeida AAFD, Jesus RMD, Oliveira RAD, Silva DDC, Costa LCDB (2015) Pb and Cd on growth, leaf ultrastructure and essential oil yield mint (*Mentha arvensis* L.). Ciência Rural 45(3):392–398

Kanazawa S, Sano S, Koshiba T, Ushimaru T (2000) Changes in antioxidative enzymes in cucumber cotyledons during natural senescence: comparison with those during dark-induced senescence. Physiol Plant 109(2):211–216

Katz SA, Salem H (1994) The biological and environmental chemistry of chromium. VCH Publishers, New York

Kavina J, Gopi R, Panneerselvam R (2011) Traditional and non-traditional plant growth regulators alter the growth and photosynthetic pigments in *Mentha piperita* Linn. Int J Environ Sci 2(1):124

Kazan K (2015) Diverse roles of jasmonates and ethylene in abiotic stress tolerance. Trends Plant Sci 20(4):219–229

Khan ZH, Khan MMA, Aftab T, Idrees M, Naeem M (2011) Influence of alginate oligosaccharides on growth, yield and alkaloid production of opium poppy (*Papaver somniferum* L.). Front Agric China 5(1):122–127

Kimbrough DE, Cohen Y, Winer AM, Creelman L, Mabuni C (1999) A critical assessment of chromium in the environment. Crit Rev Environ Sci Technol 29(1):1–46

Kokkini, S. (1992). *Essential oils as taxonomic markers in Mentha*. 325–334). Royal Botanic Gardens: Kew

Krishnamurthy A, Rathinasabapathi B (2013) Auxin and its transport play a role in plant tolerance to arsenite-induced oxidative stress in *Arabidopsis thaliana*. Plant Cell Environ 36(10):1838–1849

Kume T, Nagasawa N, Yoshii F (2002) Utilization of carbohydrates by radiation processing. Radiat Phys Chem 63(3–6):625–627

Kunwar G, Pande C, Tewari G, Singh C, Kharkwal GC (2015) Effect of heavy metals on terpenoid composition of *Ocimum basilicum* L. and *Mentha spicata* L. J Essent Oil Bear Plants 18(4):818–825

Kumar S, Lal S, Gaur R, Sahay R (2015) Interaction effect of AM fungi and heavy metals (Cd and Ni) on the growth and yield of mint species (*M. piperita* and *M. arvensis*). J Basic App Mycol 11:39–46

Lafmejani ZN, Jafari AA, Moradi P, Moghadam AL (2018) Impact of foliar application of copper sulphate and copper nanoparticles on some morpho-physiological traits and essential oil composition of peppermint (*Mentha piperita* L.). Herba Pol 64(2):13–24

Lajayer BA, Ghorbanpour M, Nikabadi S (2017) Heavy metals in contaminated environment: destiny of secondary metabolite biosynthesis, oxidative status and phytoextraction in medicinal plants. Ecotoxicol Environ Saf 145:377–390

Lange B, van der Ent A, Baker AJM, Echevarria G, Mahy G, Malaisse F et al (2017) Copper and cobalt accumulation in plants: a critical assessment of the current state of knowledge. New Phytol 213(2):537–551

Lasat MM, Pence NS, Garvin DF, Ebbs SD, Kochian LV (2000) Molecular physiology of zinc transport in the Zn hyperaccumulator *Thlaspi caerulescens*. J Exp Bot 51(342):71–79

Liu W, Lawrence BM (2007) Production of mints and mint oil in China. Mint: Genus Mentha. Taylor & Francis Group, Boca Raton, pp 165–184

López MÁ, Magnitski S (2011) Nickel: the last of the essential micronutrients. Agronomía Colombiana 29(1):49–56

Luan LQ, Hien NQ, Nagasawa N, Kume T, Yoshii F, Nakanishi TM (2003) Biological effect of radiation-degraded alginate on flower plants in tissue culture. Biotechnol Appl Biochem 38(3):283–288

Mahboubi M, Haghi G (2008) Antimicrobial activity and chemical composition of *Mentha pulegium* L. essential oil. J Ethnopharmacol 119(2):325–327

Malinowska E, Jankowski K, Wiśniewska-Kadżajan B, Skrzyczyńska J, Sosnowski J (2018) Cobalt and arsenic concentration in herbs growing in field pond areas in Poland. Appl Ecol Environ Res 16(4):3805–3814

Manikandan R, Venkatachalam P (2011) Risk assessment of mercury ion heavy metal exposure on physiological and biochemical changes and DNA damage using RAPD analysis in *Mentha arvensis* seedlings. Plant Cell Biotechnol Mol Biol 12:41–50

Manikandan R, Sahi SV, Venkatachalam P (2015) Impact assessment of mercury accumulation and biochemical and molecular response of *Mentha arvensis*: a potential hyperaccumulator plant. Sci World J 2015:1

Mazid M, Khan TA, Mohammad F (2011) Role of secondary metabolites in defense mechanisms of plants. Biol Med 3(2):232–249

McGrath SP (1995) Chromium and nickel. In: Alloway BJ (ed) Heavy metals in soils, 2nd edn. Blackie Academic and Professional, London, pp 152–178

Mishra A, Lal RK, Chanotiya CS, Dhawan SS (2017) Genetic elaborations of glandular and non-glandular trichomes in Mentha arvensis genotypes: assessing genotypic and phenotypic correlations along with gene expressions. Protoplasma 254(2):1045–1061

Misra PN, Hasan SA, Kumar S (2000) Cultivation of aromatic plants in India. Central Institute of Medicinal and Aromatic Plants, Lucknow

Mitchell P, Barre D (1995) The nature and significance of public exposure to arsenic: a review of its relevance to South West England. Environ Geochem Health 17(2):57–82

Mittler R (2002) Oxidative stress, antioxidants and stress tolerance. Trends Plant Sci 7(9):405–410

Moreno L, Bello R, Primo-Yúfera E, Esplugues J (2002) Pharmacological properties of the methanol extract from *Mentha suaveolens* Ehrh. Phytother Res 16(S1):10–13

Mostofa MG, Hossain MA, Fujita M, Tran LSP (2015) Physiological and biochemical mechanisms associated with trehalose-induced copper-stress tolerance in rice. Sci Rep 5:11433

Naeem A, Westerhoff P, Mustafa S (2007) Vanadium removal by metal (hydr) oxide adsorbents. Water Res 41(7):1596–1602

Naeem M, Khan MMA, Moinuddin, Idrees M, Aftab T (2011) Triacontanol-mediated regulation of growth and other physiological attributes, active constituents and yield of *Mentha arvensis* L. Plant Growth Regul 65(1):195–206

Naeem M, Idrees M, Aftab T, Khan MMA, Varshney L (2012a) Depolymerised carrageenan enhances physiological activities and menthol production in *Mentha arvensis* L. Carbohydr Polym 87(2):1211–1218

Naeem M, Idrees M, Aftab T, Khan MMA, Varshney L (2012b) Irradiated sodium alginate improves plant growth, physiological activities and active constituents in *Mentha arvensis* L. J Appl Pharm Sci 2(5):28

Naeem M, Idrees M, Aftab T, Alam MM, Khan M, Uddin M, Varshney L (2015a) Radiation processed carrageenan improves plant growth, physiological activities, and alkaloids production in *Catharanthus roseus* L. Adv Bot 2015:Article ID 150474, 11 pages

Naeem M, Aftab T, Ansari AA, Idrees M, Ali A, Khan MMA, Varshney L (2015b) Radiolytically degraded sodium alginate enhances plant growth, physiological activities and alkaloids production in *Catharanthus roseus* L. J Rad Res Appl Sci 8(4):606–616

Naeem M, Aftab T, Ansari AA, Shabbir S, Khan MMA, Uddin M (2019) Arsenic exposure modulates physiological attributes and artemisinin biosynthesis in Artemisia annua L. Int J Herb Med 7(2):19–26

Nagasawa N, Mitomo H, Yoshii F, Kume T (2000) Radiation-induced degradation of sodium alginate. Polym Degrad Stab 69(3):279–285

Naik PM, Al-Khayri JM (2016) Abiotic and biotic elicitors-role in secondary metabolites production through in vitro culture of medicinal plants. In: Abiotic and biotic stress in plants-recent advances and future perspectives. InTech, Rijeka, pp 247–277

Nickavar B, Alinaghi A, Kamalinejad M (2010) Evaluation of the antioxidant properties of five Mentha species. Iran J Pharm Res:203–209

Noctor G, Foyer CH (1998) Ascorbate and glutathione: keeping active oxygen under control. Annu Rev Plant Biol 49(1):249–279

Orlova I, Nagegowda DA, Kish CM, Gutensohn M, Maeda H, Varbanova M, Krichevsky A (2009) The small subunit of snapdragon geranyl diphosphate synthase modifies the chain length specificity of tobacco geranylgeranyl diphosphate synthase in planta. Plant Cell 21(12):4002–4017

Panou-Filotheou H, Bosabalidis AM, Karataglis S (2001) Effects of copper toxicity on leaves of oregano (*Origanum vulgare* subsp. hirtum). Ann Bot 88(2):207–214

Parić A, Karalija E, Čakar J (2017) Growth, secondary metabolites production, antioxidative and antimicrobial activity of mint under the influence of plant growth regulators. Acta Biol Szegediensis 61(2):189–195

Parihar P, Singh S, Singh R, Singh VP, Prasad SM (2015) Effect of salinity stress on plants and its tolerance strategies: a review. Environ Sci Pollut Res 22(6):4056–4075

Pehlivan E, Özkan AM, Dinç S, Parlayici Ş (2009) Adsorption of Cu2+ and Pb2+ ion on dolomite powder. J Hazard Mater 167(1–3):1044–1049

Peleg Z, Blumwald E (2011) Hormone balance and abiotic stress tolerance in crop plants. Curr Opin Plant Biol 14(3):290–295

Peterson PJ, Girling CA (1981) Other trace metals. In: Effect of heavy metal pollution on plants. Springer, Dordrecht, pp 213–278

Peyvandi M, AboieMehrizi Z, Ebrahimzadeh M (2016) The effect of cadmium on growth and composition of essential oils of *Mentha piperita* L. Plant Physiol 6(3):1715–1720

Pilon-Smits E, Pilon M (2000) Breeding mercury-breathing plants for environmental cleanup. Trends Plant Sci 5(6):235–236

Prajapati SK (2014) Heavy metal speciation of soil and *Calotropis procera* from thermal power plant area. Proc Int Acad Ecol Environ Sci 4(2):68

Prasad A, Singh AK, Chand S, Chanotiya CS, Patra DD (2010) Effect of chromium and lead on yield, chemical composition of essential oil, and accumulation of heavy metals of mint species. Commun Soil Sci Plant Anal 41(18):2170–2186

Rahimi Y, Taleei A, Ranjbar M (2017) Changes in the expression of key genes involved in the biosynthesis of menthol and menthofuran in *Mentha piperita* L. under drought stress. Acta Physiol Plant 39(9):203

Rai UN, Tripathi RD, Vajpayee P, Jha V, Ali MB (2002) Bioaccumulation of toxic metals (Cr, Cd, Pb and Cu) by seeds of *Euryale ferox* Salisb. (Makhana). Chemosphere 46(2):267–272

Rajput DK, Rao BR, Srivastava PC (2002) Response of cornmint (*Mentha arvensis* L. f. piperascensMalinv. ex Holmes) to micronutrients. J Hortic Sci Biotechnol 77(4):438–440

Raju K, Vijayaraghavan K, Seshachalam S, Muthumanickam J (2011) Impact of anthropogenic input on physicochemical parameters and trace metals in marine surface sediments of Bay of Bengal off Chennai, India. Environ Monit Assess 177(1–4):95–114

Ribeiro B, Shapira P (2019) Anticipating governance challenges in synthetic biology: insights from biosynthetic menthol. Technol Forecast Soc Chang 139:311–320

Rubio-Wilhelmi MM, Sanchez-Rodriguez E, Rosales MA, Begona B, Rios JJ, Romero L, Ruiz JM (2011) Effect of cytokinins on oxidative stress in tobacco plants under nitrogen deficiency. Environ Exp Bot 72(2):167–173

Sadiq Y, Khan MMA, Shabbir A, Ahmad B, Jaleel H, Uddin M, Varshney L (2017) Structural re-arrangement of depolymerized sodium alginate enriches peltate glandular trichomes and essential oil production of spearmint. Int J Biol Macromol 105:1043–1050

Šarić-Kundalić B, Fialová S, Dobeš C, Ölzant S, Tekeľová D, Grančai D et al (2009) Multivariate numerical taxonomy of Mentha species, hybrids, varieties and cultivars. Sci Pharm 77(4):851–876

Sarfaraz A, Naeem M, Shafia N, Idrees M, Tariq T, Nadeem H, Khan MMA, Varshney L (2011) An evaluation of the effects of irradiated sodium alginate on the growth, physiological activities and essential oil production of fennel (*Foeniculum vulgare* Mill.). J Medicinal Plant Res 5:15–21

Sayyed MRG, Sayadi MH (2011) Variations in the heavy metal accumulations within the surface soils from the Chitgar industrial area of Tehran. Proc Int Acad Ecol Environ Sci 1(1):36

Sharma DC (1995) Chromium accumulation by barley seedlings (*Hordeum vulgare* L.). J Exp Bot 25:241–250

Sharma I (2012) Arsenic induced oxidative stress in plants. Biologia 67(3):447–453

Sharma P, Dubey RS (2005) Lead toxicity in plants. Braz J Plant Physiol 17(1):35–52

Sharma S, Sangwan NS, Sangwan RS (2003) Developmental process of essential oil glandular trichome collapsing in menthol mint. Curr Sci Bangalore 84(4):544–550

Singh J, Kalamdhad AS (2011) Effects of heavy metals on soil, plants, human health and aquatic life. Int J Res Chem Environ 1(2):15–21

Singh R, Shushni MA, Belkheir A (2015) Antibacterial and antioxidant activities of *Mentha piperita* L. Arab J Chem 8(3):322–328

Singh M, Khan MMA, Uddin M, Naeem M, Qureshi MI (2017) Proliferating effect of radiolytically depolymerized carrageenan on physiological attributes, plant water relation parameters, essential oil production and active constituents of *Cymbopogon flexuosus* Steud. under drought stress. PloS One 12(7):e0180129

Soares C, Carvalho ME, Azevedo RA, Fidalgo F (2019) Plants facing oxidative challenges – a little help from the antioxidant networks. Environ Exp Bot 161:4–25

Sorboni AG, Bahmanyar MA, Sepanlou MG (2013) Accumulation of lead and nickel in soil and spearmint (*Mentha spicata*) growing on soil treated with municipal solid waste and sewage sludge. Int J Agric 3(1):126

Srivastava S, Srivastava AK, Suprasanna P, D'souza SF (2012) Identification and profiling of arsenic stress-induced microRNAs in *Brassica juncea*. J Exp Bot 64(1):303–315

Srivastava AK, Srivastava S, Mishra S, D'Souza SF, Suprasanna P (2014) Identification of redox-regulated components of arsenate (As V) tolerance through thiourea supplementation in rice. Metallomics 6(9):1718–1730

Suszcynsky EM, Shann JR (1995) Phytotoxicity and accumulation of mercury in tobacco subjected to different exposure routes. Environ Toxicol Chem: Int J 14(1):61–67

Syracuse Research Corporation (1993) Toxicological profile for chromium. Prepared for U.S. Dept. Health and Human Services, Public Health Service, Agency for Toxic Substances and Disease Registry, under Contract No. 205–88-0608

Száková J, Tlustoš P, Goessler W, Findenig S, Richtrová E, Balík J (2009) A comparison of arsenic mobility in *Phaseolus vulgaris*, *Mentha aquatica*, and *Pteris cretica* rhizosphere. Open Life Sci 4(1):107–116

Száková J, Tlustoš P, Goessler W, Pokorný T, Findenig S, Balík J (2011) The effect of soil contamination level and plant origin on contents of arsenic, cadmium, zinc, and arsenic compounds in *Mentha Aquatica* L. Arch Environ Prot 37:109–121

Tassou CC, Nychas GJ, Skandamis PN (2004) Herbs and spices and antimicrobials. In: Handbook of herbs and spices. Woodhead Publishing, Cambridge, pp 22–40

Tchounwou PB, Ayensu WK, Ninashvili N, Sutton D (2003) Environmental exposure to mercury and its toxicopathologic implications for public health. Environ Toxicol: Int J 18(3):149–175

Tchounwou PB, Yedjou CG, Patlolla AK, Sutton DJ (2012) Heavy metal toxicity and the environment. In: Molecular, clinical and environmental toxicology. Springer, Basel, pp 133–164

Thakur M, Bhattacharya S, Khosla PK, Puri S (2018) Improving production of plant secondary metabolites through biotic and abiotic elicitation. J Appl Res Med Aromat Plants 12:1–12

Tripathi DK, Singh S, Singh S, Mishra S, Chauhan DK, Dubey NK (2015) Micronutrients and their diverse role in agricultural crops: advances and future prospective. Acta Physiol Plant 37(7):139

Tuteja N (2007) Abscisic acid and abiotic stress signaling. Plant Signal Behav 2(3):135–138

Wagner GJ (1993) Accumulation of cadmium in crop plants and its consequences to human health. Adv Agron 51:173–212

WaLwalaba JL, Zvogbo G, Mulembo M, Mundende M, Zhang G (2017) The effect of cobalt stress on growth and physiological traits and its association with cobalt accumulation in barley genotypes differing in cobalt tolerance. J Plant Nutr 40(15):2192–2199

Wani SH, Kumar V, Shriram V, Sah SK (2016) Phytohormones and their metabolic engineering for abiotic stress tolerance in crop plants. Crop J 4(3):162–176

Wasternack C, Strnad M (2019) Jasmonates are signals in the biosynthesis of secondary metabolites – pathways, transcription factors and applied aspects – a brief review. New Biotechnol 48:1–11

Wheeler JI, Irving HR (2010) Evolutionary advantages of secreted peptide signalling molecules in plants. Funct Plant Biol 37(5):382–394

WHO (1987) Air quality guidelines for Europe. World Health Organization Regional Office for Europe, Copenhagen, p 426

Wolters H, Jürgens G (2009) Survival of the flexible: hormonal growth control and adaptation in plant development. Nat Rev Genet 10(5):305

Yadav SK (2010) Heavy metals toxicity in plants: an overview on the role of glutathione and phytochelatins in heavy metal stress tolerance of plants. S Afr J Bot 76(2):167–179

Yazaki K (2006) ABC transporters involved in the transport of plant secondary metabolites. FEBS Lett 580(4):1183–1191

Zaborowska M, Kucharski J, Wyszkowska J (2016) Biological activity of soil contaminated with cobalt, tin, and molybdenum. Environ Monit Assess 188(7):398

Zaid A, Mohammad F (2018) Methyl jasmonate and nitrogen interact to alleviate cadmium stress in *Mentha arvensis* by regulating physio-biochemical damages and ROS detoxification. J Plant Growth Regul 37(4):1331–1348

Zhou ZS, Huang SQ, Guo K, Mehta SK, Zhang PC, Yang ZM (2007) Metabolic adaptations to mercury-induced oxidative stress in roots of *Medicago sativa* L. J Inorg Biochem 101(1):1–9

Zhu XF, Jiang T, Wang ZW, Lei GJ, Shi YZ, Li GX, Zheng SJ (2012) Gibberellic acid alleviates cadmium toxicity by reducing nitric oxide accumulation and expression of IRT1 in *Arabidopsis thaliana*. J Hazard Mater 239:302–307

Zojaji F (2014) Bioaccumulation of chromium by *Zea mays* in wastewater-irrigated soil: an experimental study. Proc Int Acad Ecol Environ Sci 4(2):62

Zurayk R, Sukkariyah B, Baalbaki R, Ghanem DA (2002) Ni phytoaccumulation in *Mentha aquatica* L. and *Mentha sylvestris* L. Water Air Soil Pollut 139(1–4):355–364

Zheljazkov VD, Craker LE, Xing B (2006) Effects of Cd, Pb, and Cu on growth and essential oil contents in dill, peppermint, and basil. Environ and Exp Bot 58:9–16

Arsenic-Contaminated Soil Toxicity and Its Mitigation Through Monocot Crops

Phalguni Bhattacharyya and Md Masidur Alam

Contents

1 Introduction

Soil contamination by heavy metals, caused by rapid industrialization, urbanization, and poor management of industrial effluent, from many anthropogenic activities such as mining, improper disposal of industrial and urban waste, transportation, and unscientific agricultural activities, is becoming a serious threat to humans and animals with harmful outcomes for soil organisms and cultivated crops (Wei and Yang 2010; Yaylali 2011; Mireles 2012). Heavy metals, included in the group of metals and metalloids, have an atomic density greater than 5 g/cm^3 or five times greater than water and are toxic or poisonous even at low concentrations (Nriagu and Pacyna 1988; Hawkes 1997). Important metals and metalloids polluting our environment include arsenic (As), nickel (Ni), cadmium (Cd), lead (Pb), mercury (Hb), chromium (Cr), and selenium (Se). Soil pollution resulting from these heavy metals causes significant complications in environmental health globally (Alloway 1995a). Soils facilitate the immobilization of introduced compounds such as the ions of heavy metals. The persistence of xenobiotics is primarily the result of absorption, which is regulated by the physicochemical properties of the soil: (a) amount of clay and organic fraction, (b) pH, (c) water content, and (d) soil temperature and particulate metal ion properties (Hossain 2005). The metal content of the soil is particularly important because the occurrence of these metals influences the environment in various ways (Brookes and McGrath 1984): for example, food contamination from plants grown in polluted soil, as well as a sharp decrease in crop productivity

P. Bhattacharyya · M. M. Alam (✉)
Department of Botany, Malda College, Malda, WB, India

M. Naeem et al. (eds.), *Contaminants in Agriculture*,
https://doi.org/10.1007/978-3-030-41552-5_16

(Chaudri 1993; Rooney 2007), and through the activity of soil microbes (Akerblom 2007). Excessive concentrations of heavy metals (As, Ni, Cr, Cd, Se, Pb) have been found in the soil of farmlands near cities, industrial areas, and mines throughout the planet. Heavy metals, although found in the Earth naturally, become damaging in soils as consequences of human activities: common sources are mining and industrial wastes; lead-acid batteries; vehicle emissions; paints; fertilizers; treated woods; old deposits from water supply transportation; and microplastics found floating in the oceans. In this chapter, we concentrate on arsenic (As), which is a slow poison for human health and to plant life. The Greek word 'arsenic' is a derivative of a Persian word meaning 'yellow orpiment.' Since ancient times arsenic was available in Persia and mixed with bronze as an adulteration for hardening of the alloy. Albertus Magnus is believed to have been the first to isolate the element in 1250 C.E. The widespread use of As as a pest killer with pesticide sprays led to the contamination of the environment and thus human food with arsenic, resulting in unfavorable effects on human health. Arsenic is a solid metallic element of natural occurrence found to form inorganic or organic compounds with other elements. In the soil and groundwater, As is found as dissolved oxyanions, namely, arsenites (trivalent arsenic) or arsenates (pentavalent arsenic), or both, in addition to the organic forms. Arsenic mobility, solubility, bioavailability, and hence toxicity in the soil–crop system basically depend on its form (chemical), that is, the arsenic oxidation state (valency), and also whether the arsenic is present in inorganic or organic amalgamations.

The toxicity of As compounds in groundwater and the soil environment is in the following order: arsine (AsH; valence state of arsenic, 3) > organo-arsine compounds > arsenites and oxides (trivalent arsenic form) > arsenates (pentavalent arsenic) > arsonium metals (monovalent arsenic) > native arsenic metal. The arsenates (pentavalent form) are much less mobile, soluble, and contaminating than arsenates (trivalent form) in water and soils. Other organic forms, such as cacodylic acid or dimethyl arsinic acid (DMA), on reduction (e.g., in an anoxic soil state) include volatile di/trimethyl arsines present in the soil. Besides these organic forms present in groundwater and soil is monomethyl arsonic acid (MMA). The organic forms are nontoxic or much less toxic.

2 Soil Toxicity

Noncontaminated soils possess As concentrations in a range from 0.1 to 10 mg/kg of soil (Pendias 1995). Arsenic (As) concentration in the lands of the agricultural area of West Bengal was estimated to be about 10.7 mg/kg (Rajindiran et al. 2015) and within the range of 3.3 to 31.6 mg/kg. Arsenic toxicity in soil is the outcome of many human-based activities, including the use of pesticides and herbicides, mining and ore-processing operations, waste disposal, and coal-burning power plants. Sites formerly used for tannery purposes, with leather depositions from animals, conceal large amounts of As in the soil. Arsenic with coloring dyes, from imbibition of unlawfully distilled alcohol or arsenic-contaminated lead used for solder in the distilling equipment, or rat poison used in particle and mash stores, may be excellent sources of soil As. The maximum value of As in drinking water considered harmless was approved by the World Health Organization (WHO) to be 10 μg As/l water. The range of As contamination in the groundwater of West Bengal, Bihar, Assam, Madhya Pradesh Uttar Pradesh, Jharkhand, Punjab, Chhattisgarh, Tripura Manipur, and Nagaland is 50–3700 μg As/l, as reported by the states. The life of the element in soil is often found effective after centuries, so soil contaminated by As, especially in agricultural fields, is a serious matter. The natural occurrence of As in the groundwater does not allow safe water intake to a huge number of people in Asia and globally as it is one of the most deadly elements in the environment, liable for the highest risks of mortality and morbidity worldwide because of its toxicity and the size of the human population. In India, As is found to be naturally derived from disturbed Himalayan sediments (Polizzotto et al. 2008).

3 Arsenic-Contaminated Soil in West Bengal

The true chemical reactions and conditions that primarily cause arsenic mobilization are under debate. So far, it is in general accepted and assumed that microbial/chemical reductive disbanding of As-based iron (Fe) minerals in the aquifer (in sediments) is the chief source of As (Winkel et al. 2008). A number of investigators have reported As-based pyrite in residue samples from the West Bengal Gangetic delta (Chakraborty et al. 2013). Certain concepts suggest that As is mobilized slowly through iron arsenate or by oxyhydroxide reduction (Nriagu 2002). Das et al. (2014) reported that As discharge through sulfide-oxidation reactions is stimulated by the huge increase in dry season irrigation pumps. Intensive use of fertilizers in the agricultural fields (Acharya et al. 1998), or by carbonate formed through microbial metabolism (Appelo et al. 2002) or by transformation in the absorptive ability of ferric oxyhydroxides (Smedley and Kinniburgh 2013), may be possible causes of As contamination in soil. However, pollution of soil by heavy metals is an irreversible process and it is hard to reclaim these kinds of metal-polluted soils. The significant factors affecting arsenic (As) chemistry and mobility in the soils are soil solution (redox conditions, pH), composition of solid, As-bearing phases, adsorption/desorption, volatilization, and cycling of As in soil and biological transformations (Warren et al. 2003). Furthermore, soil fractions, organic content, and oxides of Al, Fe, and Mn also influence soil As content (Chauhan and Jain 2010). In addition to metal sulfides, salinity, temperature, distribution, and biota composition are significant factors for determining the transport of As and its fate (Ning 2002). Residential areas transformed for industrialization purposes, and a mixture of materials is released into sewage, leading to environmental pollution. Large numbers of open and covered channels carry a mixture of wastewaters generated by domestic, municipal, and industrial activities (CPCB 2009). The projected wastewater generation estimate is 122,000 million liters/day (MLD) for the country by 2050 (Bhardwaj

Table 1 Agricultural causes of soil contamination by trace elements (in mg/kg) (Kabata-Pendias and Pendias 2000)

Element	Sewage sludge	Phosphate fertilizers	Limestone	Nitrogenous fertilizers	Manures and pesticides
As	2–26	2–1200	0.1–24	2–120	3–150

2005). The cities of West Bengal all over the Ganga basin are producing 2637.7 MLD of sewage, where the treatment capacity is about 1174.4 MLD, or only about 44.2%. The leftover sewage flows into and pollutes the River Ganga because the effluent will explode without any treatment.

In farmland soils, the too large accumulation of heavy metals during wastewater irrigation not only affects soil contamination, but also increases the uptake of heavy metals by cultivated crops and hence influences food safety as well as quality (Table 1). This accumulation of heavy metals frequently leads to degradation of soil health and infectivity of the food chain, mostly through the crops and vegetables grown in such soils (Rattan 2002). The degree of soil pollution by heavy metals from various anthropogenic sources and the consequent uptake by crops depend on factors such as soil type, source, organic matter content, frequency of application, seasonal variations, major/minor nutrients, and consignment of chemical pollutants. The amount of groundwater arsenic contamination in the districts of West Bengal has reached a shocking situation because nine districts of West Bengal have been found to have groundwater arsenic levels greater than 50 μg/l water. Chakraborty et al. (2013) reported a number of communities have been found to be affected by arsenic toxicity.

In particular, during Boro (summer) rice cultivation during March to May, a huge amount of groundwater full of arsenic is used for the irrigation of agricultural crops. The majority of productive alluvial soils covering the districts of North Dinajpur, South Dinajpur Malda, Murshidabad, Burdwan, Nadia, Hoogly (North), and part of the South 24 Parganas are contaminated with As from using As-accumulated groundwater as a source of irrigation. Progressive buildup of As might result from the residual roots after harvest, contributing substantially to the accumulation of As in soils. The chief source of As in the soils is the parent resources from which the soil originated, and the deviation in As content in particular areas is frequently determined by the geological record of the area. The increase in shallow tube wells in the irrigation of West Bengal is mainly dependent on the average deep groundwater, which is the main source of As and other heavy metals in the groundwater and in agricultural soil. In West Bengal rice is grown mainly in the wetlands, which requires an enormous amount of water for farming, which leads to increased As mobility in the irrigated soil (Halder et al. 2013). Accordingly, the concentration of As in addition to other heavy metals directly affects the value of the soil and crop quality. Within contaminated areas the standard As level in the groundwater is about 122.57 μg/l, much higher than the limit prescribed by WHO, that is, 10.0 μg/l. In groundwater the high As content clearly is the reason for the high soil As contamination. Boro rice needs approximately 1000 mm irrigation water per year (season), which increases soil As concentration by 1 μg/g every year because of irrigation with water contaminated by arsenic (Meharg and Rahman 2003). Walkley and Black (1934) described a method used for the determination of percentage (%) by dry mass of organic matter in a soil sample combined with the percentage (%) of arsenic in soil.

4 Biological Approaches to Mitigation

Zhao et al. (2010) showed the efficiency of rice plants (*Oryza sativa*) in arsenite mobilization in a flooded paddy field. Arsenate enters into the plant body through phosphate transporters, and through arsenite and associated methylated As groups by the nodulin-26-like intrinsic aquaporin channels (NIP). Arsenate is gradually reduced to arsenite within the plant and detoxified by the formation of the complex with thiol-rich peptides such as phytochelatins and vacuolar sequestration. Various mitigation processes, on one side from agricultural planning to plant breeding and up to genetic alteration, may be employed to diminish As uptake through food or cultivated crops. Das (2015) described an experiment with rice, reporting two PCS genes, OsPCS1 and OsPCS2, in an *O. sativa* cultivar. These two gene transcripts are instrumental in mitigating the content of As in grains. Lakshman (2015) isolated some proteobacterial strains from the rhizosphere of rice plants that promote the growth of rice and increase the oxidizing environment near the rice plant roots. *Pantoea* sp. (EA106) strain is a microbe that may have the potential to augment food quality by reducing the accumulation of toxic arsenic species within the aboveground portions of the rice plant. Another attempt to mitigate the grain As content in rice by the expression of the rice MATE family transporter OsMATE2 has been investigated by Das (2015), who showed a unique mitigation of As with a water management system and fertilization with silicon. During the entire vegetative and reproductive stages of rice growth, the least As accumulation in rice grains and rice straw was observed in the aerobic condition. The selection of rice cultivation method also is very important to minimize As contamination.

Other biological approaches include certain plants having dicot/monocot root that act as hyperaccumulators, such as *Thlaspi caerulescens*, *Haumaniastrum robertii*, *Ipomoea alpina*, *Thlaspi rotundifolium*, *Cistus ladanifer*, *Psychotria douarret*, *Salix* sp., and *Macadamia (Virotia) neurophylla*, which are found to eliminate heavy metals from soils. These plants are tolerant of high metal levels in soil and, in comparison to other crops, can accumulate a level of metal 10 to 500 times higher. Chelates such as EDTA application have been found to boost the extraction of metal by hyperaccumulators (Abedin et al. 2002; Dotaniya 2014). Arsenic removal is accelerated by rock phosphate (Fayiga and Ma 2006) by the performance of the hyperaccumulator *Pteris vittata*. As is an omnipresent element in the environment with concentrations that vary widely from one region to another. The plants responsible for phytoremediation that are capable of accumulating elevated concentrations of metals may produce less biomass, which restricts their overall phytoextraction competence. Several high biomass-hyperaccumulating plants for which regeneration practices may be developed include Indian sunflower (*Helianthus annuus*), mustard (*Brassica juncea*), tomato (*Lycopersicon esculentum*), and yellow poplar. Certain monocot plants have also been found to remove some metals from soils, changing them by phytovolatilization or into volatile forms. Numerous plants have also been reported to use a phytostabilization function (in the rhizosphere, soluble forms of metals are inactivated) to avoid metal contamination of the groundwater and deeper soil layers (Pigna et al. 2010). Wang and Mulligan (2004) reported that some researchers demonstrated the potential of microbes for the elimination of metals from the soil ('biometal slurry reactor' technique) and indirectly through microbially generated biosurfactant. However, the utility of such strategies is yet to be proved at the field level. The presence of As is associated with natural and anthropogenic (human-governed) factors, and the higher level of toxicity frequently threatens the global ecosystems and thus human health (Nriagu et al. 2007). As has properties intermediate between metals and those of nonmetal elements, although its ionization energy and electronegativity are characteristics more related to nonmetals. This element shows a complex chemistry under natural conditions, rather like phosphorus, and tends to appear in anionic forms (Fitz and Wenzel 2002).

Soil pH-Eh conditions in the normal range are found more commonly in arsenates than arsenites (Cao et al. 2009), and the occurrence of these complexes is indispensable to control of toxicity in soils, water, or sediment (Farré and Barceló 2003). The mean As value calculated in the Earth's crust is about 1.8 mg/kg, whereas in natural soils the concentrations are generally less than 10 mg/kg (Adriano 2001; Deliyanni et al. 2003). As usually accumulates in soils because of its low mobility in the medium (Beesley and Marmiroli 2011; Basu et al. 2015). It often originates as an inorganic complex because the organic compound has less solubility in soil (Bhattacharya et al. 2007; Nriagu et al. 2007). The concentration of As in soil usually varies because the value is dependent on the initial concentration of parent material, soil type, and natural geochemical cycles (Fendorf et al. 2004). Consequently, the mean values of arsenic (As) in soils are reported to be between 0.1 and 80 mg/kg globally. Together with other micronutrients, As can be up taken up by plants through the roots, but the significant biological role of As in living organisms has not been described so far, although concentrations between 1 and 1.7 mg/kg have been found normally in plants (Yamaguchi et al. 2011). Using current techniques, several high biomass-producing plants (phytoaccumulators) were developed by relevant genes initiated from bacteria or animals as hyperaccumulators into nonaccumulators (Singh et al. 2003). Mobility, bioavailability, and consequently the potential toxicity or most element deficiencies of organisms (plants and microorganisms) are largely controlled by soil properties in terrestrial systems (Majumdar et al. 2013). Arsenic availability in the soil is affected by these criteria: calcium carbonate, iron oxides, clay content, cation-exchange capacity, pH, and organic matter content (Rooney et al. 2006; Chatterjee et al. 2013). Grasses (*Agrostis stolonifera* and *Agrostris tenuis*) were the prime species present with As content in leaves more than 1.00 g As/kg (dry weight). Wild (1973–1974) described that 72 plant species were originated on 15 Rhodesian arsenical mine dumps plus total As concentrations varying from 0.200 to 30.00 g As/kg. It is expected that the number of plant species and their density increased as As content lessened. Among the mine dumps, the survey report says the Banshee mine dump (30.00 g As/kg) was the most influenced and unable to support vegetation. In general, the weed species were identified as the most significant species on the mine dumps in terms of plant numbers. Gaika weed was often the leading or most common weed found on the mine dumps.

Among the grass species, *Cynodon dactylon* (common grass) is a key species, existing in soils with As concentrations ranging from 0.2 to 30.0 mg As/kg. Although As as well as other heavy metal concentrations at various mine dumps may slow down the stabilization of soil vegetation, some plants, viz. *Cynodon dactylon*, can tolerate elevated As content and may be useful for stabilizing the soil in mine dumps. These plants might also offer a low-cost, long-term solution for mitigating mine dumps where other remediation techniques are not practical. There are many reported occurrences of atmospheric As discharged during the smelting of Au, Pb, Cu, and Zn ores (Alloway 1995b; Acharya 1997). The use of plants that absorb large amounts of As from soils should be studied in this regard. Such hyperaccumulator plants include some ferns, mainly the brake fern (*Pteris vittata*), and Indian mustard (*Brassica* sp.). These plants, which uptake As, are generally removed from fields after the harvesting of the crops. The brake fern and mustard may be considered for application where it is not practical to provide an irrigation supply with low As content. The aquatic plant water hyacinth, *Eichhornia crassipes*, is a recognized hyperaccumulator but extends its roots only in water, so it could only be used for uptake of As from irrigation water (in settling tanks) but not from the soil. The possibility of using of hyperaccumulating plants to regenerate paddy soils needs testing, and also methods to organize the safety of huge quantities of plant-based residues with high As content to minimize health risks to children, wildlife, and livestock eating the plants (herbivores) or breathing in dust from burnt matter. The different irrigational water sources along the district subdivisions and municipalities, and their analysis for physicochemical parameters as well as heavy metals, were documented. The percolation of heavy metals through irrigation water was determined in the vegetables available daily in the market that were selected and analyzed for heavy metal occurrence. The concentrations are waiting to be analyzed especially in river-lifting irrigation (RLI) as well as the As concentration from indiscriminate use of groundwater through shallow tube wells.

Malda is a district of West Bengal, India. The district produces the country's quality production of raw silk yarn and is a good exporter of the best mango varieties with annual turnover of about Rs. 5.5 crores. To grow crops in agricultural fields, different subdivisions of the Malda district utilize three main types of irrigation facilities: river-lifting irrigation (RLI), deep tube well irrigation (DTWI), and shallow tube well irrigation (STWI). Apart from these methods, there are other sources also such as wells and ponds. The urban wastewater with pollutants and its mixing with irrigation sources as well as indiscriminate use of groundwater for agricultural purposes in the district of Malda is a matter of grave concern. Indiscriminate use of groundwater through shallow tube wells increases the threat of As contamination in food crops. We have documented the distribution of different irrigation sources in the Malda district along with the conditions of the local rivers that receive the wastewater of the locality. In another part of the study we have extracted and purified low molecular weight peptide(s) (3.0–0.5 kDa) from selected local vegetables and crops under wastewater treatment to study the biological role and also the morphological and anatomical changes in the selected crops along with an antioxidant study. In a separate experiment, rice and wheat seedlings were grown for 21 days in vitro under arsenic treatment, and were challenged by the simultaneous cultivation of *Azolla* and *Salvinia* with competition. The result was quite interesting. The sets with *Azolla* and *Salvinia* showed less contamination as evidenced by the morphological parameters of the rice and wheat.

5 Conclusion

In our country, major irrigation projects using the groundwater supply are being implemented all over As-affected areas, and there is a strong need to tackle the issue. Our investigations clearly indicate that the As level in the shallow groundwater within the study area is much higher than the WHO-recommended limit of 10 μg/l. The continuous use of unhygienic groundwater for irrigation for a long time is one of the major reasons for As buildup in the soil. Heavy metals other than As also pose a risk to various organisms for farming. Challenges arise when these metals accumulate in fields (the soil) and the contaminants might be bioaccumulated in the human body. There is no regulating group in our country to test the As content in food or food grains as our India is a developing country suffering from a food deficit. Thus, there is severe and immediate concern for all life forms including people regarding the fatal effects through drinking water as well as crops. It is also a matter of high concern that not only those who live in the As-contaminated regions at risk but also people living in other parts of India are not safe because they consume As-contaminated food produced in these communities. Several steps in this regard must be taken immediately, including institutional integration of the water supply with a proper water management system, developing real responses when contamination by As is detected in certain regions, overcoming the political wealth constraints and activities concerned with mitigation strategies, studying As in the food chain in various ecosystems, a strategic research program to address such concerns as dose–response interactions for As, along with hydrochemical and geohydrological research, and identifying or developing the low As-accumulating, less water intensive high-yielding crop varieties and cropping sequences suitable for As-contaminated areas, especially for the lean period of January to May. For example, cropping sequences could be elephant foot yam–mustard–sesame, green gram–rice–mustard, etc., instead of olitorius jute (*Corchorus olitorius*)–rice–rice and green manure–rice–rice. Further, we can irrigate with pond-based stored groundwater where partial decontamination is provided by sedimentation and dilution through rainwater, and enhance water use effectiveness (by an optimal water management system) for groundwater irrigation, particularly for summer (Boro) paddy, for example, by having recourse to judicious intermittent ponding of summer paddy during the vegetative growth period, followed by continuous ponding during the subsequent reproductive phase, which does not affect the yield significantly while decreasing the use of contaminated groundwater considerably. The high use of farmyard manure (FYM) plus other organic manures and green compost crops, as well as the use of proper inorganic supplements (Zn/Fe salts), allows identification of varieties of crops that accumulate a low level of arsenic in the consumable portion and where the inorganic:organic ratio of arsenic is too low.

References

Abedin MJ, Cotter-Howells J, Meharg AA (2002) Arsenic uptake and accumulation in rice (*Oryza sativa* L.) irrigated with contaminated water. Plant Soil 240:311–319

Acharya SK (1997) Arsenic in groundwater: geological overview. Consultation on arsenic in drinking water and resulting arsenic toxicity in India and Bangladesh. World Health Organization, New Delhi

Acharya CL, Bishoni SK, Yadavanshi HS (1998) Effect of long-term application of fertilizers and organic manures and inorganic amendments under continuous cropping on soil physical and chemical properties in an Alfisol. Indian J. Agric. Sci., 58: 509–516.

Adriano DC (2001) Trace elements in terrestrial environments. Biogeochemistry, bioavailability and risk of metals, 2nd edn. Springer, New York

Akerblom S (2007) Experimentally induced effects of heavy metals on microbial activity and community structure of forest more layers. Biol Fertil Soils 44:79–91

Alloway BJ (1995a) Heavy metals in soils. Blackie Academic & Professional, London

Alloway BJ (1995b) The origins of heavy metals in soils. In: Alloway BJ (ed) Heavy metals in soils. Blackie Academic and Professional Publication, New York, pp 38–57

Appelo CAJ, Van Der Weiden MJJ, Tournassat C, Charlet L (2002) Surface Complexation of Ferrous Iron and Carbonate on Ferrihydrite and the Mobilization of Arsenic. Environmental Science & Technology 36(14):3096–3103

Basu B, Kundu M, Hedayatullah M, Kundu CK, Bandyopadhyay P, Bhattacharya K, Sarkar S (2015) Mitigation of arsenic in rice through deficit irrigation in field and use of filtered water in kitchen. Int J Environ Sci Technol 12:2065–2207

Beesley L, Marmiroli M (2011) The immobilisation and retention of soluble arsenic, cadmium and zinc by biochar. Environ Pollut 159:474–480

Bhardwaj RM (2005) Status of wastewater generation and treatment in India. IWG-environment, international work session on water statistics, Vienna, June 20–22, 2005, p 9

Bhattacharya P, Ewlch AH, Stollenwerk KG, McMauglin MJ, Brundschuh J, Panaullah G (2007) Arsenic in the environment: biology and chemistry. Sci Total Environ 379:109–120

Brookes PC, McGrath SP (1984) The effect of metal toxicity on the size of soil microbial biomass. J Soil Sci 35:341–346

Cao Q, Hu Q-H, Baisch C, Khan S, Zhu Y-G (2009) Arsenate toxicity for wheat and lettuce in six Chinese soils with different properties. Environ Toxicol Chem 28:1946–1950

Chakraborty S, Dutta AR, Sural S, Gupta D, Sen S (2013) Ailing bones and failing kidneys: a case of chronic cadmium toxicity. Ann Clin Biochem 50:492–495

Chatterjee M, Sarkar S, Debnath S, Mukherjee A, Chakrabarti A, Bhattacharyya S (2013) Genotypic difference in temporal variation of arsenic accumulation and expression of silicon efflux transporter (LSi_2) gene in the field grown rice. Indian J Genet Pl Br 73:94–97

Chaudri AM (1993) Enumeration of indigenous *Rhizobium leguminosarum* biovar *trifolii* in soils previously treated with metal-contaminated sewage-sludge. Soil Biol Biochem 25:301–309
Chauhan A, Jain RK (2010) Biodegradation: gaining insight through proteomics. Biodegradation 21:861–879
CPCB (2009) Comprehensive environmental assessment of industrial clusters. Ecological Impact Assessment Series: EIAS/5/2009-2010. Central Pollution Control Board, Ministry of Environment and Forests, Govt. of India, New Delhi
Das N (2015) Experiment with rice reporting two PCS genes OsPCS1 OsPCS2 in *Oryza sativa* cultivar. Environ Int 242:1037–1050
Das I, Ghosh K, Das DK, Sanyal SK (2014) Transport of arsenic in some affected soils of Indian subtropics. Soil Res 52:822–832
Deliyanni EA, Bakoyannakis DN, Zouboulis AI, Peleka E (2003) Removal of arsenic and cadmium by alaganeite fixed beds. Sci Technol 38:3967–3981
Dotaniya ML (2014) Impact of tannery effluent irrigation on heavy metal build up in soil and ground water in Kanpur. Agrotechnology 2:77
Farré M, Barceló D (2003) Toxicity testing of wastewater and sewage sludge by biosensors, bioassays and chemical analysis. Trends Anal Chem 22:299–310
Fayiga AO, Ma LQ (2006) Using phosphate rock to immobilize metals in soil and increase arsenic uptake by hyperaccumulator *Pteris vittata*. Sci Total Environ 15:17–25
Fendorf S, La Force MJ, Li G (2004) Heavy metals in the environment. Temporal changes in soil partitioning and bioaccessibility of arsenic, chromium, and lead. J Environ Qual 33:2049–2055
Fitz WJ, Wenzel WW (2002) Arsenic transformations in the soil/rhizosphere/plant system: fundamentals and potential application to phytoremediation. J Biotechnol 99:259–278
Halder D, Bhowmick S, Biswas A, Mandal U, Nriagu J, Mazumdar DN, Chatterjee D, Bhattacharya P (2013) Consumption of brown rice: a potential pathway for arsenic exposure in rural Bengal. Environ Sci Technol 46:4142–4148
Hawkes JS (1997) Heavy metals. J Chem Educ 74:1374
Hossain MB (2005) Arsenic distribution in soil and water of a STW command area. In CYMMIT 2005, Ravenscroft, pp 2007
Kabata-Pendias A, Pendias H (2000) Trace elements in soil and plants, 3rd edn. CRC Press, Boca Raton
Lakshman V (2015) Proteobacterial strains promote rice growth and enhance the oxidizing environment surrounding rice root. Environ Int 74:221–230
Majumdar A, Bhattacharyya K, Kole SC, Ghosh S (2013) Efficiency of indigenous soil microbes in arsenic mitigation from contaminated alluvial soil of India. Environ Sci Pollut Res 20:5645–5653
Meharg AA, Rahman MM (2003) Arsenic contamination of Bangladesh paddy field soils: implications for rice contribution to arsenic consumption. Environ Sci Technol 37:229–234
Mireles F (2012) Assessing urban soil pollution in the cities of Zacatecas and Guadalupe, Mexico by instrumental neutron activation. Microchem J 103:158–164
Ning RY (2002) Arsenic removal by reverse osmosis. Desalination 143:237–241
Nriagu JO (2002) Arsenic poisoning through the ages. In: Frankenberger WT Jr (ed) Environmental chemistry of arsenic. Dekker, New York, pp 1–26
Nriagu JO, Pacyna J (1988) Quantitative assessment of worldwide contamination of air, water and soil by trace metals. Nature 333:134–139
Nriagu JO, Bhattacharya P, Mukherjee AB, Bundschuh J, Zevenhoven R, Loeppert RH (2007) Arsenic in soil and groundwater: an overview. In: Arsenic in Soil and Groundwater Environment: Biogeochemical Interactions, Health Effects and Remediation, P. Bhattacharya, A. B. Mukherjee, J. Bundschuh, R. Zevenhoven and R. H. Loeppert (eds), Trace Metals and other Contaminants in the Environment, Volume 9, Elsevier B. V., Amsterdam, The Netherlands, pp. 3–60.

Pendias A (1995) Agricultural problems related to excessive trace metal contents of soils. In: Salomons W, Forstner U, Mader P (eds) Heavy metals: problems and solutions. Springer-Verlag, Berlin, pp 3–18
Pigna M, Cozzolino V, Giandonato Caporale A, Mora ML, Dimeo V, Jara AA, Violante A (2010) Effects of phosphorus fertilization on arsenic uptake by wheat grown in polluted soils. J Soil Sci Plant Nutr 10:428–422
Polizzotto ML, Kocar BD, Benner SG, Sampson M (2008) Near-surface wetland sediments as a source of arsenic release to ground water in Asia. Nature 454:505–508
Rattan RK (2002) Heavy metals in environments: Indian scenario. Fertil News 47:21–40
Rooney CP (2007) Phytotoxicity of nickel in a range of European soils: influence of soil properties, Ni solubility and speciation. Environ Pollut 45:596–605
Rooney CP, Zhao FJ, McGrath SP (2006) Soil factors controlling the expression of copper toxicity to plants in a wide range of European soils. Environ Toxicol Chem 25:726–732
Rajindiran S, Dotaniya ML, Vassanda Coumar M, Panwar NR, Saha JK (2015) Heavy metal polluted soils in India: status and countermeasures. JNKVV Res J 49:320–337
Singh OV, Labana S, Pandey G, Budhiraja R, Jain RK (2003) Phytoremediation: an overview of metallic ion decontamination from soil, Appl. Microbiol. Biotech. 61:405–412
Smedley PL, Kinniburgh DG (2013) Arsenic in groundwater and the environment. In: Essentials of Medical Geology, Second Edition. Eds: Selinus, O., Alloway, B., Centeno, J.A., Finkelman, R.B., Fuge, R., Lindh, U. and Smedley, P.L. Springer, pp 279–310
Walkley A, Black IA (1934) An examination of the Detjare method for determining soil organic matter and a proposed modification of the chronic acid titration method. Soil Sci 37:29–36
Wang S, Mulligan CN (2004) Rhamnolipid foam enhanced remediation of cadmium and nickel contaminated soil. Water Air Soil Pollut 157:315–330
Warren GP, Alloway BJ, Lepp NW, Singh B, Bochereau FJM, Penny C (2003) Field trials to assess the uptake of arsenic by vegetables from contaminated soils and soil remediation with iron oxides. Sci Total Environ 311:19–33
Wei B, Yang L (2010) A review of heavy metal contaminations in urban soils, urban road dusts and agricultural soils from China. Microchem J 94:99–107
Winkel L, Berg M, Amini M, Hug SJ, Johnson CA (2008) Predicting groundwater arsenic contamination in Southeast Asia from surface parameters. Nat Geosci 1:536
Yamaguchi N, Nakamura T, Dong D, Takahashi Y, Amachi S, Makino T (2011) Arsenic release from flooded paddy soils is influenced by speciation, Eh, pH, and iron dissolution. Chemosphere 83:925–932
Yaylali AG (2011) Heavy metal contamination of surface soil around Gebze industrial area, Turkey. Microchem J 99:82–92
Zhao FJ, McGrath SP, Meharg AA (2010) Arsenic as a food chain contaminant: mechanisms of plant uptake and metabolism and mitigation strategies. Annual Review of Plant Biology. 61, pp. 535–559.

Role of Cobalt in Plants: Its Stress and Alleviation

Arshiya Akeel and Ajmat Jahan

Contents

1 Introduction

Heavy metals are one of the prime factors of abiotic stress in plants. Relatively high concentrations of heavy metals in the soil have detrimental effects on plant processes, which results in negative impact on crop growth and yield (Lanaras et al. 1993; Maksymiec 2007; Li et al. 2009). Cobalt (Co), a heavy metal and a key element of several enzymes and coenzymes, occurs naturally in the atmosphere in many distinct chemical forms. It is not classified as an essential element for plant nutrition but usually ranked among 'beneficial' elements. Plant beneficial elements (Al, Co, Na, Se and Si), are not considered essential for all crops but can be a crucial requirement for specific plants (Marschner 1995). Cobalt is regarded as a beneficial

A. Akeel (✉) · A. Jahan
Plant Physiology and Biochemistry Section, Department of Botany, Aligarh Muslim University, Aligarh, UP, India

M. Naeem et al. (eds.), *Contaminants in Agriculture*,
https://doi.org/10.1007/978-3-030-41552-5_17

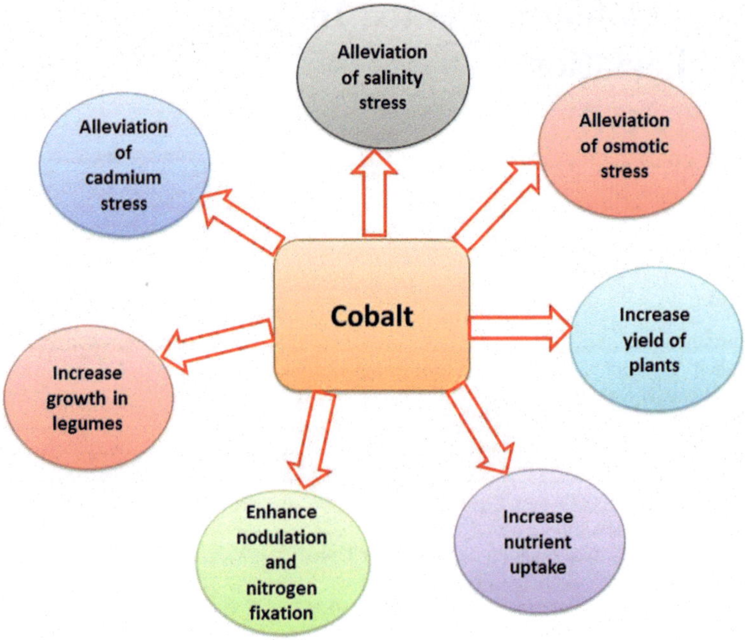

Fig. 1 Beneficial effects of cobalt

element for higher plants despite the absence of manifestation, for its direct role in plant metabolism (Gad 2012). Although cobalt is undoubtedly vital for leguminous crops as it is needed for bacterial nitrogen fixation in root nodules (Dilworth et al. 1979). Co distribution among crops depends entirely on the species. Although it is categorized as a beneficial element for both plants and animals, but at higher levels, Co is observed to produce adverse effects (Jayakumar and Vijayarengan 2006). The toxicity symptoms of Co are seen less frequently in comparison to other metals present in the soil, but its aggregation in plant crops is of considerable importance as it leads to incorporation of toxicity to the human food chain (Lison 2007). The proportions of Co-toxicity depend on the species of plants, soil and soil chemistry (Bakkaus et al. 2005). The more acidic the soil is, the higher its possibilities of Co-toxicity (Hasan et al. 2011). Accumulation of Co in plant tissues causes irretrievable harm to plant cells and its membranes, which is manifested in the form of decrease in growth and biomass, uptake of water and nutrients, chlorosis and an increased cell toxicity (Chatterjee and Chatterjee 2000, 2003; Pandey and Sharma 2002; Gopal et al. 2003; Li et al. 2009; Pandey et al. 2009). It has also been reported to suppress chlorophyll pigment synthesis by blocking the biosynthesis pathway (Mysliva-Kurdziel et al. 2004). Co is reported to have both beneficial (Fig. 1) and adverse impacts on plant growth and metabolism. It has been noted that it has a beneficial effect on crops at lower concentrations. Studies revealed that as a component of cobalamin, it has a beneficial impact on legume crops in nitrogen fixation at reduced levels (Palit et al. 1994). It has also been reported that Co favours the development of salt-stressed tomato plants and increases its leaf water potential (Rathsooriya and Nagarajah 2003). However, detrimental impacts on plant growth along with chlorosis and necrosis were recorded at higher concentrations of Co in soil (Caselles et al. 1997).

Reports revealed that it inhibits the root development by retarding the division of cells and preventing nutrient and water uptake and translocation (Jayakumar et al. 2008). Co has been reported to help in proline accumulation in counter reaction to heavy metal exposure which helps to preserve the water equilibrium in crops (Costa and Morel 1994). Cobalt salts are known to encourage numerous growth mechanisms, including stem and coleoptile elongation, leaf disc expansion, bud development and hypocotyl hook opening (Howell and Skoog 1955; Klein 1959; Miller 1951, 1954; Thimann 1956). Furthermore, when *Xanthium* plants were treated with Co^{2+}, they showed a prolonged critical night period for flowering (Salisbury 1959).

2 Cobalt Absorption and Its Transport in Plants

In plants, Co absorption from soil takes place by virtue of availability of mobile Co content present in soil and from the Co concentration given in the form of solution. In plants, Co absorption is no distinct from other heavy metals (e.g. Fe, Mn) and is transported by bonding with intricate organic compounds (MW 1000–5000), with an overall negative charge (Wiersma and Van Goor 1979). It is also evident through previous literatures that Co^{2+} is transported simultaneously (like Fe^{2+}) with citrate. In various studies, it has been observed that levels of Co in plants increased if the soil is enriched with the metal, which confirms the uptake of Co in plants through soil. Leaves are also known to uptake Co through the cuticle, which led to the conclusion that Co deficiency can be effectively overcome by the foliar applications of Co-containing solutions. The use of chemical fertilizers and liming is also a major source of Co to plants and also known to influence the solubility of Co. Liming is found to have a significant impact on phytoavailability of Co present in soil (Klessa et al. 1989).

3 Biochemical Role of Cobalt

The role of Co in fixing N_2 for leguminous plants and blue-green algae is now a well-established fact. However, the essentiality of Co for higher plants is still not evident, although studies revealed that the presence of Co is favourable for plant growth and function (Reisenauer et al. 1973; Mengel et al. 2001). Co is also known to present as a critical component of cobalamin or vitamin B_{12} (Lison 2007). Researchers have given inconclusive information on the effects of Co on nonleguminous plants, although nonlegumes show the presence of Co coenzymes which

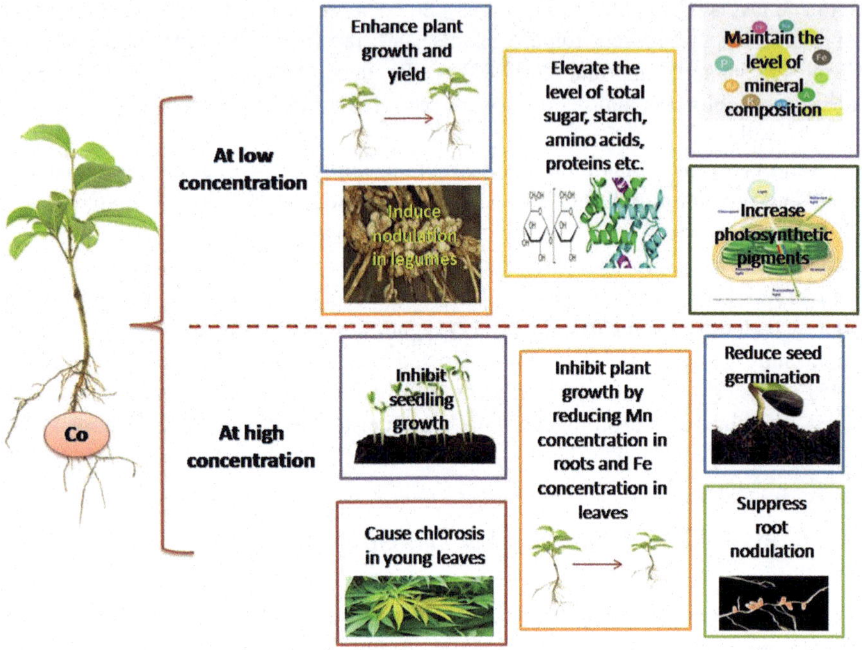

Fig. 2 Role of cobalt in plants at low and high concentration

have been detected in it, but the origin of these compounds is not clear (Smith and Carson 1981). The study on origin of these compounds suggests the role of microorganisms associated with the plant (Nicholas 1975). It is still not fully understood as how Co acts as 'beneficial' at low concentrations on plant metabolism. However, it is deduced from some studies that Co interacts with other trace metals, and the effects are cross-linked (Kabata-Pendias 2000). Structural studies of cobamide coenzyme (useful in N_2 fixation) have shown that Co is present at the centre of a porphyrin ring (Barker 1961). Nicholas (1975) critically reviewed all the studies conducted on the role of Co in N_2 fixation and revealed that cobamide coenzymes are responsible for the migration of H atom during the NH_3 formation with the help of rhizobia. Nagajyoti et al. (2010) have stated that at lower concentration Co induces crucial biochemical and physiological reactions in higher plants (Fig. 2). Cobalt is also known to regulate plant water utilization and reduce transpiration rate which in turn leads to play a significant role in many plant processes (DalCorso et al. 2014). Even though Co is recognized as vital for some bacterial activities, some studies also revealed its antimicrobial activity by inhibiting Mg uptake (Weinberg 1977). In leguminous plants, Co deficiency is reported to curb the formation of leghemoglobin which results in lowering of N_2 fixation rate (Dilworth et al. 1979). According to studies conducted by Wilson and Reisenauer (1970), 10 ppb of Co was found to be sufficient for the growth of alfalfa in the nutrient solution. Co deficiency is not reported to cause adverse effects on the growth and physiology of either nonlegumes or legumes in natural conditions. Application of Co salts to the soil can control the Co deficiency, and its effects are known to remain in soil for years (Mitchell 1972; Reuter 1975). In case the soil is rich in Mn or Fe oxides, a shorter effective period should be expected due to their role in immobilizing Co ions. However, when Co is taken up excessively by roots, it leads to decomposition of Co at the leaf margins and tip because Co is then known to follow the transpiration path (Rauser 1981). And this is the reason for white, dead leaf margins and tips in plants having Co toxicity.

4 Interactions with Other Heavy Metals

Cobalt is known to interact with several metals but has more tendency towards metals that are geochemically associated with Fe (Ti, V, Cr, Mn, Pb, etc.). However, the most relevant relationship has been noted between Co and Mn or Fe in the soil and between Co and Fe in the plant (Kabata-Pendias 2000; Talukder and Sharma 2007). Co is observed to have antagonistic relation with Cd and has been reported to limit its uptake at higher concentrations (Veltrup 1979; Kalavrouziotis et al. 2009), while some studies revealed its synergistic interactions with Ni (Anderson et al. 1973).

5 Effects of Ethylene Levels on Cobalt Function

Ethylene antagonistic relationship with Co^{2+} was observed by Lau and Yang (1976). In this study, it was revealed that Co^{2+} significantly affects ethylene-producing system in plants and acts as a potent ethylene inhibitor. Furthermore, Grover and Purves (1976) found that when ethylene was added to Petri dishes containing cucumber seedlings, the effect of Co^{2+} was considerably reduced, as it was predicted earlier that Co^{2+} promoted the hypocotyl elongation in cucumber seedlings by inhibiting the production of ethylene.

6 Cobalt Accumulation in Plants

Co level in plants is mainly dependent on its availability in soil. Plants grown on Co-rich soil like serpentine soil or soil enriched with Co ore contain higher levels of this metal (Anderson et al. 1973; Johnston and Proctor 1977). The factor which further contributes to its accumulation is the ability of a plant to uptake Co. Leguminous plants tend to accumulate more Co in comparison to other crop plants. The Co concentration in plants varies from around 8 to 100 ppm on dry weight basis. Among different plants observed, cabbage was found to have higher levels of Co, while corn and apple showed lesser level (Sillanpää and Jansson 1992). Cobalt level is also found to be organ dependent. Leaves are known to have higher deposits of Co in a specific manner. It has been previously reported by Kloke (1980) that leafy vegetables like cabbage and brassica tend to accumulate a significant amount of Co.

7 Beneficial Effects of Co on Plants

7.1 Stress Alleviation

7.1.1 Curbing of Salinity Stress

Salinity of the soil is a significant issue that adversely affects physiological and metabolic processes and ultimately reduces plant growth and yield (Ashraf and Harris 2004). Efficacy of Co in curbing salinity stress was tested by Gad and Kandil (2011) on wheat plants. The outcome of the study revealed that Co enhanced the tolerance of wheat plants to salinized soil. Also, Co concentration of 15 ppm significantly improved growth, yield and quality of treated plants over the control. In another study conducted by Gad and El-Metwally (2015), the efficiency of different cobalt concentrations (0, 15, 17.5, 20, 22.5 and 25 ppm) on two maize cultivars (Giza 310 and Hi tick 2030) was studied. They observed that increasing cobalt levels from 15 to 17.5 and 20 ppm significantly enhanced growth attributes of maize (plant height, leaves area, root length, shoot length and root fresh and dry weights) and shoots endogenous hormones such as auxins, gibberellins and cytokinins. Results also indicated that, under saline condition, all used levels of cobalt significantly enhanced the macronutrient as well as micronutrient levels of maize grains compared with untreated plants, although increasing cobalt concentration of more than 20 ppm decreased this promotive effect. Various other studies also confirmed the role of Co in alleviation of salt stress. Results of the experiment conducted by Gad (2005) on tomato plant and on cucumber plant (Gad et al. 2018) showed potentials of Co in increasing salt tolerance of treated plants. These results lead to the suggestion that Co can be used to overcome salinity stress in areas where plants are irrigated by saline water.

7.1.2 Alleviation of Cadmium Stress

Plant exposure to Cd leads to oxidative stress, suppression of respiration and photosynthetic processes and enhanced mutation rate, resulting in stunted development and reduced yield of plants. Increased production of ethylene is one of the usual responses of crops to cadmium stress, but the precise function of this hormone in plant reaction to Cd is still unknown. Chmielowska-Bąk et al. (2014) experimented to evaluate the role of cobalt (ethylene synthesis inhibitor) in curbing cadmium stress of soybean seedlings. Seedlings were subjected to various concentrations of cadmium salt with and without $CoCl_2$. The results obtained in this study showed that Co modulates the expression of a key enzyme responsible for ethylene synthesis (1-aminocyclopropane-1-carboxylic acid synthase). The expression genes were found to be slightly repressed after a few hours of treatment. So the conclusion was derived from their observation that $CoCl_2$ could be used as an ethylene inhibitor in Cd stressed conditions.

7.1.3 Alleviation of Osmotic Stress

The role of cobalt to overcome osmotic stress in potato seedlings was studied by Li et al. (2004). The findings of this experiment revealed that membrane damage caused by applying 24 hours of osmotic pressure to seedlings was considerably reduced when suitable cobalt levels were added to the growth medium. Cobalt concentration of 25 μmol L^{-1} was found to inhibit the ethylene production rate significantly, although Co 25 μmol L^{-1} had no significant effects on chlorophyll and thiobarbituric acid content in potato leaves. It was observed that cobalt helps in alleviation of declining polyamine content and the activities of anti-oxidative enzymes, and this reduces the injury to the membrane and has protective impacts on the leaves.

8 Effect of Cobalt Treatment on Leguminous Plants

Cobalt is not considered as an essential element for plant growth and metabolism, although it is categorized as a beneficial element because of its role in symbiotic N_2 fixation in legumes. It is absorbed by plants as Co^{2+} (Reisenauer 1960). Other than nitrogen fixation, Co is also known to modulate various growth and metabolic activities in legumes. A brief review of which is given below.

8.1 Effects on Growth and Yield

To study Co effects on growth attributes of leguminous plants, Kandil (2007) performed a field study on faba beans. Co was applied as various concentrations of cobalt sulphate (viz. 0, 5, 10, 15 and 20 ppm). Growth and yield attributes of treated faba beans were analysed. The results obtained in this study revealed that out of various concentrations applied, 20 ppm cobalt gave maximum value parameters studied. In a pot experiment conducted by Jaleel et al. (2008), five concentrations of Co (0, 50, 100, 150, 200 and 250 mg kg^{-1} soil) were used to assess the stimulatory role of Co on groundnut. A significant improvement was obtained by the application of 50 mg kg^{-1} Co for the growth parameters studied (plant height, number of nodules, leaf area, and shoot and root mass). Co 50 mg kg^{-1} enhances the yield characteristics. However, all yield attributes were found to decrease with an increase in the level of cobalt.

Application of Co has also been reported to enhance the growth and yield of lentil positively. In an experiment performed by Sahay and Singh (2012), cobalt application up to 6 kg ha^{-1} gave significantly maximum value for plant height, number of branches, dry biomass, pods and number of seed $plant^{-1}$ over control. The results also revealed an increase in grain and straw yield with 6 kg Co ha^{-1}. In another research on the role of cobalt in soybean crops, positive effects have been noted (Kandil et al. 2013). It has been revealed that treatment of cobalt (12 mg kg^{-1}) has a positive influence on various attributes studied. Co at 12 mg kg^{-1} produced maximum values for growth parameters, namely shoot and root length, number of nodules, total number of leaves $plant^{-1}$, number of branches per plant and fresh and dry weight of shoots and roots although a negative impact was observed with further increase of Co concentration in the soil.

Cobalt has been reported to modulate the growth and yield attributes of soybean. The results of the experiment performed by Gad et al. (2013) showed that application of Co (12 mg kg^{-1}) significantly increased the nodulation, growth and yield parameters. However, all growth yield parameters were significantly found to reduce by increasing cobalt concentrations more than 12 mg kg^{-1}.

8.2 Effects on Nodulation and Nitrogen Fixation

Nodulation in legumes is responsible for nitrogen fixation in plants, so an increase in the nodulation leads to an increment in nitrogen fixation too. It has been revealed from various studies that Co can enhance the nitrogen fixation rate among leguminous plants (Yadav and Khanna 1983; Yadav et al. 1984; Kandil 2007; Younis 2011; Gad 2012; Vijayarengan 2012). Co application of 1.5 ppm was found to fix 3.5% and 7.0% more nitrogen as compared to control, in two different varieties of cowpea. In the same study, 2.5 and 3.5 ppm of Co on soybean enhanced nitrogen fixation by 3.3% and 13.4%, respectively (Yadav and Khanna 1983). Yadav et al. (1984) further reported that cobaltous nitrate solution in combination with phosphorous significantly increased the numbers of root nodules in berseem. Cobalt is reported to increase the number of nodules in faba beans (Kandil 2007), *Lablab purpureus* (Younis 2011) and groundnut (Gad 2012). Vijayarengan (Vijayarengan 2012) conducted a pot experiment to study the effects of Co on cowpea. Plants were raised in soil applied with different concentrations of cobalt chloride (0, 50, 100, 150, 200 and 250 mg kg^{-1} soil), and the results obtained showed that the application of 50 mg kg^{-1} cobalt chloride gave maximum value for numbers of nodules with an increment of 63.25%. A number of nodules in pea plant were found to increase with treatment of Co in combination with N (Co 20 kg ha^{-1} + N 60 kg ha^{-1}). Plants treated with Co_{20} + N_{60} increased the nodules from 9 per plant in control to 16 per plant (Akbar et al. 2013).

8.3 *Effects on Nutrient Uptake*

Exogenously applied Co has been reported to increase nutrient uptake in legumes (Basu et al. 2006; Basu and Bhadoria 2008; Kandil et al. 2013; Manal et al. 2016). In a study conducted by Yadav and Khanna (1988) on groundnut, it was observed that Co, in combination with *Rhizobium*, enhanced the N, P and K uptake in plants. *Rhizobium* + phosphor bacterium inoculation + Co at 0.21 kg ha^{-1} significantly enhance the N, P and K uptake by 27.4%. Co application of 12 mg kg^{-1} was found to increase substantially the N, P and K contents of soybean plants in comparison to control (Kandil et al. 2013). Manal et al. (2016) also reported that foliar application of Co (0.24 and 0.48 gL^{-1}) showed a positive response on nutrient uptake in broad bean.

9 Effect of Cobalt Treatment on Nonleguminous Plants

The stimulatory roles of Co include increase in drought resistance in seeds, retardation of leaf senescence, inhibition of ethylene biosynthesis and governing alkaloid accumulation in medicinal plants (Palit et al. 1994). Studies revealed that lower concentrations of Co improve grain yield and quality in wheat plants (Aerya and Jagetiyaa 2000; Wen-hua et al. 2004). Jaleel et al. (2009) also studied the effects of Co on maize plants. The outcome of their study showed improvement in growth and biochemical attributes of Co-treated plants. The Co at 50 mg per kg of soil was found best, and a significant enhancement was observed in pigment content, plant height and mineral content by this treatment. Gad and El-Metwally (2015) reported a positive influence of Co treatment on growth and yield attributes of maize plants. The results of their study revealed that on increasing Co concentration from 15 to 20 ppm, a significant increase in plant height, leaf area and fresh and dry biomass was achieved. Endogenous plant hormones were also found to modulate positively by Co application.

10 Toxic Effects of Cobalt on Plants

Cobalt is reported to show cytotoxic and phytotoxic activities at higher concentrations (Palit et al. 1994). These toxic effects of Co and its compounds mainly depend on the physicochemical properties of these complexes. There is very little data on the phytotoxic impacts of Co hyper-accumulation. The study on the phytotoxicity of Co in plants like barley, oilseed rape and tomato has revealed its negative impact on shoot growth and biomass (Li et al. 2009). Besides the effects on biomass, higher levels of cobalt also reported to limit the Fe concentration, protein and chlorophyll content and antioxidant activity in leaves of cauliflower. However, diffusive capacity and relative water content in leaves were found to improve with excess Co concentration. Besides, elevated Co levels also influenced the translocation of P, S, Mn, Zn and Cu from roots to top (Chatterjee and Chatterjee 2000). Nagajyoti et al. (2010) reported that unlike excess Cu or Cr, Co considerably reduced the water potential and the rate of transpiration in plants.

Cytotoxicity caused in plants due to elevated levels of Co, include mitosis inhibition and chromosome damage, disruption of endoplasmic reticulum in root tips and disorganized phloem (Rauser 1981; Smith and Carson 1981). Although the range of Co content in plant varies, still symptoms of poisoning are not often seen. When an elevated level of Co is easily accessible, particularly in contaminated soil, it can have a severe impact on plant growth and metabolic functions. Kitagishi and Yamane (1981) reported that Co concentration of 25 and 50 ppm in soil was toxic to the rice plants. Anderson et al. (1973) found that oat crops were adversely affected when grown in a soil solution, with Co concentration of 140 μg L^{-1}. The toxicity symptoms of different levels of Co in plant tissues were recorded as follows: 43–142 ppm in bush beans (Wallace et al. 1977), 19–32 ppm in Sudan grass (Gough et al. 1979) and 6 ppm in barley seedlings (Davis et al. 1978). Findings of Kloke et al. (1984) also suggested a similar range of Co toxicity levels in plant species. However, critical Co content in crops is frequently recorded between 30 and 40 ppm (Macnicol and Beckett 1985). The toxicity symptoms of Co were observed as a gradual decrease in chlorophyll and carotenoid content and an enhanced proportion of carotenoids to chlorophyll. Other symptoms included a significant enhancement in anti-oxidative enzyme activity, which leads to indicate potent induction of oxidative stress in the growth medium due to increased cobalt level (Tewari et al. 2002).

Cereals are considered to be the most sensitive crops to Co surplus, but Anderson et al. (1973) stated that toxic impacts of Co levels at 10–20 ppm were most likely associated with excess of Ni in soil. Case et al. (1972) suggested that the level of Co in herbage should not extend beyond 60 ppm due to risk for animal health. Plants are known to accumulate bulk quantities of Co and create a Co-tolerance mechanism that is substantially comparable to that found in any metalliferous plant species. Several species of plants, mostly from the Boraginaceae, Caryophyllaceae, Cruciferae, Leguminosae, Myrtaceae, Nyssaceae and Violaceae families, are renowned for their elevated accumulation of Co and are also suggested as biogeochemical markers.

11 Cobalt Toxicity Symptoms

As Co concentration is raised in growth medium, the toxicity symptom becomes clearly visible. These symptoms tend to get worse with increase in concentration and duration of Co application in medium. The concentration of Co in plant tissues tends to increase with an increase in supply of this metal. White, dead edges and leaf tips are widespread symptoms of Co toxicity (Table 1). However, interveinal

Table 1 Cobalt toxicity symptoms in crops

Crops	Visible symptoms	References
French bean	Chlorosis of young leaves from the apex towards the base which changed to necrosis, dried and withered	Chatterjee et al. (2006)
Groundnut	Chlorosis with pale white colour and necrosis of young leaves	Singh et al. (2004)
Mung bean	Chlorosis of thc younger leaves	Liu et al. (2000)
Tomato	Diffused chlorosis of young leaves from base, necrotic spots and marginal scorching, distorted leaves that appeared hook-like with rudimentary leaflets at the top	Gopal et al. Gopal et al. (2003)
Tomato	Interveinal chlorosis of young leaves, fruits developed black patches	Chatterjee and Chatterjee (2005)

chlorosis of fresh leaves is the primary response of crops to excess of Co, which is strongly connected to Fe chlorosis (Kabata-Pendias 2000).

12 Cobalt Stress in Plants and Its Alleviation

The effect of some heavy metal stress was studied on cauliflower (*Brassica oleracea* L.). Plants were grown in medium containing 0.5 mM each of Co, Cr and Cu. The observed effects showed that symptoms of Co toxicity appeared first and were most pronounced. Excess of each heavy metal retards the biomass of cauliflower, chlorophylls *a* and *b*, concentrations of Fe, protein and antioxidant enzyme activity in leaves in the order Co > Cu > Cr. It was further noted that there was minimal translocation of Cr from roots to tops and maximum translocation of Co when cauliflower was supplied individually with excess Co, Cu or Cr. The translocation of elements like P, S, Mn, Zn and Cu from roots to tops of cauliflower plants was affected most significantly by Co and least by Cr. Unlike hyper-accumulation of Cu or Cr, Co reduced water potential and transpiration rates considerably and enhanced relative water content in leaves (Chatterjee and Chatterjee 2000). Furthermore, Chatterjee and Chatterjee (2003) ameliorated Co stress by exogenous application of phosphorus. The visible symptoms of excess cobalt were less pronounced in tomato but early to appear at high (3 mM) than sufficient (1.5 mM) phosphorus when cultivated in refined sand. They studied various parameters like fresh and dry weight, chlorophyll content, fruit volume and size, ascorbic acid content, lycopene and reducing sugars, that were found to increase by application of phosphorus whereas, a decrease in phenol and starch content was recorded by the same.

Cobalt at higher levels is reported to inhibit chlorophyll pigment synthesis by blocking the pathway of its biosynthesis (Mysliva-Kurdziel et al. 2004). Zengin (2006) investigated the role of heavy metal ions (Co^{2+} and Zn^{2+}) on bean plants (*Phaseolus vulgaris*). Effects were analysed in terms of proline, total protein content, chlorophyll and abscisic acid (ABA) levels in leaves. The results revealed a significant increase of ABA contents in roots and leaves of treated plants. The proline content in leaves of Co^{2+} and Zn^{2+} applied plants was also found to enhance as compared to control, although the contents of chlorophyll (a + b) and total protein were found inhibited with the concentration of both metals. The results of this study showed that Co stress was more pronounced than Zn. The response of exogenous supply of Co was studied on two annual high-biomass yielding crops, *Sorghum bicolor* and *Helianthus annuus*. Plants were grown in a polymetallic soil with Co 51 ppm in soil, in combination with mineral fertilization and organic amendment. The results showed an increased biomass yield but did not heighten the concentration of metals in the harvestable tissue of the plants during the crop cycle. However, these crops showed high removal of metal making them good potential for phytoremediation (Marchiol et al. 2007).

It has also shown inhibition of root development by retarding the division of cells and preventing nutrient and water uptake and translocation (Jayakumar et al. 2008). Li et al. (2009) studied phytotoxicity and bioavailability of Co in three plants, namely, barley (*Hordeum vulgare* L.), oilseed rape (*Brassica napus* L.) and tomato (*Lycopersicon esculentum* L.). The results revealed that growth retardation and inhibition of shoot biomass in all plants with Co added in soil. The Co toxicity due to higher concentrations of Co (100, 200, 400 ppm) in chickpea (*Cicer arietinum*) results in negative impacts on growth and biochemical attributes. The results revealed a significant decrease in seed germination percentage, growth, biomass and leaf chlorophyll content. Excess Co also caused injury to shoot and root, leaf chlorosis, suppression of root nodules and lower yield of the plant (Khan and Khan 2010).

Arora et al. (2012) reported that Co toxicity in *Brassica juncea* L. plants could be ameliorated by the application of epibrassinolide (EBL). Out of various concentrations applied, 10^{-8} M of 24-EBL was most effective in alleviating Co stress. Plants were subjected to different levels of Co toxicity (viz. 0, 5×10^{-4}, 10^{-3}, 1.5×10^{-3} and 2×10^{-3} M). It was observed that 24-EBL foliar application was able to relieve cobalt ion stress and considerably improve growth and biochemical parameters. In leaves of crops treated with 24-EBL alone, the activity of antioxidant enzymes (SOD, CAT, POD, GR, APOX) and protein content was also significantly controlled. It was also noted that the application of 24-EBL to plants under stress showed positive impacts. Plant's antioxidant defence mechanism was further enhanced and reinforced by stimulating the activity of various antioxidant enzymes, 10^{-8} M being the most effective. The toxicity of Co in *Medicago sativa* was reported to be successfully ameliorated by giving the pretreatment of 0.01 M EDTA (ethylenediaminetetraacetic acid) in combination with heavy metal solutions. This pretreatment reduced the negative impact on growth and the metabolic activities of the plant due to Co hyper-accumulation (Zeid et al. 2013).

The effect of Co (100 M) on Indian mustard (*Brassica juncea* L.) was investigated by Karuppanapandian and Kim (2013). A hydroponic experiment was conducted, and effects were assessed in terms of plant growth, biochemical parameters and oxidative stress. The responses of antioxidant enzymes were evaluated. Co-toxicity was correlated with a rise in palisade and spongy cell quantity and a decrease in chloroplasts per cell. It was noted that DNA fragmentation and a 36 kDa DNase activity was the primary reason for Co-induced cell death. The current findings indicate that excess Co decreases seedling development by causing oxidative stress related to lipid peroxidation and overproduction of O^{2-} and H_2O_2.

Sharma et al. (2014) conducted an experiment to observe Co toxicity in wheat plants. Various levels of Co (100–500 ppm) were applied to wheat (*Triticum aestivum* L.) plants in a sand culture medium using Hoagland solution. Seed germination percentage was studied in Petri dishes using the same treatments. The results showed that treatment with Co-200 reduced germination rate; however, seedling vigour index decreased with increasing Co concentration. There were no adverse effects of Co on the germination index, whereas more inhibition of wheat germination has been observed beyond Co-300 treatment. A positive impact on plant height, leaf number and area and dry matter accumulation was recorded up to 200 ppm of Co, but with higher levels, detrimental effects were observed for the same attributes. They summarized that Co has a positive impact on wheat plant development at reduced levels (up to 200 ppm).

Response of *Anisopappus chinensis* to metalliferous ions was studied by Lange et al. (2016). Two concentrations of Co in the form of Co (II) sulphate (20 and 100 mg kg^{-1}) were supplied in soil along with foliar application of 0.028 M concentration. The results revealed an increased accumulation of Co in treated plants. Toxicity symptoms observed as leaf chlorosis and plants also exhibited inhibition of growth.

According to Lwalaba et al.'s (2017) study, Co toxicity could be ameliorated by the addition of calcium (Ca) in the growth medium. Ca is reported to decrease the Co toxicity by inhibiting Co uptake and increasing the antioxidant enzyme activity in barley plants. The Ca ion plays a key role in improving abiotic stress tolerance of plants (Liang et al. 2005); it is because of its nature to react with calmodulin proteins, which further regulate metal ion transport and gene expression (Yang and Poovaiah 2003). Various other reports also confirmed the toxicity effects of Co. It was observed that inhibition of seedling growth in mung beans occurred at 5 μM Co (Liu et al. 2000). In tomato, excess Co (0.5 mM Co) was reported to reduce the biomass (Gopal et al. 2003), weight, volume and size and quality of fruits (Chatterjee and Chatterjee 2005). The Co (II)-Glychelate was more toxic than the Co(II)-EDTA chelate and had adverse effects on the lucerne crop dry matter, nitrogen, protein and essential amino acid content and also decreased the physiological efficiency of the symbiotic system of lucerne and caused changes in its structural organization (Molas 2008). Under excess cobalt concentrations (Co >0.0001 mM cobalt sulphate) in French bean (cv. 'Anupama'), the biomass was decreased, the flowers produced were fewer and smaller and many failed to mature, leading to lower seed yields (Chatterjee et al. 2006). In a study conducted by Liu (1998), Co toxicity was alleviated by 1.25 mM of Ca in mung bean plants. Their study showed a positive effect on the growth and photosynthetic activity of Co-treated plants.

Prajapati et al. (2012) used *Pistia stratiotes*, for phytoremediation of water bodies polluted with heavy metals like Cr and Co. Their study provides a sustainable way to overcome heavy metal stress. The effects of various concentrations of four heavy metals including Cd, Cr, Co and Ni were studied in two cultivars of wheat, bean and lucerne. The outcome of this study was assessed on the basis of plant biomass, metal content and percent recovery values. Their study showed that the bean plant was the most effective crop in removing heavy metals from the medium. The lucerne, though, had higher tissue concentration, because of a low biomass, and wheat was not an active species (Hajiboland 2005).

The role of salicylic acid (SA) in response to various stresses by modulating biochemical processes in plants and decreasing the adverse effects of stress is well established. A study was conducted by Sinha et al. (2015) to investigate the impacts of SA in curbing heavy metal stress in cauliflower cv. Different concentrations of SA were applied to plants. Heavy metal toxicity produced adverse impact on dry weight and specific activity of catalase and increased the concentration of lipid peroxidation, proline, non-protein thiol, electrolyte leakage percentage and specific activity of peroxidase and superoxide dismutase. Application of salicylic acid reverted all negative effects of heavy metal stress that were found on all the parameters. Application of SA (100 mM) helps in reducing adverse effects of HMs to some extent showing antagonism in cauliflower. Further, they suggested that other methods may be employed for alleviation of Co stress, such as the use of tolerant crop genotype, phytoremediation by tolerant crops and inoculation by beneficial microorganisms.

13 Conclusions

It has been concluded that Co is beneficial element for higher plants, but it plays a vital role in leguminous plants as it is required for nitrogen fixation and nodule formation. Cobalt is a transition element and is an essential component of several enzymes and coenzymes. It has also been shown to affect growth and metabolism of nonleguminous plants, in different degrees, when applied exogenously. So, it was concluded that Co in lower concentrations can be used to increase productivity of plants and also in alleviation of various stress conditions. The toxicity of cobalt seems to be more common than that of cobalt deficiency although the prevalence depends on the area. However, Co toxicity can be alleviated by the application of mineral nutrients and PGRs.

References

Aerya NC, Jagetiyaa BL (2000) Effect of cobalt treatments on dry matter production of wheat and DTPA extractable cobalt content in soils. Commun Soil Sci Plant Anal 31:1275–1286

Akbar FM, Zafar M, Hamid A, Ahmed M, Khaliq A, Khan MR, Rehman ZU (2013) Interactive effect of cobalt and nitrogen on growth, nodulation, yield and protein content of field grown pea. Hort Environ Biotechnol 54:465–474

Anderson AJ, Meyer DR, Mayer FK (1973) Heavy metal toxicities: levels of nickel, cobalt and chromium in the soil and plants associated with visual symptoms and variation in growth of an oat crop. Aust J Agric Res 24:557–571

Arora P, Bhardwaj R, Kanwar MK (2012) Effect of 24-epibrassinolide on growth, protein content and antioxidative defense system of *Brassica juncea* L. subjected to cobalt ion toxicity. Acta Physiol Plant 34:2007–2017

Ashraf M, Harris PJC (2004) Potential biochemical indicators of salinity tolerance in plants. Plant Sci 166:3–6

Bakkaus E, Gouget B, Gallien JP, Khodja H, Carrot H, Morel JL, Collins R (2005) Concentration and distribution of cobalt in higher plants: the use of micro-PIXE spectroscopy. Nucl Instrum Methods Phys Res B 23:350–356

Barker HA (1961) Structure and function of cobamide coenzymes. Fed Proc 20:956

Basu M, Bhadoria PBS (2008) Performance of groundnut under nitrogen fixing and phosphorous microbial inoculates with different levels of cobalt in alluvial soils of eastern India. Agron Res 6:15–25

Basu M, Bhadoria PBS, Mahapatra SC (2006) Influence of microbial culture in combination with micronutrient in improving the groundnut productivity under alluvial soil of India. Acta Agric Slov 87:435–444

Case AA, Selby LA, Hutcheson DP, Ebens RJ, Erdman JA, Feder GL (1972) Infertility and growth suppression in beef cattle associated with abnormalities in their geochemical environment. Trace Subst Environ Health 6:15

Caselles JM, Pérez-Espinosa A, Pérez-Murcia MD, Moral R, Gómez I (1997) Effect of increased cobalt treatments on cobalt concentration and growth of tomato plants. J Plant Nutr 20:805–811

Chatterjee J, Chatterjee C (2000) Phytotoxicity of cobalt, chromium and copper in cauliflower. Environ Pollut 109:69–74

Chatterjee J, Chatterjee C (2003) Management of phytotoxicity of cobalt in tomato by chemical measures. Plant Sci 164:793–801

Chatterjee J, Chatterjee C (2005) Deterioration of fruit quality of tomato by excess cobalt and its amelioration. Commun Soil Sci Plant Anal 36:1931–1945

Chatterjee C, Gopal R, Dube BK (2006) Physiological and biochemical responses of French bean to excess cobalt. J Plant Nutr 29:127–136

Chmielowska-Bąk J, Lefèvre I, Lutts S, Kulik A, Deckert J (2014) Effect of cobalt chloride on soybean seedlings subjected to cadmium stress. Acta Soc Bot Pol 83:201–207

Costa G, Morel JL (1994) Water relations, gas exchange and amino acid content in Cd-treated lettuce. Plant Physiol and Biochem 32:561–570

DalCorso G, Manara A, Piasentin S, Furini A (2014) Nutrient metal elements in plants. Metallomics 6:1770–1788

Davis RD, Beckett PHT, Wollan E (1978) Critical levels of twenty potentially toxic elements in young spring barley. Plant Soil 49:395–408

Dilworth MJ, Robson AD, Chatel DL (1979) Cobalt and nitrogen fixation in *Lupinus angustifolius* L. II. Nodule formation and function. New Phytol 83:63–79

Gad N (2005) Interactive effect of cobalt and salinity on tomato plants II- some physiological parameters as affected by cobalt and salinity. Res J Agri Bio Sci Pakistan 1:270–276

Gad N (2012) Physiological and chemical response of groundnut (*Arachis hypogaea*) to cobalt nutrition. World Appl Sci J 20:327–335

Gad N, El-Metwally IM (2015) Chemical and physiological response of maize to salinity using cobalt supplement. Int J ChemTech Res 8:45–52

Gad N, Kandil H (2011) Maximizing the tolerance of wheat plants to soil salinity using cobalt. 1- Growth and mineral composition. J Appl Sci Res 7:1569–1574

Gad N, El-Moez A, Bekbayeva LK, Karabayeva AA, Surif M (2013) Effect of cobalt supplement on the growth and productivity of soybean (*Glycine max* L. Merrill). World Appl Sci J 26:926–933

Gad N, Abdel-Moez MR, Fekry Ali ME, Abou-Hussein SD (2018) Increasing salt tolerance in cucumber by using cobalt. Middle East J Appl Sci 08:345–354
Gopal R, Dube BK, Sinha P, Chatterjee C (2003) Cobalt toxicity effects on growth and metabolism of tomato. Commun Soil Sci Plant Anal 34:619–628
Gough LP, Shacklette HT, Case AA (1979) Element concentrations toxic to plants, animals and man. LiSGS Bull 1466:80–84
Grover S, Purves WK (1976) Cobalt and plant development: interactions with ethylene in hypocotyl growth. Plant Physiol 57:886–889
Hajiboland R (2005) An evaluation of the efficiency of cultural plants to remove heavy metals from growing medium. Plant Soil Environ 51:156–164
Hasan SA, Hayat S, Wani AS, Ahmad A (2011) Establishment of sensitive and resistant variety of tomato on the basis of photosynthesis and antioxidative enzymes in the presence of cobalt applied as shotgun approach. Braz J Plant Physiol 23:175–185
Howell RW, Skoog F (1955) Effect of adenine and other substances on growth of excised *Pisum* epicotyls cultured in vitro. Am J Bot 42:356–360
Jaleel CA, Jayakumar K, Chang-xing Z, Azooz MM (2008) Effect of soil applied cobalt on activities of antioxidant enzymes in *Arachis hypogaea*. Global J Mole Sci 3:42–45
Jaleel CA, Jayakumar K, Chang-xing Z, Iqbal M (2009) Low concentration of cobalt increases growth, biochemical constituents, mineral status and yield in *Zea Mays*. J Sci Res 1:128–137
Jayakumar K, Jaleel CA, Azooz MM (2008) Impact of cobalt on germination and seedling growth of *Eleusine coracana* L. and *Oryza sativa* L. under hydroponic culture. Global J Mol Sci 3:18–20
Jayakumar K, Vijayarengan P (2006) Influence of cobalt on seed germination and seedling growth of *Vigna mungo* (L.) Hepper. Plant Archives 6:681–2
Johnston WR, Proctor J (1977) Metal concentrations in plants and soils from two British serpentine sites. Plant Soil 46:275–278
Kabata-Pendias A (2000) Trace elements in soils and plants, 3rd edn. CRC Press, Boca Raton
Kalavrouziotis IK, Koukoulakis PH, Manouris G, Papadopoulos AH (2009) Interactions between cadmium, lead, cobalt, and nickel in broccoli, irrigated with treated municipal wastewater. Eur Water 25:13–23
Kandil H (2007) Effect of cobalt fertilizer on growth, yield and nutrients status of faba bean (*Vicia faba* L.) plants. J Appl Sci Res 3:867–872
Kandil H, Ihab MF, El-Maghraby A (2013) Effect of cobalt level and nitrogen source on quantity and quality of soybean plant. J Basic Appl Sci Res 3:185–192
Karuppanapandian T, Kim W (2013) Cobalt-induced oxidative stress causes growth inhibition associated with enhanced lipid peroxidation and activates antioxidant responses in Indian mustard (*Brassica juncea* L.) leaves. Acta Physiol Plant 35:2429–2443
Khan MR, Khan MM (2010) Effect of varying concentration of nickel and cobalt on the plant growth and yield of chickpea. Aust J Basic Appl Sci 4:1036–1046
Kitagishi K, Yamane I (1981) Heavy metal pollution in soils of Japan. Japan Science Society Press, Tokyo, pp 65–80
Klein WH (1959) Interaction of growth factors with photoprocess in seedling growth. In: Withrow RB (ed) Photoperiodism and related phenomena in plants and animals. American Association for the Advancement of Science, D. C, pp 207–215
Klessa DA, Dixon J, Voss RC (1989) Soil and agronomic factors influencing the cobalt content of herbage. Res Develop Agric 6:25–35
Kloke A, Sauerbeck DR, Vetter H (1984) The contamination of plants and soils with heavy metals and the transport of metals in terrestrial food chains. In: Changing metal cycles and human health. Springer, Berlin, Heidelberg, pp 113–141
Kloke A (1980) Immissionsbelastete Landwirtschaftliche Standorte. Leistungen von Landwirtschaft und Landschaft zur Wohlfart in Verdichtungsräumen. Verlag M &H Schaper, Hannover 84 pp. 24–44

Lanaras T, Moustakas M, Symeonidis L, Diamatoglou S, Karataglis S (1993) Plant metal content, growth responses and some photosynthetic measurements on field cultivated wheat growing on ore bodies enriched in cu. Physiol Plant 88:307–314

Lange B, Pourret O, Meerts P, Jitaru P, Cancès B, Grison C, Faucon MP (2016) Copper and cobalt mobility in soil and accumulation in a metallophyte as influenced by experimental manipulation of soil chemical factors. Chemosphere 146:75–84

Lau OL, Yang SF (1976) Inhibition of ethylene production by cobaltous ion. Plant Physiol 58:114–117

Li CZ, Wang D, Wang GX (2004) The protective effects of cobalt on potato seedling leaves during osmotic stress. Bot Bull Acad Sin 46:119–125

Li HF, Gray C, Mico C, Zhao FJ, McGrath SP (2009) Phytotoxicity and bioavailability of cobalt to plants in a range of soils. Chemosphere 75:979–986

Liang Y, Wong J, Wei L (2005) Silicon-mediated enhancement of cadmium tolerance in maize (*Zea mays* L.) grown in cadmium contaminated soil. Chemosphere 58:475–483

Lison D (2007) Cobalt. In: Handbook on the toxicology of metals, 3rd edn. Academic Press, Amsterdam, pp 511–528

Liu J (1998) Cobalt. In: Physiological effects and uptake mechanisms in plants. Doctoral dissertation

Liu J, Reid RJ, Smith FA (2000) The mechanism of cobalt toxicity in mung beans. Physiol Plant 110:104–110

Lwalaba JL, Zvobgo G, Fu L, Zhang X, Mwamba TM, Muhammad N, Mundende RP, Zhang G (2017) Alleviating effects of calcium on cobalt toxicity in two barley genotypes differing in cobalt tolerance. Ecotoxicol Environ Safe 139:488–495

Macnicol RD, Beckett PH (1985) Critical tissue concentrations of potentially toxic elements. Plant Soil 85:107–129

Maksymiec W (2007) Signalling responses in plants to heavy metal stress. Acta Physiol Plant 29:177–187

Manal AA, Manal A, Alla A, Allam SMM (2016) Effect of zinc and cobalt applied with different methods and rates on the yield components of *Vicia faba* L. World Wide J Multidiscip Res Develop 2:52–58

Marchiol L, Fellet G, Perosa D, Zerbi G (2007) Removal of trace metals by *Sorghum bicolor* and *Helianthus annuus* in a site polluted by industrial wastes: a field experience. Plant Physiol Biochem 45:379–387

Marschner H (1995) Mineral nutrition of higher plants, 2nd edn. Academic Press, London

Mengel K, Kirkby EA, Kosegarten H, Appel T (2001) Principles of plant nutrition, 5th edn. Springer, Kluwer Academic Publishers, Dordrecht

Miller CO (1951) Promoting effect of cobalt and nickel ions on expansion of etiolated bean leaf disks. Arch Biochem Biophys 32:216–218

Miller CO (1954) The influence of cobalt and sugars upon the elongation of etiolated pea stem segments. Plant Physiol 29:79–82

Mitchell RL (1972) Cobalt in soil and its uptake by plants. 9th SimposioInt di Agrochimica, vol 2, Punta Alta

Molas J (2008) Comparison of the toxicity of Co(II)-Gly and Co(II)-EDTA chelates to lucerne (*Medicago sativa* L.). Proc ECOpole 2:71–75

Mysliva-Kurdziel B, Prasad MNV, Strzalka K (2004) Photosynthesis in heavy metal stressed plants. In: Heavy metal stress in plants. Springer, Berlin, Heidelberg, pp 146–181

Nagajyoti PC, Lee KD, Sreekanth TVM (2010) Heavy metals, occurrence and toxicity for plants: a review. Environ Chem Lett 8:199–216

Nicholas DJD (1975) The functions of trace elements. In: Trace elements in soil-plant-animal systems. Academic Press, New York, pp 181–198

Palit S, Sharma A, Talukder G (1994) Effects of cobalt on plants. Bot Rev 60:149–181

Pandey N, Sharma CP (2002) Effect of heavy metals Co^{2+}, Ni^{2+}, and Cd^{2+} on growth and metabolism of cabbage. Plant Sci 163:753–758

Pandey N, Pathak GC, Pandey DK, Pandey R (2009) Heavy metals, Co, Ni, Cu, Zn and Cd, produced oxidative damage and evoke differential antioxidative responses in spinach. Braz J Plant Physiol 21:103–111

Prajapati SK, Meravi N, Singh S (2012) Phytoremediation of chromium and cobalt using *Pistia stratiotes*: a sustainable approach. Proc Int Acad Ecol Environ Sci 2:136

Rathsooriya GB, Nagarajah S (2003) Studies effect of salinity with cobalt on tomato plants. Plant Soil 44:158

Rauser WE (1981) Entry of sucrose into minor veins of bean seedlings exposed to phytotoxic burdens of Co, Ni or Zn. J Plant Nutr 3:319–328

Reisenauer HM (1960) Cobalt in nitrogen fixation by a legume. Nature 186:375–376

Reisenauer HM, Walsh LM, Hoeft RG (1973) Testing soil for sulphur, boron, molybdenum, and chlorine. In: Soil testing and plant analysis. Soil Science Society of America, Madison, pp 173–200

Reuter DJ (1975) The recognition and correction of trace element deficiencies. In: Trace elements in soil plant- animal systems. Academic Press, New York, pp 291–324

Sahay N, Singh SP (2012) Effect of cobalt application on growth, yield attributes and yield of lentil cultivars. Ann Plant Soil Res 14:39–41

Salisbury FB (1959) Growth regulators and flowering. II. The cobaltous ion. Plant Physiol 34:598–604

Sharma B, Devi P, Gogoi N, Devi YM (2014) Effects of cobalt induced stress on *Triticum aestivum* L. crop. Asian J Agri Biol 2:137–147

Sillanpää M, Jansson H (1992) Status of cadmium, lead, cobalt and selenium in soils and plants of thirty countries, vol 65. Food & Agriculture Organization of the United Nations, Rome

Singh AL, Basu MS, Singh NB (2004) Mineral disorders of groundnut. National Research Center for groundnut (ICAR), Junagadh, p 85

Sinha P, Shukla AK, Sharma YK (2015) Amelioration of heavy-metal toxicity in cauliflower by application of salicylic acid. Commun Soil Sci Plant Anal 46:1309–1319

Smith HI, Carson BL (1981) Cobalt. In: An appraisal of environmental exposure, Trace metals in the environment, vol 6. Ann Arbor Science, Ann Arbor

Talukder G, Sharma A (2007) Cobalt. In: Handbook of plant nutrition. CRC Taylor and Francis, Boca Raton, pp 449–514

Tewari RK, Kumar P, Sharma PN, Bisht SS (2002) Modulation of oxidative stress responsive enzymes by excess cobalt. Plant Sci 162:381–388

Thimann KV (1956) Studies on the growth and inhibition of isolated plant parts V. The effect of cobalt and other metals. Am J Botany 43:241–250

Veltrup W (1979) The effect of Ni^{2+}, Cd^{2+}, and Co^{2+} on the uptake of copper by intact barley roots. Z Pflanzenphysiol 93:1–9

Vijayarengan P (2012) Changes in growth, biochemical constitution and antioxidant potential in cowpea (*Vigna unguiculata* L.) under cobalt stress. Int J Res Environ Sci Tech 2:74–82

Wallace A, Alexander GV, Chaudhry FM (1977) Phytotoxicity of cobalt, vanadium, titanium, silver and chromium. Commun Soil Sci Plant Anal 8:751–756

Weinberg ED (1977) Microorganisms and minerals. Marcel Dekker, New York, p 492

Wen-hua F, Xiao-guang X, Li-fang Y (2004) Effect of cobalt on yield and grain quality of winter wheat. J Plant Nutr Fert 10:429–432

Wiersma D, Van Goor BJ (1979) Chemical forms of nickel and cobalt in phloem of *Ricinuscommunis*. Physiol Plant 45:440–442

Wilson DO, Reisenauer HM (1970) Effects of some heavy metals on the cobalt nutrition of *Rhizobium meliloti*. Plant Soil 32:81–89

Yadav DV, Khanna SS (1983) Effect of cobalt and phosphorous on their availability and nitrogen fixation by legumes. Haryana Agric Univ J Res 13:396–401

Yadav RP, Yadav DV, Khanna SS (1984) Improved nodulation and nitrogen fixation in berseem with the application of cobalt and phosphorous. Indian J Agric Sci 54:506–508

Yang T, Poovaiah BW (2003) Calcium/calmodulin-mediated signal network in plants. Trends Plant Sci 8:505–512

Younis M (2011) Soil applied cobalt and copper alter the *Rhizobium* symbiosis and growth status of *Lablab purpureus*. J Plant Nutri 34:1736–1745

Yadav DV and Khanna SS (1988) Role of cobalt in nitrogen fixation a review. Review 9: 180–182

Zeid IM, Ghazi SM, Nabawy DM (2013) Alleviation of Co and Cr toxic effects on alfalfa. Int J Agron Plant Prod 4:984–993

Zengin FK (2006) The effects of Co^{2+} and Zn^{2+}on the contents of protein, abscisic acid, proline and chlorophyll in bean (*Phaseolus vulgaris* cv. Strike) seedlings. J Environ Biol 27:441–448

Part IV
Approaches and Challenges for Crop Protection and Production Under Contaminated Soils

Bioremediation of Toxic Pollutants: Features, Strategies, and Applications

Arti Yadav, Deepika Goyal, Mrinalini Prasad, Teg Bahadur Singh, Preksha Shrivastav, Akbar Ali, and Prem Kumar Dantu

Contents

1 Introduction

Any substance which has no further economic use for living beings and exists in the environment (open fields, water or air) is supposed to be a pollutant (Megharaj et al. 2011). Due to fast population growth, urbanization, and industrialization, the amount of hazardous waste is increasing annually. The dumping of hazardous waste, namely, rubber, plastics, pesticides, heavy metals, and industrial waste, into the environment is unsafe. The harmful effects of these substances on natural or man-made resources are due to their physicochemical and biological properties which led to the pollution, and as a result the resources become unfit for use and are of concern to the environmentalists (Fulekar 2010). The foremost sources of such hazardous substances are chemical industries. Approximately 6×10^6 chemical compounds have been synthesized, with 1000 new chemicals being synthesized yearly. Almost 60,000 to 95,000 chemicals are in commercial use. According to Third World Network Reports, more than one billion pounds (450 million kilograms) of toxins are released worldwide in air, water, and land (Shukla et al. 2010).

A. Yadav · D. Goyal · M. Prasad · T. B. Singh · P. Shrivastav · A. Ali · P. K. Dantu (✉)
Department of Botany, Dayalbagh Educational Institute (Deemed University), Agra, Uttar Pradesh, India

M. Naeem et al. (eds.), *Contaminants in Agriculture*,
https://doi.org/10.1007/978-3-030-41552-5_18

The chemical compounds causing ecological problems leading to environmental imbalance are of global concern now Garima and Singh (2014). The hazardous wastes from chemical industries and household sewage comprise noxious organic and inorganic chemicals containing heavy metals, high pH solvents, and salts. The oil spills and long-term use of fertilizers led to the accumulation of heavy metals in the soil and water affecting health of human beings and other organisms. After the green revolution, soil fertility as well as microbial flora and fauna was devastated due to accelerated use of chemical fertilizers and pesticides in order to increase agricultural productivity (Henis 1997).

Agricultural pollutants are classified in four broad groups, namely, fertilizers, pesticides, animal waste, and fossil fuels and its combustion products. Among them, fertilizers and animal wastes are easily recycled by nature itself; however, recycling of fuel and pesticides needs more time and energy due to their xenobiosis, recalcitrance, and potential toxicity hence are long-term persistent pollutants (Henis 1997).

There are various physical, chemical, and thermal methods adopted for the treatment of persistent toxic contaminants; however, they are not found effective to remove pollutants from all sites at satisfactory level of the Environmental Protection Acts. Further, cost of their renewal run into billions of US dollars (Roseberg 1993; Glass 2000; McIntyre 2003; Kuiper et al. 2004). Therefore, the biological approach popularly known as bioremediation received a special interest due to its low cost, high public acceptance, and eco-friendly nature. "Bioremediation" as the word itself explains is the process of treatment of environmental contaminants using living organisms. Bioremediation uses bacteria, fungi, algae, higher plants, or their enzymes to recover the environmental disturbances to its original state altered by hazardous wastes or pollutants (Glazer and Nikaido 1995).

Bioremediation is not a new process on Earth and perhaps has been there since the beginning of life (Okpokwasili 2007). As the waste is dumped into soil or water, the microbes get adopted for that changed environment, start degradation of that waste material, and use it for their own carbon and energy source. However, due to fast industrialization, the amount of hazardous waste increased beyond permissible limit; therefore, the same process has become challenging to microorganisms. Bioremediation is the most effective and promising method to resolve this environmental problem. Bioremediation can occur at the site of contamination or other places after excavation of contaminated soil or water (Sharma 2012). In bioremediation process, several technologies have been used to reduce contamination from the environment, such as bioventing, biopiling, bioaugmentation, biosparging, compositing, land farming, biopiles, bioleaching, bioreactor, composting, etc. Microbes have potential metabolic activities to remediate toxic wastes; however, the process is very slow, and satisfactory level of remediation does not take place. Hence, recent research is being focused toward molecular approaches in the development of transgenic microbes or consortia, engineered protein, metabolic engineering, whole-transcriptome profiling, and proteomics for remediation of environmental contaminants (Wood 2008). Cell surface expression of specific proteins allows the engineered microorganisms to transport, bioaccumulate, and/or detoxify heavy metals and to degrade xenobiotics (Arshad et al. 2007).

2 Salient Features of Bioremediation

- Bioremediation is a living organism (bacteria, fungi, yeasts, algae, and green plants) based natural environment cleaning technique that requires less time, cost, and efforts.
- When the pollutants are present in huge amount, microbial population increases using polluting substances as nutrient or energy source, and once the pollutants are degraded or imbibed, their population declines. The effectiveness of bioremediation depends on metabolic potential of microorganisms (Antizar-Ladislao 2010) and the environmental conditions which allows the microbial growth and activity.
- Bioremediation can occur at the site of contamination avoiding the human health risk which may be possible due to transportation of hazardous waste. It is generally carried out under aerobic condition though it is also possible under anaerobic conditions.
- Bioremediation is based on the principle of elimination, mineralization, attenuation, or transformation of toxic substances by the use of biological processes (Shannon and Unterman 1993).
- Bioremediation is the most promising method of pollution treatment than other traditional ones as it has the ability to completely abolish or render the pollutants from site of contamination.
- Conventional method may not completely destruct pollutants; rather, they only convert them into new waste such as incineration. On the other hand, bioremediation along with degradation transforms the toxic compounds to harmless products (water and harmless gases) eliminating the jeopardy of future liability related to the treatment and disposal of hazardous waste.
- Bioremediation is an eco-friendly and inexpensive technique. It does not use toxic chemicals for treatment of contaminants and hence has high environmental acceptance.
- Bioremediation process is affected by several factors such as nature of pollutants, pH and moisture content of soil, nutritional (contaminants) concentration and bioavailability, microbial density and diversity, oxygen content, temperature, and redox potential.
- Bioremediation activity can be enhanced by the addition of nutrients such as nitrogen and phosphorus.
- The microbial populations exposed to a specific pollutant (e.g., hydrocarbons) become adapted to that and develop genetic changes. When adapted microbial populations are used to treat hydrocarbon-contaminated sites, they respond very quickly (within hours) resulting in higher biodegradation rates than non-adapted ones (Leahy and Colwell 1990; Atlas and Bartha 1998).

3 Natural and Living Agents of Bioremediation

Environmental toxicants imposed on plants are a serious concern in most countries. These heavy metals imbalance the ecological harmony and cause disturbance to animals as well as to plants (Table 1). Removal of heavy metals from a contaminated site is known as remediation. Before the knowledge of microbes for remediation, there were some conventional methods which were being widely used for contaminant removal. These conventional methods include dredging (physical removal of the contaminated sediment layers), capping (covering the contaminated sediment surface with clean material, thus isolating the sediments), and incineration (waste treatment technology which involves the combustion of organic substances contained in waste materials). Nowadays, there are many strategies adopted by researchers to remediate heavy metals. There are certain biotechnological approaches that require the use of living organisms (Table 2) and cell manipulation to develop alternative and innovative methods to maintain natural environment. Living organisms that are used for remediation of contaminants from soil and water could be algae, bacteria, fungi, or plants. On the basis of types of biological organisms used for remediation processes, they are categorized as phycoremediation, bioremediation, mycoremediation, and phytoremediation (Fig. 1).

3.1 Bioremediation

Bioremediation is made up of two words: "bios" that means life and refers to living organisms and "to remediate" that means to solve a problem. Bioremediation is a biological process of the decontamination of contaminated environment. Microbes produce some enzymes which have the ability to degrade organic contaminants into nontoxicants. There are some microbes which are being widely used in remediation process as *Pseudomonas putida, Dechloromonas aromatica, Deinococcus radiodurans, Methylibium petroleiphilum,* and *Alcanivorax borkumensis*. However, there are some drawbacks/limitations in this process. One is that microbes (bacteria and fungi) do not act on a broad range of organic compounds. No organism is reported till now which can destroy a large percentage of the natural chemicals that exist. Another hurdle of bioremediation is that it takes a long period of time to act and impose its effect.

3.2 Mycoremediation

Besides the use of bacteria, fungal species as *Aspergillus niger*, *Aureobasidium pullulans*, *Ganoderma lucidum*, and *Cladosporium resinae* are found to be capable in mycoremediation (Mani and Kumar 2014). Fungi secrete more potent enzymes

Table 1 Living organisms involved in bioremediation

Living organism	Species	Metals	References
Bacteria	*Pseudomonas veronii*	Cd, Zn, Cu	Vullo et al. (2008)
	Burkholderia species	Cd, Pb	Jiang et al. (2008)
	Bacillus sp.	Cd, Pb, Cu	Guo et al. (2010)
	Kocuria flava	Cu	Achal et al. (2011)
	Serratia marcescens	U	Kumar et al. (2011)
	Pseudomonas aeruginosa	U	Choudhary and Sar (2011)
	Bacillus cereus	Cd, Zn	Hrynkiewicz and Baum (2012)
	Halomonas sp.	Sr	Achal et al. (2012b)
	Sporosarcina ginsengisoli	As	Achal et al. (2012a)
Fungi	*Penicillium canescens*	Cr	Say et al. (2003)
	Ganoderma lucidum	Ar	Loukidou et al. (2003)
	Aspergillus fumigates	Pb	Ramasamy et al. (2011)
Algae	*Chlorella pyrenoidosa*	U	Singhal et al. (2004)
	Chlorella fusca	Pb, Zn, Cd, Cr, Cu, Ni	Ahluwalia and Goyal (2007)
	Spirogyra sp.	Pb, Cu	Lee and Chang (2011)
	Spirulina sp.	Cr, Cu, Fe, Mn, Se, Zn	Mane and Bhosle (2012)
	Hydrodictyon sp.	V, As	Saunders et al. (2012)
	Oedogonium sp.	V, As	Saunders et al. (2012)
Lichen	*Cladonia rangiformis*	Pb	Ekmekyapar et al. (2012)
Plants	*Pteris vittata*	Cu, Ni, Zn, As	Ma et al. (2001)
	Brassica juncea	Se, Cd	Banuelos et al. (2005)
	Helianthus annuus	Cd	Mani and Kumar (2014)
	Populus sp.	Hg	Lyyra et al. (2007)
	Brassica napus	Cd	Selvam and Wong (2008)
	Typha latifolia	Pb	Tiwari et al. (2008)
	Nelumbo nucifera	Zn, Cu, Pb, Ni	Kumar et al. (2008)
	Amaranthus viridis	Cr	Liu et al. (2008)
	Helianthus annuus	Cu, Zn, Pb, Hg, As, Cd, Ni	Mani et al. (2012)
	Trifolium pratense	Cs	Wu and Tang (2009)
	Spinacea oleracea	Pb, Zn	Mani et al. (2012)
	Vetiveria zizanioides	Cd, Pb	Danh et al. (2009)
	Nicotiana tabacum	Cd	Wojas et al. (2009)
	Brassica juncea	Pb	Zarei et al. (2010)
	Pistia stratiotes	Cd, Pb, Zn	Vesely et al. (2012)
	Populus tremula	Zn, Cd, Cu	Ruiz et al. (2011)
	Gmelina arborea	Al	Dudhane et al. (2012)

Table 2 Developmental methods of bioremediation

Techniques	Principle requirement	Advantages	Disadvantages	References
Bioventing	Air (O_2) supply	Effective for treatment of petroleum-, diesel-, phenanthrene-, and hydrocarbon-contaminated sites Used in both aerobic and anaerobic bioremediation	Prolonged treatment; soil vapors of volatile compounds are formed due to high airflow rate which require additional off-gas treatment	Azubuike et al. (2016), Höhener and Ponsin (2014), Thome' et al. (2014), Frutos et al. (2010), Rayner et al. (2007), Mihopoulos et al. (2000, 2002), Shah et al. (2001), and Burgess et al. (2001)
Biopiling	Treatment bed, aeration system, nutrient or irrigation system, and leachate collection system	It transforms contaminants to carbon dioxide and water and completed in 3 to 6 months	Forced aeration system and excessive heating of air causes drying of contaminated soil resulting to inhibition of microbial activities which in turn promotes volatilization rather than biodegradation	Azubuike et al. (2016), Garima and Singh (2014), Niu et al. (2009), Wu and Crapper (2009), and Sanscartier et al. (2009)
Bioaugmentation	Metabolically more active microorganism	Natural attenuation processes Treats soil and water contaminated with chlorinated ethane such as tetrachloroethylene and trichloroethylene pollutants transformed to harmless ethylene and chloride	System that is difficult to monitor	Plangklang and Alissara (2010); Sei et al. (2001) and Bouwer and Zehnder (1993)
Biosparging	Air (O_2) supply under high pressure	Readily available, easy to install, noninvasive to operation site, and completed in short time Used for treatment of petroleum products like gasoline, diesel fuel, jet fuel, and kerosene Effectively treats benzene-, toluene-, ethylbenzene- and xylene (BTEX)-contaminated groundwater	It has environmental constraints as it requires uniform air sparging, permeable soil and unconfined aquifer, etc. No field and laboratory data available to support design consideration. Prediction of airflow direction is difficult	Azubuike et al. (2016), Garima and Singh (2014), Hajabbasi et al. (2011); Lambert et al. (2009), Kao et al. (2008), and Philp and Atlas (2005)

Compositing	Elevated temperature (55 to 65 °C°C) Straw, alfalfa, manure, agricultural waste, and wood chips used as bulking agents and supplementary carbon source	More convenient and inexpensive, self-heating by microorganisms useful to treat hydrocarbon-polluted soils, to make plants healthier, it is good alternative to land filling or incinerating practical	Extended treatment time (incubation periods range from months to years). Requires nitrogen supplementation. Requires periodic turning of piled polluted soil for effective treatment	Garima and Singh (2014), Blanca et al. (2007, 2008), and Barr (2002)
Land farming	Tillage (for aeration), nutrients, and irrigation Excavated polluted soil is sandwiched between clean soil (at bottom) and clay and/or concrete soil (at upper most layer)	Simple, self-heating, and cost-effective Requires less energy and has no negative impact on the environment Operative to any climate and site Useful to treat diesel- and hydrocarbon-polluted (polyaromatic hydrocarbons) site	Large operating space requirement, slow degradation rates, and requires long incubation periods, additional cost due to excavation and reduced efficacy in inorganic pollutant removal, not suitable for toxic volatiles due to its design and mechanism of pollutant removal (volatilization), especially in tropical climate regions, less efficient	Garima and Singh (2014), Cerqueira et al. (2014), Silva-Castro et al. (2015, 2012), Besaltatpour et al. (2011), Paudyn et al. (2008), Blanca et al. (2007, 2008), Khan et al. (2004), and Maila and Colete (2004)
Biopiles	Aboveground pilling of excavated polluted soil, aeration, irrigation, nutrient, leachate collection system, and treatment bed	Economic, requires less space, reduces volatilization of low molecular weight pollutants, operative to extreme cold climate and clay and sandy soils, used to treat total petroleum hydrocarbon, applicable to agriculture to municipal waste	Mass transfer problem, bioavailability limitation, robust engineering, maintenance and operation cost, lack of power supply at remote areas, need to control abiotic loss, excessive air heating causes soil drying which affects microbial activity, promotes volatilization rather than biodegradation	Dias et al. (2015), Whelan et al. (2015), Gomez and Sartaj (2014), Akbari and Ghoshal (2014), Chemlal et al. (2013), Sanscartier et al. (2009), and Blanca et al. (2007, 2008)

(continued)

Table 2 (continued)

Techniques	Principle requirement	Advantages	Disadvantages	References
Slurry bioreactors	Bioaugmenting or biostimulating agents (sewage sludge)	Fast degradation kinetics, optimized environmental parameters, toxicity of amendments Applicable for crude oil (total petroleum and polyaromatic hydrocarbons)-polluted sediment and soil polluted with volatile organic compounds such as BTEX (benzene, toluene, ethylbenzene, and xylene)	Soil requires excavation which can disrupt soil structure, due to ex situ treatment; it requires more manpower, cost inefficient due to transportation of pollutant to treatment site	Chikere et al. (2012, 2016), Zangi-Kotler et al. (2015), Mustafa et al. (2015), Plangklang and Alissara Reungsang (2010), Antizar-Ladislao et al. (2007, 2008); Arsam et al. (2007), Philp and Atlas (2005), Fuller et al. (2003)
Aqueous bioreactors	Bioaugmenting or biostimulating agents (sewage sludge)	Enhances mass transfer, effective use of inoculants and surfactant, effective to reduce toxic concentrations of contaminants. Used to treat laundry wastewater polluted with linear alkylbenzene sulfonate, BTEX-contaminated water and coal gasification wastewater	Relatively high cost capital and high operating cost; transportation may cause environmental problems	Delforno et al. (2015), Firmino et al. (2015), Xu et al. (2015), and Arsam et al. (2007)
Precipitations/ flocculations	It requires nondirected physicochemical complexation reaction between dissolved contaminants and charged cellular components (dead biomass)	Cost-effective and advantageous for removal of heavy metals	Yet to be exploited commercially	Natrajan (2008)

Microfiltrations	Microfiltration membranes and constant pressure	Remove dissolved solids rapidly, effective for wastewater treatment; more than 90% of original wastewater can be recovered and reused	Yet to be exploited commercially	Shukla et al. (2010)
Electrodialysis	Requires cation and anion exchange membrane pairs	Electrodialysis is reusable and can tolerate high temperature Useful to eliminate dissolved solids efficiently	Yet to be exploited commercially	Shukla et al. (2010)

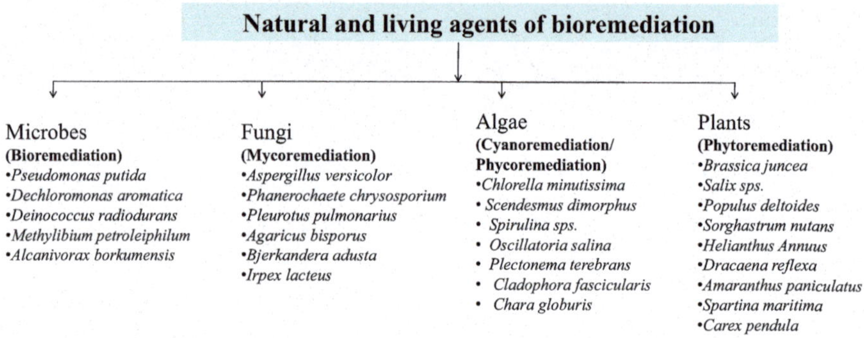

Fig. 1 Living organisms involved in bioremediation

even in nutrient-deficient conditions, and these enzymes act on a broad category of natural chemicals. Remediation through fungus may proceed faster than bacterial degradation, with hurdle suggested as the main mechanism of calcium mobilization (Gadd 2010). Many fungal species are reported to metabolize hydrocarbons, and some of them may be used in bioremediation of oil-polluted regions. These fungal genera include *Acremonium, Aspergillus, Aureobasidium, Candida, Cephalosporium, Cladosporium, Cunninghamella, Fusarium, Geotrichum, Gliocladium, Graphium, Hansenula, Mortierella*, etc. Few of the fungus, as *Trichoderma*, increases biomass of plant acting as a biocontrol agent as well as remediates agricultural waste (Pakdaman and Goltapeh 2006). *Lentinus edodes*, the gourmet mushroom, has potential of removing more than 60% of pentachlorophenol from soil (Pletsch et al. 1999). Such a potent fungus is being used as a boon in oil industries and refineries. *Phanerochaete chrysosporium* and other white-rot fungi degrade some xenobiotics as DDT and lindane (Kirk et al. 1992).

3.3 *Phycoremediation or Cyanoremediation*

Phycoremediation is defined as the “use of algae to treat solid wastes or wastewaters.” There are few microalgae and macroalgae such as more commonly known as the seaweeds that have the ability of removing soil and water toxicants such as heavy metals, hydrocarbons, and pesticides through various mechanisms, ranging from biosorption, bioconcentration, biotransformation, to volatilization. The most common examples of microalgae are *Chlorella pyrenoidosa*, *Chlorella fusca*, *Spirogyra sp.*, *Spirulina sp.*, *Hydrodictyon sp.*, and *Oedogonium sp.* Microalgae are reported for potent remediation of pollutants from environments (Phang et al. 2015).

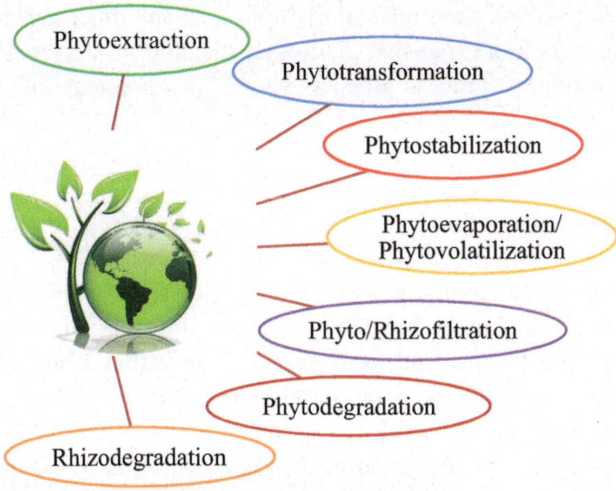

Fig. 2 Various processes used by plants for bioremediation

3.4 Phytoremediation

When plants or plant parts are involved in the removal of environmental toxicant, the process is called phytoremediation. Modern technology of phytoremediation includes phytoextraction, phytotransformation, phytostabilization, phytoevaporation or phytovolatilization, phyto–/rhizofiltration, phytodegradation, and rhizodegradation (Mahar et al. 2016), as depicted in Fig. 2.

3.4.1 Phytoextraction

Phytoextraction is a remediation process where pollutants are taken up by plant roots or algae from the contaminated soil, sediments, and/or water, and then they are accumulated in the shoots (harvestable plant biomass) (Sekara et al. 2005; Rafati et al. 2011; Razzaq 2017). Since the last two decades or so, phytoremediation technique has become more popular worldwide for extracting heavy metals from soil or water (Sulmon et al. 2007). Plants absorb pollutants from soil or water through roots and store them in root biomass or transport them up to shoot biomass or leaves. Plants continuously absorb pollutants until it is harvested. At the time of plant biomass harvesting, it was reported that plant concentrates the pollutants to much smaller volume than they were initially present in the polluted site. After the harvest, the level of pollutants is generally reduced in the soil which can be further removed through repeated process of plantation of several crops, and pollutant-free soil could become suitable for other vegetation. It was also reported that plants along with fungus (*T. atroviride*) showed more effective phytoextraction of Cd and Ni than without fungus (Cao et al. 2008). Phytoextraction is more advantageous than other traditional methods of bioremediation in several ways such as it is more eco-friendly process and prevents soil disruption or any other harm to soil quality; phytoremediation is less expensive than other cleanup processes. However, it is a more time-consuming process due to direct involvement of plants (Shukla et al. 2010).

3.4.2 Phytotransformation

In phytotransformation, complex organic molecules are converted into the simpler form through degradation or breakdown, and simple organic molecules can be retained in the plant tissues, soil, or water (Razzaq 2017). Thus, complete breakdown of the compound does not occur in phytotransformation. The complex organic pollutants such as pesticides, explosives, solvents, industrial chemicals, and other xenobiotic substances are metabolized to nontoxic forms by several plants (e.g., *Canas*), or sometimes microorganisms associated with plant roots may metabolize them in soil or water (Shukla et al. 2010). Hence, the term "green liver model" is used to explain phytotransformation, as plants behave analogously to the human liver when dealing with these xenobiotic substances (pollutant). The phytotransformation process completes in two phases of metabolism: In Phase I, the polarity of pollutants is increased by nitroreductase enzymes (Yoon et al. 2008), followed by phase II where glucose and amino acids are added to the polarized pollutants to further increase polarity (also called conjugation) (Mendez and Maier 2008); thus, the plants reduce toxicity and sequester the xenobiotics. Trinitrotoluene phytotransformation has been extensively researched, and a transformation pathway has been proposed (Vanderford et al. 1997).

3.4.3 Phytostabilization

Phytostabilization is a kind of phytoimmobilization technique in which plants are used for immobilization of soil or water pollutants (Singh 2012; Shukla et al. 2010). In this technique, pollutants are generally absorbed and accumulated in roots, adsorbed on roots, or precipitated in the rhizosphere which reduces contaminant mobility to groundwater or air, thus decreasing the bioavailability and preventing spread through the food chain (Yoon et al. 2008; Erakhrumen 2007; Ghosh 2010; Shukla et al. 2010; Wuana and Okieimen 2011). The major limitation of this technique is that it does not remove pollutants from soil or water completely, but it reduces only pollutant mobility to water stream or soil. Therefore, this technique alone is not sufficient for removal of contaminants; however, this technique can be used along with other bioremediation processes to manage the polluted sites (Vangronsveld et al. 2009; Razzaq 2017).

3.4.4 Phytovolatilization

Phytovolatilization refers to the removal of pollutants in volatile form where plants uptake water-soluble pollutants from the soil along with minerals that convert them into volatile form and then release them into atmosphere as they transpire water (Danika et al. 2005; Shukla et al. 2010; Razzaq 2017). The degree of success varies with plant as phytovolatilizers with one study showing poplar trees to volatilize up to 90% of the trichloroethylene (TCE) they absorb (Danika et al. 2005).

3.4.5 Phytofiltration or Rhizofiltration

Phytofiltration is the process where plants absorb or adsorb organic pollutants from wastewater in order to prevent its mixing with groundwater (Danika et al. 2005). Phytofiltration is slightly different in concept to phytoextraction as the former is related to the remediation of contaminated groundwater rather than polluted soils. Since plant roots are used in this technique, the term rhizofiltration is generally used in place of phytofiltration. However, rhizofiltration can be called blastofiltration when young seedlings are used or caulofiltration when excised plant shoots are used (Macek et al. 2000; Razzaq 2017). Earlier studies revealed that movement of toxic pollutants can be reduced in groundwater using this technique (Memon et al. 2001; Sakai et al. 2012). In rhizofiltration, acclimatized plants are used for remediation of contaminants (Marcia et al. 1999).

3.4.6 Rhizodegradation

Rhizodegradation refers to the degradation of organic pollutants in the soil by soil living microorganisms where the enzymatic activity of soil microbes is enhanced by plant root exudates (Razzaq 2017; Shukla et al. 2010). The plant root exudates such as sugars, alcohols, and other organic acids act as carbohydrate sources for soil microbes for enhancing their growth and activity. Few of these exudates also act as chemotactic signals for microflora. Since the biodegradation activity of soil microbes is stimulated by plant-derived exudates, the process is also called enhanced rhizosphere biodegradation, phytostimulation, and plant-assisted bioremediation (KudjoDzantor 2007).

3.4.7 Phytodegradation

It is the process by which plant-driven breakdown or degradation of toxic organic pollutants such as herbicides or trichloroethylene occurs. Degradation can take place by internal or external metabolic processes (Razzaq 2017). In external processes, complex organic compounds are hydrolyzed to simple and small units by plant enzymes. The simpler forms of contaminant can be absorbed by plants which can be incorporated and used as metabolites by the plant as it grows (Singh and Jain 2003).

4 Techniques Involved in Bioremediation

Bioremediation is broadly classified in to two groups (Fig. 3): in situ and ex situ which are further categorized into several techniques on the basis of amenability of the contaminants to biological transformation (biochemistry), availability of the pollutant to microorganisms (bioavailability), and opportunity for optimization of biological activity (bioactivity).

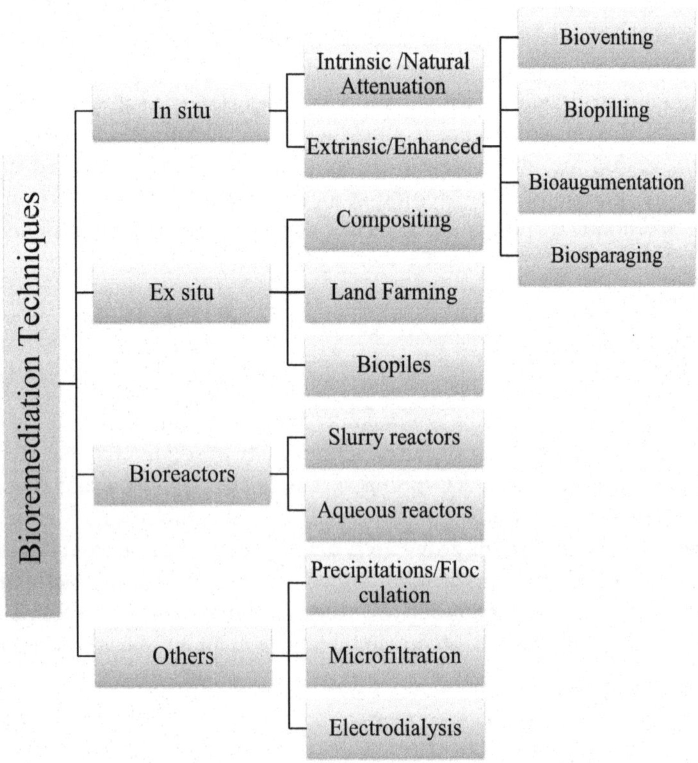

Fig. 3 Techniques of bioremediation

5 Recent Developments in Remediation Technology

Besides conventional methods as landfilling and leaching, excavation, and burial or soil washing which are time-consuming and less efficient, some advanced methods like the use of nanoparticles, nonliving biomass, and genetically modified plants are in trends for remediation Dhermendra et al. (2008). Nanoparticles are being used due to their small size and large surface area which can interact with heavy metals. High surface-to-volume ratio of nanoparticles makes them more suitable for adsorption of heavy metals. Nowadays, superparamagnetic iron oxide nanoparticles (SPION) are also being used for the separation of contaminants from soil and aquatic wastes due to their ultrafine structure and high competence and prepared iron nanoparticles for the remediation of heavy metals as Cd, Cr, Cu, and Ni. In another study, chitosan nanoparticles were formulated for the treatment of Cu from aquatic system (Yuwei and Jianlong 2011). Another approach is using nonliving biomass where no media or chemicals are required; thus, it is economical. *Cladonia rangiformis* (a nonliving lichen) is being used for the accumulation and remediation of lead from aqueous solution (Mohamad et al. 2012). Some other dead cells are also reported as *Mesorhizobium amorphae* and *Spirulina sp.* for the remediation of heavy metals as Cu and Pb, respectively (Aneja et al. 2010). Genetic engineering of plants is done to improve phytoaccumulation, phytoextraction, and phytosequestration. Recently, *Arabidopsis thaliana* was developed transgenically to increase the tolerance and accumulation of arsenic and cadmium by overexpression of AsPCS1 and YCF1 genes. These genes are derived from garlic and baker's yeast (Gaur et al. 2013). A metallothionein gene is transferred from yeast to *Nicotiana tabacum* to accumulate Cd in the roots of this transgenic plant (Krystofova et al. 2012).

6 Applications of Bioremediation

There are several advantages of bioremediation making this technique a preferred technology to remediate polluted sites:

- Bioremediation is a scientifically accepted natural process, which uses microorganisms and higher plants to remediate a wide range of organic and inorganic compounds and metabolize them to harmless products or into carbon dioxide and water.
- The complete elimination of contaminants reduces any chance of future liability associated with treatment and disposal of contaminated material.
- Microbes increase their numbers when a huge amount of contaminant is present, and once the contaminant is degraded, their population declines.
- Bioremediation can be employed on the site of contamination (in situ) without any environmental distraction. In situ bioremediation reduces the chance of environmental expose of pollutants, while transportation eliminates the threats to human health.

- On-site bioremediation with natural attenuation and fewer inputs makes it a less expensive technique for cleaning of toxic wastes (Hussain et al. 2009; Kumar et al. 2011).

7 Limitations of Bioremediation

Although bioremediation seems to be a good alternative for toxic contaminant removal, it is not fully developed/established method. Further, it requires continuous research due the involvement of microorganisms and toxic chemical compounds. Few limitations of bioremediation are as follows:

- Bioremediation is in general labor intensive and can take several months for the remediation of toxic waste to achieve at satisfactory levels.
- Bioremediation is limited to biodegradable compounds only; further complete degradation of all pollutants is not possible.
- Bioremediation involves degradation of hazardous wastes that possess a huge number of contaminants and toxicity which can inhibit the growth of microorganism or sometimes kill them.
- The pollutants which are converted to another form of chemical compound during the process of bioremediation may be more persistent or hazardous.
- Bioremediation is highly specific process that requires potentially active microorganisms, proper aeration, nutrients, irrigation, favorable pH, and temperature 20 °C to 30 °C (Vidali 2001).
- In order to enhance the activity of bacterium, fungi, or any other microorganisms, additives are supplemented which may be disruptive to other creatures inhabiting in same environment when done in situ (Vidali 2001). Thus, there is chance of more damage by bioremediation than the actual pollutant itself.
- The factors such as chemical composition, solubility, oxidation–reduction, and microbial interaction of waste likewise affect bioremediation process.
- It is time-consuming process as compared to excavation and removal of soil or incineration Kumar et al. (2011).
- It is difficult to extrapolate from bench and pilot-scale studies to full-scale field operations Hussain et al. (2009).
- Bioremediation is still a developing technology, and continuous research is needed to develop and engineer bioremediation technologies (genetically modified microorganisms) that are appropriate for sites with complex mixtures of contaminants that are not evenly dispersed in the environment Sharma (2012).
- Further, there is a problem after release of genetically engineered microorganism into environment because as time will pass it becomes difficult to remove them Garima and Singh (2014).

8 Conclusions

Organic and inorganic toxic pollutants are major problems to the environment and human health. Worldwide research on chemical contaminants helps to understand its recalcitrance and toxicity (Alcock et al. 2011). Although a variety of physical and chemical methods are used for the removal of these toxic wastes, the biological method (bioremediation) is the only one which is economic and eco-friendly technology for better and safe future (Uqab et al. 2016). A diverse group of metabolically active microorganisms are involved for in situ and ex situ bioremediation. However, response to environmental pollutants varies within a microbial guild (Ramakrishnan et al. 2010), and the presence of co-contaminants can elicit variable responses (Ramakrishnan et al. 2011). Reports reveal that nutrient supplement promotes microbial growth as well as pollutant degradation (Adams et al. 2015). Besides microorganisms, plants are also helpful to extract, degrade, transform, and store pollutants. Site characterization is the crucial step for effective bioremediation so that suitable technique (ex situ or in situ) can be employed. Geological characteristics of polluted site(s) including soil type, pollutant depth and type, site location relative to human habitation, and performance characteristics of each bioremediation technique should be incorporated in deciding the most suitable and efficient method to effectively treat polluted sites (Azubuike et al. 2016). Researchers are conducting pilot-scale bioremediation research which helps one to understand applications and limitations of this strategy. In this chapter, research finding on successful use of bioremediation to treat a variety of toxic waste has been discussed. Though bioremediation is recommended as an effective alternative for pollutant treatment, it has several practical limitations which need more research regarding soil–microbe–plant–contaminant interactions to translate effectively the bench- and pilot-scale findings to field scale (Hussain et al. 2009). However, the advantages of this technology generally compensate the disadvantages making it more reliable (Kumar et al. 2011) and have proved again and again its potential to degrade variety of pollutants (Garima and Singh 2014; Megharaj et al. 2011).

References

Achal V, Pan X, Zhang D (2011) Remediation of copper-contaminated soil by Kocuriaflava CR1, based on microbially induced calcite precipitation. Ecol Eng 37(10):1601–1605

Achal V, Pan X, Fu Q, Zhang D (2012a) Biomineralization based remediation of as (III) contaminated soil by Sporosarcina ginsengisoli. J Hazar Mater 201–202:178–184

Achal V, Pan X, Zhang D (2012b) Bioremediation of strontium (Sr) contaminated aquifer quartz sand based on carbonate precipitation induced by Sr resistant Halomonas sp. Chemosphere 89:764–768

Adams GO, Fufeyin PT, Okoroz SE, Ehinomen I (2015) Bioremediation, biostimulation and bioaugmentation: a review. Int J Envt Bioremed Biodegred 3(1):28–39

Ahluwalia SS, Goyal D (2007) Microbial and plant derived biomass for removal of heavy metals from wastewater. Bioresource Technol 98(12):2243–2257

Akbari A, Ghoshal S (2014) Pilot-scale bioremediation of a petroleum hydrocarbon-contaminated clayey soil from a sub-Arctic site. J Hazard Mater 280:595–602
Alcock RE, MacGilliray BH, Busby JS (2011) Understanding the mismatch between the demands of risk assessment and practice of scientists — the case of Deca-BDE. Environ Int 37:216–225
Aneja RK, Chaudhary G, Ahluwalia SS, Goyal D (2010) Biosorption of Pb and Zn by Non-Living Biomass of Spirulina sp. Indian J Microbiol 50:438–442
Antizar-Ladislao B (2010) Bioremediation: working with bacteria. Elements 6:389–394
Antizar-Ladislao B, Beck AJ, Spanova K, Lopez-Real J, Russell NJ (2007) The influence of different temperature programmes on the bioremediation of polycyclic aromatic hydrocarbons (PAHs) in a coal-tar contaminated soil by in-vessel composting. J Hazard Mater 14:340–347
Antizar-Ladislao B, Spanova K, Beck AJ, Russell NJ (2008) Microbial community structure changes during bioremediation of PAHs in an aged coal-tar contaminated soil by in-vessel composting. Int Biodeterior Biodegrad 61:357–364
Arsam B, Romain L, Laurent S, Rachid O, Badie IM (2007) Gas holdup and bubble size behavior in a large-scale slurry bubble column reactor operating with an organic liquid under elevated pressures and temperatures. Chem Eng J 128:69–84
Arshad M, Saleem M, Hussain S (2007) Perspectives of bacterial ACC deaminase in phytoremediation. Trends Biotechnol 25:356–362
Atlas R, Bartha R (1998) Microbial Ecology: Fundamentals and Applications, Benjamin/Cummings Sci Pub, Menlo Park, CA. 99–103
Azubuike CC, Chikere CB, Okpokwasili (2016) Bioremediation techniques-classification based on site of application: principles, advantages, limitations and prospects. World J Microbiol Biotechnol 32:180
Blanca A, Angus JB, Katerina S, Joe L, Nicholas JR (2007) The influence of different temperature programmes on the bioremediation of polycyclic aromatic hydrocarbons (PAHs) in a coal-tar contaminated soil by in-vessel composting. J Hazard Mater 14:340–347
Blanca A, Katerina S, Angus JB, Nicholas JR (2008) Microbial community structure changes during bioremediation of PAHs in an aged coal-tar contaminated soil by in-vessel composting. Int Biodeteriorat & Biodegrad 61:357–364
Banuelos G, Terry N, Leduc DL, Pilon-Smits EAH, Mackey B (2005) Field trial of transgenic Indian mustard plants shows enhanced phytoremediation of selenium contaminated sediment. Environ Sci Technol 39:1771–1777
Barr D (2002) Biological methods for assessment and remediation of contaminated land: case studies. Construction Industry Research and Information Association, London
Bouwer EJ, Zehnder AJB (1993) Bioremediation of organic compounds putting microbial metabolism to work. Trends Biotechnol 11:287–318
Burgess JE, Parsons SA, Stuetz RM (2001) Developments in odour control and waste gas treatment biotechnology: a review. Biotechnol Adv 19:35–63
Cao L, Jiang M, Zeng Z, Du A, Tan H, Liu Y (2008) *Trichodermaatroviride*F6 improves phytoextraction efficiency of mustard (*Brassica juncea* (L.) Coss. var. *foliosa*Bailey) in Cd, Ni contaminated soils. Chemosphere 71:1769–1173
Cerqueira VS, Peralba MR, Camargo FAO, Bento FM (2014) Comparison of bioremediation strategies for soil impacted with petrochemical oily sludge. Int Biodeterior Biodegrad 95:338–345
Chemlal R, Abdi N, Lounici H, Drouiche N, Pauss A, Mameri N (2013) Modeling and qualitative study of diesel biodegradation using biopile process in sandy soil. Int Biodeterior Biodegrad 78:43–48
Chikere CB, Chikere BO, Okpokwasili GC (2012) Bioreactor-based bioremediation of hydrocarbon-polluted Niger Delta marine sediment, Nigeria. 3 Biotech 2:53–66
Chikere CB, Okoye AU, Okpokwasili GC (2016) Microbial community profiling of active oleophilic bacteria involved in bioreactor based crude-oil polluted sediment treatment. World J Microbiol Biotechnol 32:180
Choudhary S, Sar P (2011) Uranium biomineralization by a metal resistant Pseudomonas aeruginosa strain isolated from contaminated mine waste. J Hazard Mater 186(1):336–343

Danh LT, Truong P, Mammucari R, Tran T, Foster N (2009) Vetiver grass, Vetiveria zizanioides: a choice plant for phytoremediation of heavy metals and organic wastes. Int J Phytoremediation 11:664–691

Danika L, LeDuc, Norman T (2005) Phytoremediation of toxic trace elements in soil and water. J Ind Microbiol Biotechnol 32:514–520

Delforno TP, Moura AGL, Okada DY, Sakamoto IK, Varesche MBA (2015) Microbial diversity and the implications of sulfide levels in an anaerobic reactor used to remove an anionic surfactant from laundry wastewater. Bioresour Technol 192:37–45

Dhermendra KT, Behari J, Prasenjit S (2008) Application of nanoparticles in waste water treatment. World Appl Sci J 3(3):417–433

Dias RL, Ruberto L, Calabro' A, Balbo AL, Del Panno MT, Mac Cormack WP (2015) Hydrocarbon removal and bacterial community structure in on-site biostimulated biopile systems designed for bioremediation of diesel-contaminated Antarctic soil. Polar Biol 38:677–687

Dudhane M, Borde M, Jite PK (2012) Effect of aluminium toxicity on growth responses and antioxidant activities in Gmelina arborea Roxb inoculated with AM Fungi. Int J Phytoremediation 14(7):643–655

Ekmekyapar F, Aslan A, Bayhan YK, Cakici A (2012) Biosorption of Pb(II) by Nonliving Lichen Biomass of Cladonia rangiformis Hoffm. Int J Environ Res 6(2):417–424

Erakhrumen AA (2007) Phytoremediation: an environmentally sound technology for pollution prevention, control and remediation in developing countries. Edu Res Rev 2:151–156

Firmino PIM, Farias RS, Barros AN, Buarque PMC, Rodrı'guez E, Lopes AC, dos Santos AB (2015) Understanding the anaerobic BTEX removal in continuous-flow bioreactors for ex situ bioremediation purposes. Chem Eng J 281:272–280

Frutos FJG, Escolano O, Garcı'a S, Mar Babı'n M, Ferna'ndez MD (2010) Bioventing remediation and ecotoxicity evaluation of phenanthrene-contaminated soil. J Hazard Mater 183:806–813

Fulekar MH (2010) Bioremediation Technology for Hazardous Wastes-Recent Advances. In Bioremediation Technology (135–166). Springer, Dordrecht

Fuller ME, Kruczek J, Schuster RL, Sheehan PL, Arienti PM (2003) Bioslurry treatment for soils contaminated with very high concentrations of 2,4,6-trinitrophenylmethylnitramine (tetryl). J Hazard Mater 100:245–257

Gadd GM (2010) Metals, minerals and microbes: geomicrobiology and bioremediation. Microbiol 156:609–643

Garima T, Singh SP (2014) Application of bioremediation on solid waste management: a review. J Bioremed Biodegr 5:248–256

Gaur N, Flora G, Yadav M, Archana Tiwari (2013) A review with recent advancements on bioremediation-based abolition of heavy metals. Environ Sci Processes & Impacts 16(2):180–193

Ghosh S (2010) Wetland macrophytes as toxic metal accumulators. Int J Environ Sci 1:523–528

Glass DJ (2000) Economic potential of phytoremediation. In: hytoremediation of toxic metals – using plants to clean up the environment, vol 7. Wiley, New York, pp 15–33

Glazer AN, Nikaido H (1995) Microbial biotechnology: fundamentals of applied microbiology. Freeman, New York

Gomez F, Sartaj M (2014) Optimization of field scale biopiles for bioremediation of petroleum hydrocarbon contaminated soil at low temperature conditions by response surface methodology (RSM). Int Biodeterior Biodegrad 89:103–109

Guo H, Luo S, Chen L, Xiao X, Xi Q, Wei W, He Y (2010) Bioremediation of heavy metals by growing hyper accumulator endophytic bacterium Bacillus sp. L14. Bioresource Tech 101(22):8599–8605

Hajabbasi AM, Khoshgoftarmanesh A, Dorostkar V (2011) Landfarming process effects on biochemical properties of petroleum-contaminated soils. Soil Sediment Contam Int J 20:234–248

Henis Y (1997) Bioremediation in agriculture: dream or reality? In Modern Agriculture and the Environment (481–489). Springer, Dordrecht

Höhener P, Ponsin V (2014) In situ vadose zone bioremediation. Curr Opin Biotechnol 27:1–7

Hrynkiewicz K, Baum C (2012) The potential of rhizosphere microorganisms to promote the plant growth in disturbed soils. In Environmental protection strategies for sustainable development (35–64). Springer, Dordrecht

Hussain S, Siddique T, Arshad M, Saleem M (2009) Bioremediation and phytoremediation of pesticides: recent advances. Critic Rev Environ Sci Tech 39:843–907
Jiang CY, Sheng XF, Qian M, Wang QY (2008) Isolation and characterization of a heavy metal-resistant Burkholderia sp. from heavy metal-contaminated paddy field soil and its potential in promoting plant growth and heavy metal accumulation in metal-polluted soil. Chemosphere 72(2):157–164
Kao CM, Chen CY, Chen SC, Chien HY, Chen YL (2008) Application of in situ biosparging to remediate a petroleum hydrocarbon spill site: field and microbial evaluation. Chemosphere 70:1492–1499
Khan FI, Husain T, Hejazi R (2004) An overview and analysis of site remediation technologies. J Environ Manag 71:95–122
Kirk TK, Lamar RT, Glaser JA (1992) The potential of white-rot fungi in bioremediation. In: Mongkolsuk S, Lovett PS, Trempy JE (eds) Biotechnology and environmental science – molecular approaches. Proced. Int. Conf. Biotechnol. Environ. Sci. Mol. Approach, New York, pp 131–138
Krystofova O, Zitka O, Krizkova S, Hynek D, Shestivska V, Adam V, Hubalek J, Mackova M, Macek T, Zehnalek J (2012) Int. J Electrochem Sci 7:886–907
KudjoDzantor E (2007) Phytoremediation: the state of rhizosphere engineering for accelerated rhizodegradation of xenobiotic contaminants. J Chem Technol Biotechnol 82:228–232
Kuiper I, Lagendijk EL, Bloemberg GV, Lugtenberg BJJ (2004) Rhizoremediation: a beneficial plant-microbe interaction. Mol Plant-Microbe Interact 7:6–15
Kumar JIN, Soni H, Kumar RN, Bhatt I (2008) Macrophytes in phytoremediation of heavy metal contaminated water and sediments in Pariyej community reserve, Gujarat. India Turk J Aquat Fish Sci 8:193–200
Kumar A, Bisht BS, Joshi VD, Dhewa T (2011) Review on bioremediation of polluted environment: a management tool. Int J Environ Sci 1(6):1079–1093
Lambert JM, Yang T, Thomson NR, Barker JF (2009) Pulsed biosparging of a residual fuel source emplaced at CFB borden. Int J Soil Sedi Water 2(3):6
Lee YC, Chang SP (2011) The biosorption of heavy metals from aqueous solution by Spirogyra and Cladophora filamentous macroalgae. Bioresour Technol 102(9):5297–5304
Leahy JG, Colwell RR (1990) Microbial degradation of hydrocarbons in the environment. Microbial Rev 53(3):305–315
Liu D, Zou J, Wang M, Jiang W (2008) Hexavalent chromium uptake and its effects on mineral uptake, antioxidant defence system and photosynthesis in Amaranthus viridis L. Bioresour Technol 99(7):2628–2636
Loukidou MX, Matis KA, Zouboulis AI, Liakopoulou-Kyriakidou M (2003) Removal of As(V) from wastewaters by chemically modified fungal biomass. Water Res 37(18):4544–4552
Lyyra S, Meagher RB, Kim T, Heaton A, Montello P, Balish RS, Merkle SA (2007) Coupling two mercury resistance genes in Eastern cottonwood enhances the processing of organomercury. Plant Biotechnol J 5(2):254–262
Ma LQ, Komar KM, Tu C, Zhang W, Cai Y, Kennelley ED (2001) A fern that hyper accumulates arsenic. Nature 409(6820):579
Macek T, Mackova M, Kas J (2000) Exploitation of plants for the removal of organics in environmental remediation. Biotechnol Adv 18:23–34
Maila MP, Colete TE (2004) Bioremediation of petroleum hydrocarbons through land farming: are simplicity and cost-effectiveness the only advantages? Rev Environ Sci Biotechnol 3:349–360
Mahar A, Wang, P, Ali A, Awasthi MK, Lahori AH, Wang Q, Zhang Z (2016) Challenges and opportunities in the phytoremediation of heavy metals contaminated soils: a review. Ecotoxicol Environl Safety 26:111–121
Mane PC, Bhosle AB (2012) Bioremoval of Some Metals by Living Algae Spirogyra sp. and Spirullina sp. from aqueous solution. Int J Environ Res 6(2):571–576
Mani D, Kumar C (2014) Biotechnological advances in bioremediation of heavy metals contaminated ecosystems: an overview with special reference to phytoremediation. Int J Environ Sci Technol 11:843–872

Mani D, Sharma B, Kumar C, Balak S (2012) Cadmium and lead bioaccumulation during growth stages alters sugar and vitamin C content in dietary vegetables. Proc Natl Acad Sci India Sect B Biol Sci 82(4):477–488
Marcia P, Brancilene SA, Charlwood BV (1999) Novel biotechnological approaches in environmental remediation research. Biotechnol Adv 17:679–687
McIntyre T (2003) Phytoremediation of heavy metals from soils. Adv Biochem Engg Biotechnol 78:97–123
Megharaj M, Ramakrishnan B, Venkateswarlu K, Sethunathan N, Naidu R (2011) Bioremediation approaches for organic pollutants: a critical perspective. Environ Int 37:1362–1375
Memon AR, Aktoprakligil D, Ozdemir A, Vertii A (2001) Heavy metal accumulation and detoxification mechanisms in plants. Turk J Bot 25:111–121
Mendez MO, Maier RM (2008) Phytostabilization of mine tailings in arid and semiarid environments—an emerging remediation technology. Environ Health Perspect 116(3):278–283
Mihopoulos PG, Suidan MT, Sayles GD (2000) Vapor phase treatment of PCE by lab-scale anaerobic bioventing. Water Res 34:3231–3237
Mihopoulos PG, Suidan MT, Sayles GD, Kaskassian S (2002) Numerical modeling of oxygen exclusion experiments of anaerobic bioventing. J Contam Hydrol 58:209–220
Mohamad OA, Hao X, Xie P, Hatab S, Lin Y, Wei G (2012) Microbes Environ 27:234–241
Mustafa YA, Abdul-Hameed HM, Razak ZA (2015) Biodegradation of 2, 4-dichlorophenoxyacetic acid contaminated soil in a roller slurry bioreactor. Clean-Soil Air Water 43:1115–1266
Natrajan KA (2008) Microbial aspects of acid mine drainage and its bioremediation. Trans Nonferrous Metals Soc China 18:1352–1360
Niu GL, Zhang JJ, Zhao S, Liu H, Boon N et al (2009) Bioaugmentation of a 4-chloronitrobenzene contaminated soil with Pseudomonas putida ZWL73. Environ Pollut 157:763–771
Okpokwasili GC (2007) Biotechnology and clean environment, in Proceedings of the 20th Annual conference of the Biotechnology Society of Nigeria (BSN) (Abakaliki: Ebonyi State University)
Pakdaman BS, Goltapeh EM (2006) An in vitro study on the possibility of rapeseed white stem rot disease control through the application of prevalent herbicides and Trichoderma species. Pak J Biol Sci 10:7–12
Paudyn K, Rutter A, Rowe RK, Poland JS (2008) Remediation of hydrocarbon contaminated soils in the Canadian Arctic by land farming. Cold Reg Sci Technol 53:102–114
Phang SM, Chu WL, Rabiei R (2015) Phycoremediation. In: Sahoo D, Seckbach J (eds) The algae world. Cellular origin, life in extreme habitats and astrobiology, vol 26. Springer, Dordrecht
Philp JC, Atlas RM (2005) Bioremediation of contaminated soils and aquifers. In: Atlas RM, Philp JC (eds) Bioremediation: applied microbial solutions for real-world environmental cleanup. American Society for Microbiology (ASM) Press, Washington, D.C., pp 139–236
Plangklang P, Alissara RA (2010) Bioaugmentation of carbofuran by Burkholderia cepacia PCL3 in a bioslurry phase sequencing batch reactor. Proc Chem 45:230–238
Pletsch M, de Araujo B, Charlwood B (1999) Novel biotechnological approaches in environmental remediation research. Biotechnol Adv 17:679–687
Rafati M, Khorasani N, Moattar F, Shirvany A, Moraghebi F et al (2011) Phytoremediation potential of Populus alba and Morus alba for cadmium, chromium and nickel absorption from polluted soil. Int J Environ Res 5:961–970
Ramakrishnan B, Megharaj M, Venkateswarlu K, Naidu R, Sethunathan N (2010) The impacts of environmental pollutants on microalgae and cyanobacteria. Crit Rev Environ Sci Technol 40:699–821
Ramakrishnan B, Megharaj M, Venkateswarlu K, Sethunathan N, Naidu R (2011) Mixtures of environmental pollutants: effects on microorganisms and their activities. Rev Environ Contam Toxicol 211:63–120
Ramasamy RK, Congeevaram S, Thamaraiselvi K (2011) Evaluation of isolated fungal strain from e-waste recycling facility for effective sorption of toxic heavy metals Pb(II) ions and fungal protein molecular characterization-a Mycoremediation approach. Asian J Exp Biol 2(2):342–347

Rayner JL, Snape I, Walworth JL, Harvey PM, Ferguson SH (2007) Petroleum–hydrocarbon contamination and remediation by microbioventing at sub-Antarctic Macquarie Island. Cold Reg Sci Technol 48:139–153

Razzaq (2017) Phytoremediation: an environmental friendly technique–a review. J Environ Anal Chem 4(2):2380–2391

Roseberg E (1993) Exploring microbial growth on hydrocarbons- new markets. Tibtech 11:419–424

Ruiz ON, Alvarez D, Gonzalez-Ruiz G, Torres C (2011) Characterization of mercury bioremediation by transgenic bacteria expressing metallothionein and polyphosphate kinase. BMC Biotechnol 11:82–89

Sakai Y, Ma Y, Xu C, Wu H, Zhu W (2012) Phytodesalination of a salt affected soil with four halophytes in China. J Arid Land Stud 22:17–20

Sanscartier D, Zeeb B, Koch I, Reimer K (2009) Bioremediation of diesel-contaminated soil by heated and humidified biopile system in cold climates. Cold Reg Sci Technol 55:167–173

Saunders RJ, Paul NA, Hu Y, de Nys R (2012) Sustainable sources of biomass for bioremediation of heavy metals in waste water derived from coal-fired power generation. PloS one 7(5):36470

Say R, Yimaz N, Denizli A (2003) Removal of heavy metal ions using the fungus Penicillium canescens. Adsorpt Sci Technol 21:643–650

Sei K, Nakao M, Mori KM, Ike M, Kohno T, Fujita M (2001) Design of PCR primers and a gene probe for extensive detection of poly (3-hydroxybutyrate) (PHB)-degrading bacteria possessing fibronectin type III linker type- PHB depolymerases. Appl Microbiol Biotechnol 55:801–806

Sekara A, Poniedzialeek M, Ciura J, Jedrszczyk E (2005) Cadmium and lead accumulation and distribution in the organs of nine crops: implications for phytoremediation. Pol J Environ Stud 14:509–516

Selvam A, Wong JW (2008) Phytochelatin synthesis and cadmium uptake of Brassica napus. Environ Technol 29:765–773

Shah JK, Sayles GD, Suidan MT, Mihopoulos PG, Kaskassian SR (2001) Anaerobic bioventing of unsaturated zone contaminated with DDT and DNT. Water Sci Technol 43:35–42

Shannon MJ, Unterman R (1993) Evaluating bioremediation: distinguishing fact from fiction. Anu Rev Microbiol 47:24

Sharma S (2012) Bioremediation: features, strategies and applications. Asian J Pharmac Life Sci 2(2):202–213

Shukla KP, Singh NK, Sharma S (2010) Bioremediation: developments, current practices and perspectives. Genetic Eng Biotech J 3:1–20

Silva-Castro GA, Uad I, Go'nzalez-Lo'pez J, Fandin˜o CG, Toledo FL, Calvo C (2012) Application of selected microbial consortia combined with inorganic and oleophilic fertilizers to recuperate oil-polluted soil using land farming technology. Clean Technol Environ Policy 14:719–726

Silva-Castro GA, Uad I, Rodrı'guez-Calvo A, Gonza'lez-Lo'pez J, Calvo C (2015) Response of autochthonous microbiota of diesel polluted soils to land- farming treatments. Environ Res 137:49–58

Singh S (2012) Phytoremediation: a sustainable alternative for environmental challenges. Int J Gr Herb Chem 1:133–139

Singh OV, Jain RK (2003) Phytoremediation of toxic aromatic pollutants from soil. Appd Microbiol Biotechnol 63:128–135

Singhal RK, Joshi S, Tirumalesh K, Gurg RP (2004) Reduction of uranium concentration in well water by Chlorella (Chlorella pyrenoidosa) a fresh water algae immobilized in calcium alginate. J Radioanal Nuclear Chem 261:73–78

Sulmon C, Gouesbet G, Binet F, Martin-Laurent F, Amrani AE, Couée I (2007) Sucrose amendment enhances phytoaccumulation of the herbicide atrazine in Arabidopsis thaliana. Environ Pollut 145:507–515

Thome' A, Reginatto C, Cecchin I, Colla LM (2014) Bioventing in a residual clayey soil contaminated with a blend of biodiesel and diesel oil. J Environ Eng 140:1–6

Tiwari S, Kumari B, Singh SN (2008) Evaluation of metal mobility/ immobility in fly ash induced by bacterial strains isolated from rhizospheric zone of Typha latifolia growing on fly ash dumps. Bioresour Technol 99:1305–1310

Uqab B, Mudasir S, Nazir R (2016) Review on bioremediation of pesticides. J Bioremed Biodegr 7(3):343–347

Vanderford M, Shanks JV, Hughes JB (1997) Phytotransformation of trinitrotoluene (TNT) and distribution of metabolic products in *Myriophyllum aquaticum*. J Biotechnol Lett 19:277–280

Vangronsveld J, Herzig R, Weyens N, Boulet J, Adriaensen K (2009) Phytoremediation of contaminated soils and groundwater: lessons from the field. Environ Sci Pollut Res 16:765–794

Vesely T, Tlustos P, Szakova J (2012) Organic acid enhanced soil risk element (Cd, Pb and Zn) leaching and secondary bioconcentration in water lettuce (*Pistia stratiotes* L.) in the rhizofiltration process. Int J Phytoremediat 14(4):335–349

Vidali M (2001) Bioremediation: an overview. Pure Appl Chem 73:1163–1172

Vullo DL, Ceretti HM, Daniel MA, Ramírez SA, Zalts A (2008) Cadmium, zinc and copper biosorption mediated by Pseudomonas veronii 2E. Bioresource Technol 99(13): 5574–5581

Whelan MJ, Coulon F, Hince G, Rayner J, McWatters R, Spedding T, Snape I (2015) Fate and transport of petroleum hydrocarbons in engineered biopiles in polar regions. Chemosphere 131:232–240

Wojas S, Hennig J, Plaza S, Geisler M, Siemianowski O et al (2009) Ectopic expression of Arabidopsis ABC transporter MRP7 modifies cadmium root-to-shoot transport and accumulation. Environ Pollut 157(10):2781–2789

Wood TK (2008) Molecular approaches in bioremediation. Curr Opin Biotechnol 19:572–578

Wu T, Crapper M (2009) Simulation of biopile processes using a hydraulics approach. J Hazard Mater 171(1–3):1103–1111

Wu HB, Tang SR (2009) Using CO_2 to increase the biomass of a Sorghum vulgare 9 Sorghum vulgare var. Sudanese hybrid and *Trifolium pratense* L and to trigger hyperaccumulation of cesium. J Hazard Mater 170:861–870

Wuana RA, Okieimen FE (2011) Heavy metals in contaminated soils: a review of sources, chemistry, risks and best available strategies for remediation. ISRN Ecol 2011:1–20

Xu P, Ma W, Han H, Jia S, Hou B (2015) Isolation of a naphthalene- degrading strain from activated sludge and bioaugmentation with it in a MBR treating coal gasification wastewater. Bull Environ Contam Toxicol 94:358–364

Yoon JM, Oliver DJ, Shanks JV (2008) Phytotransformation of 2, 4-Dinitrotoluene in *Arabidopsis thaliana*: toxicity, fate, and gene expression studies in vitro. Biotech Prog 19:1524–1531

Yuwei C, Jianlong W (2011) Chem Eng J 168:286–292

Zangi-Kotler M, Ben-Dov E, Tiehm A, Kushmaro A (2015) Microbial community structure and dynamics in a membrane bioreactor supplemented with the flame retardant dibromoneopentyl glycol. Environ Sci Pollut Res Int 22:17615–17624

Zarei M, Hempel S, Wubet T, Schafer T, Savaghebi G et al (2010) Molecular diversity of arbuscular mycorrhizal fungi in relation to soil chemical properties and heavy metal contamination. Environ Pollut 158:2757–2765

Utilisation of Sericulture Waste by Employing Possible Approaches

R. N. Manjunath, Amit Kumar, and K. P. Arun Kumar

Contents

1 Introduction

Sericulture is basically the rearing of silkworms for the production of natural silk fibre. The production of silk cocoons is the main source of income for the farmers involved in sericulture. The chain of activities in sericulture revolves around the cultivation of food plants that can be fed to silkworms which further spin to form silk cocoons and reeling of such cocoons for unwinding the silk filament for making value-added products. As the end-product users are consumed by higher economic people, the money flows from rich to poor, thereby creating a sectorial value addition for rural households. It has now turned out to be an important cottage industry in various countries like China, Uzbekistan, Thailand, India, Brazil, Korea, etc. Table 1 lists the ten major silk-producing countries in the world. Today, India and China are the major silk producers in the world, producing almost 98% of the

R. N. Manjunath · A. Kumar (✉) · K. P. Arun Kumar (✉)
Central Muga Eri Research & Training Institute, Central Silk Board,
Lahdoigarh, Assam, India

M. Naeem et al. (eds.), *Contaminants in Agriculture*,
https://doi.org/10.1007/978-3-030-41552-5_19

Table 1 Global silk production

Sl No.	Country	Silk production (metric tonnes)
01	China	1,42,000
02	India	31,906
03	Uzbekistan	1200
04	Thailand	680
05	Brazil	600
06	Vietnam	520
07	North Korea	365
08	Iran	120
09	Bangladesh	41
10	Turkey	30

Source: International sericulture commission-International sericulture commission (2019)

world's total silk production. Sericulture is a labour-intensive industry, and in India, it provides employment to almost eight million people in rural areas, thereby securing their economy and preventing them from migrating to big cities.

India has the unique distinction that produces four types of silk, namely, Mulberry, Eri, Tasar and Muga. The mulberry silkworm *Bombyx mori* produces a major chunk of silk globally. Muga silkworms, *Antheraea assama* Westwood (syn. *Antheraea assamensis* Helf., *Anthereame jankori* Moore), are naturally reared outdoors on host plants like Som and Soalu trees to produce Muga silk. These silkworms are basically multivoltine having five to six generations per year, and these days they are also commercially reared on semi-domestic conditions. Tasar silkworms *Antheraea mylitta* are reared on host plants like *Terminalia arjuna* and *Terminalia tomentosa* in wild conditions to produce Tasar silk, and Eri silkworms *Samia ricini* are domestically reared by feeding with castor leaves to produce Eri silk. Among these four types of silks, mulberry silkworm contributes to about 75% of the total silk production, while the rest of the portion is shared by wild silks (popularly called as Vanya silks in India). Therefore, the term silk is generally referred to as 'Mulberry silk', while the other three are categorised as 'non-Mulberry silk' or 'Vanya silk'.

In sericulture, the production of host plants is interwoven with cocoon production, where the output of plant cultivation is the input for cocoon production. The continuous strive for cocoon production in all the silk varieties has led to the rise of various mutually dependent technologies between agriculture and sericulture that in turn generates numerous wastes and secondary by-products. The sustainability of sericulture can be strongly reinforced in people's mind when it concerns the efficient use of such waste materials to high economic useful products. Therefore, it is necessary to work on approaches for proper utilisation of secondary wastes and by-products that can add up to the farmers' income. The recent trends considering the waste management and by-product utilisation in sericulture are revived in the further sections.

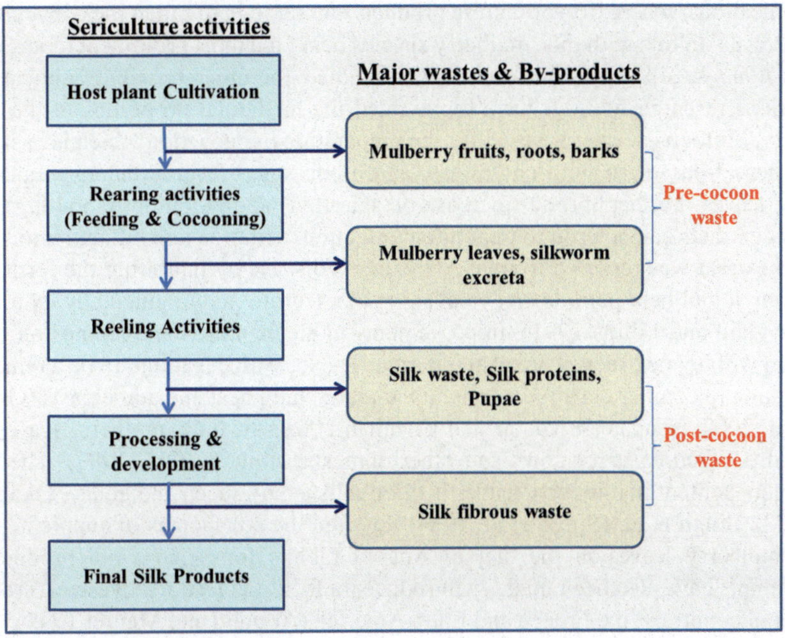

Fig. 1 Waste generation in sericulture

2 Waste Generation in Sericulture

Sericulture activities can be widely categorised into pre-cocoon and post-cocoon activities. The pre-cocoon activities basically involve the cultivation of host plants and silkworm-rearing activities till the cocoon formation. Silk wastes generated from these activities are mainly silkworm litter, leaf residue, dead larvae, moths, etc. The post-cocoon activities involve reeling activities (silk extraction) and product development through various processing methods like weaving, dyeing, finishing, etc. The possible wastes generated from these activities would be pupa, damaged cocoons or unusable cocoons (floss, double cocoons, pierced cocoons, stained cocoons, etc.) and fibrous silk waste (Chaubey et al. 2019; Kamili and Mosoodi 2000). The types of waste generated throughout the sericulture activities are shown in Fig. 1.

3 Utilisation of Wastes from the Pre-cocoon Sector

The proper utilisation of by-products from sericulture can generate extra income to farmers in addition to silk production. Silk is a major textile fibre, and the approach of by-product utilisation can facilitate the search for various non-textile applications, where the sericulture products can serve better than the conventional materials. In line with this, mulberry species being the most prominent food source for *Bombyx* silkworms have also been studied for other possible applications. Various research findings have highlighted the medicinal properties of the mulberry. Mulberry leaves are used for the preparation of decoction which is basically an herbal-based tea with cholesterol and blood sugar level-reducing properties. The leaves are diaphoretic (increase perspiration) and emollient (soothing), and their extracts can be used to treat throat infection, irritation and inflammation. The leaf extract was reported to treat Alzheimer's disease by inhibiting the formation of amyloidal beta-peptide and weakening the neurotoxicity induced by its formation (Jain and Fillips 1991). Improvements in elephantiasis and tetanus on treatment with leaf extracts were also reported. The leaf juice is found to be a remedial solution to counter diarrhoea, endemic, malaria, lung heat and amoebiasis (Grover et al. 2002; Venkatesh Kumar and Chauhan 2008). In Italy, mulberry leaves are used for feeding dairy cows and other domestic animals (FAO 1993). The high protein content in mulberry leaves is often relished by sheep and goats (Takahashi 1998). Singh et al. (Singh et al. 1984) reported the advantages of supplementing of mulberry leaves in the diet of Angora rabbits for better wool production. Attempts have also been made to introduce mulberry as feed for livestock production in countries like France and Latin America (Armand and Meuret 1995).

3.1 The Potential of Biogas Generation

Biogas (a mixture of different gases) is produced by the breakdown of organic matter in anaerobic fermentation process. The biogas production is an economic, sustainable, eco-friendly process which also provides homogeneous manure for agro-ecosystems. All agricultural waste, manure, municipal waste, plant material, sewage, green waste or food waste can serve as organic raw materials and can be successfully transformed into biogas. The silkworm waste is a mixture of food residue (leaf) with excreta in mulberry sericulture serving as high-quality feedstock for biogas production. Wastes like mulberry silkworm defatted pupae have been favourably proposed as feedstock for biogas production. Silkworm waste contains biodegradable organic matter with optimal C/N ratio of 15–35 (Mao et al. 2015). The other important advantage is that it is also devoid of inhibitory compounds, such as antibiotics, antiseptics and detergents (Dobre et al. 2014). The inhibitory substances are toxic to bacteria and thus effectively hamper the anaerobic fermentation process of biogas generation. Besides substrate quality, biogas production is also subject to other factors such as an effective hydraulic retention time (HRT), the optimum organic loading rate (OLR) and optimum process temperature (Mao et al. 2015).

The silkworm waste and its excreta can ferment relatively faster, due to its simple organic matter composition (Wang et al. 2016), and also this feedstock contains low amounts of macromolecular compounds (Uzakova et al. 1987) which is advantageous for better methane yield (Kiran et al. 2014; Niu et al. 2013). The silkworm breeding material has been found to have high energy potential than silkworm excrement (Łochyn'ska and Frankowski 2018). Łochyn'ska and Frankowski (2018) studied the composition and biogas yield of the substrates generated from mulberry sericulture and concluded that the high content of dry matter and methane in the silkworm waste and its excreta considerably increased the calorific value of biogas production. Karthikeyan and Shivakumar (2007) and Patil et al. (2013) successfully utilised pupal waste as a medium for mass cultivation of *B. thuringiensis,* a familiar bio-pesticide that is used against various lepidopteron pests worldwide. The production of biogas from silkworm excreta (through anaerobic fermentation process) will not just be a financial gain, but the homogeneity of the farm manure will also be significantly improved.

3.2 Potential Composting Materials

Compost is the biologically decomposed organic matter under controlled aerobic condition rich in nutrients and can function as soil conditioner, a fertiliser and also a natural pesticide for soil. Nutrients contained in organic manures are released more slowly (Bhatia et al. 2013) and stored for a long time in the soil, ensuring a long residual effect (Kumar et al. 2013; Mukherjee et al. 2018; Sharma and Mittra 2007). At ecosystem level, compost controls erosion, reclaims land and stream and also increases the plant's immunity to diseases, pests, etc. Compost functions as an absorbent material that holds moisture and soluble minerals, porous, providing the support and nutrients for plant growth. It is used as mixture with soil, sand, grit, bark chips, vermiculite and perlite.

The waste from pre-cocoon sector like larval excreta, faeces and residue leaves can be utilised as the substrate for composting materials. This substrate can be digested by utilising the fungi *Pleurotus florida* and *Pleurotus os treatus* and can be used for production of quality compost (Naik et al. 1992). Composting is completed in 50–60 days in vermicomposting, whereas in anaerobic composting it takes 120–150 days (Kalaiyarasan et al. 2015). The non-decomposed waste is not recommended to be utilised directly into the field, as the heat generation during the decomposition process can negatively affect the seed germination, root growth and non-capability to supply the actual nutrient demand of the crop immediately. Besides this, direct application of silk waste especially the dead larval can also increase the chance of infectious disease. Hence, the conversion of sericulture waste into usable form is very important. Further, the manure resulted out of silk waste has been shown to have better nutrient characteristics than farmyard manure (Kalaiyarasan et al. 2015). Therefore, the potential of sericulture waste-based compost to supply the micronutrients is higher than the farmyard manure (Sinha et al. 2005).

3.3 *Vermicompost*

Vermicompost is the compost derived from specific species of worms, especially red wigglers (*Eisenia foetida* or *Eisenia andrei*), European night crawlers (*Eisenia hortensis* or *Dendrobaena veneta*), white worms and earthworms. The potential of the worm species is measured on the fecundity and decomposition rate. However, the nature of the substrate and other physical parameters such as temperature, humidity, etc., also play an important role in the preparation of vermicompost. The redworms (*Eisenia andrei, Eisenia foetida* and *Lumbricus rubellus*) feed most rapidly at temperatures of 15–25 °C and can even survive at 10 °C. Temperatures above 30 °C can hamper the production of vermicompost due to harming of earthworms (Rathore and Srinivasulu 2018). *Eisenia foetida* are frequently used because of their high fecundity and decomposition rates.

Vermicompost is an end product (excreta) of the earthworms' digestive system. It is rich in humus and water-soluble nutrients (Rathore et al. 2007) and used as organic fertiliser and soil conditioner. It contains nitrogen, 1.5–2.5%; calcium, 0.5–1.0%, phosphorus, 0.9–1.7%; magnesium, 0.2–0.3%; potash, 1.5–2.4%; sulphur, 0.4–0.5%; and other micronutrients with vitamins, enzymes and plant growth regulators like auxins and gibberellins. It is rich in almost all essential plant nutrients and free from toxic elements, pathogens, weed seeds, etc. Vermicompost enhances the growth of new leaves and shoots and also improves the produce quality and shelf life. Vermicompost is easy to handle, store and apply in field. In addition to these uses, it also improves the soil texture, soil structure, aeration process and water-holding capacity of soil and prevents soil from its erosion. Vermicompost is also rich in beneficial microbiota such as fixers, cellulose-decomposing microflora, P-solubiliser microbiota, etc. Vermicompost prevents nutrient losses and also increases the N use efficiency.

The process of vermicomposting of sericulture waste was carried out by introducing mixed culture of juvenile earthworms *Eudrilus eugeniae, Eisellia felida and Perionyx excavatus* at the rate of 1.5 kwmt of wastes in each trench (Singhal et al. 2001). Vermicompost quality can be higher when sericulture waste is supplemented with farmyard manure (Sinha et al. 2005). Fortification with microbial inoculum like *Azotobacter*, phosphate-solubilising microbes and single superphosphate (SSP) will also improve the quality of produce (Dandin et al. 2000).

4 Waste Generation from Post-cocoon Sector

The post-cocoon sector involves extraction of raw silk from the cocoons and conversion of fibrous silk to useful valuable products. The process of extracting (unwinding) silk filament from the cocoon is called 'reeling', in which a number of cocoons are reeled together to produce single thread. Among the four types of silkworm cocoons, Eri cocoons are open ended and cannot be reeled due to the

Fig. 2 Post-cocoon waste and their useful products

nonavailability of continuous filament. It is therefore used for manufacturing spun yarns (machine spun or handspun) with 100% utilisation of fibrous material. Hence, the waste generated during the conversion of Eri cocoons to silk yarns is almost negligible. However, Mulberry, Tasar and Muga cocoons are constructed with continuous silk filaments, and hence they are reeled to extract continuous filament yarn. Other than silk production, the major by-products generated during the reeling process are 'pupal waste' and 'fibrous silk waste', which is unavoidable. Figure 2 shows the waste generated in the post-cocoon sector and their possible applications, which will be briefly discussed in the following sections.

4.1 Utilisation of Pupal Waste

In general, the pupa contains crude 50–60% proteins; 25–35% fats; 5–8% free amino acids; 8–10% sugars; E, B1 and B2 vitamins; calcium; and phosphorous. The list of constituent amino acids in the silkworm pupae is listed in Table 2. The

Table 2 Amino acid composition of silkworm pupae

Amino acid	Silkworm pupae			References
	Mulberry	Tasar	Eri	
Alanine	5.5	6.3	6.1	Rao (1994), Zhou and Han (2006a, b), Longvah et al. (2011), Kumar et al. (2015)
Arginine	6.8	12.2	4.4	
Aspartic acid	10.9	6.4	9.9	
Cysteine	1.4	1.5	0.5	
Glutamic acid	14.9	12.7	12.9	
Glycine	4.6	4.4	4.9	
Histidine	2.5	2.9	2.7	
Isoleucine	5.7	9.9	4.4	
Leucine	8.3	3.2	6.6	
Lysine	7.5	4.5	6.5	
Methionine	4.6	1.5	2.3	
Phenylalanine	5.1	8.1	5.2	
Proline	4.0	–	6.5	
Serine	4.7	4.6	5.3	
Threonine	5.4	4.6	4.6	
Tyrosine	5.4	2.6	6.4	
Valine	5.6	6.6	5.4	

silkworm pupae with good source of fat content (over 30%) are used as chrysalis oil in the manufacture of cosmetic products (cream, soap, lotion, emulsion) and as protein powder for valuable animal fodder.

The edible pupae are the by-product of commercial silk production, and any insect that can produce two or more useful products simultaneously increases its economic value. The Eri silk produced by *Samia ricini* is known for its unique thermal properties. Eri culture has previously been mostly confined to the Brahmaputra valley in the tribal-inhabited districts. It is also practised in a few districts of neighbouring states, mainly Manipur, Nagaland, Meghalaya, Mizoram and Arunachal Pradesh. However, due to its ease of cultivation and increased use in different applications, its culture is spread to other parts of India. Eri culture has been in practice in Northeast India since long time ago and closely linked with the culture and tradition of region. The tribes as well as general Eri farmers of Northeast India consume Eri pupa. While most of the tribal communities of Northeast India consume prepupal stage, the people of the Ahom community consume Eri pupa in the mature stage. In these areas, Eri silkworm is mainly reared for its consumption rather than for its silk. The pupae of mulberry silkworm are also used as food in various Asian countries (Sarmah 2011).

Recent findings show that the silkworm pupae have excellent antioxidant potential and good anti-tyrosinase activity (Kwon et al. 2012). Pupal oil is extracted by the process of drying, deacidification, bleaching and molecular refinement of the pupal waste. These oil extracts are good source of vitamin B2 which can be used to treat vitamin B2 deficiency (Harris et al. 1997). They can effectively reduce triglycerides and therefore are suitable to treat fatty lever, improve blood quality, lower blood pressure and prevent arteriosclerosis. It can reduce blood glucose levels by improving the functions of insulin-producing beta cells (Gavia et al. 2003). The natural steroids present in the oil can improve fertility, and the fatty acids in them can improve the vitality of immune cells. The α-linolenic acid in pupal oil can effectively prevent symptoms like wrinkles, pigmentation, sallow skins and premature aging effects.

Silkworm pupae (a waste of silk industry and breeding experiments) are rich in proteins and lipids and are comparatively low-cost nutritional feed for livestock and poultry contains protein (65–75%) and lipid (Iyengar 2002). Several studies have shown that silkworm pupa–based animal feed is a better alternative in meeting nutritional needs of farm animals. The de-oiled pupae feed improves hens' egg-laying capacity and improved colour of egg yolk. The dried pupal feed has also enhanced growth rate and egg quality in hens. The fatless silkworm pupae were used as feed of fish and give better yields (Buhroo et al. 2018a, b). Pupae were used as food in piggery, poultry and dog feed due to their richness in protein and fatty acids. The silkworm pupae were also profitable in magur fish production (Ghosh 2005). The dried pupal feed is also reported effective to improve survival rate, feed conversion rate and specific growth rate in fish (Buhroo et al. 2018a, b). Pupae were used as food in piggery, poultry and dog feed due to their richness in protein and fatty acids. The silkworm pupae were found to be very cost effective in the production of magur fish as well (Ghosh 2005). The de-oiled feed of pupae made rabbits to gain better weight and growth of fur (Velayudhan et al. 2008).

4.2 *Utilisation of Fibrous Silk Waste*

The fibrous silk wastes are usually generated from the following sources:

- Cut and pierced cocoons from breeding and silkworm seed production centres.
- Double cocoons (two cocoons spun together) that cannot be reeled.
- Extreme outside (floss) and extreme inside (palades) layers of cocoon.
- Waste left over during reeling, winding, re-reeling, silk throwing and other manufacturing processes.

These fibrous wastes are collectively utilised for spun yarn production called 'spun silk', which is much cheaper than the reeled silk. The sequence of operations involved in conversion of silk waste for spun silk is shown in Fig. 3.

The first process of a spun silk mill would be to remove the gum (sericin) which is naturally coated on to the filament's surface for its protection. The process of gum removal is called 'degumming', which is basically carried out by boiling the cocoons in hot water with added alkalis. The process of degumming softens the silk filaments and eases material handling in the eventual processes. The degummed cocoons are dried and opened into fibrous chunks using a mechanical cocoon opener. The fibrous chunks are taken to fillers, where further opening

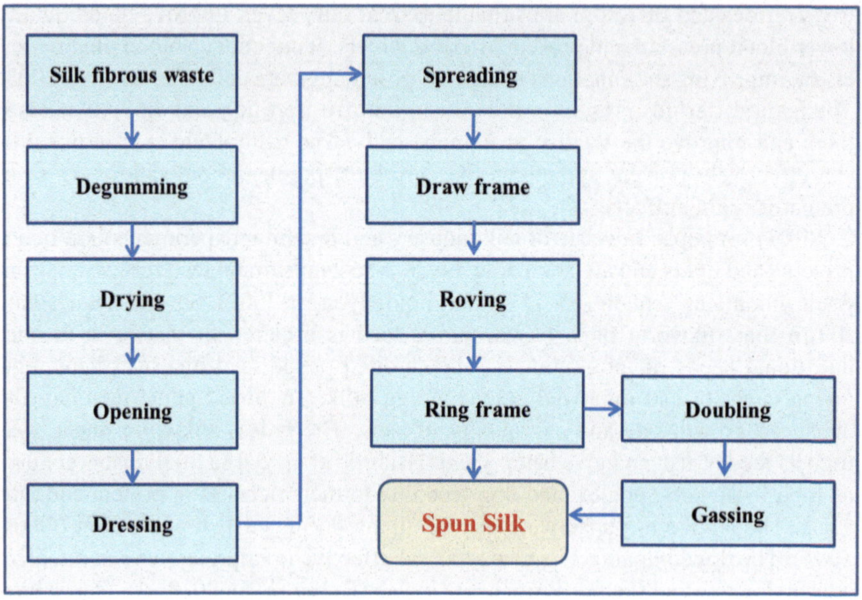

Fig. 3 Sequence of process involved in spun silk production

and cleaning of the fibrous chunks takes place. Further, the fibrous chunks are taken on to dressing machines where the material is combed to remove embedded foreign materials, neps (knots among the fibres) and short fibres. In this process, only those fibres with average staple are retained, and the fibrous material is converted into laps (consolidation of webs). These laps are taken to set frames where the proper distribution and parallelisation of fibres take place within the material to form slivers (long strips). Various such slivers (6–8 nos) are combined together on a draw frame and redrafted again to form slivers with perfect parallel orientation of fibres. These slivers are drawn further on a roving machine along with a slight twist to form rovings which are further spun into yarn on a ring frame by inserting sufficient twist and strength. These yarns can be further doubled and twisted to form twofold or threefold yarns as per commercial requirement. Since the spun yarn is formed by drafting and twisting of short-stapled fibres, the fibres' ends would be protruding outwards from the surface of the yarn. These protruding fibres are removed in a gassing process, where the yarn is passed through a burning flame. The smoothly finished yarn is wounded and collected on to a package suitable for shipping.

The twisting and weaving units of export-oriented industries produce hard silk waste to the extent of 300 MT which are mostly utilised in the manufacturing of coarser yarns, noil, throwster and carpet yarns. In view of further product diversification and waste management, short stapled silk filaments from the silk waste can be more effectively utilised in the development of nonwoven fabrics for diversified applications. The technique of non-fabric formation through mechanical bonding is illustrated in Fig. 4.

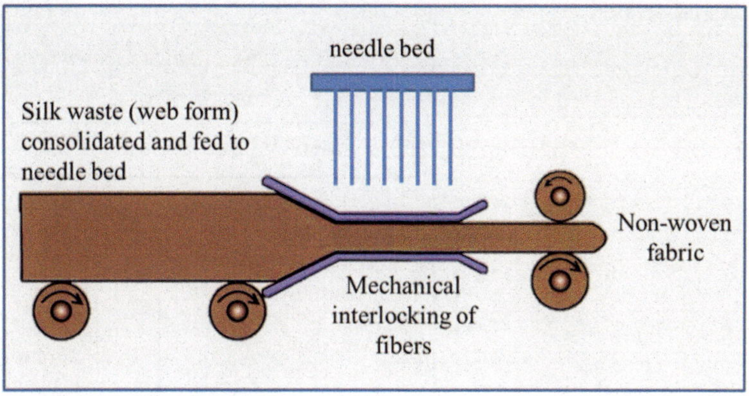

Fig. 4 Principle of needle-punching technique

Firstly, the fibrous silk waste materials are collectively processed through degumming and bleaching for removal of sericin and colouring matter. The lumps of processed waste material are further cleaned, opened and carded to form thin fibre webs. Many such webs are consolidated and mechanically bonded (orientation and interlocking of fibres achieved by a set of barbed needles passing in and out of the web) through needle-punching technique. Based on the requirement of end uses, nonwoven fabrics of specific weight and dimension can be produced. Owing to their excellent properties, these nonwoven can be used in a wide range of applications, like geotextiles for soil reinforcement, filtration, automobile, aerospace industries and medical applications.

Silk is the constituent of two major proteins, namely, fibroin and sericin. The unique structures of these proteins are featured by large domains of hydrophobic amino acids segregated by hydrophilic regions. Silk proteins being biocompatible and biodegradable with controllable morphology are most favoured materials for tissue engineering and drug delivery. When compared to other natural materials, silk proteins possess excellent mechanical properties. They exhibit high encapsulation efficiency and their drug release kinetics can be easily tuned due to their controllable crystalline β-sheet formation. The structural attributes of both the proteins are listed in Fig. 5. The silk waste can effectively be utilised for extraction of fibroin and sericin that can be used in various biomedical, cosmetic and pharmaceutical applications.

5 Summary and Outlook

Proper utilisation of secondary by-products and wastes from sericulture can generate additional income to farmers in addition to cocoon production. Silk wastes basically include agriculture waste, silkworm waste, pupal wastes and fibrous wastes. Agricultural wastes find their major route of reutilisation in biogas, composting and

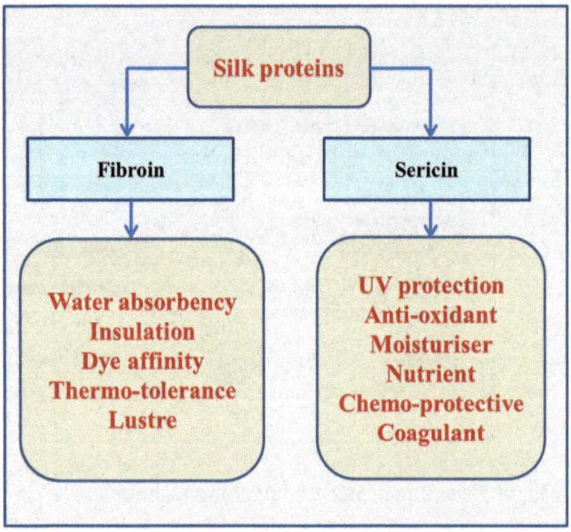

Fig. 5 Attributes of silk proteins

fodder for animals. Some of the mulberry herbs have been studied for their medicinal properties, while the studies on wild silk host plants are yet to be conducted. The silkworms are being widely exploited for the production of recombinant proteins. They can be the major source of supplement to treat protein deficiency in livestock and poultry. Silkworm pupae being the major by-product of the silk industry are effectively utilised by the biomedical science and pharmaceutical industries.

In north-eastern states of India, Eri culture is mainly practised for rearing of silkworm pupae for human consumptions. Recent spread of Eri culture into the rest of India has called for the development of pupal preservation techniques. The Eri pupa is not considered food in the rest of India and so is a sericulture waste. Any processing and preservation method that preserves the nutritional value and flavour of Eri pupa for at least 2 weeks will be helpful in utilising the sericulture waste in an appropriate manner. Attempts are being made to develop preservation methods such as vacuum packing, pickling, canning, dry snacks, etc. If successful, such an exercise will save tonnes of prepupa of Eri silkworm from going into waste in the rest of Indian states.

Most of the research findings reported to date are involved in the exploration and attributes of Mulberry silk. Considering the potential of wild or Vanya sericulture resources, the future research should focus more on utilising them to create new and valuable products that can actually benefit small-scale farmers in their social and economic upliftment. The development of sericulture in this way may indirectly facilitate reaching the target of global silk production.

Acknowledgement We are thankful for the Central Muga Eri Research and Training Institute (CMERTI), Central Silk Board (CSB), Lahdoigarh, Jorhat, Assam, for providing necessary support for this work.

References

Armand D, Meuret M (1995) Selection and utilization of cultivated fodder trees and shrubs in Mediterranean extensive livestock production systems. Avignon, France, Institut National de la Recherche Agronomique. Rapport Final, 1991–1994

Bhatia A, Kumar A, Kumar V, Jain N (2013) Low carbon technologies for sustainable agriculture. Indian Farming 63(2):18–22

Buhroo ZI, Bhat MA, Kamili AS, Ganai NA, Bali GK, Khan IL, Aziz A (2018a) Trends in development and utilization of sericulture resources for diversification and value addition. J Entomol Zool Stud 6(4):601–615

Buhroo NZI, Mushtaq N, Javid K, Rasool S, Mir GM (2018b) Medicinal values of products and by products of sericulture. J Pharmacog Phytochem 6(5):1388–1392

Chaubey R, Singh J, Baig MM, Kumar A (2019) Recent advancement and the way forward for *Cordyceps*. In: Yadav A, Singh S, Mishra S, Gupta A (eds) Recent advancement in white biotechnology through fungi: fungal biology. Springer, Cham, pp 441–474. https://doi.org/10.1007/978-3-030-14846-1_15

Dandin SB, Jayaswal J, Giridhar K (2000) Handbook of sericulture technologies. Central Silk Board, India, p S5-60

Dobre P, Nicolae F, Matei F (2014) Main factors affecting biogas production – an overview. Rom Biotechnol Lett 19:9283–9296

FAO (1993) Possibility of combined utilization of *Morus alba* and *Trifolium subterraneum* in the Tuscan Maremma (Italy). In Management of mediterranean shrublands and related forage resources. REUR technical series Rome, vol 28, Rome: Food and Agriculture Organization of the United Nations pp 206–209

Gavia MH, Couto RC, Oyama LM et al (2003) Diets rich in polyunsaturated fatty acids effect on hepatic metabolism in rats. Nutrition 19:144–149

Ghosh MK (2005) Utilization of silkworm pupae in pisciculture. Indian Silk 44:6–11

Grover JK, Yadav S, Vats V (2002) Medicinal plants of India. A review. J Ethnopharmacol 81:81–100

Harris WS, Ginsberg HN, Arunakul N et al (1997) Safety and efficacy of Omacor in severe hypertriglyceridemia. J Cardiovasc Risk 4:385–391

International sericulture commission (2019). https://inserco.org/en/statistics. Accessed 15 May 2019

Iyengar MNS (2002) Recycled silk wastes as feed integrated for poultry. Indian Silk. 41(5):30

Jain SK, De Fillips A (1991) Medicinal plants of India, Reference Publication, Inc., Algonac Michigan 2:438

Kalaiyarasan V, Nandhini UD, Udhayakumar K (2015) Seri waste vermicompost- a trend of new sustainable generation – a review. Agric Rev 36(2):159–163

Kamili AS, Mosoodi AM (2000) Principle of temperate sericulture. Kalyani Publishers, New Delhi, pp 105–116

Karthikeyan A, Sivakumar N (2007) Sericulture pupal waste — a new production medium for mass cultivation of *Bacillus thuringiensis*. Indian J Biotechnol 6:557–559

Kiran EU, Trzcinski AP, Ng WJ, Liu Y (2014) Bioconversion of food waste to energy: a review. Fuel 134:389–399

Kumar A, Gupta DK, Kumar M (2013) Green manure crops: a boon for agricultural soil. Int J Agric Environ Biotechnol 6(1):193–198

Kumar D, Dev P, Kumar RV (2015) Biomedical applications of silkworm pupae proteins. In: Kumar D, Kundapur RR (eds) Biomedical applications of natural proteins. Springer briefs in biochemistry and molecular biology. Springer, New Delhi, pp 41–49. https://doi.org/10.1007/978-81-322-2491-4_3

Kwon MG, Kim DS, Lee JH, Park SW, Choo Y et al (2012) Isolation and analysis of natural compounds from silkworm pupae and effect of its extracts on alcohol detoxification. Entomol Res 42:55–62

Łochyn'ska M, Frankowski J (2018) The biogas production potential from silkworm waste. Waste Manag 79:564–570

Longvah T, Mangthya K, Ramulun P (2011) Nutrient composition and protein quality evaluation of eri silkworm (Samia ricinii) prepupae and pupae. Food Chem 128:400–403

Mao C, Feng Y, Wang X, Ren G (2015) Review on research achievements of biogas from anaerobic digestion. Renew Sust Energy Rev 45:540–555

Mukherjee J, Mridha N, Mondal S, Chakraborty D, Kumar A (2018) Identifying suitable soil health indicators under variable climate scenarios: a ready reckoner for soil management. In: Bal S, Mukherjee J, Choudhury B, Dhawan A (eds) Advances in crop environment iInteraction. Springer, Singapore, pp 205–227

Naik SV, Hariraj G, Lakshmipattiah BN, Nandiger GS (1992) The silk waste raw material for spun silk industry. Indian Silk 31(8):15–17

Niu Q, Qiao W, Qiang H, Hojo T, Li YY (2013) Mesophilic methane fermentation of chicken manure at a wide range of ammonia concentration: stability, inhibition and recovery. Bioresour Technol 137:358–367

Patil SR, Amena S, Vikas A, Rahul P, Jagadeesh K, Praveen K (2013) Utilization of silkworm litter and pupal waste-an eco-friendly approach for mass production of *Bacillus thuringiensis*. Bioresour Technol 131:545–547. https://doi.org/10.1016/j.biortech.2012.12.153

Rao PU (1994) Chemical composition and nutritional evaluation of spent silkworm pupae. J Agric Food Chem 42:2201–2203

Rathore MS, Srinivasulu Y (2018) Vermicomposting bed types for recycling of seri cultural waste. Int Res J Eng Technol 5(8):1484–1488

Rathore MS, Singh M, Panwar D, Shekhawat NS (2007) Vermicomposting: towards clean and green farming. Green Farming 1(1):28–30

Sarmah MC (2011) Eri pupa: a delectable dish of North East India. Curr Sci 100:279

Sharma AR, Mittra BN (2007) Effect of different rates of application of organic security in harmony with nature. In: Shirashankar K (ed) Held at UAS, Bangalore during 1st–4th Dec, p 97

Singh B, Goel GC, Negi SS (1984) Effect of supplementing mulberry (Morus alba) leaves ad libitum to concentrate diets of Angora rabbits on wool production. J Appl Rabbit Res 7:156–160

Singhal BK, Dhara, Tripathy PM, Gadry SM, Bindroo BB, Ahsan MM (2001) Medicinal utilities of mulberry and non-mulberry food plants of the silk worm. In: Recent progress in medicinal plants. Research periodicals and book publishing house, Texas

Sinha PS, Angina R, Singh BD, Griyaghey UP, Sharatchandra B, Sinha BRRP (2005) Studies on the vermiculture technique and efficacy of Vermicompost in substituting NPK and FYM requirements of *Morus alba* L. Indian J Agric Res 39(4):235–241

Takahashi R (1998) Sericicultura: Amoreira (Morusalba L.) Bicho da Seda (Bombyx mori L.). Jaboticabal: FCAV UNESP, 135 (Apostila)

Uzakova DU, Kolesnik AA, Zherebin YL, Evstigneeva RP, Sarycheva IK (1987) Lipids of mulberry leaves and of mulberry silkworm excreta. Chem Nat Compd 23:419–422

Velayudhan K, Balachandran N, Sinha RK, Kamble KC (2008) Utility of silkworm pupae: a new dimension as food and medicine. Indian Silk 47(1):11–18

Venkatesh Kumar R, Chauhan S (2008) Mulberry: life enhancer. J Med Plant Res 2:271–278

Wang X, Zhang J, Li Y, Liangyu P, Tang T (2016) The status of quantitative analysis and utilization of various components in silkworm excrement. Sci Sericult 42:918–925. (in Chinese)

Zhou J, Han D (2006a) Proximate, amino acid and mineral composition of pupae of the silkworm Antheraea pernyi in China. J Food Comp Anal 19:850–853

Zhou J, Han D (2006b) Safety evaluation of protein of silkworm (Antheraea pernyi) pupae. Food Chem Toxi 44:1123–1130, 11

Inimical Effects of Arsenic on the Plant Physiology and Possible Biotechnological Solutions to Mitigate Arsenic-Induced Toxicity

Pinki Sharma, Monika, Khushboo Goyal, Tarun Kumar, and Nar Singh Chauhan

Contents

P. Sharma · Monika · K. Goyal · T. Kumar · N. S. Chauhan (✉)
Department of Biochemistry, Maharshi Dayanand University, Rohtak, Haryana, India
e-mail: nschauhan@mdurohtak.ac.in

M. Naeem et al. (eds.), *Contaminants in Agriculture*,
https://doi.org/10.1007/978-3-030-41552-5_20

1 Introduction

Arsenic is one of most abundant elements on Earth with an approximate concentration of 1–2 mg/kg in the Earth's crust. Even though it is present in traces, arsenic is broadly found in nature along with the other metal ores like gold, iron, lead, copper, etc. (Meliker et al. 2008). Along with this, it may be naturally produced through volcanic eruptions and weathering and may be found as aerosols in water and air or as suspended particulates (Bhattacharya et al. 2007). Anthropogenic resources such as coal combustion, dyes and paints, tanning wastes, etc., are also responsible for arsenic presence in various forms such as copper chromate arsenate. In the 1930s, the introduction of arsenic-fed pesticides like calcium arsenate and dimethylarsonate contributed to environmental arsenic load with an average of 10,000 metric tons annually (Tollestrup et al. 1995). Till now, many arsenic-based insecticides and pesticides are being used in agriculture. Traces of free arsenic are present in soil, water, and living communities.

Arsenic may be found as As (-III), As (0), As (III), and As (V). So, it is also named as a metalloid. Arsenic is also present in the gases produced by the anoxygenic conditions. The anoxic and aerobic environments accommodate different arsenic forms with the prevalence of inorganic form As (V) as $H_2AsO_4^-$ and $HAsO_4^{2-}$ in aerobic and aqueous conditions and As (III) as H_3AsO_3 and $H_2AsO_3^-$ in anoxic environment. The presence of arsenic in water is affected by pKa, salinity, pH, and redox potential. The arsenic forms can be interconverted through oxidation-reduction reactions. The inorganic arsenic may be transformed through biomethylation to form organo-arsenic which is majorly present in living organisms including plants and humans.

Due to the water contamination by arsenic, majority of the population is facing arsenic toxicity. Arsenic affects the cell metabolic activities and overall cellular development. The chemical form determines the severity of toxic effects. As arsenate is similar to phosphate, thus, it competitively inhibits energy-generating reactions such as oxidative phosphorylation. It enters the cell through phosphate transporters. As As (III) has no charge at pH less than 9.2, it makes its entry to the cell through aquaglyceroporins. Arsenite is comparatively more noxious as it reacts with sulfhydryl groups and alters protein functioning. It may also bind vicinal thiol moieties present in 2-oxoglutarate dehydrogenase and pyruvate dehydrogenase and thereby alter respiratory process. Arsenite can also deteriorate DNA and affect the replication and repair processes. Thus, arsenic affects many cellular aspects through effect on various pathways and processes and may lead to inhibition of growth, differentiation, apoptosis, and induction of apoptosis.

Biomolecules like proteins, lipids, etc., are rapidly destroyed by arsenic (Gunes et al. 2009). As (V) exposure increases certain stress in plants such as growth inhibition (Stoeva and Bineva 2003) and physiological deterioration (Stoeva et al. 2005) leading to death as As (V) clamps down the P_i/arsenic uptake system (Smith et al. 2010). In addition, the reaction of arsenite with the thiol group results in the modification of the structure of proteins (Wang and Wai 2004), which leads to the loss of activity of most of the enzymes (Akter et al. 2005). Even at a very low concentration, arsenic causes many changes at biological and molecular level and affects the productivity of plants (Chandrakar et al. 2016a). In soil–plant system, arsenic toxicity level modifies with the plant species. At the subcellular level, arsenic leads to free radical formation and ROS generation (Rafiq et al. 2017). These ROS species affect the plant metabolism by deteriorating the macromolecules, including DNA, proteins, lipids, and carbohydrates. The ROS production is because of formation of As (III) (Chandrakar et al. 2016a).

Moreover, As (III) causes deterioration of morphological aspects of the plants (Adriano 2001), reduction of root hair growth, cortical region deterioration (Singh et al. 2007), and thylakoid membrane rupture (Li et al. 1994), thus causing cell death. As (V) competitively binds to biomolecules, and As (III) disturbs the various biochemical reactions through the –SH moieties of enzymes (Ullrich-Eberius et al. 1989). In this chapter, we have discussed about arsenic mobilization, distribution, accumulation, toxicity, and mitigation strategies in detail.

2 Arsenic Uptake in Plants

2.1 Mechanism of Arsenic Uptake

Terricolous plants primarily uptake arsenic with the help of absorption through roots, although some of the submerged plants absorb arsenic dissolved in water by their leaves (Wolterbeek and Meer 2002). Plants usually have a less concentration of arsenic (~1.0 mg/kg dry weight) (Adriano 2001). This concentration gradient facilitates arsenic uptake in plants. Arsenic uptake mechanism in plants varies with arsenic species (Fig. 1). Plants possess both high- and low-affinity transporters for arsenic (Mirza et al. 2014). Arsenic translocation occurs through the xylem tissues (Kumar et al. 2015). Within plant cell, arsenic reductase reduces As (V) (Bleeker et al. 2006).

2.2 Arsenate Uptake

Plant cells uptake As (V) by the P_i channels (Lei et al. 2012) as phosphate is chemically analogous to oxyanion As (V). These P-channels possess different P_i transporter proteins (PHT) for As (V) uptake (Nussaume et al. 2011; LeBlanc et al. 2013). Many studies reported that plants have both high- and low-affinity transporters of phosphate. One of the high-affinity phosphate transporters is PHT1 which may persuade low-affinity transportation, while low-affinity transport proteins are still unidentified. The arsenic transporter expression in roots causes the uptake of P_i (Zhao et al. 2009). These transporters may also be expressed in the plasma

Fig. 1 Arsenic uptake and distribution in plants

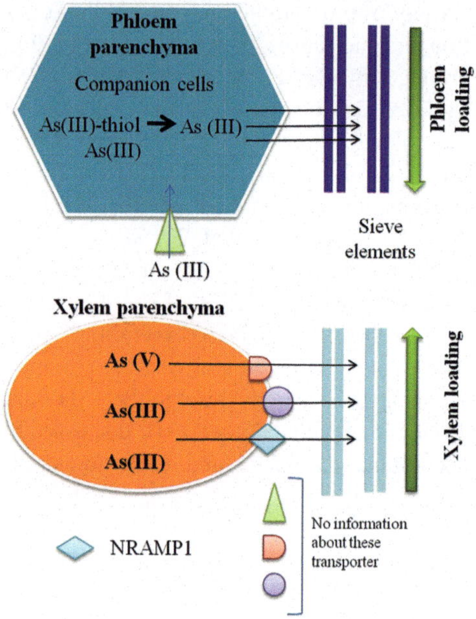

membrane like the P_i transporters, AtPHT in *Arabidopsis thaliana* (Catarecha et al. 2007; LeBlanc et al. 2013), P_i transporters of *Oryza sativa* (Wu et al. 2011; Jia et al. 2011), and P_i transporters in *Lotus japonicus* (Bienert et al. 2008) or may be present in the tonoplast such as the P_i transporters in *Pteris vittata* (Indriolo et al. 2010).

The dicarboxylate carriers present in the inner mitochondrial membrane also transport As (V) (Palmieri et al. 2008). In *A. thaliana*, two isoforms of high-affinity Pht1 transporters, AtPht1:1 and AtPht1:4, are present, both of which are responsible for transport of P_i and arsenate (Shin et al. 2004).

2.3 Uptake of Arsenite

Majority of the arsenite is found in submerged soil. Arsenite is taken up by plants through various nodulin-26-like intrinsic proteins (NIPs). Many studies account that the PHT transporters work unidirectionally (transport molecules in one direction), whereas the NIP transporters transport molecules in both directions. So, movement of As (III) occurs bidirectionally between the plant cells and growth. Some of the plant aquaporins and the membrane channels, which are responsible for water transport and movement of neutral molecules, also play a role in the As (III) transport (Li et al. 2014; Mukhopadhyay et al. 2014). NIP5;1 and NIP6;1 are present in *A. thaliana*, and NIP2;1 is present in *O. sativa* (Bienert et al. 2008). These are expressed in the roots. Suspension cells and anthers are known to express NIP3;2, while *Lotus japonicus* shows expression of NIP5;1 and NIP6;1 on the cell membrane. In addition, in developing anthers, AtNIP7;1 is present, while AtNIP1;1 expresses more in roots, and seeds express AtNIP1;2. AtNIP3;1 also expresses in roots and is responsible for As (III) transport (Xu et al. 2015).

OsLsi1 reported in *O. sativa* is a type of NIP2;1 transporter. It is located distal to the exodermis and endodermis which influx As (III) through roots (Ma et al. 2006, 2008). OsNIP2;2 (OsLsi6), OsNIP1;1, and OsNIP3;1 are expressed less as compared to OsNIP2;1, thus making them less prone to As (III) uptake by roots as reported in rice (Ma et al. 2008). MIP family including OsPIP2;4, OsPIP2;6, and OsPIP2;7 which are the members of intrinsic protein of rice plasma membrane also play a role in As (III) transport (Mosa et al. 2012).

2.4 Uptake of Methylated Arsenic Species

The organic arsenic forms are slowly taken up than the inorganic forms of arsenic, while it is transported fast from root to shoot (Chandrakar et al. 2016a). The Si influx and efflux transporters, namely, Lsi1 and Lsi2, also transport one of the organic species of arsenic (Mitani-Ueno et al. 2011). Recently, it has been reported that the transport of DMA is more than the transport of As (V) and MMA (Raab et al. 2007). By the usage of arsenic pesticide and herbicide, there is a reduction in the methylated arsenic species, namely, monomethyl arsenic (MMA) and dimethyl arsenic (DMA). Many studies revealed that methylated arsenic originated from microorganisms which lie in the rhizosphere of plants (Lomax et al. 2012). There is no clear information about the transport and the uptake of methylated arsenic. The aquaporin NIP2;1 is vital for taking up undissociated methylated form via roots (Li et al. 2009). Moreover, monomethyl and dimethyl forms of arsenic enter the same way as glycerol, that is, by roots, followed by further transport in the plant cells with the help of aquaporins (Rahman et al. 2011).

3 Arsenic Impact on Plant Physiology

Arsenic causes several physiological changes in the plants (Smith et al. 2010) such as metabolic disturbance, plant growth inhibition, and ultimate damage to the crop yield (Miteva and Merakchiyska 2002; Stoeva and Bineva 2003) (Table 1 and Fig. 2).

Table 1 Physiological and biochemical changes in plants after arsenic exposure

Plant species	Effects	References
Cicer arietinum L.	Reduction in essential and nonessential amino acids and Fe concentrations. Overexpression of dehydration-responsive genes (MIPS, bPGIP, and DRE). Reduction in antioxidant enzyme activities (GR, CAT, SOD, APX, and GPX)	Tripathi et al. (2017)
Zea mays L.	Exposure of arsenic leads to increase the CAT activity and glutathione-S-transferase	Mylona et al. (1998)
Zea mays L.	Increased shoot arsenic and P concentrations, reduction in pigment concentration (chlorophyll a, chlorophyll b, and total chlorophyll), and gas exchange attributes	Mehmood et al. (2017)
Brassica napus and *Brassica juncea*	Reduction in leaf area, plant height, number of leaves, shoot and root dry biomass, photosynthetic rate, transpiration rate, stomatal conductance, photosynthetic pigments, and water use efficiency	Niazi et al. (2017)
Vigna mungo L.	Decrease in Chlorophyll a, chlorophyll b, total chlorophyll, and carotenoids; increased lipid peroxidation. The activities of antioxidative enzymes (SOD, POD, and APX), except CAT, were increased	Srivastava et al. (2017)
Glycine max	Changed expression of a phosphatidic acid via phospholipase D and phospholipase C.	Talukdar (2017)

3.1 Photosynthesis

Photosynthesis is a vital process in plants. It serves as the largest solar source. Arsenic reduces the photosynthetic rate in plants (Gusman et al. 2013). Due to arsenic toxicity, there is a shift in the adjustment of both the photosystems due to which chlorophyll content declines. It causes injury to the chloroplast membrane, thereby altering the photosynthesis (Rafiq et al. 2017). Due to arsenic, there is a reduction in carbon dioxide fixation, and the PS-II activity also reduces to a greater extent (Stoeva and Bineva 2003), which in turn blocks the flow of electrons destined to generate electrochemical potential for ATP and NADPH production through the thylakoid membrane. As (V) inhibits ribulose-1,5-bisphosphate carboxylase/oxygenase (RuBisCo) activity (Ahsan et al. 2010) by retarding the plastid-encoded large subunit (Bock 2007) such as in rice but in other cases like in *Arabidopsis*, the small subunit of RuBisCo gets affected by As (V) exposure. This increase or decrease in the amount of RuBisCo has a negative impact on plant photosynthesis (Abercrombie et al. 2008). Arsenic causes a change in the rate of gaseous exchange as it decreases the heat dissipation from plants and negatively affects the photosynthetic efficiency (Debona et al. 2017).

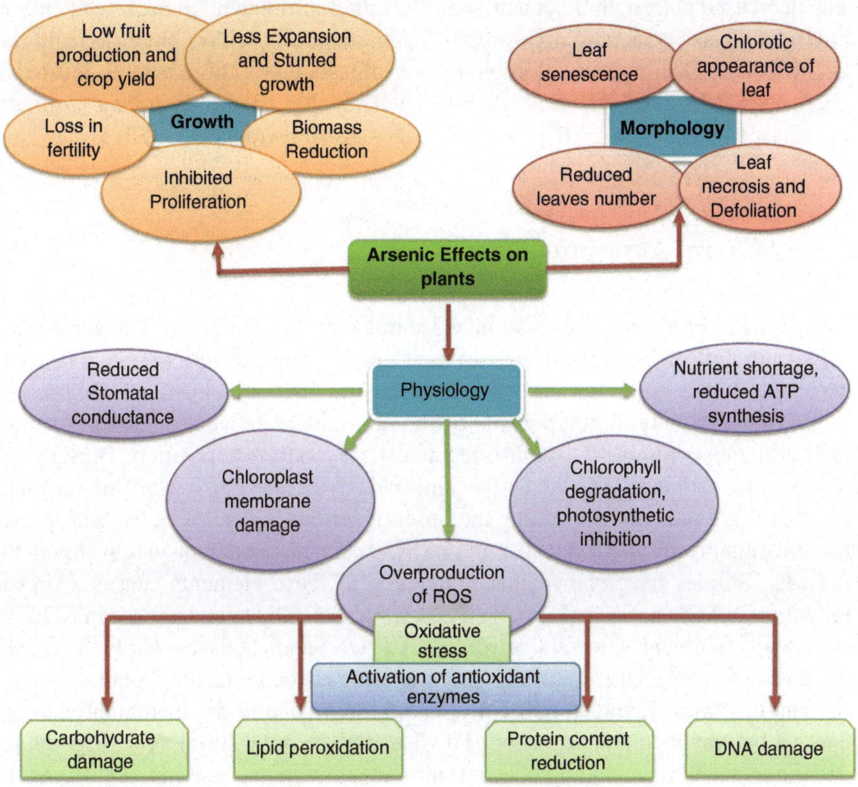

Fig. 2 Physiological and biochemical effects of arsenic in plants

3.2 *Chlorophyll Content*

The chloroplast is the photosynthetic pigment which plays a major role in photosynthesis. Chloroplast has two different chlorophylls, chl-a and chl-b. Both types of chlorophylls differ from each other due to different absorption wavelengths. Some factors such as conductance in stomata, amount of chlorophyll, and specific leaf weight affect the photosynthetic rate per leaf (Peng 2000). The soil arsenic concentration varies inversely to the chl-a and chl-b, and this relationship exists among all rice varieties. There is an inverse relationship between arsenic concentration and photosynthesis. Arsenic surge may alter the shape of the chloroplast, manifesting its round shape by decreasing the longitude of the cell. Overall arsenic toxicity forms the concave cell membrane, thereby hindering the flow and accumulation of molecules altering the chlorophyll amount in rice leaf (Miteva and Merakchiyska 2002). As arsenic concentration increases, the photosynthetic activity of plant decreases. The declined chlorophyll concentration owes to decreased activity of δ-aminolevulinic acid dehydrogenase facilitating the biosynthesis of chlorophyll.

The decreased chlorophyll content could be either attributable to decreased chlorophyll synthesis or increased chlorophyll degradation by chlorophyllase (Jain and Gadre 2004). Decrease in chlorophyll biosynthesis due to high arsenic levels has also been observed in *Lactuca sativa* (Suneja 2014), *Trifolium pratense* L. (Hasanuzzaman et al. 2017), and *Zea mays* (Emamverdian et al. 2015).

3.3 Mineral Absorption

The presence of arsenic causes an increase in the absorption of Cu, Mn, and Fe, but the accumulation of arsenic decreases their uptake (Päivöke and Simola 2001). Cu, Mn, and Fe are micronutrients that have several functions in plants (Marschner 1995), being structural components of the antioxidant enzymatic system of plant cells. The most prominent arsenic effect to cells is the ROS generation. Thus, upregulation of enzymes involved in the antioxidant metabolism occurs to scavenge excess of free radicals, preventing the onset of serious injuries. Cu, Fe, and Mn are the constituents of superoxide dismutase (SOD), the main enzyme in scavenging the radicals. Studies have shown that deficiency in these elements causes a drastic reduction in SOD activity (Leao et al. 2013), which leads to an increase in ROS. Fe is also a constituent of several peroxidases (POX) and catalases (CAT). Thus, the observed increase in the uptake of these nutrients is due to the upregulation of antioxidant enzymes. Reduced nutrient uptake may be due to the membrane damage caused by the metalloid toxicity (Päivöke and Simola 2001). The Mg uptake decreases when the metalloid concentration in solution increases as As disturbs the oxidative phosphorylation and the overall chlorophyll composition which is one of the main toxic influences of the arsenic (Marschner 1995).

3.4 ATP Synthesis

As As (V) is chemically similar to P_i molecules, the processes involving phosphate get disrupted during ATP synthesis (Singh 2005). Although the affinity of P_i transporters for phosphate is more, yet As (V) competes with P_i to bind these transporters (Tawfik and Viola 2011). Thus, higher amount of As (V) than the P_i binds to the root surfaces, resulting in the formation of unstable adducts of As (V). The As (V) forms a complex with ADP, thus uncoupling the oxidative phosphorylation and the photophosphorylation due to which the capability of cells to produce the ATP reduces (Singh 2005; Tawfik and Viola 2011). Metabolism of nucleic acids and phospholipids and other processes like oxidative phosphorylation, protein phosphor/dephosphorylation, and glycolysis are also affected by As (V) (Finnegan and Chen 2012). ATP formation from ADP requires Pi. Due to the arsenic abundance and deficiency of P_i within the cell, the mitochondrial enzymes form ADP-arsenic (Gresser 1981). On the other hand, some enzymes of mitochondria, that is, F_1F_o-type ATP synthase, show equal compatibility to both As (V) and Pi (Watling-Payne and Selwyn 1974); therefore, As (V) has a crucial role in energy and ATP production of the cell.

3.5 *Membrane Damage*

Under arsenic stress, plasma membrane undergoes severe damage (Sharma et al. 2012). This damage occurs due to reduced uptake of minerals and water which alters stomatal conductance owing to the alteration in transpiration (De Oliveira et al. 2013). Arsenic toxicity retards transpiration in *A. sativa* (Evangelou et al. 2007). Cellular membrane destabilizes due to arsenic stress in *Pteris ensiformis L.* as well as *Pteris vittata L.* seedlings (Singh et al. 2006). In arsenic-contaminated seedlings of *Phaseolus aureus*, electrolyte leakage in the membrane and an increased lipid peroxidation have been observed (Singh et al. 2007).

3.6 *Plant Growth*

Arsenic disturbs the normal functioning of the plants, thereby reducing the productivity and leading to the stunted growth of plants (Shahid et al. 2015). Suitable amount of arsenic proves to be beneficial in promoting plant growth, but high concentration may overcome the benefits (Azam and Gousul 2015). The application of a high concentration of arsenic to *Helianthus annuus* leads to plant death because of altering critical metabolic and biochemical processes (Imran et al. 2013). Plant height also decreases with arsenic-contaminated irrigation water (Monteiro et al. 2012). *Oryza sativa* L. and *Cicer arietinum* L. seedlings also show stunted shoot and roots (Vromman et al. 2013; Malik et al. 2011). Arsenic also causes the decrease in fresh and dry mass of plants and overall number and area of leaf surface (Nath et al. 2014). It causes necrosis, wilting, and curling of leaves (Finnegan and Chen 2012) and lowers the rate of photosynthesis and fruit yield (Duman et al. 2010). The solubility and toxicity of arsenite exceed arsenate (Sachs and Michael 1971). Arsenic treatment makes the plant deficient in phosphorylated compounds, while arsenite treatment can cause the sudden loss of principal enzymes involved in oxidative phosphorylation. Furthermore, arsenite leads to cell damage and ultimately cell death as it is prone to react with the sulfhydryl group of enzymes. The growth reduction of plants caused by the arsenic application has been observed by many investigators (Woolson and Kearney 1973).

4 Arsenic and Cellular Metabolism

Disturbance in cellular biochemical and metabolic pathways (e.g., hampered absorption of nutrient, the retarded rate of photosynthesis, interruption in plant water repute, and substituent of essential ions from ATP (Khalid et al. 2017, Rafiq et al. 2017)) is due to arsenic toxicity (Fig. 2).

4.1 *Oxidative Stress*

Arsenite binds adjacent to sulfhydryl/thiol (SH) groups present in proteins and disturbs their main catalytic structure and functions. Toxic properties of arsenate (AsV) is due to its reduction in oxidation forms i.e. arsenite (AsIII) (Hughes 2002). Arsenate exposure creates harmful reactive ROS in plant tissues and induces oxidative stress (OS) such as genotoxicity and peroxidation of lipid in cell membrane. A number of antioxidants, enzymatic and nonenzymatic, are involved in the antioxidant responses upregulated in the plants on exposure of arsenite (AsIII). Reduction of GSH in the cell may be the cause of arsenic-triggered OS (Mylona et al. 1998). Toxic ROS formed as a result of arsenic stress prerequisite to be scavenged to maintain proper development of a normal growing cell.

4.2 *Arsenic Aggravates ROS Generation*

Under arsenic stress, plants can generate oxidative stress, which makes disturbance between formation of ROS and scavenging of toxic ROS. ROS are very toxic free radical oxygen-containing particles, are unstable, are chemically reactive, and possess single electron in their valence orbital shell. Inside plant cellular organelles mitochondria, a series of complexes, among them complex I and complex III of electron transport pathway, are well-known sites of reactive free radical (O^-) generation (Keshavkant and Naithani 2001). Here, primarily two pathways are involved in O_2 consumption: (i) respiratory enzymes, among them cytochrome oxidase/complex IV, used up 95% of overall oxygen to produce water; and (ii) NADH dehydrogenase segment containing iron–sulfur (Fe–S) center or flavoprotein reduced oxygen directly.

4.3 *Vacuolar Sequestration of Toxic Metal Complexes*

The phytochelatin–arsenite complexes (PC–arsenite complexes) are expected to be stored in enclosed compartments of plant cell known as vacuoles. The uptake of arsenite-reduced glutathione (As(III)-$(GS)_3$) complexes occurs with the help of vesicles that are formed from *H. lanatus* plant roots in a neutral charge condition, and ATP-dependent process occurs via ATP-binding cassette (ABC) family transporter system. Additionally transport also occurs by ABC protein, that is for PC–arsenite complexes transport into the vacuoles. *P. vittata* fronds store arsenic in arsenite form. Arsenite (AsV) transportation across the vacuolar membrane probably involves an ATP-dependent active transport mechanism due to concentration gradient difference between the vacuoles and cytoplasm that resides within the cell.

4.4 Lipid Metabolism

Arsenic stress has detrimental effect on the integrity of the cell membranes through oxidation of membranous lipids (Jomova and Valko 2011; Clemens and Ma 2016). If level of ROS generation in cell is above a tolerable limit, it leads to peroxidation of lipid moieties (Parkhey et al. 2012). This causes the formation of lipid-derived oxidative products such as cytotoxic peroxide radicals, which affect the proper functioning of cell organelles or cell tissue. Mostly, initiation, propagation, and termination are three main steps that occur in lipid peroxidation process. In cellular membranes, OH radical generation leads to the peroxidation of lipid molecules (Gill and Tuteja 2010). Lipid peroxidation disturbs cell membrane structure and results in increase in leakiness and decrease in the membrane fluidity, causing serious damage to membrane proteins. Various toxic oxidative derivatives such as 4-hydroxynonenal (4-HNE) and malondialdehyde (MDA) occur as enol, keto, and epoxy fatty acids upon lipid peroxidation, and these harmful oxidative products can form complexes with macromolecules DNA and proteins and cause cell death (Das and Roy 2014).

4.5 Carbohydrate Metabolism in Plants

Plant exposure to arsenic has a deleterious effect on prime simple carbohydrate metabolism of sugars and starch (Chandrakar et al. 2016b). Soluble sugars can accumulate inside the plant in response to arsenic stress. It is observed that in arsenic-exposed plant *Oryza sativa* (rice), the ratio of both reducing and nonreducing carbohydrates decreases in shoots (Jha and Dubey 2004). Mainly sucrose non-reducing sugar that transforms into reducing sugars has been detected during arsenic exposure. It can also imitate sucrose synthesis suppression compared with available hexose monophosphate. Additionally, carbohydrate-metabolizing enzymes like amylases (α-amylase and β-amylase) and starch phosphorylase that is starch-degrading enzymes showed strong inhibition in their activities under arsenic-exposed conditions and caused impairment in carbohydrate-metabolizing pathway. Furthermore, it was examined that the activities of sucrose-hydrolyzing enzymes, sucrose synthase and acid invertase, upregulated with the suppression of the activity of sucrose phosphate synthase, under in situ (subsurface) arsenic toxicity.

4.6 Protein Metabolism

Both forms of arsenic such as arsenite and arsenate are toxic to plant and can interrupt with metabolism of plant via the distinct mechanism, but reduction product of arsenate, arsenite, is 100-fold more lethal than arsenate. The excessive toxicity of arsenite is due to its affinity of binding with thiol (SH) groups present in the proteins, resulting in membrane destruction, and subsequently cell damage occurs (Meharg and Hartley-Whitaker 2002). In *P. ensiformis*, *P. vittata*, *Oryza sativa*, *Zea mays*, *Trifolium pretense*, and *Vigna radiata* plant species exposed to arsenic, there was a reduction in total protein. Within the cell, protein hydrolysis is the most important and major part, which provides different turnovers of free amino acids of individual proteins. It has been observed in *Oryza sativa* plant that both nitrate and nitrite reductase enzyme activities suppressed as a result of arsenic toxicity. Generally, proteins are hydrolyzed into free amino acids or fragment into short peptides by the action of protease and peptidase enzymes. In arsenic-exposed plants, reduction of enzymes especially proteases has been observed (Mascher et al. 2002). In addition to protease based protein hydrolysis, lipid peroxidation-derived lipid peroxide moieties affects adjacent proteins within the cell and also make them susceptible to ROS attack. The ROS produced in response to arsenic stress can modify proteins, by forming more toxic oxidation products such as carbonyls (Parkhey et al. 2014). Due to arsenic stress, free carbonyl groups are formed upon amino acid oxidation, specifically arginine (R), histidine (H), lysine (L), proline (P), threonine (T), and tryptophan (W), which may alter, inhibit, or impair the protein activities and protein catalytic structure. In turn, protein fragmentizes into shorter peptides through proteolytic attack due to arsenic stress in a plant (Parkhey et al. 2014).

4.7 DNA Structure

It has been reported that exposure of excessive arsenic on plant induces genotoxic responses due to some harmful oxidative reactions inside the cell. Various studies have proposed that initially during biotransformation reactions, ROS species impair with the base pair of DNA and cause genotoxicity due to arsenic stress in the plant. Arsenic genotoxicity is associated with the production toxic reactive oxygen species during its biotransformation (Dalle-Donne et al. 2006). Thus, the generated toxic ROS can form adducts of DNA–protein and cause oxidative nitrogenase base damage of DNA, a breakup or exchange of chromosome and chromatid, the formation of apurinic/apyrimidinic (AP/abasic) sites, and DNA lesions (Faita et al. 2013). DNA–protein conjugate formation, chromosomal abnormalities, exchange of chromatid, aneuploidy, and deletion result in DNA repair mechanism fails, DNA lesions, DNA mutations, and DNA damage (Kitchin and Wallace 2008). Another important impact of ROS attack on plant DNA is its base modification from the DNA structure (Ziech et al. 2010). Base modification within the DNA structure, for example, 8-oxoguanine (8-OHdG), is highly miscoding that leads to mutagenic lesions and is responsible for nitrogenous base (G:C to T:A) transversion mutations. The accumulation of 8-hydroxyguanine/8-oxoguanine (8-OHdG) adducts, in response to arsenic exposure, has been assessed in numerous plant tissues (Vizcaya-Ruiz et al. 2009). Likewise, single- and double-strand nick in DNA can be produced directly by toxic ROS or indirectly during base excision repair (BER) mechanism (Kligerman et al. 2010). In addition, telomere shortening and inhibition in DNA repair methods, such as nucleotide and base excision repair, are also probable causes of arsenic-induced genotoxic effects in exaggerated plant tissues.

5 Proteomic Responses of Plants to Arsenic Stress

Until now, not much information is available for the proteomic studies as regards the effects of arsenic on the plants. The proteomic study is limited to only a few species of plants such arsenic maize roots, shoots, etc., and pseudo-metallophytes and arsenic accumulators. The studies have found that approximately 11 proteins play their role in cellular homeostasis using the redox reactions. Thus, arsenic is mainly responsible for oxidative toxicity in plants. One of the attempts to find the toxicity mechanism of arsenic on plants was done on rice roots (Ahsan et al. 2008), where 23 differentially expressed proteins were compared. The comparative analysis showed that these proteins are involved in several processes such as protein biosynthesis, cell signaling, defense and development, detoxification, and stress. Most of the protein activities are related to metabolism of energy, like malic enzyme which is NADP-dependent. The under-regulation of RuBisCO enzyme and ribonucleoproteins is the possible outcome for the decrease in the rate of photosynthesis.

6 Genomic Analysis of Plants on Exposure to Arsenic

Under arsenic exposure, analysis of the genome of the plants has been done to identify some of the common pathways. Genomic analysis shows the number of the various genes involved in cellular differentiation and downregulated genes involved in cellular development during arsenic toxicity. In the rice genome, lesser expression of some genes was observed on higher exposure to arsenic, and these genes were two tubulin genes (Os03g56810 and Os03g45920), one actin gene (Os01g64630), two microtubule genes (Os03g27700 and Os09g13460), and two expansion genes (Os01g46650 and Os04g14660) (Norton et al. 2008). The microarray studies show that arsenic stress sways the different pathways that are related to plant defense strategies, photosynthesis, and signaling of molecules. As (V) shows the direct impact on cell wall, primary and secondary metabolites, germination of the seedlings, and abscisic acid metabolism. On the other hand, accumulation of As (III) influences plant hormonal and signaling pathways (Chakrabarty et al. 2009).

7 Defense Mechanisms Against Arsenic in Plants

The well-organized defense system is found in plants to mitigate metal stress (Fig. 3). Plants use certain enzymes like catalase, dismutase, peroxidase, and compounds like ascorbate, glutathione, and tocopherol for subcellular and cellular protection (Gunes et al. 2009). Transporters playing an important role in mineral transport also facilitate the transport and metabolism of arsenic (Hussain et al. 2013) using various pathways and transporters for distinct arsenic forms.

To overcome the arsenic toxicity, plant system facilitates binding of metalloid with phytochelatins (PCs) (derived from glutathione). Arsenic toxicity in plant also reduces on exposure to arsenic via the reduction of cytoplasmic free arsenic by the vacuolar sequestration and addition of certain groups to AsPC complex (Zhao et al.

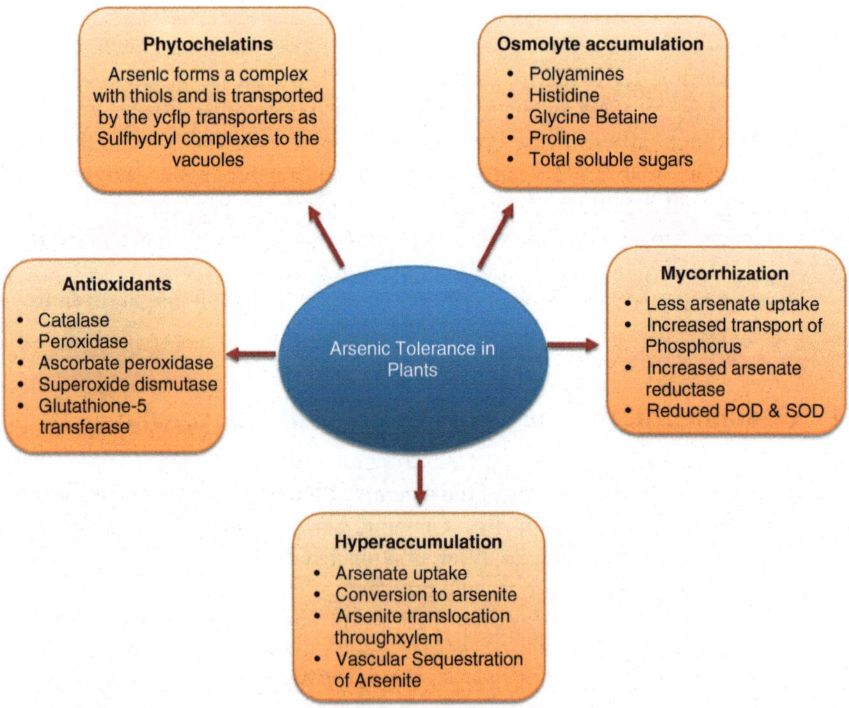

Fig. 3 Arsenic toxicity tolerance mechanisms in plants

2009; Liu et al. 2010). Plant arsenic uptake relies on the bioavailability of arsenic in the soil (Rafiq et al. 2017). Proline is an osmo-protectant as it shows maximum accumulation in the arsenic stress condition. It acts by serving cell wall protection and regulating the moisture conditions in the plant cell. Thus, it becomes important to explore the assimilation and metabolism of arsenic so as to design mitigation strategies for its removal.

8 Potential Biotechnological Strategies to Mitigate Arsenic Toxicity

8.1 Bioremediation

Bioremediation is the process of cleaning the toxicants from the environment by using wild and recombinant microbes. It facilitates better detoxification as well as the degradation of toxic pollutants either with the help of sequestration or through the conversion to nontoxic compounds (Brar et al. 2006). Most of the transgenic or naturally occurring organisms have the ability to transform and degrade toxicants so they can be used to overcome the environment pollution. Genetic, biochemical, and physiological properties of plants are used for soil as well as water pollutant remediation. New genetic tools must be designed to better understand the plant gene structures and functions.

8.1.1 Biosorption

It is the process where a biomass is used to absorb the arsenic from the aqueous medium. It is a physical and chemical interaction-based process where polysaccharide, protein, and lipid content of the cell wall acts as biosorbents and provides binding sites to arsenic (Prasad et al. 2013). The sensitivity in detecting toxic metal, biomaterial reuse, low operating cost, high efficiency, and short operating time are some of the benefits of these biomaterials (Bhargavi and Savitha 2014). The algae are more efficient in biosorption than other chemical substances. The amino, carbonyl, carboxyl, hydroxyl, and sulfhydryl moieties in its cell wall act as the binding sites for metals and metalloids (Wang et al. 2015). Ion exchange is also crucial in binding of the metals to the biosorbents (Sulaymon et al. 2013). Bacteria also have the potential to be used as biosorbents. Dead cells extracted from fermentation broth are also one of the good biosorbents. The biosorption capacity of fungi varies with the fungal species, but the functional groups responsible for metal binding are amine, hydroxyl, imidazole, sulfhydryl, sulfate, and phosphate (Dhankhar and Hooda 2011). Physical and chemical adsorption, precipitation, ion exchange, and metal-ion complex formation are involved in biosorption (Othman et al. 2013). Chemical modification may also facilitate adsorption (Vinh et al. 2015).

8.2 Phytoremediation

Phytoremediation comprises of various plant-based strategies including wild or genetically manipulated plants which are used to detoxify the environment (Ellis et al. 2006). The methods that are based on biotechnology are used in phytoremediation of most of the heavy metals and some of the metalloids like arsenic, selenium, and lead. Phytoremediation comprises of two main strategies in case of metal and metalloid toxicity.

8.2.1 Manipulation of Genes and Metal Uptake System

Metal forbearance can be accomplished by the manipulation of their transporters. It has been shown in recent studies that in plants aquaporins are also involved in arsenite transport (Bienert et al. 2008). Phytoremediation is achieved by reducing the uptake of metals by plants, enhancing metal tolerance, or pacing up the methylation process for greater food safety (Zhu and Rosen 2009). The genome-wide As (V) analysis in rice roots exposed to metalloids revealed several differentially expressed genes belonging to various transporter families, for example, genes for major facilitator superfamily (MFS) antiporter (Dubey et al. 2014).

8.2.2 Enrichment of Metalloid Ligands

There are some reports on the research of using cysteine-rich peptides as the ligands for metals to get rid of toxic substances and heavy metal accumulation. Similarly, modification of certain cysteine peptides such as GSH may enhance the tolerance to arsenic forms. For example, in *Arabidopsis*, genes expressed for γ-glutamylcysteine synthetase (*γ-ECS*) and arsenate reductase (*arsC*) show significant tolerance to arsenic (Dhankher et al. 2006). Likewise, arsenate tolerance was increased when AtPCS1 was constitutively expressed in *A. thaliana*. Similar AtPCS1 expression in *B. juncea* enhanced arsenate endurance (Garsenicic and Korban 2007). Thus, the manipulation of genes provides a platform to increase the tolerance to metals and metalloids and accumulation in plants.

8.3 Genetic Engineering

Genetic manipulation has different strategies depending on the aim of manipulation such as enhanced recognition, tolerance, etc.

8.3.1 Arsenate Reduction by Genetic Manipulation

Reduction of arsenic is the crucial phase in arsenic metabolism in plants. Phytoremediation for arsenic has been done by manipulating the arsenate reductase genes. The overexpression of *arsC* gene coding for arsenate reductase in *E. coli* along with γ-ECS gene for increased GSH biosynthesis enhanced arsenic tolerance (Dhankher et al. 2006). The arsenic is translocated slowly from roots to shoot because of arsenate reductase so the studies have been done to slow down the expression of gene coding for endogenous arsenate reductase. The silencing of arsenate reductase coding gene AtACR2 leads to the shoot arsenic hyperaccumulation (Dhankher et al. 2006). Even the phytoextraction can be enhanced by inoculating plants with certain specific bacteria which can reduce arsenic and promote the plant growth as reported in *P. vittata* (Lampis et al. 2006).

8.3.2 Genetic Modification for Vacuolar Sequestration

As discussed earlier, arsenic can be detoxified by complexing arsenic peptides rich in cysteine group like phytochelatins (Lu et al. 1997; Tommasini et al. 1998). Thus, arsenic phytoremediation is done by increasing the synthesis or accumulation of PCs and GSH. It can be done by improving phytochelatin synthase expression. Overexpression of AtPCS1 and GSH1 in *A. thaliana* resulted in increased arsenic endurance (Guo et al. 2008). Although the arsenite transport whether in complex or free form across the tonoplast affects the arsenic accumulation, no such report of genetic engineering of tonoplast transport has been shown.

8.3.3 Genetic Engineering for Volatilization

Plants have methylated arsenic formed through endogenous methylation (Wu et al. 2002), end product of which is a volatile gas trimethylarsine (TMAs(III)). The cloning and expression of *arsM* gene from *Rhodopseudomonas palustris* encoding As(III)-S-adenosylmethionine methyltransferase (*arsM*) in an *E. coli* strain sensitive to arsenic resulted in more methylated forms of arsenic and TMAs(III) (Qin et al. 2006). Thus, a single gene is enough for volatilization and tolerance to arsenic. One of the ArsM homologous genes is found in the eukaryotic alga *Cyanidioschyzon merolae* (Qin et al. 2009). Whether it is possible in higher plants or not is not clear.

8.3.4 Genetic Modification for Enhanced Translocation Between Shoot and Root

Varying arsenic translocation between root and shoot indicates the genetic regulation. The most important phase during translocation is the xylem loading. Lsi2 is an efflux protein responsible for xylem loading (Ma et al. 2002, 2006). Lsi2 mutation causes less arsenic accumulation by about 50% (Meng et al. 2004). Whether the overexpression of Lsi2 results in increased translocation between root and shoot is still not clear.

References

Abercrombie JM, Halfhill MD, Ranjan P, Rao MR, Saxton AM, Yuan JS, Stewart CN Jr (2008) Transcriptional responses of *Arabidopsis thaliana* plants to As(V) stress. BMC Plant Biol 8:87–102. https://doi.org/10.1186/1471-2229-8-8

Adriano DC (2001) Trace elements in terrestrial environments. In: Biogeochemistry bioavailability and risks of metals, 2nd edn. Springer, New York, p 867

Ahsan N, Lee DG, Alam I, Kim PJ, Lee JJ, Ahn YO, Kwak SS, Lee IJ, Bahk JD, Kang KY, Renaut J, Komatsu S, Lee BH (2008) Comparative proteomic study of arsenic-induced differentially expressed proteins in rice roots reveals glutathione plays a central role during As stress. Proteomics 8(17):3561–3576. https://doi.org/10.1002/pmic.200701189

Ahsan N, Lee DG, Kim KH, Alam I, Lee SH, Lee KW, Lee H, Lee BH (2010) Analysis of arsenic stress-induced differentially expressed proteins in rice leaves by two-dimensional gel electrophoresis coupled with mass spectrometry. Chemosphere 78(3):224–231. https://doi.org/10.1016/j.chemosphere.2009.11.004

Akter KF, Owens G, Davey DE, Naidu R (2005) Arsenic speciation and toxicity in biological systems. Rev Environ Contam Toxicol 184:97–149

Azam SM, Gousul G (2015) Arsenic removal from contaminated waters by Fe-based (hydr) oxides and its phytoavailability in soil-plant system. PhD thesis, University of Naples Federico II

Bhargavi SD, Savitha J (2014) Arsenate resistant *Penicillium coffeae*: a potential fungus for soil bioremediation. Bull Environ Contam Toxicol 92(3):369–373. https://doi.org/10.1007/s00128-014-1212-y

Bhattacharya P, Mukherjee A, Bundschuh J, Zevenhoven R, Loeppert R (2007) Arsenic in soil and groundwater environment. Elsevier Press, Amsterdam

Bienert GP, Thorsen M, Schüssler MD, Nilsson HR, Wagner A, Tamás MJ et al (2008) A subgroup of plant aquaporins facilitate the bi-directional diffusion of Arsenic$(OH)_3$ and $Sb(OH)_3$ across membranes. BMC Biol 6:26. https://doi.org/10.1186/1741-7007-6-26

Bleeker PM, Hakvoort HW, Bliek M, Souer E, Schat H (2006) Enhanced arsenate reduction by a CDC25-like tyrosine phosphate arsenic explains increased arsenic phytochelatin accumulation in arsenate-tolerant *Holcus lanatus*. Plant J 45(6):917–929. https://doi.org/10.1111/j.1365-313X.2005.02651.x

Bock R (2007) Structure, function, and inheritance of plastid genomes. In: Topics in current genetics, Cell and molecular biology of plastids, 19th edn. Springer, Berlin, Heidelberg, pp 29–62

Brar SK, Verma M, Surampalli RY, Misra K, Tyagi RD, Meunier N, Blias JF (2006) Bioremediation of hazardous wastes—a review. Pract Period Hazard Toxic Radioact Waste Manag 10(2):59–72. https://doi.org/10.1061/(ASCE)1090-025X(2006)10:2(59)

Catarecha P, Segura MD, Franco-Zorrilla JM, García-Ponce B, Lanza M, Solano R, Paz-Ares J, Leyva A (2007) A mutant of the Arabidopsis phosphate transporter PHT1;1 displays enhanced arsenic accumulation. Plant Cell 19(3):1123–1133. https://doi.org/10.1105/tpc.106.041871

Chakrabarty D, Trivedi PK, Misra P, Tiwari M, Shri M, Shukla D, Kumar S, Rai A, Pandey A, Nigam D, Tripathi RD, Tuli R (2009) Comparative transcriptome analysis of arsenate and arsenite stresses in rice seedlings. Chemosphere 74(5):688–702. https://doi.org/10.1016/j.chemosphere.2008.09.082

Chandrakar V, Dubey A, Keshavkant S (2016a) Modulation of antioxidant enzymes by salicylic acid in arsenic exposed Glycine max L. J Soil Sci Plant Nutr 16(3):662–676. https://doi.org/10.4067/S0718-95162016005000048

Chandrakar V, Naithani SC, Keshavkant S (2016b) Arsenic-induced metabolic disturbances and their mitigation mechanisms in crop plants: a review. Biologia 71(4):367–377. https://doi.org/10.1515/biolog-2016-0052

Clemens S, Ma JF (2016) Toxic heavy metal and metalloid accumulation in crop plants and foods. Annu Rev Plant Biol 67:489–512. https://doi.org/10.1146/annurev-arplant-043015-112301

Dalle-Donne I, Rossi R, Colombo R, Giustarini D, Milzani A (2006) Biomarkers of oxidative damage in human disease. Clin Chem 52:601–623. https://doi.org/10.1373/clinchem.2005.061408

Das K, Roy CA (2014) Reactive oxygen species (ROS) and response of antioxidants arsenic ROS-scavengers during environmental stress in plants. Front Environ Sci 2. https://doi.org/10.3389/fenvs.2014.00053

De Oliveira AB, Alencar NLM, Gomes-Filho E (2013) Comparison between the water and salt stress effects on plant growth and development. In responses of organisms to water stress. Intech, Rijeka. https://doi.org/10.5772/54223

Debona D, Rodrigues FA, Datnoff LE (2017) Silicon's role in abiotic and biotic plant stresses. Annu Rev Phytopathol 55(4):85–107. https://doi.org/10.1146/annurev-phyto-080516-035312

Dhankhar R, Hooda A (2011) Fungal biosorption–an alternative to meet the challenges of heavy metal pollution in aqueous solutions. Environ Technol 32(5–6):467–491. https://doi.org/10.1080/09593330.2011.572922

Dhankher OP, Rosen BP, McKinney EC, Meagher RB (2006) Hyperaccumulation of arsenic in the shoots of Arabidopsis silenced for arsenate reductase (ACR2). Proc Natl Acad Sci U S A 103:5413–5418. https://doi.org/10.1073/pnas.0509770102

Dubey S, Shri M, Misra P, Lakhwani D, Bag SK, Asif MH, Trivedi PK, Tripathi RD, Chakrabarty D (2014) Heavy metals induce oxidative stress and genome-wide modulation in transcriptome of rice root. Funct Integr Genomics 14(2):401–417. https://doi.org/10.1007/s10142-014-0361-8

Duman F, Ozturk F, Aydin Z (2010) Biological responses of duckweed (*Lemna minor* L.) exposed to the inorganic arsenic species Arsenic(III) and Arsenic(V): effects of concentration and duration of exposure. Ecotoxicology 19:983–993. https://doi.org/10.1007/s10646-010-0480-5

Ellis DR, Gumaelius L, Indriolo E, Pickering IJ, Banks JA, Salt DE (2006) A novel arsenate reductase from the arsenic hyperaccumulating fern *Pteris vittata*. Plant Physiol 141:1544–1554. https://doi.org/10.1104/pp.106.084079

Emamverdian A, Ding Y, Mokhberdoran F, Xie Y (2015) Heavy metal stress and some mechanisms of plant defense response. Sci World J 18:1. https://doi.org/10.1155/2015/756120

Evangelou MW, Ebel M, Schaeffer A (2007) Chelate assisted phytoextraction of heavy metals from soil. Effect, mechanism, toxicity, and fate of chelating agents. Chemosphere 68(6):989–1003. https://doi.org/10.1016/j.chemosphere.2007.01.062

Faita F, Cori L, Bianchi F, Andreassi MG (2013) Arsenic-induced genotoxicity and genetic susceptibility to arsenic-related pathologies. Int J Environ Res Public Health 10:1527–1546. https://doi.org/10.3390/ijerph10041527

Finnegan PM, Chen W (2012) Arsenic toxicity: the effects on plant metabolism. Front Physiol 3:182. https://doi.org/10.3389/fphys.2012.00182

Garsenicic K, Korban SS (2007) Transgenic Indian mustard (*Brassica juncea*) plants expressing an Arabidopsis phytochelatin synthase (AtPCS1) exhibit enhanced Arsenic and Cd tolerance. Plant Mol Biol 64:361–369

Gill SS, Tuteja N (2010) Reactive oxygen species and antioxidant machinery in abiotic stress tolerance in crop plants. Plant Physiol Biochem 48(12):909–930. https://doi.org/10.1016/j.plaphy.2010.08.016

Gresser MJ (1981) ADP-arsenate. Formation by submitochondrial particles under phosphorylating conditions. J. Biol. Chem. 256(12):5981–5983

Gunes A, Pilbeam DJ, Inal A (2009) Effect of arsenic-phosphorous interaction on arsenic-induced oxidative stress in chick pea plants. Plant Soil 314(1–2):211–220. https://doi.org/10.1007/s11104-008-9719-9

Guo JB, Dai XJ, Xu WZ, Ma M (2008) Overexpression of GSH1 and AsPCS1 simultaneously increase the tolerance and accumulation of cadmium ad arsenic in Arabidopsis thaliana. Chemosphere 72(7):1020–1026. https://doi.org/10.1016/j.chemosphere.2008.04.018

Gusman GS, Oliveira JA, Farnese FS, Cambraia J (2013) Mineral nutrition and enzymatic adaptation induced by arsenate and arsenite exposure in lettuce plants. Plant Physiol Biochem 71:307–314. https://doi.org/10.1016/j.plaphy.2013.08.006

Hasanuzzaman M, Nahar K, Rahman A, Al Mahmud J, Hossain S, Alam K, Oku H, Fujita M (2017) Actions of biological trace elements in plant abiotic stress tolerance. In: Essential plant nutrients. Springer, Berlin, pp 213–274

Hughes MF (2002) Arsenic toxicity and potential mechanisms of action. Toxicol Lett 133(1):1–16. https://doi.org/10.1016/S0378-4274(02)00084-X

Hussain S, Ashraf U, Khan I, Schroeder JI, Delhaize E, Frommer WB, Guerinot ML, Harrison MJ, Herrera-Estrella L, Horie T, Kochian LV, Munns R, Nishizawa NK, Tsay YF, Sanders D (2013) Using membrane transporters to improve crops for sustainable food production. Nature 497(7447):60–66. https://doi.org/10.1038/nature11909

Imran MA, Khan RM, Ali Z, Mahmood T (2013) Toxicity of arsenic (as) on seed germination of sunflower (*Helianthus annuus* L.). Int J Phys Sci 8(17):840–847. https://doi.org/10.5897/IJPS2013.3894

Indriolo E, Na G, Ellis D, Salt DE, Banks JA (2010) A vacuolar arsenite transporter necessary for arsenic tolerance in the arsenic hyperaccumulating fern *Pteris vittata* is missing in flowering plants. Plant Cell 22:2045–2057. https://doi.org/10.1105/tpc.109.069773

Jain M, Gadre RP (2004) Inhibition of 5-amino levulinic acid dehydratase activity by arsenic in excised etiolated maize leaf segments during greening. J Plant Physiol 161(3):251–255. https://doi.org/10.1078/0176-1617-00879

Jha A, Dubey R (2004) Carbohydrate metabolism in growing rice seedlings under arsenic toxicity. J Plant Physiol 161(7):867–872. https://doi.org/10.1016/j.jplph.2004.01.004

Jia HF, Ren HY, Gu M, Zhao JN, Sun SB, Zhang X (2011) The PHOSPHATE TRANSPORTER gene OsPht1;8 is involved in phosphate homeostasis in rice. Plant Physiol 156(3):1164–1175. https://doi.org/10.1104/pp.111.175240

Jomova K, Valko M (2011) Advances in metal-induced oxidative stress and human disease. Toxicology 283:65–87. https://doi.org/10.1016/j.tox.2011.03.001

Keshavkant S, Naithani S (2001) Chilling-induced oxidative stress in young sal (*Shorea robusta*) seedlings. Acta Physiol Plant 23(4):457–466. https://doi.org/10.1007/s11738-001-0056-3

Khalid S, Shahid M, Niazi NK, Murtaza B, Bibi I, Dumat C (2017) A comparison of technologies for remediation of heavy metal contaminated soils. J Geochem Explor 182:247–268. https://doi.org/10.1016/j.gexplo.2016.11.021

Kitchin KT, Wallace K (2008) Evidence against the nuclear in situ binding of arsenicals–oxidative stress theory of arsenic carcinogenesis. Toxicol Appl Pharmacol 232(2):252–257. https://doi.org/10.1016/j.taap.2008.06.021

Kligerman AD, Malik SI, Campbell JA (2010) Cytogenetic insights into DNA damage and repair of lesions induced by a monomethylated trivalent arsenical. Mutat Res 695(1–2):2–8. https://doi.org/10.1016/j.mrgentox.2009.09.007

Kumar S, Dubey RS, Tripathi RD, Chakrabarty D, Trivedi PK (2015) Omics and biotechnology of arsenic stress and detoxification in plants: current updates and prospective. Environ Int 74:221–230. https://doi.org/10.1016/j.envint.2014.10.019

Lampis S, Santi C, Ciurli A, Andreolli M, Vallini G (2006) Promotion of arsenic phytoextraction efficiency in the fern *Pteris vittata* by the inoculation of As-resistant bacteria: a soil bioremediation perspective. Front Plant Sci 6:80. https://doi.org/10.3389/fpls.2015.00080

Leao GA, de Oliveira JA, Felipe RTA, Farnese FS, Gusman GS (2013) Anthocyanins, thiols, and antioxidant scavenging enzymes are involved in *Lemna gibba* tolerance to arsenic. J Plant Interact 9(1):143–151

LeBlanc MS, McKinney EC, Meagher RB, Smith AP (2013) Hijacking membrane transporters for arsenic phytoextraction. J Biotechnol 163(1):1–9. https://doi.org/10.1016/j.jbiotec.2012.10.013

Lei M, Wan X-M, Huang Z-C, Chen T-B, Li X-W, Liu Y-R (2012) First evidence on different transportation modes of arsenic and phosphorus in arsenic hyperaccumulator *Pteris vittata*. Environ Pollut 161:1–7. https://doi.org/10.1016/j.envpol.2011.09.017

Li HM, Kaneko Y, Keegstra K (1994) Molecular cloning of a chloroplastic protein associated with both the envelope and thylakoid membranes. Plant Mol Biol 25(4):619–632

Li RY, Ago Y, Liu WJ, Mitani N, Feldmann J, McGrath SP, Ma JF, Zhao FJ (2009) The Rice Aquaporin Lsi1 mediates uptake of methylated arsenic species. Plant Physiol 150(4):2071–2080. https://doi.org/10.1104/pp.109.140350

Li GW, Santoni V, Maurel C (2014) Plant aquaporins: roles in plant physiology. Biochim Biophs Acta 1840(5):1574–1582. https://doi.org/10.1016/j.bbagen.2013.11.004

Liu WJ, Wood BA, Raab A, McGrath SP, Zhao FJ, Feldmann J (2010) Complexation of arsenite with phytochelatins reduces arsenite efflux and translocation from roots to shoots in Arabidopsis. Plant Physiol 152:2211–2221. https://doi.org/10.1104/pp.109.150862

Lomax C, Liu WJ, Wu L, Xue K, Xiong J, Zhou J, McGrath SP, Meharg AA, Miller AJ, Zho FJ (2012) Methylated arsenic species in plants originate from soil microorganisms. New Phytol 193(3):665–672. https://doi.org/10.1111/j.1469-8137.2011.03956.x

Lu YP, Li ZS, Rea PA (1997) *AtMRP1* gene of *Arabidopsis* encodes a glutathione S-conjugate pump: isolation and functional definition of a plant ATP-binding cassette transporter gene. Proc Natl Acad Sci U S A 94(15):8243–8248

Ma JF, Tamai K, Ichii M, Wu GF (2002) A rice mutant defective in Si uptake. Plant Physiol 130(4):2111–2117. https://doi.org/10.1104/pp.010348

Ma JF, Tamai K, Yamaji N, Mitani N, Konishi K, Katsuhara M, Ishiguro M, Murata Y, Yano M (2006) A silicon transporter in rice. Nature 440(7084):688–691. https://doi.org/10.1038/nature04590

Ma JF, Yamaji N, Mitani N, Xu XY, Su YH, McGrath SP, Zhao FJ (2008) Transporters of arsenite in rice and their role in arsenic accumulation in rice grain. Proc Natl Acad Sci U S A 105(105):9931–9935. https://doi.org/10.1073/pnas.0802361105

Malik JA, Goel S, Sandhir R, Nayyar H (2011) Uptake and distribution of arsenic in chickpea: effects on seed yield and seed composition. Commun Soil Sci Plant Anal 42:1728–1738. https://doi.org/10.1080/00103624.2011.584593

Marschner H (1995) Mineral nutrition of higher plants, 2nd edn. Academic Press, San Diego, p 889

Mascher R, Lippmann B, Holzinger S, Bergmann H (2002) Arsenate toxicity: effects on oxidative stress response molecules and enzymes in red clover plants. Plant Sci 163:961–969

Meharg AA, Hartley-Whitaker J (2002) Arsenic uptake and metabolism in arsenic resistant and non-resistant plant species. New Phytol 154:29–43. https://doi.org/10.1046/j.1469-8137.2002.00363.x

Mehmood T, Bibi I, Shahid M, Niazi NK, Murtaza B, Wang H, Ok YS, Sarkar B, Javed MT, Murtaza G (2017) Effect of compost addition on arsenic uptake, morphological and physiological attributes of maize plants grown in contrasting soils. J Geo chem Explor 178:83–91. https://doi.org/10.1016/j.gexplo.2017.03.018

Meliker JR, AvRuskin GA, Slotnick MJ (2008) Validity of spatial models of arsenic concentrations in private well water. Environ Res 106(1):42–50. https://doi.org/10.1016/j.envres.2007.09.001

Meng YL, Liu Z, Rosen BP (2004) As(III) and Sb(III) uptake by GlpF and efflux by ArsB in Escherichia coli. J Biol Chem 279(18):18334–18341. https://doi.org/10.1074/jbc.M400037200

Mirza N, Mahmood Q, Maroof Shah M, Pervez A, Sultan S (2014) Plants arsenic useful vectors to reduce environmental toxic arsenic content. Sci World J 11:1. https://doi.org/10.1155/2014/921581

Mitani-Ueno N, Yamaji N, Zhao FJ, Ma JF (2011) The aromatic/arginine selectivity filter of NIP aquaporins plays a critical role in substrate selectivity for silicon, boron, and arsenic. J Exp Bot 62(12):4391–4398. https://doi.org/10.1093/jxb/err158

Miteva E, Merakchiyska M (2002) Response of chloroplasts and photosynthetic mechanism of bean plants to excess arsenic in soil. Bulg J Agric Sci 8:151–156

Monteiro C, Santos CAO, Pinho SN, Oliveira H, Pedrosa T, Diarsenic MC (2012) Cadmium-induced cyto- and genotoxicity are organ-dependent in lettuce. Chem Res Toxicol 25(7):1423–1434. https://doi.org/10.1021/tx300039t

Mosa KA, Kumar K, Chhikara S, Mcdermott J, Liu Z, Musante C, White JC, Dhankher OP (2012) Members of rice plasma membrane intrinsic proteins subfamily are involved in arsenite permeability and tolerance in plants. Transgenic Res 21(6):1265–1277. https://doi.org/10.1007/s11248-012-9600-8

Mukhopadhyay R, Bhattacharjee H, Rosen BP (2014) Aquaglyceroporins: generalized metalloid channels. Biochim Biophys Acta 1840(5):1583–1591. https://doi.org/10.1016/j.bbagen.2013.11.021

Mylona PV, Polidoros AN, Scandalios JG (1998) Modulation of antioxidant responses by arsenic in maize. Free Radic Biol Med 25(4–5):576–585. https://doi.org/10.1016/S0891-5849(98)00090-2

Nath S, Panda P, Mishra S, Dey M, Choudhury S, Sahoo L, Panda SK (2014) Arsenic stress in rice: redox consequences and regulation by iron. Plant Physiol Biochem 80:203–210. https://doi.org/10.1016/j.plaphy.2014.04.013

Niazi NK, Bibi I, Fatimah A, Shahid M, Javed MT, Wang H, Ok YS, Bashir S, Murtaza B, Saqib ZA, Shakoor MB (2017) Phosphate-assisted phytoremediation of arsenic by *Brassica napus* and *Brassica juncea*: morphological and physiological response. Int J Phytoremediation 19(7):670–678. https://doi.org/10.1080/15226514.2016.1278427

Norton GJ, Lou-Hing DE, Meharg AA, Price AH (2008) Rice–arsenate interactions in hydroponics: whole genome transcriptional analysis. J Exp Bot 59(8):2267–2276. https://doi.org/10.1093/jxb/ern097

Nussaume L, Kanno S, Javot H, Marin E, Pochon N, Ayadi A, Nakanishi TM, Thibaud MC (2011) Phosphate import in plants: focus on the PHT1 transporters. Front Plant Sci 2:83. https://doi.org/10.3389/fpls.2011.00083

Othman N, Mohd-Asharuddin S, Azizul-Rahman MFH (2013) An overview of fruit waste as sustainable adsorbent for heavy metal removal. Appl Mech Mater 389:29–35. https://doi.org/10.4028/www.scientific.net/AMM.389.29

Päivöke AE, Simola LK (2001) Arsenate toxicity to Pisum sativum: mineral nutrients, chlorophyll content, and phytase activity. Ecotoxicol Environ Saf 49(2):111–121. https://doi.org/10.1006/eesa.2001.2044

Palmieri L, Picault N, Arrigoni R, Besin E, Palmieri F, Hodges M (2008) Molecular identification of three *Arabidopsis thaliana* mitochondrial dicarboxylate carrier isoforms: organ distribution, bacterial expression, reconstitution into liposomes and functional characterization. Biochem J 410(3):621–629. https://doi.org/10.1042/BJ20070867

Parkhey S, Naithani SC, Keshavkant S (2012) ROS production and lipid catabolism in desiccating *Shorea robusta* seeds during aging. Plant Physiol Biochem 57:261–267. https://doi.org/10.1016/j.plaphy.2012.06.008

Parkhey S, Naithani S, Keshavkant S (2014) Protein metabolism during natural ageing in desiccating recalcitrant seeds of *Shorea robusta*. Acta Physiol Plant 36(7):1649–1659. https://doi.org/10.1007/s11738-014-1540-x

Peng S (2000) Single-leaf and canopy photosynthesis of rice. In: Sheehy JE et al (eds) Redesigning rice photosynthesis to increase yield. Elsevier, Amsterdam, pp 213–228

Prasad KS, Ramanathan AL, Paul J, Subramanian V, Prasad R (2013) Biosorption of Arsenite (As3þ) and Arsenate (As5þ) from aqueous solution by *Arthrobacter sp*. Biomass Environ Technol 34(17–20):2701–2708. https://doi.org/10.1080/09593330.2013.786137

Qin J, Rosen BP, Zhang Y, Wang G, Franke S, Rensing C (2006) Arsenic detoxification and evolution of trimethylarsine gas by a microbial arsenite S-adenosylmethionine methyltransferase. Proc Natl Acad Sci U S A 103(7):2075–2080. https://doi.org/10.1073/pnas.0506836103

Qin J, Lehr CR, Yuan CG, Le XC, McDermott TR, Rosen BP (2009) Biotransformation of arsenic by a yellow stone thermos acidophilic eukaryotic alga. Proc Natl Acad Sci U S A 106(13):5213–5217. https://doi.org/10.1073/pnas.0900238106

Raab A, Wright SH, Jaspars M, Meharg AA, Feldmann J (2007) Pentavalent arsenic can bind to biomolecules. Angew Chem Int Ed 46(15):2594–2597. https://doi.org/10.1002/anie.200604805

Rafiq M, Shahid M, Abbas G, Shamshad S, Khalid S, Niazi NK, Dumat C (2017) Comparative effect of calcium and EDTA on arsenic uptake and physiological attributes of *Pisum sativum*. Int J Phytoremediation 19(7):662–669. https://doi.org/10.1080/15226514.2016.1278426

Rahman MA, Kadohashi K, Maki T, Hasegawa H (2011) Transport of DMAA and MMAA into rice (*Oryza sativa* L.) roots. Environ Exp Bot 72:41–46

Sachs RM, Michael JL (1971) Comparative phytotoxicity among four arsenical herbicides. Weed Sei 19:558–564

Shahid M, Khalid S, Abbas G, Shahid N, Nadeem M, Sabir M, Aslam M, Dumat C (2015) Heavy metal stress and crop productivity. In: Crop production and global environmental issue. Springer, Cham, pp 1–25. https://doi.org/10.1007/978-3-319-23162-4_1

Sharma P, Jha AB, Dubey RS, Pessarakli M (2012) Reactive oxygen species, oxidative damage, and antioxidative defense mechanism in plants under stressful conditions. J Bot 24(2012):217037. https://doi.org/10.1155/2012/217037

Shin H, Shin HS, Dewbre GR, Harrison MJ (2004) Phosphate transport in Arabidopsis: Pht1;1 and Pht1;4 play a major role in phosphate acquisition from both low- and high-phosphate environments. Plant J 39(4):629–642. https://doi.org/10.1111/j.1365-313X.2004.02161.x

Singh VP (2005) Toxic metals and environmental issues. Sarup & Sons, New Delhi

Singh N, Ma LQ, Srivastava M, Rathinabapathi B (2006) Metabolic adaptations to arsenic-induced oxidative stress in *Pteris vittata* L. and *Pteris ensiformis* L. Plant Sci 170(2):274–282. https://doi.org/10.1016/j.plantsci.2005.08.013

Singh HP, Batish DR, Kohli RK, Arora K (2007) Arsenic-induced root growth inhibition in mung bean (*Phaseolus aureus* Roxb.) is due to oxidative stress resulting from enhanced lipid peroxidation. Plant Growth Regul 53(1):65–73. https://doi.org/10.1007/s10725-007-9205-z

Smith SE, Christophersen HM, Pope S, Smith FA (2010) Arsenic uptake and toxicity in plants: integrating mycorrhizal influences. Plant Soil 327(1–2):1–21. https://doi.org/10.1007/s11104-009-0089-8

Srivastava S, Sinha P, Sharma YK (2017) Status of photosynthetic pigments, lipid peroxidation and anti-oxidative enzymes in Vigna mungo in presence of arsenic. J Plant Nutr 40(3):298–306. https://doi.org/10.1080/01904167.2016.1240189

Stoeva N, Bineva T (2003) Oxidative changes and photosynthesis in oat plants grown in As-contaminated soil. Bulg J Plant Physiol 29(1–2):87–95. https://doi.org/10.2478/s11756-012-0024-y

Stoeva N, Berova M, Vassilev A, Zlatev Z (2005) Effect of arsenic on some physiological parameters in bean plants. Biol Planta 49(2):293–296

Sulaymon AH, Mohammed AA, Al-Musawi TJ (2013) Competitive biosorption of lead, cadmium, copper, and arsenic ions using algae. Environ Sci Pollut Res 20(5):3011–3023. https://doi.org/10.1007/s11356-012-1208-2

Suneja Y (2014) Physio-biochemical responses and allelic diversity for water deficit tolerance related traits in *Aegilops tauschii* and *Triticum dicoccoides*. PhD thesis, Punjab Agricultural University, Ludhiana

Talukdar D (2017) Balancing roles of reactive oxygen species in plants' response to metalloid exposure: boon or bane—revisiting the role of ROS. Reactive oxygen species in plants. Wiley, Hobokem. https://doi.org/10.1002/9781119324928.ch3

Tawfik DS, Viola RE (2011) Arsenate replacing phosphate: alternative life chemistries and ion promiscuity. Biochemistry 50(7):1128–1134. https://doi.org/10.1021/bi200002a

Tollestrup K, Daling JR, Allard J (1995) Mortality in a cohort of orchard workers exposed to lead arsenate pesticide spray. Arch Environ Health 50(3):221–229. https://doi.org/10.1080/00039896.1995.9940391

Tommasini R, Vogt E, Fromenteau M, Hortensteiner S, Matile P, Amrhein N, Martinoia E (1998) An ABC-transporter of *Arabidopsis thaliana* has both glutathione-conjugate and chlorophyll catabolite transport activity. Plant J 13(6):773–780

Tripathi P, Singh PC, Mishra A, Srivastava S, Chauhan R, Awasthi S, Mishra S, Dwivedi S, Tripathi P, Kalra A (2017) Arsenic tolerant *Trichoderma sp*. reduces arsenic induced stress in chickpea (*Cicer arietinum*). Environ Pollut 223:137–145. https://doi.org/10.1016/j.envpol.2016.12.073

Ullrich-Eberius CI, Sanz A, Novacky AJ (1989) Evaluation of arsenate and vanadate-associated changes of electrical membrane potential and phosphate transport in *Lemna gibba* G1. J Expn Bot 40(210):119–128

Vinh NV, Zafar M, Behera SK, Park HS (2015) Arsenic(III) removal from aqueous solution by raw and zinc loaded pine cone biochar: equilibrium, kinetics, and thermodynamics studies. Int J Environ Sci Technol 12(4):1283–1294. https://doi.org/10.1007/s13762-014-0507-1

Vizcaya-Ruiz AD, Barbier O, Ruiz-Ramos R, Cebrian ME (2009) Biomarkers of oxidative stress and damage in human populations exposed to arsenic. Mutat Res 674(1–2):85–92

Vromman D, Lutts S, Lefèvre I, Somer L, De Vreese O, Šlejkovec Z, Quinet M (2013) Effects of simultaneous arsenic and iron toxicities on rice (*Oryza sativa* L.) development, yield-related parameters and Arsenic and Fe accumulation in relation to Arsenic speciation in the grains. Plant Soil 371(1–2):199–217. https://doi.org/10.1007/s11104-013-1676-2

Wang JS, Wai CM (2004) Arsenic in drinking water—a global environmental problem. J Chem Educ 81:207–213

Wang Y, Wang S, Xu P, Liu C, Liu M, Wang Y, Wang C, Zhang C, Ge Y (2015) Review of arsenic speciation, toxicity and metabolism in microalgae. Rev Environ Sci Biotechnol 14(3):427–451. https://doi.org/10.1007/s11157-015-9371-9

Watling-Payne AS, Selwyn MJ (1974) Inhibition and uncoupling of photophosphorylation in isolated chloroplasts by organotin, organomercury and diphenyleneiodonium compounds. Biochem J 142(1):65–74

Wolterbeek HT, Van der Meer AJGM (2002) Transport rate of arsenic, cadmium, copper and zinc in *Potamogeton pectinatus* L.: radiotracer experiments with 76As, 109,115Cd, 64Cu and 65,69 m Zn. Sci Total Environ 287:13–30. https://doi.org/10.1016/S0048-9697(01)00989-5

Woolson EA, Kearney PC (1973) Persistence and reactions of 14C-cacodylic acid in soils. Environ Sci Technol 7(1):47–50. https://doi.org/10.1021/es60073a008

Wu JH, Zhang R, Lilley RM (2002) Methylation of arsenic in vitro by cell extracts from the bentgrass (*Agrostis tenuis*): effect of acute exposure of plants to arsenate. Funct Plant Biol 29:73–80. https://doi.org/10.1071/PP01022

Wu Z, Ren H, McGrath SP, Wu P, Zhao FJ (2011) Investigating the contribution of the phosphate transport pathway to arsenic accumulation in rice. Plant Physiol 157(1):498–508. https://doi.org/10.1104/pp.111.178921

Xu W, Dai W, Yan H, Li S, Shen H, Chen Y, Xu H, Sun Y, He Z, Ma M (2015) Arabidopsis NIP3;1 plays an important role in arsenic uptake and root-to-shoot translocation under arsenite stress conditions. Mol Plant 8(5):722–733. https://doi.org/10.1016/j.molp.2015.01.005

Zhao F, Ma J, Meharg A, McGrath S (2009) Arsenic uptake and metabolism in plants. New Phytol 181:777–794. https://doi.org/10.1111/j.1469-8137.2008.02716.x

Zhu YG, Rosen BP (2009) Perspectives for genetic engineering for the phytoremediation of arsenic-contaminated environments: from imagination to reality? Curr Opin Biotechnol 20(2):220–224. https://doi.org/10.1016/j.copbio.2009.02.011

Ziech D, Franco R, Georgakilas AG, Georgakila S, Malamou-Mitsi V, Schoneveld O, Pappa A, Panayiotidis MI (2010) The role of reactive oxygen species and oxidative stress in environmental carcinogenesis and biomarker development. Chem Biol Interact 188(2):334–339. https://doi.org/10.1016/j.cbi.2010.07.010

Vermicompost Leachate Use for the Irrigation of Two South American Woody Plant Species

Gabriel Basílico, Mauricio Cogollo-Rueda, Romina López, María Luz Montenegro, Anahí Magdaleno, and Laura de Cabo

Contents

G. Basílico (✉)
Museo Argentino de Ciencias Naturales "Bernardino Rivadavia" – Consejo Nacional de Investigaciones Científicas y Técnicas, Buenos Aires, Argentina

Laboratorio de Bioindicadores y Remediación – Universidad de Flores, Buenos Aires, Argentina
e-mail: gabriel.basilico@uflo.edu.ar

M. Cogollo-Rueda · R. López · M. L. Montenegro
Laboratorio de Bioindicadores y Remediación – Universidad de Flores, Buenos Aires, Argentina

A. Magdaleno
Facultad de Farmacia y Bioquímica – Universidad de Buenos Aires, Buenos Aires, Argentina

L. de Cabo
Museo Argentino de Ciencias Naturales "Bernardino Rivadavia" – Consejo Nacional de Investigaciones Científicas y Técnicas, Buenos Aires, Argentina

M. Naeem et al. (eds.), *Contaminants in Agriculture*,
https://doi.org/10.1007/978-3-030-41552-5_21

1 Introduction

Inadequate management of municipal solid waste (MSW) generates negative environmental impacts on soil, surface water, groundwater, air, ecosystems, and human health. The global production of MSW continues to grow, and by the year 2025 a total daily production of more than 3.5 million of tons is expected (Zorpas et al. 2015). With some exceptions, in both developed and developing countries, only a small part of the MSW is reused or recycled, while the rest is buried in landfills or dumping sites or incinerated. These management modalities are linear and represent millionaire costs for local governments, which often have limited budgets.

The composition of the MSW varies considerably in relation to the climate, the frequency of collection, social customs and consumption habits, per capita income, the degree of urbanization and industrialization of the area, etc. (Henry 1999). The MSW is composed of materials of vegetable and animal origin, such as food and pruning waste, plastics, paper, cardboard, glass, textiles, metals, diapers and dressings, hazardous and pathogenic waste, construction materials, ashes, and other materials.

Classifying MSW in different categories allows the reuse or subsequent recycling of part of the materials and continues to be one of the main actions that contribute to improving waste management, both in urban and rural areas. Although there is a wide variety of categories for the segregation of the MSW, the most basic alternative is to separate them into a wet fraction (organic materials), a dry fraction (plastics, paper, cardboard, glass, and metals), and a third fraction that includes other waste whose reuse or recycling is complex, such as electronic waste. The wet fraction usually represents the highest percentage of the MSW in developing countries. For example, in the Metropolitan Area of Buenos Aires (Argentina), the wet fraction corresponds to 50.4% w/w of MSW (CEAMSE and FIUBA 2011). Taking into account that most of this fraction is currently destined for landfill disposal, the application of recycling methodologies by local governments would allow a significant reduction in costs derived from transportation and final disposal of MSW.

Recycling wet fraction of the MSW and other agricultural and industrial wastes can be carried out through aerobic or anaerobic processes. Among the aerobic processes, composting and vermicomposting stand out, while anaerobic digestion allows the production of biogas, a methane-rich mixture of gases. Composting of organic matter consists basically in the partial oxidation of carbohydrates, proteins, fats and other biomolecules, and their transformation into "compost." This material serves as fertilizer and soil improver (Fornes et al. 2012) and can be prepared both industrially and at a domestic scale. The main organisms involved in the stabilization of organic matter during composting are bacteria and fungi, although the process can be accelerated by *Eisenia foetida* (Das et al. 2016) and other earthworms (Lumbricidae), larvae of *Hermetia illucens* (Salomone et al. 2017), and other higher organisms. The process based on the use of Lumbricidae is called "vermicomposting" and is used mainly in the treatment of plant materials.

An important part of the water contained in the organic material is removed by evaporation and leaching during vermicomposting. The amount and composition of the produced leachates are variable and depend on the type and composition of the source material and the humidity content of material, among other factors. However, leachate usually contains inorganic nutrients, suspended solids, soluble organic molecules, and other substances (Sall et al. 2019). On the other hand, leachates contain plant growth regulators (PGR) and other phytochemicals that contribute favorably to the growth of plants (Aremu et al. 2015).

The volume of leachates produced during vermicomposting can be high. Then, several possible uses of leachates have been proposed, such as the irrigation of horticultural (Ávila-Juárez et al. 2015) and tree seedlings (Kandari et al. 2011), in order to take advantage of the substances contained in this by-product. This alternative also allows reducing the use of chemical fertilizers. However, due to the large amount of dissolved substances contained in the leachate, its electrical conductivity (EC) is usually high. On the other hand, there is evidence of phytotoxicity of the leachates generated in the vermicomposting process (Gutiérrez-Miceli et al. 2008). These characteristics make dilution of the leachate necessary.

The objective of this study was to evaluate the use of vermicompost leachate dilutions in the irrigation of the shrub *Austroeupatorium inulifolium* (Asteraceae) and tree *Erythrina crista-galli* (Fabaceae), two native, ornamental, woody plant species of South America. In order to establish the appropriate dilutions of use, phytotoxicity and genotoxicity of vermicompost leachate was evaluated by *Allium cepa* test.

2 Materials and Methods

2.1 *Vermicomposting and Leachate Production*

Vermicomposting was carried out in the Laboratory of Bioindicators and Remediation of the Flores University (Autonomous City of Buenos Aires, Argentina). A plastic container with a volume of 50 L was used for the elaboration of vermicompost. Vegetable material (food waste from a grocery store and coffee grounds) and pieces of cardboard were placed in the container, in an approximate volumetric ratio of 10:1. Along with the vegetable material and cardboard, small amounts of mature compost and approximately 60 individuals of *E. foetida* were incorporated. Perforations were made in the base of the plastic container, in order to collect the leachate produced during the process (90 days). Fresh vegetable waste and/or cardboard were incorporated periodically, in order to keep the vermicompost humidity content approximately constant and to obtain a sufficient amount of leachate.

2.2 *Leachate Stabilization and Sampling*

The produced leachate was collected periodically and stored in a plastic container with a capacity of 5 L. In order to stabilize the collected leachate and allow its subsequent use (Romero-Tepal et al. 2014), air was bubbled continuously for a period of 30 days, by using an aquarium aerator.

At the end of the leachate stabilization period, the pH and EC were measured using a Hanna® multiparameter sensor. In addition, a sample was taken for total nitrogen (TN), total phosphorus (TP), and potassium (K) determination, according to standardized methodologies (APHA et al. 2012).

2.3 A. inulifolium *and* E. crista-galli *Bioassays*

A. inulifolium and *E. crista-galli* bioassays were carried out in quadruplicate, considering as an experimental unit (EU) each plastic pot containing 400 g of commercial substrate, with its corresponding seedling. Bioassays were started from seeds (*E. crista-galli*) or seedlings (*A. inulifolium*) of homogeneous size. Each assay had a different duration due to the growth rate characteristic of tested species. Dilutions of leachate were prepared with deionized water and used to irrigate seedlings. In the *A. inulifolium* bioassay, two treatments were established: control (deionized water) and L10% (10% v/v leachate dilution), while the bioassay with the species *E. crista-galli* consisted of five treatments: control, L5% (5% v/v leachate dilution), L10%, CF1, and CF2, where CF1 and CF2 were two commercial liquid fertilizers (Table 1). These fertilizers were diluted with deionized water and applied every 15 days according to the product recommendations.

The volume of irrigation solution applied in each EU was equivalent to the volume necessary to obtain the saturation of the substrate and fluctuated between 20 and 60 mL/EU. Irrigation was carried out every 2 or 3 days. When the application of commercial fertilizers CF1 and CF2 did not correspond, the EUs assigned to these treatments were irrigated with deionized water. In addition to the usual inorganic nutrients (N, P, and K), common to both fertilizers, the CF1 fertilizer

Table 1 Bioassays characteristics

Species	Bioassay duration	Treatment	Replicates	Description
A. Inulifolium	51 days	Control	4	Irrigation with deionized water
		L10%	4	Irrigation with leachate, 10% dilution
E. crista-galli	107 days	Control	4	Irrigation with deionized water
		L5%	4	Irrigation with leachate, 5% dilution
		L10%	4	Irrigation with leachate, 10% dilution
		CF1	4	Commercial fertilizer 1
		CF2	4	Commercial fertilizer 2

contained growth phytohormones, and CF2 fertilizer included in its formulation bacterial inoculum to promote the absorption of nutrients.

2.4 *Plant Growth*

The effect of irrigation with different solutions of leachate, fertilizers, and deionized water (control) was evaluated by measuring the height, stem diameter and biomass. In both species, height measurement was performed every 4–7 days after planting (DAP) with a measuring tape. In the case of *E. crista-galli*, the first measurement was made at 23 DAP, due to the germination time of the seeds. At the end of each assay, stem diameter and root, aerial tissues and total fresh weight (FW) of plants were measured.

2.5 *Substrate Characterization*

Samples of substrate were taken at the beginning and at the end of each test in order to perform the corresponding analytical determinations. The characterization of the substrate included the pH, the EC, and the content of organic matter (OM) by the semiquantitative method of hot wet oxidation with 30% hydrogen peroxide (Schumacher 2002). The determination of pH and EC was made in the supernatant of a 1:2.5 w/v suspension of substrate: water (Islam et al. 2016).

2.6 *Phytotoxicity*

Organically grown seeds of *A. cepa* (2n = 16), variety Valcatorce, with more than 90% germination, were used for evaluate the phytotoxicity and genotoxicity of vermicompost leachate. The seeds were genetically and physiologically homogeneous. The assays were performed according to a modified version of Grant's protocol (Matsumoto et al. 2006). One hundred (100) onion seeds were germinated in 90-mm-diameter Petri dishes lined with filter paper containing 4 mL of pure leachate and dilutions L5% and L10%. Distilled water and methyl methanesulfonate (MMS, 12 mg/L) were used as negative and positive controls, respectively. The Petri dishes were kept in darkness in an incubator at 22 ± 2 °C for 96 h. After this period, the seeds were collected and the roots fixed in alcohol–acetic acid (3:1) for 24 h. The fixed roots were stored in 70% ethyl alcohol until microscopic analysis. To prepare the slides, the meristematic regions were covered with cover slips and carefully squashed in a drop of 2% acetic orcein solution. The mitotic index (MI) was calculated by counting all stages of mitotic cells with respect to the total number of cells. For the chromosome aberration (CA) analyses, several aberrations such as fragments, vagrants, and bridges in the anaphase and telophase were analyzed. All these categories were placed into one category in order to evaluate the CA as a single endpoint, following the criteria used by Hoshina and Marin-Morales (2009). The micronuclei (MN) induction was recorded by observing the interphase cells. The analyses were performed by scoring 5000 cells per treatment, that is, 1000 cells per slide and a total of five slides. Phytotoxicity was evaluated based on the seed germination index, which was calculated as the ratio of the number of germinated seeds to the total seeds allowed to germinate.

2.7 Statistical Analysis

Substrate data were analyzed by means of parametric (Student *t*-test; ANOVA) or nonparametric tests (Mann–Whitney *U*-test; Kruskal–Wallis test). Growth data were statistically analyzed by means of the Student *t*-test (*A. inulifolium* bioassay) or ANOVA with Tukey's (between all treatments) and Dunnett's (between each treatment and control) (*E. crista-galli* bioassay) post hoc comparisons (Zar 1999). Phytotoxicity was evaluated by means of nonparametric tests (Kruskal–Wallis test and pairwise comparisons by Mann–Whitney *U*-test) (Basílico et al. 2017).

3 Results and Discussion

3.1 Leachate, Irrigation Solutions, and Substrate Characterization

Undiluted leachate had an alkaline pH (>8.00), high EC (22.3 mS/cm), and concentrations of TN, TP, and K of 144 mg/l, 50.8 mg/l, and 5430 mg/l, respectively. The high EC reflects the high salt content, limiting its use in irrigation (Klapperich et al. 2014). The pH of the irrigation dilutions was in the range of 7.05 (CF2) and 8.47 (L10%), while the average EC fluctuated between 0.001 mS/cm (control) and 2.23 mS/cm (L10%). The pH, TP, and K were in the ranges reported in the literature; however, TN was lower than the minimum reported value of 247 mg/L (Sall et al. 2019). The low TN content can be attributed to source material poor in nitrogen and ammonia volatilization during composting (He et al. 2018).

The initial pH of commercial substrate was slightly acidic in both tests, and the EC was 929 μS/cm (0.93 mS/cm) (Figs. 1 and 2). After 51 DAP, the pH and the EC of the substrates increased markedly with respect to the control, as a result of the irrigation with the dilution of the 10% leachate (L10%, Fig. 1) in *A. inulifolium*. At the end of *E. crista-galli* bioassay (107 DAP), pH and EC also increased in the leachate treatments (L5% and L10%, Fig. 2). Differences in pH and EC were statistically significant ($P < 0.05$) in both bioassays (Figs. 1 and 2). Increases in substrate

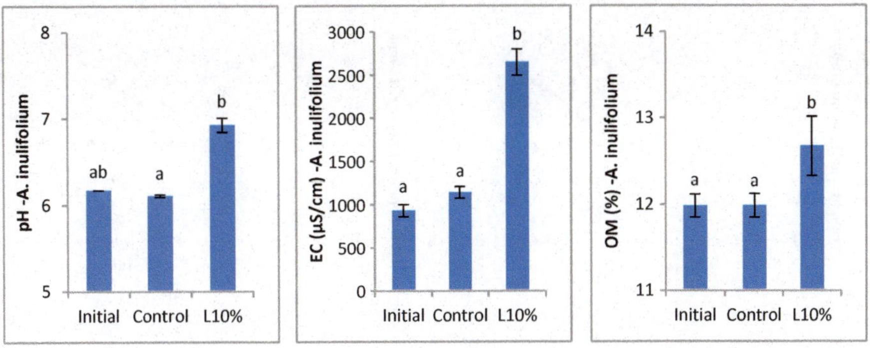

Fig. 1 Values of pH, EC, and OM in the substrates used in *A. inulifolium* bioassay (mean and SD). The same letters (**a, b**) indicate membership to homogeneous groups ($P < 0.05$). Different tests were used according to data set characteristics

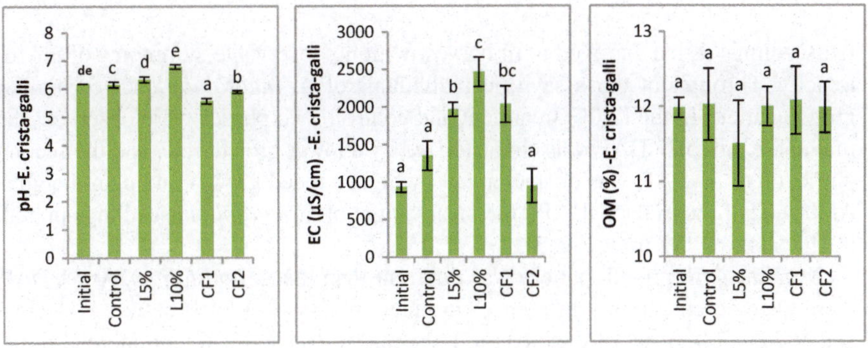

Fig. 2 Values of pH, EC, and OM in the substrates used in *E. crista-galli* bioassay (mean and SD). The same letters (**a–e**) indicate membership to homogeneous groups ($P < 0.05$). Different tests were used according to data set characteristics

pH in both bioassays were due to leachate irrigation which contains NH_4^+ and other base-forming cations such as Ca^{+2} and Mg^{+2} (Cheng and Chu 2007). Moreover, it is well documented that leachate irrigation that contains elevated salt concentration leads to soil salinization (Hernández et al. 1999). In the *E. crista-galli* bioassay, the irrigation with commercial fertilizers (CF1 and CF2) causes the lowering of pH, with respect to the initial values and control. The EC of the substrate irrigated with CF1 increased with respect to the initial value and was superior to the control (Fig. 2).

Substrate organic matter content (OM) in L10% treatment was higher than in the initial and the control for *A. inulifolium*, and the differences were statistically significant ($P < 0.05$) (Fig. 1). This tendency was not observed in *the E. crista-galli* assay (Fig. 2).

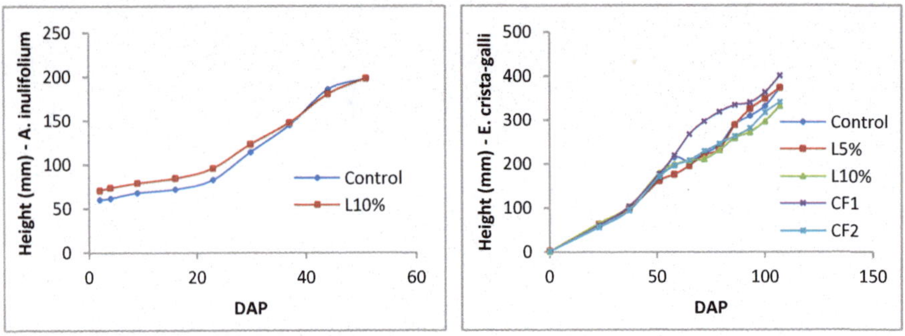

Fig. 3 Mean heights of *A. inulifolium* (left) and *E. crista-galli* (right) seedlings throughout the bioassay

3.2 A. inulifolium *and* E. crista-galli *Growth*

All seedlings tested survived until harvest with no observable symptoms of phytotoxicity. Throughout the assay, the individuals of *A. inulifolium* showed similar growth pattern in the L10% treatment and control. Two phases were observed: the first phase, up to 23 DAP, was characterized by a lower growth rate, and the second phase, up to 51 DAP, was characterized by higher speed growth and plant heights. At the end of the trial (51 DAP), the mean height of *A. inulifolium* seedlings in both treatments was 197 mm (Fig. 3).

E. crista-galli growth pattern was different than the *A. inulifolium*. At 51 DAP, there were practically no differences in plant heights. However, between 51 DAP and 107 DAP (the end of the bioassay), the plants irrigated with commercial fertilizer (CF1 treatment) had a higher height than the other treatments and the control. The mean height of the *E. crista-galli* seedlings, at 107 DAP, varied between 332 mm (L10%) and 400 mm (CF1) (Fig. 3).

At the end of the test with *A. inulifolium* (51 DAP), no statistically significant differences were observed in any of the growth variables considered (*t*-test, $P > 0.05$). However, both the stem diameter and the total fresh weight and aerial tissues were higher in the plants irrigated with the leachate dilution (Tables 2 and 3).

E. crista-galli growth showed statistically significant differences among treatments (ANOVA, $P < 0.05$), at the end of the trial (107 DAP) in the height (Table 2) and total and aerial tissue fresh weight (Table 3). For these variables, the highest mean values corresponded to the CF1 treatment. The lowest average values of height and fresh weight of aerial tissues corresponded to the L10% treatment, while the lowest total fresh weight observed corresponded to the L5% treatment (Tables 2 and 3).

No statistically significant differences were observed in stem diameter and root fresh weight, in both species (Tables 2and 3).

In Tables 2 and 3, it can also be observed that the height, the diameter of the stem, the total fresh weight, and the fresh weight of the aerial tissues of *the E. crista-galli* seedlings were lower in the L10% treatment with respect to control. Similar

Table 2 Height and stem diameter of *A. inulifolium* and *E. crista-galli* seedlings at the end of each bioassay

Species	DAP	Treatment	Height (mm)	Stem diameter (mm)
A. Inulifolium	**51**	Control	197 ± 12 a	4.73 ± 0.38 a
		L10%	197 ± 15 a	4.95 ± 0.92 a
E. crista-galli	**107**	Control	372 ± 29 ab	10.23 ± 1.03 a
		L5%	372 ± 28 ab	8.73 ± 1.15 a
		L10%	332 ± 26 a	9.69 ± 0.78 a
		CF1	400 ± 17 b	10.78 ± 1.27 a
		CF2	340 ± 42 ab	9.95 ± 1.29 a

The same letters (a, b) indicate membership to homogeneous groups according to Tukey's post hoc comparisons ($P < 0.05$)

Table 3 Total roots and aerial tissue fresh weight of *A. inulifolium* (left) and *E. crista-galli* (right) seedlings at the end of each bioassay

Species	DAP	Treatment		Fresh weight (g)	
			Total	Roots	Aerial tissues
A. inulifolium	51	Control	5.94 ± 0.93 a	2.06 ± 0.57 a	3.88 ± 0.37 a
		L10%	6.16 ± 0.77 a	1.76 ± 0.41 a	4.40 ± 0.41 a
E. crista-galli	107	Control	15.96 ± 1.61 ab	4.95 ± 0.61 a	11.01 ± 1.37 ab
		L5%	14.42 ± 3.26 a	5.21 ± 1.28 a	9.21 ± 2.48 a
		L10%	14.53 ± 1.87 a	5.51 ± 0.73 a	9.02 ± 1.23 a
		CF1	20.35 ± 0.18 b	6.17 ± 0.85 a	14.18 ± 0.91b
		CF2	16.86 ± 2.11 ab	6.26 ± 0.84 a	10.60 ± 1.44 a

The same letters (a, b) indicate membership to homogeneous groups according to Tukey's post hoc comparisons ($P < 0.05$)

results were obtained when irrigating the seedlings with L5%. In this case, no differences were observed in height, which had the same mean value as control treatment seedlings (372 mm). The fresh weight of the roots was slightly higher in seedlings irrigated with both dilutions of leachate (5% and 10%), in comparison with control. It is noted that those differences indicate a trend, but they were not statistically significant according to Dunnett's post hoc contrasts ($P > 0.05$).When comparing the growth of the leachate-irrigated plants (both dilutions) with respect to CF1, it can be observed that total and aerial tissue fresh weight was significantly lower ($P < 0.05$) in the leachate treatments. Height of L10% plants was also significantly lower ($P < 0.05$) than CF1 plants.

Leachate irrigation could lead to yield reduction (Menser et al. 1983) or to enhance growth (Cureton et al. 1991; Liang et al. 1999) depending on plant species, leachate source, and application method. *A. inulifolium* and *E. crista-galli* are native woody plants that grow in highly disturbed sites. Also, *E. crista-galli* has proved to be tolerant to the presence of heavy metals in soils and has been commonly used in local restoration programs (Basílico et al. 2018).

Table 4 *A. cepa* test results

Treatment		Indexes	
	Micronuclei (%)	Chromosomal aberrations (%)	Mitotic index
Control	0.28 ± 0.20 a	0.03 ± 0.07 a	59.16 ± 8.14 a
MMS	4.20 ± 1.50 d	0.55 ± 0.35 b	55.26 ± 2.33 a
L5%	0.90 ± 0.29 b	0.16 ± 0.14 a	59.98 ± 3.87 a
L10%	1.79 ± 0.96 c	0.07 ± 0.12 a	65.44 ± 16.83 a

The same letters (a, b, c, d) indicate membership to homogeneous groups according Mann–Whitney *U*-test pairwise comparisons ($P < 0.05$)

3.3 *Leachate Phytotoxicity*

Chlorosis and leaf burn are common symptoms of soil salinization (Cheng and Chu 2007). However, both species showed no symptoms of phytotoxicity or significant growth decrease with the dilutions of leachate applied. The results obtained in this study incorporate a new factor which these species are tolerant.

The germination percentage of *A. cepa* seeds was 93 ± 1% in the negative control and 92 ± 8% in the positive control. Undiluted vermicompost leachate was highly toxic to *A. cepa*, almost completely inhibiting the germination of the seeds, whose percentage was 5 ± 3%. In this way, it was not possible to carry out the mutagenicity test with the pure leachate, and only the L5% and L10% dilutions were tested. The germination of the seeds in these two dilutions was also lower than in the controls (L5% = 86 ± 3% and L10% = 83 ± 4%), which suggests a certain degree of toxicity. The genotoxicity test using these two dilutions showed statistically significant differences between the control treatments, MMS, L5%, and L10%, according to the nonparametric Kruskal–Wallis test (Table 4). The percentage of MN in the seeds of *A. cepa* exposed to both dilutions of leachate was higher than in the control, and this genotoxic effect was highly significant (Mann–Whitney *U*-test, $P < 0.01$). On the other hand, no significant differences were found in CA with respect to the control (Table 4).

Similar results were observed by Datta et al. (2018) using *A. cepa* test. These authors found that genotoxicity depends on the vermicompost leachate concentration. At 10% there were no increases in chromosomal aberrations with respect to the control.

The growth promotion obtained for *A. inulifolium* was due to nutrients in leachate solution in spite of the presence of toxic substances that produced genotoxicity in the *A. cepa* assay. It is necessary to develop additional research on the genotoxicity of vermicomposting leachates.

4 Conclusions

Substrate pH and EC increased as a result of application of diluted vermicompost leachate in both species; however, OM content increased only in *A. inulifolium* bioassay. Increasing the organic content in the substrate is a potential benefit of irrigation with leachate, although this enrichment depends on plant species. Therefore, irrigation with diluted vermicompost leachate should be spaced out over time, discouraging its continuous use for irrigation.

Both plant species, *A. inulifolium* and *E. crista-galli*, were tolerant to irrigation with diluted vermicompost leachate. Irrigation of plants with 10% leachate dilution has no statistically significant effects on the plant growth, compared to those irrigated with deionized water. However, it was found that seedlings of *A. inulifolium* irrigated with leachate had a higher total biomass than those of the control treatment. For *E. crista-galli*, the total and aerial fresh weights were sensitive variables when comparing the growth of leachate-irrigated plant against a commercial fertilizer. Stem diameter and root fresh weight of both species were not affected by application of any irrigation solution.

The low germination percentage of seeds of *A. cepa* indicates that undiluted leachate is phytotoxic. The two dilutions of leachate were genotoxic for *A. cepa*, suggesting the presence of mutagenic substances in the complex mixture of vermicompost. However, leachate solution induced growth promotion in *A. inulifolium*, probably due to the presence of nutrients that overlapped the effect of substances with a genotoxic effect.

Acknowledgments The authors wish to thank Flores University for the funding received. We are grateful to Mr. Ricardo J. Piccolo for kindly providing the *Allium cepa* seeds for performing the genotoxicity assay.

References

APHA, AWWA, & WEF (2012) Standard methods for the examination of water and wastewater, 22nd edn. American Public Health Association, Washington, D.C.

Aremu AO, Stirk WA, Kulkarni MG, Tarkowská D, Turečková V, Gruz J, Šubrtová M, Pěnčík A, Novák O, Doležal K, Strnad M, Van Staden J (2015) Evidence of phytohormones and phenolic acids variability in garden-waste-derived vermicompost leachate, a well-known plant growth stimulant. Plant Growth Regul 75(2):483–492

Ávila-Juárez L, Rodríguez González A, Rodríguez Piña N, Guevara González RG, Torres Pacheco I, Ocampo Velázquez RV, Moustapha B (2015) Vermicompost leachate as a supplement to increase tomato fruit quality. J Soil Sci Plant Nutr 15(1):46–59

Basílico G, Magdaleno A, Paz M, Moretton J, Faggi A, de Cabo L (2017) Sewage pollution: genotoxicity assessment and phytoremediation of nutrients excess with *Hydrocotyle ranunculoides*. Environ Monit Assess 189(4):182

Basílico G, Faggi A, de Cabo L (2018) Tolerance to metals in two species of Fabaceae grown in riverbank sediments polluted with chromium, copper, and lead. In: Phytoremediation. Springer, Cham, pp 169–178

CEAMSE & FIUBA (2011) Estudio de calidad de los residuos sólidos urbanos del Área Metropolitana de Buenos Aires – Tercer informe de avance. Coordinación Ecológica Área Metropolitana Sociedad del Estado (CEAMSE). Facultad de Ingeniería de la Universidad de Buenos Aires (FIUBA). http://www.ceamse.gov.ar/wp-content/uploads/2012/06/Tercer-Informe-ECRSU-AMBA.pdf

Cheng CY, Chu LM (2007) Phytotoxicity data safeguard the performance of the recipient plants in leachate irrigation. Environ Pollut 145(1):195–202

Cureton PM, Groenevelt PH, McBride RA (1991) Landfill leachate recirculation: effects on vegetation vigor and clay surface cover infiltration. J Environ Qual 20(1):17–24

Das D, Bhattacharyya P, Ghosh BC, Banik P (2016) Bioconversion and biodynamics of *Eisenia foetida* in different organic wastes through microbially enriched vermiconversion technologies. Ecol Eng 86:154–161

Datta S, Singh J, Singh J, Singh S, Singh S (2018) Assessment of genotoxic effects of pesticide and vermicompost treated soil with *Allium cepa* test. Sustain Environ Res 28:171–178

Fornes F, Mendoza-Hernández D, García-de-la-Fuente R, Abad M, Belda RM (2012) Composting versus vermicomposting: a comparative study of organic matter evolution through straight and combined processes. Bioresour Technol 118:296–305

Gutiérrez-Miceli FA, García-Gómez RC, Rosales RR, Abud-Archila M, Angela OLM, Cruz MJG, Dendooven L (2008) Formulation of a liquid fertilizer for sorghum (*Sorghum bicolor* (L.) Moench) using vermicompost leachate. Bioresour Technol 99(14):6174–6180

He Z, Lin H, Hao J, Kong X, Tian K, Bei Z, Tian X (2018) Impact of vermiculite on ammonia emissions and organic matter decomposition of food waste during composting. Bioresour Technol 263:548–554

Henry JG (1999) Residuos sólidos. In: Henry JG, Hinke GW (eds) Ingeniería Ambiental. Prentice Hall, México

Hernández AJ, Adarve MJ, Gil A, Pastor J (1999) Soil salivation from landfill leachates: effects on the macronutrient content and plant growth of four grassland species. Chemosphere 38(7):1693–1711

Hoshina MM, Marin-Morales MA (2009) Micronucleus and chromosome aberrations induced in onion (*Allium cepa*) by a petroleum refinery effluent and by river water that receives this effluent. Ecotoxicol Environ Saf 72(8):2090–2095

Islam MK, Yaseen T, Traversa A, Kheder MB, Brunetti G, Cocozza C (2016) Effects of the main extraction parameters on chemical and microbial characteristics of compost tea. Waste Manag 52:62–68

Kandari LS, Kulkarni MG, Van Staden J (2011) Vermicompost leachate improves seedling emergence and vigour of aged seeds of commercially grown *Eucalyptus* species. J For Sci 73(2):117–122

Klapperich RJ, Liu G, Stepan DJ, Jensen MD, Steadman EN, Harju JA, Nakles DV (2014) IEAGHG investigation of extracted water from CO2 storage: potential benefits of water extraction and lesson learned. Energy Procedia 63:7173–7186

Liang J, Zhang J, Wong MH (1999) Landfill leachate used as irrigation water on landfill sites during dry seasons. In: Remediation and management of degraded lands. Lewis Publishers, Boca Raton, pp 305–317

Matsumoto ST, Mantovani MS, Malaguttii MIA, Dias AL, Fonseca IC, Marin-Morales MA (2006) Genotoxicity and mutagenicity of water contaminated with tannery effluents, as evaluated by the micronucleus test and comet assay using the fish *Oreochromis niloticus* and chromosome aberrations in onion root-tips. Genet Mol Biol 29(1):148–158

Menser HA, Winant WA, Bennett OL (1983) Spray irrigation with landfill leachate. BioCycle J Waste Recycl 24(3):22–25

Romero-Tepal EM, Contreras-Blancas E, Navarro-Noya YE, Ruíz-Valdiviezo VM, Luna-Guido M, Gutiérrez-Miceli FA, Dendooven L (2014) Changes in the bacterial community structure in stored wormbed leachate. J Mol Microbiol Biotechnol 24(2):105–113

Sall PM, Antoun H, Chalifour FP, Beauchamp CJ (2019) Potential use of leachate from composted fruit and vegetable waste as fertilizer for corn. Cogent Food Agric 5:1580180

Salomone R, Saija G, Mondello G, Giannetto A, Fasulo S, Savastano D (2017) Environmental impact of food waste bioconversion by insects: application of life cycle assessment to process using *Hermetia illucens*. J Clean Prod 140:890–905

Schumacher BA (2002) Methods for the determination of total organic carbon (TOC) in soils and sediments. United States Environmental Protection Agency

Zar JH (1999) Biostatistical analysis. Pearson Education

Zorpas AA, Lasaridi K, Voukkali I, Loizia P, Chroni C (2015) Household waste compositional analysis variation from insular communities in the framework of waste prevention strategy plans. Waste Manag 38:3–11

Index

M. Naeem et al. (eds.), *Contaminants in Agriculture*,
https://doi.org/10.1007/978-3-030-41552-5

D

Printed by Printforce, the Netherlands